Universitext

For further volumes:
http://www.springer.com/series/223

Giulia Di Nunno · Bernt Øksendal
Frank Proske

Malliavin Calculus for Lévy Processes with Applications to Finance

Springer

Giulia Di Nunno
Bernt Øksendal
Frank Proske
Department of Mathematics
University of Oslo
0316 Oslo
Blindern
Norway
giulian@math.uio.no
oksendal@math.uio.no
proske@math.uio.no

ISBN 978-3-540-78571-2   e-ISBN 978-3-540-78572-9
DOI 10.1007/978-3-540-78572-9
Springer Heidelberg Dordrecht London New York

Library of Congress Control Number: 2009934515

Mathematics Subject Classification (2000): 60H05, 60H07, 60H40, 91B28, 93E20, 60G51, 60G57

Cover design: WMXDesign GmbH, Heidelberg

The cover design is based on a graph provided by F.E. Benth, M. Groth, and O. Wallin

Printed on acid-free paper

Springer is part of Springer Science+Business Media (www.springer.com)

To Christian. To my parents.

G.D.N.

To Eva.

B.Ø.

To Simone, Paulina and Siegfried.

F.P.

Preface

There are already several excellent books on Malliavin calculus. However, most of them deal only with the theory of Malliavin calculus for Brownian motion, with [35] as an honorable exception. Moreover, most of them discuss only the application to regularity results for solutions of SDEs, as this was the original motivation when Paul Malliavin introduced the infinite-dimensional calculus in 1978 in [158]. In the recent years, Malliavin calculus has found many applications in stochastic control and within finance. At the same time, Lévy processes have become important in financial modeling. In view of this, we have seen the need for a book that deals with Malliavin calculus for Lévy processes in general, not just Brownian motion, and that presents some of the most important and recent applications to finance.

It is the purpose of this book to try to fill this need. In this monograph we present a general Malliavin calculus for Lévy processes, covering both the Brownian motion case and the pure jump martingale case via Poisson random measures, and also some combination of the two. We also present many of the recent applications to finance, including the following:

- The Clark–Ocone theorem and hedging formulae
- Minimal variance hedging in incomplete markets
- Sensitivity analysis results and efficient computation of the "greeks"
- Optimal portfolio with partial information
- Optimal portfolio in an anticipating environment
- Optimal consumption in a general information setting
- Insider trading

To be able to handle these applications, we develop a general theory of anticipative stochastic calculus for Lévy processes involving the Malliavin derivative, the Skorohod integral, the forward integral, which were originally introduced for the Brownian setting only. We dedicate some chapters to the generalization of our results to the white noise framework, which often turns out to be a suitable setting for the theory. Moreover, this enables us to prove results that are general enough for the financial applications, for example, the generalized Clark–Ocone theorem.

This book is based on a series of courses that we have given in different years and to different audiences. The first one was given at the Norwegian School of Economics and Business Administration (NHH) in Bergen in 1996, at that time about Brownian motion only. Other courses were held later, every time including more updated material. In particular, we mention the courses given at the Department of Mathematics and at the Center of Mathematics for Applications (CMA) at the University of Oslo and also the intensive or compact courses presented at the University of Ulm in July 2006, at the University of Cape Town in December 2006, at the Indian Institute of Science (IIS) in Bangalore in January 2007, and at the Nanyang Technological University in Singapore in January 2008.

At all these occasions we met engaged students and attentive readers. We thank all of them for their active participation to the classes and their feedback. Our work has benefitted from the collaboration and useful comments from many people, including Fred Espen Benth, Maximilian Josef Butz, Delphine David, Inga Baadshaug Eide, Xavier Gabaix, Martin Groth, Yaozhong Hu, Asma Khedher, Paul Kettler, An Ta Thi Kieu, Jørgen Sjaastad, Thilo Meyer-Brandis, Farai Julius Mhlanga, Yeliz Yolcu Okur, Olivier Menoukeu Pamen, Ulrich Rieder, Goncalo Reiss, Steffen Sjursen, Alexander Sokol, Agnès Sulem, Olli Wallin, Diane Wilcox, Frank Wittemann, Mihail Zervos, Tusheng Zhang, and Xunyu Zhou. We thank them all for their help. Our special thanks go to Paul Malliavin for the inspiration and continuous encouragement he has given us throughout the time we have worked on this book. We also acknowledge with gratitude the technical support with computers of the Drift-IT at the Department of Mathematics at the University of Oslo.

Oslo,
June 2008.

Giulia Di Nunno
Bernt Øksendal
Frank Proske

Contents

Part II The Discontinuous Case: Pure Jump Lévy Processes

Introduction

The mathematical theory now known as *Malliavin calculus* was first introduced by Paul Malliavin in [158] as an infinite-dimensional integration by parts technique. The purpose of this calculus was to prove the results about the smoothness of densities of solutions of stochastic differential equations driven by Brownian motion. For several years this was the only known application. Therefore, since this theory was considered quite complicated by many, Malliavin calculus remained a relatively unknown theory also among mathematicians for some time. Many mathematicians simply considered the theory as too difficult when compared with the results it produced. Moreover, to a large extent, these results could also be obtained by using Hörmander's earlier theory on hypoelliptic operators. See also, for example, [20, 114, 225, 230].

This was the situation until 1984, when Ocone in [173] obtained an explicit interpretation of the Clark representation formula [46, 47] in terms of the Malliavin derivative. This remarkable result later became known as the *Clark–Ocone formula*. Sometimes also called *Clark–Haussmann–Ocone formula* in view of the contribution of Haussmann in 1979, see [98]. In 1991, Ocone and Karatzas [174] applied this result to finance. They proved that the Clark–Ocone formula can be used to obtain explicit formulae for replicating portfolios of contingent claims in complete markets.

Since then, new literature helped to distribute these results to a wider audience, both among mathematicians and researchers in finance. See, for example, the monographs [53, 161, 169, 212, 216] and the introductory lecture notes [178]; see also [206].

The next breakthrough came in 1999, when Fournié et al. [81] obtained numerically tractable formulae for the computation of the so-called *greeks* in finance, also known as *parameters of sensitivity*. In the recent years, many new applications of the Malliavin calculus have been found, including partial information optimal control, insider trading and, more generally, anticipative stochastic calculus.

At the same time Malliavin calculus was extended from the original setting of Brownian motion to more general Lévy processes. This extensions were at first motivated by and taylored to the original application within the study of smoothness of densities (see e.g. [12, 35, 37, 38, 44, 141, 142, 143, 163, 189, 190, 218, 219]) and then developed largely targeting the applications to finance, where Lévy processes based models are now widely used (see, e.g., [25, 29, 64, 69, 148, 171, 181]). Within this last direction, some extension to random fields of Lévy type has also been developed, see, for example, [61, 62]. Other extension of Malliavin calculus within quantum probability have also appeared, see, for example, [84, 85].

One way of interpreting the Malliavin derivative of a given random variable $F = F(\omega)$, $\omega \in \Omega$, on the given probability space $(\Omega, \mathcal{F}, P)$ is to regard it as a derivative with respect to the random parameter ω. For this to make sense, one needs some mathematical structure on the space Ω. In the original approach used by Malliavin, for the Brownian motion case, Ω is represented as the *Wiener space* $C_0([0, T])$ of continuous functions $\omega : [0, T] \longrightarrow \mathbb{R}$ with $\omega(0) = 0$, equipped with the uniform topology. In this book we prefer to use the representation of Hida [99], namely to represent Ω as the *space $\mathcal{S}'$ of tempered distributions* $\omega : \mathcal{S} \longrightarrow \mathbb{R}$, where $\mathcal{S}$ is the Schwartz space of rapidly decreasing smooth functions on $\mathbb{R}$ (see Chap. 5). The corresponding probability measure P is constructed by means of the Bochner–Minlos theorem. This is a classical setting of white noise theory. This approach has the advantage that the Malliavin derivative $D_t F$ of a random variable $F : \mathcal{S}' \longrightarrow \mathbb{R}$ can simply be regarded as a *stochastic gradient*.

In fact, if γ is deterministic and in $L^2(\mathbb{R})$ (note that $L^2(\mathbb{R}) \subset \mathcal{S}'$), we define the *directional derivative of F in the direction γ, $D_\gamma F$*, as follows:

$$D_\gamma F(\omega) = \lim_{\varepsilon \to 0} \frac{F(\omega + \varepsilon \gamma) - F(\omega)}{\varepsilon}, \qquad \omega \in \mathcal{S}',$$

if the limit exists in $L^2(P)$. If there exists a process $\Psi(\omega, t) : \Omega \times \mathbb{R} \longrightarrow \mathbb{R}$ such that

$$D_\gamma F(\omega) = \int_{\mathbb{R}} \Psi(\omega, t) \gamma(t) dt, \quad \omega \in \mathcal{S}'$$

for all $\gamma \in L^2(\mathbb{R})$, then we say that F is *Malliavin–Hida differentiable* and we define

$$D_t F(\omega) := \Psi(\omega, t), \qquad \omega \in \mathcal{S}'$$

as the *Malliavin–(Hida) derivative* (or *stochastic gradient*) of F at t.

This gives a simple and intuitive interpretation of the Malliavin derivative in the Brownian motion case. Moreover, some of the basic properties of calculus such as *chain rule* follow easily from this definition. See Chap. 6.

Alternatively, the Malliavin derivative can also be introduced by means of the *Wiener–Itô chaos expansion* [120]:

$$F = \sum_{n=0}^{\infty} I_n(f_n)$$

of the random variable F as a series of iterated Itô integrals of symmetric functions $f_n \in L^2(\mathbb{R}^n)$ with respect to Brownian motion. In this setting, the Malliavin derivative gets the form

$$D_t F = \sum_{n=1}^{\infty} n I_{n-1}(f_n(\cdot, t)),$$

see Chap. 3, cf. [169]. This form is appealing because it has some resemblance to the derivative of a monomial:

$$\frac{d}{dx} x^n = n x^{n-1}.$$

Moreover, the chaos expansion approach is convenient because it gives easy proofs of the Clark–Ocone formula and several basic properties of the Malliavin derivative.

The chaos expansion approach also has the advantage that it carries over in a natural way to the *Lévy process* setting (see Chap. 12). This provides us with a relatively unified approach, valid for both the continuous and discontinuous case, that is, for both Brownian motion and Lévy processes/Poisson random measures. See, for example, the proof of the Clark–Ocone theorem in the two cases. At the same time it is important to be aware of the differences between these two cases. For example, in the continuous case, we base the interpretation of the Malliavin derviative as a stochastic gradient, while in the discontinuous case, the Malliavin derivative is actually a difference operator.

How to use this book

It is the purpose of this book to give an introductory presentation of the theory of Malliavin calculus and its applications, mainly to finance. For pedagogical reasons, and also to make the reading easier and the use more flexible, the book is divided into two parts:

Part I. The Continuous Case: Brownian Motion
Part II. The Discontinuous Case: Pure Jump Lévy Processes

In both parts the emphasis is on the topics that are most central for the applications to finance. The results are illustrated throughout with examples. In addition, each chapter ends with exercises. Solutions to some selection of exercises, with varying level of detail, can be found at the back of the book.

We hope the book will be useful as a graduate text book and as a source for students and researchers in mathematics and finance. There are several possible ways of selecting topics when using this book, for example, in a graduate course:

Alternative 1. If there is enough time, all eighteen chapters could be included in the program.

Alternative 2. If the interest is only in the continuous case, then the whole Part I gives a progressive overview of the theory, including the white noise approach, and gives a good taste of the applications.

Alternative 3. Similarly, if the readers are already familiar with the continuous case, then Part II is self-contained and provides a good text choice to cover both theory and applications.

Alternative 4. If the interest is in an introductory overview on both the continuous and the discontinuous case, then a good selection could be the reading from Chaps. 1 to 4 and then from Chaps. 9 to 12. This can be possibly supplemented by the reading of the chapters specifically devoted to applications, so according to interest one could choose among Chaps. 8, 15, 16, and also Chaps. 17 and 18.

Part I

The Continuous Case: Brownian Motion

1

The Wiener–Itô Chaos Expansion

The celebrated Wiener–Itô chaos expansion is fundamental in stochastic analysis. In particular, it plays a crucial role in the Malliavin calculus as it is presented in the sequel. This result which concerns the representation of square integrable random variables in terms of an infinite orthogonal sum was proved in its first version by Wiener in 1938 [227]. Later, in 1951, Itô [120] showed that the expansion could be expressed in terms of *iterated Itô integrals* in the Wiener space setting.

Before we state the theorem we introduce some useful notation and give some auxiliary results.

1.1 Iterated Itô Integrals

Let $W = W(t) = W(\omega, t)$, $\omega \in \Omega$ $t \in [0,T]$ $(T > 0)$, be a one-dimensional Wiener process, or equivalently Brownian motion, on the complete probability space $(\Omega, \mathcal{F}, P)$ such that $W(0) = 0$ P-a.s.

For any t, let $\mathcal{F}_t$ be the σ-algebra generated by $W(s)$, $0 \leq s \leq t$, augmented by all the P-zero measure events. We denote the corresponding filtration by

$$\mathbb{F} = \{\mathcal{F}_t, \ t \in [0,T]\}. \tag{1.1}$$

Note that this filtration is both left- and right-continuous, that is,

$$\mathcal{F}_t = \lim_{s \nearrow t} \mathcal{F}_s := \sigma\Big\{ \bigcup_{s < t} \mathcal{F}_s \Big\},$$

respectively,

$$\mathcal{F}_t = \lim_{u \searrow t} \mathcal{F}_u := \bigcap_{u > t} \mathcal{F}_u.$$

See, for example, [129] or [207].

G.Di Nunno et al., *Malliavin Calculus for Lévy Processes with Applications to Finance*,
© Springer-Verlag Berlin Heidelberg 2009

Definition 1.1. *A real function* $g : [0,T]^n \to \mathbb{R}$ *is called* symmetric *if*

$$g(t_{\sigma_1}, \ldots, t_{\sigma_n}) = g(t_1, \ldots, t_n) \tag{1.2}$$

for all permutations $\sigma = (\sigma_1, \ldots, \sigma_n)$ *of* $(1, 2, \ldots, n)$.

Let $L^2([0,T]^n)$ be the standard space of square integrable Borel real functions on $[0,T]^n$ such that

$$\|g\|^2_{L^2([0,T]^n)} := \int\limits_{[0,T]^n} g^2(t_1, \ldots, t_n) dt_1 \cdots dt_n < \infty. \tag{1.3}$$

Let $\widetilde{L}^2([0,T]^n) \subset L^2([0,T]^n)$ be the space of *symmetric* square integrable Borel real functions on $[0,T]^n$. Let us consider the set

$$S_n = \{(t_1, \ldots, t_n) \in [0,T]^n : \ 0 \leq t_1 \leq t_2 \leq \cdots \leq t_n \leq T\}.$$

Note that this set S_n occupies the fraction $\frac{1}{n!}$ of the whole n-dimensional box $[0,T]^n$. Therefore, if $g \in \widetilde{L}^2([0,T]^n)$ then $g_{|S_n} \in L^2(S_n)$ and

$$\|g\|^2_{L^2([0,T]^n)} = n! \int\limits_{S_n} g^2(t_1, \ldots, t_n) dt_1 \ldots dt_n = n! \|g\|^2_{L^2(S_n)}, \tag{1.4}$$

where $\|\cdot\|_{L^2(S_n)}$ denotes the norm induced by $L^2([0,T]^n)$ on $L^2(S_n)$, the space of the square integrable functions on S_n.

If f is a real function on $[0,T]^n$, then its *symmetrization* $\widetilde{f}$ is defined by

$$\widetilde{f}(t_1, \ldots, t_n) = \frac{1}{n!} \sum_\sigma f(t_{\sigma_1}, \ldots, t_{\sigma_n}), \tag{1.5}$$

where the sum is taken over all permutations σ of $(1, \ldots, n)$. Note that $\widetilde{f} = f$ if and only if f is symmetric.

Example 1.2. The symmetrization of the function

$$f(t_1, t_2) = t_1^2 + t_2 \sin t_1, \quad (t_1, t_2) \in [0,T]^2,$$

is

$$\widetilde{f}(t_1, t_2) = \frac{1}{2} \left[t_1^2 + t_2^2 + t_2 \sin t_1 + t_1 \sin t_2 \right], \quad (t_1, t_2) \in [0,T]^2.$$

Definition 1.3. *Let* f *be a deterministic function defined on* S_n $(n \geq 1)$ *such that*

$$\|f\|^2_{L^2(S_n)} := \int\limits_{S_n} f^2(t_1, \ldots, t_n) dt_1 \cdots dt_n < \infty.$$

Then we can define the n-fold iterated Itô integral as

$$J_n(f) := \int\limits_0^T \int\limits_0^{t_n} \cdots \int\limits_0^{t_3} \int\limits_0^{t_2} f(t_1, \ldots, t_n) dW(t_1) dW(t_2) \cdots dW(t_{n-1}) dW(t_n). \tag{1.6}$$

Note that at each iteration $i = 1, ..., n$ the corresponding Itô integral with respect to $dW(t_i)$ is well-defined, being the integrand $\int_0^{t_i} \cdots \int_0^{t_2} f(t_1, ..., t_n) dW(t_1)...dW(t_{i-1})$, $t_i \in [0, t_{i+1}]$, a stochastic process that is $\mathbb{F}$-adapted and square integrable with respect to $dP \times dt_i$. Thus, (1.6) is well-defined.

Thanks to the construction of the Itô integral we have that $J_n(f)$ belongs to $L^2(P)$, that is, the space of square integrable random variables. We denote the norm of $X \in L^2(P)$ by

$$\|X\|_{L^2(P)} := \left(E\left[X^2\right] \right)^{1/2} = \left(\int_\Omega X^2(\omega) P(d\omega) \right)^{1/2}.$$

Applying the Itô isometry iteratively, if $g \in L^2(S_m)$ and $h \in L^2(S_n)$, with $m < n$, we can see that

$$E\left[J_m(g)J_n(h)\right] = E\left[\left(\int_0^T \int_0^{s_m} \cdots \int_0^{s_2} g(s_1, ..., s_m) dW(s_1) \cdots dW(s_m) \right) \right.$$

$$\left. \cdot \left(\int_0^T \int_0^{s_m} \cdots \int_0^{t_2} h(t_1, ..., t_{n-m}, s_1, ..., s_m) dW(t_1) \cdots dW(t_{n-m}) dW(s_1) \cdots dW(s_m) \right) \right]$$

$$= \int_0^T E\left[\left(\int_0^{s_m} \cdots \int_0^{s_2} g(s_1, ..., s_{m-1}, s_m) dW(s_1) \cdots dW(s_{m-1}) \right) \right. \tag{1.7}$$

$$\left. \cdot \left(\int_0^{s_m} \cdots \int_0^{t_2} h(t_1, ..., s_{m-1}, s_m) dW(t_1) \cdots dW(s_{m-1}) \right) \right] ds_m = \ldots$$

$$= \int_0^T \int_0^{s_m} \cdots \int_0^{s_2} g(s_1, s_2, ..., s_m) E\left[\int_0^{s_1} \cdots \int_0^{t_2} h(t_1, ..., t_{n-m}, s_1, ..., s_m) \right.$$

$$\left. \cdot dW(t_1) \cdots dW(t_{n-m}) \right] ds_1 \cdots ds_m = 0$$

because the expected value of an Itô integral is zero. On the other hand, if both g and h belong to $L^2(S_n)$, then

$$E\left[J_n(g)J_n(h)\right] = \int_0^T E\left[\int_0^{s_n} \cdots \int_0^{s_2} g(s_1, ..., s_n) dW(s_1) \cdots dW(s_{n-1}) \right.$$

$$\left. \cdot \int_0^{s_n} \cdots \int_0^{s_2} h(s_1, ..., s_n) dW(s_1) \cdots dW(s_{n-1}) \right] ds_n = \ldots \tag{1.8}$$

$$= \int_0^T \cdots \int_0^{s_2} g(s_1, ..., s_n) h(s_1, ..., s_n) ds_1 \cdots ds_n = (g, h)_{L^2(S_n)}$$

We summarize these results as follows.

Proposition 1.4. *The following relations hold true:*

$$E[J_m(g)J_n(h)] = \begin{cases} 0 & , n \neq m \\ (g,h)_{L^2(S_n)} & , n = m \end{cases} \qquad (m,n = 1,2,...), \qquad (1.9)$$

where

$$(g,h)_{L^2(S_n)} := \int\limits_{S_n} g(t_1,\ldots,t_n)h(t_1,\ldots,t_n)dt_1 \cdots dt_n$$

is the inner product of $L^2(S_n)$. In particular, we have

$$\|J_n(h)\|_{L^2(P)} = \|h\|_{L^2(S_n)}. \qquad (1.10)$$

Remark 1.5. Note that (1.9) also holds for $n = 0$ or $m = 0$ if we define $J_0(g) = g$, when g is a constant, and $(g,h)_{L^2(S_0)} = gh$, when g, h are constants.

Remark 1.6. It is straightforward to see that the n-fold iterated Itô integral

$$L^2(S_n) \ni f \implies J_n(f) \in L^2(P)$$

is a linear operator, that is, $J_n(af + bg) = aJ_n(f) + bJ_n(g)$, for $f, g \in L^2(S_n)$ and $a, b \in \mathbb{R}$.

Definition 1.7. *If $g \in \widetilde{L}^2([0,T]^n)$ we define*

$$I_n(g) := \int\limits_{[0,T]^n} g(t_1,\ldots,t_n)dW(t_1)\ldots dW(t_n) := n!J_n(g). \qquad (1.11)$$

We also call n-fold iterated Itô integrals the $I_n(g)$ here above.

Note that from (1.9) and (1.11) we have

$$\|I_n(g)\|^2_{L^2(P)} = E[I_n^2(g)] = E[(n!)^2 J_n^2(g)]$$
$$= (n!)^2 \|g\|^2_{L^2(S_n)} = n! \|g\|^2_{L^2([0,T]^n)} \qquad (1.12)$$

for all $g \in \widetilde{L}^2([0,T]^n)$. Moreover, if $g \in \widetilde{L}^2([0,T]^m)$ and $h \in \widetilde{L}^2([0,T]^n)$, we have

$$E[I_m(g)I_n(h)] = \begin{cases} 0 & , n \neq m \\ (g,h)_{L^2([0,T]^n)} & , n = m \end{cases} \qquad (m,n = 1,2,...),$$

with $(g,h)_{L^2([0,T]^n)} = n!(g,h)_{L^2(Sn)}$.

There is a useful formula due to Itô [120] for the computation of the iterated Itô integral. This formula relies on the relationship between Hermite polynomials and the Gaussian distribution density. Recall that the *Hermite polynomials* $h_n(x)$, $x \in \mathbb{R}$, $n = 0,1,2,\ldots$ are defined by

$$h_n(x) = (-1)^n e^{\frac{1}{2}x^2} \frac{d^n}{dx^n}(e^{-\frac{1}{2}x^2}), \qquad n = 0,1,2,\ldots, \qquad (1.13)$$

Thus, the first Hermite polynomials are

$$h_0(x) = 1, \; h_1(x) = x, \; h_2(x) = x^2 - 1, \; h_3(x) = x^3 - 3x,$$
$$h_4(x) = x^4 - 6x^2 + 3, \; h_5(x) = x^5 - 10x^3 + 15x, \ldots .$$

We also recall that the family of Hermite polynomials constitute an orthogonal basis for $L^2(\mathbb{R}, \mu(dx))$ if $\mu(dx) = \frac{1}{\sqrt{2\pi}} e^{\frac{x^2}{2}} dx$ (see, e.g., [215]).

Proposition 1.8. *If* $\xi_1, \xi_2, \ldots$ *are orthonormal functions in* $L^2([0,T])$, *we have that*

$$I_n\big(\xi_1^{\otimes \alpha_1} \hat{\otimes} \cdots \hat{\otimes} \xi_m^{\otimes \alpha_m}\big) = \prod_{k=1}^{m} h_{\alpha_k}\Big(\int_0^T \xi_k(t) W(t)\Big), \qquad (1.14)$$

with $\alpha_1 + \cdots + \alpha_m = n$. *Here* $\otimes$ *denotes the* tensor power *and* $\alpha_k \in \{0, 1, 2, \ldots\}$ *for all* k.

See [120]. In general, the *tensor product* $f \otimes g$ of two functions f, g is defined by

$$(f \otimes g)(x_1, x_2) = f(x_1)g(x_2)$$

and the *symmetrized tensor product* $f \hat{\otimes} g$ is the symmetrization of $f \otimes g$. In particular, from (1.14), we have

$$n! \int_0^T \int_0^{t_n} \cdots \int_0^{t_2} g(t_1)g(t_2) \cdots g(t_n) dW(t_1) \cdots dW(t_n) = \|g\|^n h_n\Big(\frac{\theta}{\|g\|}\Big), \quad (1.15)$$

for the tensor power of $g \in L^2([0,T])$. Here above we have used $\|g\| = \|g\|_{L^2([0,T])}$ and $\theta = \int_0^T g(t) dW(t)$.

Example 1.9. Let $g \equiv 1$ and $n = 3$, then we get

$$6 \int_0^T \int_0^{t_3} \int_0^{t_2} 1 \, dW(t_1)dW(t_2)dW(t_3) = T^{3/2} h_3\Big(\frac{W(T)}{T^{1/2}}\Big) = W^3(T) - 3T W(T).$$

1.2 The Wiener–Itô Chaos Expansion

Theorem 1.10. The Wiener–Itô chaos expansion. *Let* ξ *be an* $\mathcal{F}_T$-*measurable random variable in* $L^2(P)$. *Then there exists a unique sequence* $\{f_n\}_{n=0}^\infty$ *of functions* $f_n \in \widetilde{L}^2([0,T]^n)$ *such that*

$$\xi = \sum_{n=0}^{\infty} I_n(f_n), \qquad (1.16)$$

where the convergence is in $L^2(P)$. Moreover, we have the isometry

$$\|\xi\|_{L^2(P)}^2 = \sum_{n=0}^{\infty} n!\|f_n\|_{L^2([0,T]^n)}^2. \tag{1.17}$$

Proof By the Itô representation theorem there exists an $\mathbb{F}$-adapted process $\varphi_1(s_1)$, $0 \leq s_1 \leq T$, such that

$$E\left[\int_0^T \varphi_1^2(s_1)ds_1\right] \leq E\left[\xi^2\right] \tag{1.18}$$

and

$$\xi = E[\xi] + \int_0^T \varphi_1(s_1)dW(s_1). \tag{1.19}$$

Define

$$g_0 = E[\xi].$$

For almost all $s_1 \leq T$ we can apply the Itô representation theorem to $\varphi_1(s_1)$ to conclude that there exists an $\mathbb{F}$-adapted process $\varphi_2(s_2, s_1)$, $0 \leq s_2 \leq s_1$, such that

$$E\left[\int_0^{s_1} \varphi_2^2(s_2, s_1)ds_2\right] \leq E[\varphi_1^2(s_1)] < \infty \tag{1.20}$$

and

$$\varphi_1(s_1) = E[\varphi_1(s_1)] + \int_0^{s_1} \varphi_2(s_2, s_1)dW(s_2). \tag{1.21}$$

Substituting (1.21) in (1.19) we get

$$\xi = g_0 + \int_0^T g_1(s_1)dW(s_1) + \int_0^T \int_0^{s_1} \varphi_2(s_2, s_1)dW(s_2)dW(s_1), \tag{1.22}$$

where

$$g_1(s_1) = E[\varphi_1(s_1)].$$

Note that by (1.18), (1.20), and the Itô isometry we have

$$E\left[\left(\int_0^T \int_0^{s_1} \varphi_2(s_2, s_1)dW(s_2)dW(s_1)\right)^2\right] = \int_0^T \int_0^{s_1} E[\varphi_2^2(s_2, s_1)]ds_2ds_1 \leq E[\xi^2].$$

Similarly, for almost all $s_2 \leq s_1 \leq T$, we apply the Itô representation theorem to $\varphi_2(s_2, s_1)$ and we get an $\mathbb{F}$-adapted process $\varphi_3(s_3, s_2, s_1)$, $0 \leq s_3 \leq s_2$, such that

$$E\left[\int_0^{s_2} \varphi_3^2(s_3, s_2, s_1)ds_3\right] \le E[\varphi_2^2(s_2, s_1)] < \infty \tag{1.23}$$

and

$$\varphi_2(s_2, s_1) = E[\varphi_2(s_2, s_1)] + \int_0^{s_2} \varphi_3(s_3, s_2, s_1)dW(s_3). \tag{1.24}$$

Substituting (1.24) in (1.22) we get

$$\xi = g_0 + \int_0^T g_1(s_1)dW(s_1) + \int_0^T \int_0^{s_1} g_2(s_2, s_1)dW(s_2)dW(s_1)$$

$$+ \int_0^T \int_0^{s_1} \int_0^{s_2} \varphi_3(s_3, s_2, s_1)dW(s_3)dW(s_2)dW(s_1),$$

where

$$g_2(s_2, s_1) = E[\varphi_2(s_2, s_1)], \qquad 0 \le s_2 \le s_1 \le T.$$

By (1.18), (1.20), (1.23), and the Itô isometry we have

$$E\left[\left(\int_0^T \int_0^{s_1} \int_0^{s_2} \varphi_3(s_3, s_2, s_1)dW(s_3)dW(s_2)dW(s_1)\right)^2\right] \le E\left[\xi^2\right].$$

By iterating this procedure we obtain after n steps a process $\varphi_{n+1}(t_1, t_2, \ldots, t_{n+1})$, $0 \le t_1 \le t_2 \le \cdots \le t_{n+1} \le T$, and $n+1$ deterministic functions $g_0, g_1, \ldots, g_n$, with g_0 constant and g_k defined on S_k for $1 \le k \le n$, such that

$$\xi = \sum_{k=0}^n J_k(g_k) + \int_{S_{n+1}} \varphi_{n+1}dW^{\otimes(n+1)},$$

where

$$\int_{S_{n+1}} \varphi_{n+1}dW^{\otimes(n+1)} := \int_0^T \int_0^{t_{n+1}} \cdots \int_0^{t_2} \varphi_{n+1}(t_1, \ldots, t_{n+1})dW(t_1)\cdots dW(t_{n+1})$$

is the $(n+1)$-fold iterated integral of φ_{n+1}. Moreover,

$$E\left[\left(\int_{S_{n+1}} \varphi_{n+1}dW^{\otimes(n+1)}\right)^2\right] \le E\left[\xi^2\right].$$

In particular, the family

$$\psi_{n+1} := \int_{S_{n+1}} \varphi_{n+1} dW^{\otimes(n+1)}, \qquad n = 1, 2, \ldots$$

is bounded in $L^2(P)$ and, from the Itô isometry,

$$(\psi_{n+1}, J_k(f_k))_{L^2(P)} = 0 \tag{1.25}$$

for $k \leq n$, $f_k \in L^2([0, T]^k)$. Hence we have

$$\|\xi\|^2_{L^2(P)} = \sum_{k=0}^{n} \|J_k(g_k)\|^2_{L^2(P)} + \|\psi_{n+1}\|^2_{L^2(P)}.$$

In particular,

$$\sum_{k=0}^{n} \|J_k(g_k)\|^2_{L^2(P)} < \infty, \; n = 1, 2, \ldots$$

and therefore $\sum_{k=0}^{\infty} J_k(g_k)$ is convergent in $L^2(P)$. Hence

$$\lim_{n \to \infty} \psi_{n+1} =: \psi$$

exists in $L^2(P)$. But by (1.25) we have

$$(J_k(f_k), \psi)_{L^2(P)} = 0$$

for all k and for all $f_k \in L^2([0, T]^k)$. In particular, by (1.15) this implies that

$$E\left[h_k\left(\frac{\theta}{\|g\|}\right) \cdot \psi\right] = 0$$

for all $g \in L^2([0, T])$ and for all $k \geq 0$, where $\theta = \int_0^T g(t)dW(t)$. But then, from the definition of the Hermite polynomials,

$$E[\theta^k \cdot \psi] = 0$$

for all $k \geq 0$, which again implies that

$$E[\exp \theta \cdot \psi] = \sum_{k=0}^{\infty} \frac{1}{k!} E[\theta^k \cdot \psi] = 0.$$

Since the family

$$\{\exp \theta : \quad g \in L^2([0, T])\}$$

is total in $L^2(P)$ (see [179, Lemma 4.3.2]), we conclude that $\psi = 0$. Hence, we conclude

$$\xi = \sum_{k=0}^{\infty} J_k(g_k) \qquad (1.26)$$

and

$$\|\xi\|_{L^2(P)}^2 = \sum_{k=0}^{\infty} \|J_k(g_k)\|_{L^2(P)}^2. \qquad (1.27)$$

Finally, to obtain (1.16)–(1.17) we proceed as follows. The function g_n is defined only on S_n, but we can extend g_n to $[0,T]^n$ by putting

$$g_n(t_1,\ldots,t_n) = 0, \qquad (t_1,\ldots,t_n) \in [0,T]^n \setminus S_n.$$

Now define $f_n := \widetilde{g}_n$ to be the symmetrization of g_n - cf. (1.5). Then

$$I_n(f_n) = n! J_n(f_n) = n! J_n(\widetilde{g}_n) = J_n(g_n)$$

and (1.16) and (1.17) follow from (1.26) and (1.27), respectively. □

Example 1.11. What is the Wiener–Itô expansion of $\xi = W^2(T)$? From (1.15) we get

$$2 \int_0^T \int_0^{t_2} 1\, dW(t_1) dW(t_2) = T h_2\Big(\frac{W(T)}{T^{1/2}}\Big) = W^2(T) - T,$$

and therefore

$$\xi = W^2(T) = T + I_2(1).$$

Example 1.12. Note that for a fixed $t \in (0,T)$ we have

$$\int_0^T \int_0^{t_2} \chi_{\{t_1 < t < t_2\}}(t_1, t_2) dW(t_1) dW(t_2) = \int_t^T W(t) dW(t_2) = W(t)\big(W(T) - W(t)\big).$$

Hence, if we put

$$\xi = W(t)(W(T) - W(t)), \qquad g(t_1, t_2) = \chi_{\{t_1 < t < t_2\}}$$

we can see that

$$\xi = J_2(g) = 2 J_2(\widetilde{g}) = I_2(f_2),$$

where

$$f_2(t_1, t_2) = \widetilde{g}(t_1, t_2) = \frac{1}{2}\big(\chi_{\{t_1 < t < t_2\}} + \chi_{\{t_2 < t < t_1\}}\big).$$

Here and in the sequel we denote the *indicator function* by

$$\chi = \chi_A(x) = \chi_{\{x \in A\}} := \begin{cases} 1, & x \in A, \\ 0, & x \notin A. \end{cases}$$

1.3 Exercises

Problem 1.1. (*) Let $h_n(x)$, $n = 0, 1, 2, \ldots$, be the Hermite polynomials defined in (1.13).

(a) Prove that

$$\exp\left\{tx - \frac{t^2}{2}\right\} = \sum_{n=0}^{\infty} \frac{t^n}{n!} h_n(x).$$

[*Hint.* Write $\exp\{tx - \frac{t^2}{2}\} = \exp\{\frac{1}{2}x^2\} \cdot \exp\{-\frac{1}{2}(x - t)^2\}$ and apply the Taylor formula on the last factor.]

(b) Show that if $\lambda > 0$ then

$$\exp\left\{tx - \frac{t^2\lambda}{2}\right\} = \sum_{n=0}^{\infty} \frac{t^n \lambda^{\frac{n}{2}}}{n!} h_n\left(\frac{x}{\sqrt{\lambda}}\right).$$

(c) Let $g \in L^2([0, T])$. Put

$$\theta = \int_0^T g(s) dW(s).$$

Show that

$$\exp\left\{\int_0^T g(s) dW(s) - \frac{1}{2}\|g\|^2\right\} = \sum_{n=0}^{\infty} \frac{\|g\|^n}{n!} h_n\left(\frac{\theta}{\|g\|}\right),$$

where $\|g\| = \|g\|_{L^2([0,T])}$.

(d) Let $t \in [0, T]$. Show that $\exp\{W(t) - \frac{1}{2}t\} = \sum_{n=0}^{\infty} \frac{t^{n/2}}{n!} h_n\left(\frac{W(t)}{\sqrt{t}}\right)$.

Problem 1.2. Let ξ and ζ be F_T-measurable random variables in $L^2(P)$ with Wiener–Itô chaos expansions $\xi = \sum_{n=0}^{\infty} I_n(f_n)$ and $\zeta = \sum_{n=0}^{\infty} I_n(g_n)$, respectively. Prove that the chaos expansion of the sum $\xi + \zeta = \sum_{n=0}^{\infty} I_n(h_n)$ is such that $h_n = f_n + g_n$ for all $n = 1, 2, \ldots$

Problem 1.3. (*) Find the Wiener–Itô chaos expansion of the following random variables:

(a) $\xi = W(t)$, where $t \in [0, T]$ is fixed,

(b) $\xi = \int_0^T g(s) dW(s)$, where $g \in L^2([0, T])$,

(c) $\xi = W^2(t)$, where $t \in [0, T]$ is fixed,

(d) $\xi = \exp\{\int_0^T g(s) dW(s)\}$, where $g \in L^2([0, T])$ [*Hint.* Use (1.15).],

(e) $\xi = \int_0^T g(s) W(s) ds$, where $g \in L^2([0, T])$.

Problem 1.4. (*) The *Itô representation theorem* states that if $F \in L^2(P)$ is $\mathcal{F}_T$-measurable, then there exists a unique $\mathbb{F}$-adapted process $\varphi = \varphi(t), 0 \leq t \leq T$, such that

$$F = E[F] + \int_0^T \varphi(t)dW(t).$$

This result only provides the *existence* of the integrand φ, but from the point of view of applications it is important also to be able to find the integrand φ more explicitly. This can be achieved, for example, by the *Clark–Ocone formula* (see Chap. 4), which says that, under some suitable conditions,

$$\varphi(t) = E[D_t F | \mathcal{F}_t], \qquad 0 \leq t \leq T,$$

where $D_t F$ is the *Malliavin derivative* of F. We discuss this topic later in the book. However, for certain random variables F it is possible to find φ directly, by using the Itô formula. For example, find φ when

(a) $F = W^2(T)$
(b) $F = \exp\{W(T)\}$
(c) $F = \int_0^T W(t)dt$
(d) $F = W^3(T)$
(e) $F = \cos W(T)$ [*Hint.* Check that $N(t) := e^{\frac{1}{2}t} \cos W(t), t \in [0, T]$, is a martingale.]

Problem 1.5. (*) This exercise is based on [108]. Suppose the function F of Problem 1.4 has the form

$$F = f(X(T)),$$

where $X = X(t), t \in [0, T]$, is an Itô diffusion given by

$$dX(t) = b(X(t))dt + \sigma(X(t))dW(t); \qquad X(0) = x \subset \mathbb{R}.$$

Here $b : \mathbb{R} \to \mathbb{R}$ and $\sigma : \mathbb{R} \to \mathbb{R}$ are given Lipschitz continuous functions of at most linear growth, so there exists a unique strong solution $X(t) = X^x(t), t \in [0, T]$. Then there is a useful formula for the process φ in the Itô representation theorem. This formula is achieved as follows. If g is a real function such that

$$E[|g(X^x(t))|] < \infty,$$

then we define

$$u(t, x) := P_t g(x) := E[g(X^x(t))], \qquad t \in [0, T], \quad x \in \mathbb{R}.$$

Suppose that there exists $\delta > 0$ such that

$$|\sigma(x)| \geq \delta \quad \text{for all } x \in \mathbb{R}. \tag{1.28}$$

Then $u(t, x) \in C^{1,2}(\mathbb{R}^+ \times \mathbb{R})$ and

$$\frac{\partial u}{\partial t} = b(x)\frac{\partial u}{\partial x} + \frac{1}{2}\sigma^2(x)\frac{\partial^2 u}{\partial x^2}$$

(this is the Kolmogorov backward equation, see, for example, [74, Volume 1, Theorem 5.11, p. 162 and Volume 2, Theorem 13.18, p. 53], [177, Theorem 8.1] for details on this issue).

(a) Use the Itô formula for the process

$$Y(t) = g(t, X(t)), \quad t \in [0, T], \quad \text{with} \quad g(t, x) = P_{T-t}f(x)$$

to show that

$$f(X(T)) = P_T f(x) + \int\limits_0^T \left[\sigma(\xi)\frac{\partial}{\partial \xi}P_{T-t}f(\xi)\right]_{|\xi=X(t)} dW(t), \qquad (1.29)$$

for all $f \in C^2(\mathbb{R})$. In other words, with the notation of Problem 1.4, we have shown that if $F = f(X(T))$, then

$$E[F] = P_T f(x) \quad \text{and} \quad \varphi(t) = \left[\sigma(\xi)\frac{\partial}{\partial \xi}P_{T-t}f(\xi)\right]_{|\xi=X(t)}. \qquad (1.30)$$

(b) Use (1.30) to compute $E[F]$ and find φ in the Itô representation of the following random variables:
(b.1) $F = W^2(T)$
(b.2) $F = W^3(T)$
(b.3) $F = X(T)$, where $X(t), t \in [0, T]$, is the geometric Brownian motion, that is,

$$dX(t) = \rho X(t)dt + \alpha X(t)dW(t); \quad X(0) = x \in \mathbb{R} \qquad (\rho, \alpha \text{ constants}).$$

(c) Extend formula (1.30) to the case when $X(t) \in \mathbb{R}^n$, $t \in [0, T]$, and $f : \mathbb{R}^n \to \mathbb{R}$. In this case, condition (1.28) must be replaced by the *uniform ellipticity* condition

$$\eta^T \sigma^T(x)\sigma(x)\eta \geq \delta|\eta|^2 \quad \text{for all } x \in \mathbb{R}^n, \eta \in \mathbb{R}^n, \qquad (1.31)$$

where $\sigma^T(x)$ denotes the transposed of the $m \times n$-matrix $\sigma(x)$.

2

The Skorohod Integral

The Wiener–Itô chaos expansion is a convenient starting point for the introduction of several important stochastic concepts. In this chapter we focus on the *Skorohod integral*. This stochastic integral, introduced for the first time by A. Skorohod in 1975 [217], may be regarded as an extension of the Itô integral to integrands that are not necessarily $\mathbb{F}$-adapted, see also, for example, [30, 31]. The Skorohod integral is also connected to the Malliavin derivative, which is introduced with full detail in Chap. 3.

As for other extensions of the Itô integral closely related to the Skorohod integral, we can mention the *noncausal integral* (also called *Ogawa integral*) and refer to [175, 176]; see also [86].

2.1 The Skorohod Integral

Let $u = u(t, \omega)$, $t \in [0, T], \omega \in \Omega$, be a measurable stochastic process such that, for all $t \in [0, T]$, $u(t)$ is a $\mathcal{F}_T$-measurable random variable and

$$E[u^2(t)] < \infty.$$

Then, for each $t \in [0, T]$, we can apply the Wiener–Itô chaos expansion to the random variable $u(t) = u(t, \omega)$, $\omega \in \Omega$, and thus there exist the symmetric functions $f_{n,t} = f_{n,t}(t_1, \ldots, t_n)$, $(t_1, \ldots, t_n) \in [0, T]^n$, in $\widetilde{L}^2([0, T]^n)$, $n = 1, 2, \ldots$, such that

$$u(t) = \sum_{n=0}^{\infty} I_n(f_{n,t}).$$

Note that the functions $f_{n,t}$, $n = 1, 2, \ldots$, depend on the parameter $t \in [0, T]$, and so we can write

$$f_n(t_1, \ldots, t_n, t_{n+1}) = f_n(t_1, \ldots, t_n, t) := f_{n,t}(t_1, \ldots, t_n)$$

G.Di Nunno et al., *Malliavin Calculus for Lévy Processes with Applications to Finance,*
© Springer-Verlag Berlin Heidelberg 2009

and we may regard f_n as a function of $n+1$ variables. Since this function is symmetric with respect to its first n variables, its *symmetrization* $\widetilde{f}_n$ is given by

$$\widetilde{f}_n(t_1,\ldots,t_{n+1}) = \frac{1}{n+1}\Big[f_n(t_1,\ldots,t_{n+1})$$
$$+f_n(t_2,\ldots,t_{n+1},t_1)+\cdots+f_n(t_1,\ldots,t_{n-1},t_{n+1},t_n)\Big], \quad (2.1)$$

see (1.5).

Example 2.1. Let us consider

$$f_{2,t}(t_1,t_2) = f_2(t_1,t_2,t) = \frac{1}{2}\big[\chi_{\{t_1<t<t_2\}} + \chi_{\{t_2<t<t_1\}}\big].$$

Then the symmetrization $\widetilde{f}_2$ of f_2 is given by

$$\widetilde{f}_2(t_1,t_2,t_3) = \frac{1}{3}\Big[\frac{1}{2}(\chi_{\{t_1<t_3<t_2\}} + \chi_{\{t_2<t_3<t_1\}})$$
$$+\frac{1}{2}(\chi_{\{t_1<t_2<t_3\}} + \chi_{\{t_3<t_2<t_1\}}) + \frac{1}{2}(\chi_{\{t_3<t_1<t_2\}} + \chi_{\{t_2<t_1<t_3\}})\Big],$$

which gives

$$\widetilde{f}_2(t_1,t_2,t_3) = \frac{1}{6}. \qquad (2.2)$$

Definition 2.2. *Let* $u(t)$, $t \in [0,T]$, *be a measurable stochastic process such that for all* $t \in [0,T]$ *the random variable* $u(t)$ *is* $\mathcal{F}_T$-*measurable and* $E[\int_0^T u^2(t)dt] < \infty$. *Let its Wiener–Itô chaos expansion be*

$$u(t) = \sum_{n=0}^{\infty} I_n(f_{n,t}) = \sum_{n=0}^{\infty} I_n(f_n(\cdot,t)).$$

Then we define the Skorohod integral of u *by*

$$\delta(u) := \int_0^T u(t)\delta W(t) := \sum_{n=0}^{\infty} I_{n+1}(\widetilde{f}_n) \qquad (2.3)$$

when convergent in $L^2(P)$. *Here* $\widetilde{f}_n$, $n = 1,2,...$, *are the symmetric functions* (2.1) *derived from* $f_n(\cdot,t)$, $n = 1,2,....$ *We say that* u *is Skorohod integrable, and we write* $u \in Dom(\delta)$ *if the series in* (2.3) *converges in* $L^2(P)$ *(see also Problem 2.1).*

Remark 2.3. By (1.17) a stochastic process u belongs to $Dom(\delta)$ if and only if

$$E[\delta(u)^2] = \sum_{n=0}^{\infty}(n+1)!\|\widetilde{f}_n\|_{L^2([0,T]^{n+1})}^2 < \infty. \qquad (2.4)$$

Example 2.4. Let us verify that

$$\int_0^T W(T)\delta W(t) = W^2(T) - T.$$

The Wiener–Itô chaos expansion of the integrand $u(t) = W(T) = \int_0^T 1\, dW(s)$, $t \in [0, T]$, is given by $f_{0,t} = 0$, $f_{1,t} = 1$, and $f_{n,t} = 0$ for $n \geq 2$. Hence

$$\delta(u) = I_2(\widetilde{f_1}) = I_2(1) = 2 \int_0^T \int_0^{t_2} 1\, dW(t_1)dW(t_2) = W^2(T) - T.$$

Note that, even if the integrand does not depend on t, we have

$$\int_0^T W(T)\delta W(t) \neq W(T) \int_0^T \delta W(t).$$

This last remark illustrates that, *for* $u \in Dom(\delta)$, *even if* G *is an* $\mathcal{F}_T$-*measurable random variable such that* $Gu \in Dom(\delta)$, *we have in general that*

$$\int_0^T Gu(t)\delta W(t) \neq G \int_0^T u(t)\delta W(t). \tag{2.5}$$

Example 2.5. What is $\int_0^T W(t)\big[W(T) - W(t)\big]\delta W(t)$? Note that

$$\int_0^T \int_0^{t_2} \chi_{\{t_1 < t < t_2\}}(t_1, t_2)dW(t_1)dW(t_2) = \int_0^T W(t)\chi_{\{t < t_2\}}(t_2)dW(t_2)$$

$$= W(t)\big[W(T) - W(t)\big].$$

Hence

$$u(t) = W(t)\big[W(T) - W(t)\big] = I_2(f_2(\cdot, t)),$$

where

$$f_{2,t}(t_1, t_2) = f_2(t_1, t_2, t) = \frac{1}{2}\Big(\chi_{\{t_1 < t < t_2\}} + \chi_{\{t_2 < t < t_1\}}\Big).$$

Hence by Example 2.1 we have

$$\delta(u) = I_3(\widetilde{f_2}) = I_3(\frac{1}{6}) = \frac{1}{6}I_3(1) = \frac{1}{6}\Big[W^3(T) - 3T\, W(T)\Big].$$

2.2 Some Basic Properties of the Skorohod Integral

The reader accustomed with classical analysis and Itô stochastic integration may find (2.3) to be just a formal definition for an operator, which can hardly be matched with the general meaning of integral. The purpose of the two following sections is to motivate Definition 2.2, showing that the operator (2.3) is a meaningful stochastic integral having strong links with the Itô stochastic integral itself. In the forthcoming Chaps. 3 and 5, more will be said about the properties of the Skorohod integral.

First of all we recognize that, just like any integral in classical analysis, the Skorohod integral (2.3) is a *linear* operator:

$$L^2(P \times \lambda) \supseteq Dom(\delta) \ni u \implies \delta(u) \in L^2(P).$$

See Problem 2.2.

Another typical property of integrals is the additivity on adjacent intervals of integration. This also holds for the Skorohod integral.

Proposition 2.6. *For any fixed* $t \in [0, T]$ *and* $u \in Dom(\delta)$ *we have* $\chi_{(0,t]}u \in Dom(\delta)$ *and* $\chi_{(t,T]}u \in Dom(\delta)$ *and*

$$\int_0^t u(s)\delta W(s) = \int_0^T \chi_{(0,t]}(s)u(s)\delta W(s) \text{ and } \int_t^T u(s)\delta W(s) = \int_0^T \chi_{(t,T]}(s)u(s)\delta W(s),$$

with

$$\int_0^T u(s)\delta W(s) = \int_0^t u(s)\delta W(s) + \int_t^T u(s)\delta W(s).$$

Proof The proof, based on the Wiener-Itô chaos expansions and (2.4), is left as an exercise. See Problem 2.3.   □

Proposition 2.7. *For any* $u \in Dom(\delta)$ *the Skorohod integral has zero expectation, that is,*

$$E[\delta(u)] = 0. \tag{2.6}$$

Proof This is a trivial consequence of the fact that Itô integrals and thus also iterated Itô integrals have zero expectation.   □

Here we address all those who associate the name of "integral" to the operators resulting from the classical construction, which defines the integral as some limit of certain finite sums derived from simple functions (e.g., Rieman integral, Lebesgue integral, and Itô integral). In some sense also the Skorohod integral can be regarded as such (see, e.g., [169]). A full characterization in this sense can be given in the white noise framework, see Theorem 5.20 and Corollary 5.21.

2.3 The Skorohod Integral as an Extension of the Itô Integral

As mentioned earlier, the Skorohod integral is an extension of the Itô integral. More precisely, if the integrand u is $\mathbb{F}$-adapted, then the two integrals coincide as elements of $L^2(P)$. To prove this, we need a characterization of adaptedness with respect to $\mathbb{F}$ in terms of the functions $f_n(\cdot, t)$, $n = 1, 2, \ldots$, in the chaos expansion.

Lemma 2.8. *Let* $u = u(t)$, $t \in [0, T]$, *be a measurable stochastic process such that, for all* $t \in [0, T]$, *the random variable* $u(t)$ *is* $\mathcal{F}_T$-*measurable and* $E[u^2(t)] < \infty$. *Let*

$$u(t) = \sum_{n=0}^{\infty} I_n(f_n(\cdot, t))$$

be its Wiener-Itô chaos expansion. Then u *is* $\mathbb{F}$-*adapted if and only if*

$$f_n(t_1, \ldots, t_n, t) = 0 \qquad if \quad t < \max_{1 \le i \le n} t_i. \tag{2.7}$$

The above equality is meant a.e. in $[0, T]^n$ *with respect to Lebesgue measure.*

Proof First note that for any $g \in \widetilde{L}^2([0, T]^n)$ we have

$$
\begin{aligned}
E[I_n(g)|\mathcal{F}_t] &= n! E[J_n(g)|\mathcal{F}_t] \\
&= n! E\Big[\int_0^T \int_0^{t_n} \cdots \int_0^{t_2} g(t_1, \ldots, t_n) dW(t_1) \cdots dW(t_n) \Big| \mathcal{F}_t \Big] \\
&= n! \int_0^t \int_0^{t_n} \cdots \int_0^{t_2} g(t_1, \ldots, t_n) dW(t_1) \cdots dW(t_n) \\
&= n! J_n(g(t_1, \ldots, t_n) \cdot \chi_{\{\max t_i < t\}}) \\
&= I_n(g(t_1, \ldots, t_n) \cdot \chi_{\{\max t_i < t\}}).
\end{aligned}
\tag{2.8}
$$

Now, u is $\mathbb{F}$-adapted if and only if $E[u(t)|\mathcal{F}_t] = u(t)$. Namely, if and only if $\sum_{n=0}^{\infty} I_n(f_n(\cdot, t)) = \sum_{n=0}^{\infty} E[I_n(f_n(\cdot, t))|\mathcal{F}_t] = \sum_{n=0}^{\infty} I_n(f_n(\cdot, t) \cdot \chi_{\{\max t_i < t\}})$. And thus if and only if $f_n(t_1, \ldots, t_n, t) \cdot \chi_{\{\max t_i < t\}} = f_n(t_1, \ldots, t_n, t)$ a.e. in $[0, T]^n$ with respect to Lebesgue measure. By uniqueness of the sequence of deterministic functions in the Wiener-Itô chaos expansion and since the last identity is equivalent to (2.7), the lemma is proved. $\square$

Theorem 2.9. *Let* $u = u(t)$, $t \in [0, T]$, *be a measurable* $\mathbb{F}$-*adapted stochastic process such that*

$$E\Big[\int_0^T u^2(t) dt \Big] < \infty.$$

Then $u \in Dom(\delta)$ and its Skorohod integral coincides with the Itô integral

$$\int_0^T u(t)\delta W(t) = \int_0^T u(t)dW(t). \qquad (2.9)$$

Proof Let $u(t) = \sum_{n=0}^{\infty} I_n(f_n(\cdot, t))$ be the chaos expansion of $u(t)$. First note that by (2.1) and Lemma 2.8 we have

$$\widetilde{f}_n(t_1, \ldots, t_n, t_{n+1}) = \frac{1}{n+1} f_n(t_1, \cdots, t_{j-1}, t_{j+1}, \ldots, t_{n+1}, t_j),$$

where

$$j := \mathrm{argmax}_{1 \le i \le n+1} t_i.$$

Hence

$$\|\widetilde{f}_n\|_{L^2([0,T]^{n+1})}^2 = (n+1)! \int_{S_{n+1}} \widetilde{f}_n^2(t_1, \ldots, t_{n+1})dt_1 \cdots dt_{n+1}$$

$$= \frac{(n+1)!}{(n+1)^2} \int_{S_{n+1}} f_n^2(t_1, \ldots, t_{n+1})dt_1 \cdots dt_{n+1}$$

$$= \frac{n!}{n+1} \int_0^T \int_0^t \int_0^{t_n} \cdots \int_0^{t_2} f_n^2(t_1, \ldots, t_n, t)dt_1 \cdots dt_n dt$$

$$= \frac{n!}{n+1} \int_0^T \int_0^T \int_0^{t_n} \cdots \int_0^{t_2} f_n^2(t_1, \ldots, t_n, t)dt_1 \cdots dt_n dt$$

$$= \frac{1}{n+1} \int_0^T \|f_n(\cdot, t)\|_{L^2([0,T]^n)}^2 dt,$$

again by using Lemma 2.8. Hence, by (1.17),

$$\sum_{n=0}^{\infty}(n+1)!\|\widetilde{f}_n\|_{L^2([0,T]^{n+1})}^2 = \sum_{n=0}^{\infty} n! \int_0^T \|f_n(\cdot, t)\|_{L^2([0,T]^n)}^2 dt$$

$$= \int_0^T \sum_{n=0}^{\infty} n!\|f_n(\cdot, t)\|_{L^2([0,T]^n)}^2 dt$$

$$= E\left[\int_0^T u^2(t)dt\right] < \infty.$$

This proves that $u \in Dom(\delta)$, see (2.4). Finally, we prove the relationship (2.9):

$$\int_0^T u(t)dW(t) = \sum_{n=0}^{\infty} \int_0^T I_n(f_n(\cdot, t))dW(t)$$

$$= \sum_{n=0}^{\infty} \int_0^T n! \int_{0 \leq t_1 \leq \cdots \leq t_n \leq t} f_n(t_1, \ldots, t_n, t)dW(t_1) \cdots dW(t_n)dW(t)$$

$$= \sum_{n=0}^{\infty} \int_0^T n!(n+1) \int_{0 \leq t_1 \leq \cdots \leq t_n \leq t_{n+1}} \widetilde{f}_n(t_1, \ldots, t_n, t_{n+1})dW(t_1)$$

$$\cdots dW(t_n)dW(t_{n+1})$$

$$= \sum_{n=0}^{\infty} (n+1)!J_{n+1}(\widetilde{f}_n) = \sum_{n=0}^{\infty} I_{n+1}(\widetilde{f}_n) = \int_0^T u(t)\delta W(t).$$

By this the proof is complete.   □

2.4 Exercises

Problem 2.1. Let $u(t), 0 \leq t \leq T$, be a measurable stochastic process such that

$$E\left[\int_0^T u^2(t)dt\right] < \infty.$$

Show that there exists a sequence of deterministic measurable kernels $f_n(t_1, \ldots, t_n, t)$ on $[0, T]^{n+1}$ $(n \geq 0)$, with

$$\int_{[0,T]^{n+1}} f_n^2(t_1, \ldots, t_n, t)dt_1 \ldots dt_n dt < \infty$$

such that all f_n are symmetric with respect to the variables $t_1, \ldots, t_n$ and such that

$$u(t) = u(\omega, t) = \sum_{n=0}^{\infty} I_n(f_n(\cdot, t))(\omega), \qquad \omega \in \Omega, t \in [0, T],$$

with convergence in $L^2(P \times \lambda)$. [*Hint.* Consider approximations of $u(t)$, $t \in [0, T]$, in $L^2(P \times \lambda)$ of the form $\sum_{i=1}^{m} a_i(\omega)b_i(t)$, $m = 1, 2, \ldots$, where $a_i \in L^2(P)$ and $b_i \in L^2([0, T])$.]

Problem 2.2. Prove the linearity of the Skorohod integral. [*Hint.* See Problem 1.2.]

Problem 2.3. Prove Proposition 2.6.

Problem 2.4. (*) Compute the following Skorohod integrals:

(a) $\int\limits_0^T W(t)\delta W(t)$,

(b) $\int\limits_0^T \left(\int\limits_0^T g(s)dW(s) \right)\delta W(t)$,   for a given function $g \in L^2([0,T])$,

(c) $\int\limits_0^T W^2(t_0)\delta W(t)$,   where $t_0 \in [0,T]$ is fixed,

(d) $\int\limits_0^T \exp\{W(T)\}\delta W(t)$ [*Hint.* Use Problem 1.3.],

(e) $\int_0^T F\delta W(t)$, where $F = \int_0^T g(s)W(s)ds$, with $g \in L^2([0,T])$ [*Hint.* Use Problem 1.3].

3

Malliavin Derivative via Chaos Expansion

3.1 The Malliavin Derivative

The Malliavin calculus (see [158], see also, for example, [53, 72, 160, 169, 212]) was originally created as a tool for studying the regularity of densities of solutions of stochastic differential equations. Subsequently, partly due to the papers [173] and [174], the significance of Malliavin calculus in finance became clear. This triggered a tremendous interest in the subject, also among economists. Today the range of applications has extended even further to include numerical methods, stochastic control, and insider trading, not just for systems driven by Brownian motion, but for systems driven by general Lévy processes. These applications will be covered later in this book.

There are many ways of introducing the Malliavin derivative. The original construction was given on the Wiener space $\Omega = C_0([0,T])$ consisting of all continuous functions $\omega : [0,T] \longrightarrow \mathbb{R}$ with $\omega(0) = 0$. This construction is outlined in Appendix A.

In this book, we mainly use an approach based on chaos expansions. We give a presentation in this chapter. In the Brownian motion case this approach is basically equivalent to the construction of the Malliavin derivative as a stochastic gradient on the space $\Omega = \mathcal{S}'(\mathbb{R})$. This last approach has the advantage of being more intuitive. Moreover, it opens for a useful combination with Hida white noise calculus, which turns out to be a useful framework for both Malliavin calculus, Skorohod integrals, and anticipative calculus in general. We discuss this in Chap. 6.

Definition 3.1. *Let $F \in L^2(P)$ be $\mathcal{F}_T$-measurable with chaos expansion*

$$F = \sum_{n=0}^{\infty} I_n(f_n),$$

where $f_n \in \widetilde{L}^2([0,T]^n)$, $n = 1, 2, \ldots$.

G.Di Nunno et al., *Malliavin Calculus for Lévy Processes with Applications to Finance,*
© Springer-Verlag Berlin Heidelberg 2009

(i) We say that $F \in \mathbb{D}_{1,2}$ if

$$\|F\|_{\mathbb{D}_{1,2}}^2 := \sum_{n=1}^{\infty} nn! \|f_n\|_{L^2([0,T]^n)}^2 < \infty. \tag{3.1}$$

(ii) If $F \in \mathbb{D}_{1,2}$ we define the Malliavin derivative $D_t F$ of F at time t as the expansion

$$D_t F = \sum_{n=1}^{\infty} n I_{n-1}(f_n(\cdot, t)), \qquad t \in [0,T], \tag{3.2}$$

where $I_{n-1}(f_n(\cdot,t))$ is the $(n-1)$-fold iterated integral of $f_n(t_1, ..., t_{n-1}, t)$ with respect to the first $n - 1$ variables $t_1, ..., t_{n-1}$ and $t_n = t$ left as parameter.

Remark 3.2. Note that if (3.1) holds, then

$$\|D.F\|_{L^2(P \times \lambda)}^2 = E\Big[\int_0^T (D_t F)^2 dt \Big] = \sum_{n=1}^{\infty} \int_0^T n^2(n-1)! \|f_n(\cdot,t)\|_{L^2([0,T]^n)}^2 dt \tag{3.3}$$

$$= \sum_{n=1}^{\infty} nn! \|f_n\|_{L^2([0,T]^n)}^2 = \|F\|_{\mathbb{D}_{1,2}}^2 < \infty,$$

so $D.F = D_t F$, $t \in [0,T]$, is well defined as an element of $L^2(P \times \lambda)$.

We first establish the following fundamental result.

Theorem 3.3. Closability of the Malliavin derivative. *Suppose* $F \in L^2(P)$ *and* $F_k \in \mathbb{D}_{1,2}$, $k = 1, 2, ...$, *such that*

(i) $F_k \longrightarrow F$, $k \to \infty$, *in* $L^2(P)$
(ii) $\{D_t F_k\}_{k=1}^{\infty}$ *converges in* $L^2(P \times \lambda)$.

Then $F \in \mathbb{D}_{1,2}$ *and* $D_t F_k \longrightarrow D_t F$, $k \to \infty$, *in* $L^2(P \times \lambda)$.

Proof Let $F = \sum_{n=0}^{\infty} I_n(f_n)$ and $F_k = \sum_{n=0}^{\infty} I_n(f_n^{(k)})$, $k = 1, 2, ...$. Then by (i)

$$f_n^{(k)} \longrightarrow f_n, \quad k \to \infty, \quad \text{in } L^2(\lambda^n)$$

for all n. By (ii) we have

$$\sum_{n=1}^{\infty} nn! \|f_n^{(k)} - f_n^{(j)}\|_{L^2(\lambda^n)}^2 = \|D_t F_k - D_t F_j\|_{L^2(P \times \lambda)}^2 \longrightarrow 0, \quad j, k \to \infty.$$

Hence by the Fatou lemma,

$$\lim_{k \to \infty} \sum_{n=1}^{\infty} nn! \|f_n^{(k)} - f_n\|_{L^2(\lambda^n)}^2 \le \lim_{k \to \infty} \lim_{j \to \infty} \sum_{n=1}^{\infty} nn! \|f_n^{(k)} - f_n^{(j)}\|_{L^2(\lambda^n)}^2 = 0.$$

This implies that $F \in \mathbb{D}_{1,2}$ and

$$D_t F_k \longrightarrow D_t F, \quad k \to \infty, \quad \text{in } L^2(P \times \lambda). \qquad \square$$

3.2 Computation and Properties of the Malliavin Derivative

In this section we proceed presenting a collection of results that constitute the rules of calculus of the Malliavin derivatives.

3.2.1 Chain Rules for Malliavin Derivative

We proceed to prove a useful chain rule for Malliavin derivatives. First let us consider the case when $f_n = f^{\otimes n}$ for some $f \in L^2([0,T])$, that is,

$$f_n(t_1, ..., t_n) = f(t_1) \cdots f(t_n).$$

Then by (1.15) we have

$$I_n(f_n) = \|f\|^n h_n\Big(\frac{\theta}{\|f\|}\Big), \qquad (3.4)$$

where $\|f\| = \|f\|_{L^2([0,T])}$, $\theta = \int_0^T f(t)dW(t)$ and h_n is the Hermite polynomial of order n. Then by (3.2) we have

$$\begin{aligned}
D_t I_n(f_n) &= nI_{n-1}(f_n(\cdot, t)) \\
&= nI_{n-1}(f^{\otimes(n-1)})f(t) \\
&= n\|f\|^{n-1} h_{n-1}\Big(\frac{\theta}{\|f\|}\Big)f(t).
\end{aligned} \qquad (3.5)$$

A basic property of the Hermite polynomials is that

$$h_n'(x) = nh_{n-1}(x). \qquad (3.6)$$

Combining this with (3.4) and (3.5) we get

$$D_t h_n\Big(\frac{\theta}{\|f\|}\Big) = h_n'\Big(\frac{\theta}{\|f\|}\Big)\frac{f(t)}{\|f\|}. \qquad (3.7)$$

In particular, choosing $n = 1$, we get

$$D_t \int_0^T f(s)dW(s) = f(t). \qquad (3.8)$$

Similarly, by (3.6) and induction, for $n - 2, 3, ...$, we have

$$D_t\Big(\int_0^T f(s)dW(s)\Big)^n = n\Big(\int_0^T f(s)dW(s)\Big)^{n-1}f(t). \qquad (3.9)$$

Let $\mathbb{D}_{1,2}^0$ be the set of all $F \in L^2(P)$ whose chaos expansion has only finitely many terms. Then we have the following result.

Theorem 3.4. Product rule for Malliavin derivative. *Suppose $F_1, F_2 \in \mathbb{D}_{1,2}^0$. Then $F_1, F_2 \in \mathbb{D}_{1,2}$ and also $F_1 F_2 \in \mathbb{D}_{1,2}$ with*

$$D_t(F_1 F_2) = F_1 D_t F_2 + F_2 D_t F_1. \tag{3.10}$$

Proof Being $F_1, F_2 \in \mathbb{D}_{1,2}^0$, clearly $F_1, F_2 \in \mathbb{D}_{1,2}$ and, since the Gaussian random variables have all finite moments, we also have that $F_1 F_2 \in L^2(P)$. First of all let us consider the random variables $F_k^{(n)}$ ($n = 1, 2, ..., k = 1, 2$) as linear combination of iterated integrals of tensor products of functions ξ_i in an orthogonal basis $\{\xi_j\}_{j=1}^\infty$ of $L^2([0, T])$. Thanks to the structure of the Hermite polynomials, the argument above together with (1.14) shows that $F_1^{(n)}, F_2^{(n)}$ and $F_1^{(n)} F_2^{(n)}$ are in $\mathbb{D}_{1,2}$ for all n, with

$$D_t(F_1^{(n)} F_2^{(n)}) = F_1^{(n)} D_t F_2^{(n)} + F_2^{(n)} D_t F_1^{(n)}. \tag{3.11}$$

We can choose the two sequences so that $F_k^{(n)} \longrightarrow F_k$ in $L^2(P)$ and $D_t F_k^{(n)} \longrightarrow D_t F_k$ in $L^2(P \times \lambda)$, for $n \to \infty$ ($k = 1, 2$). Then, being $F_1 F_2 \in \mathbb{D}_{1,2}^0$, we have that $F_1^{(n)} F_2^{(n)} \longrightarrow F_1, F_2$ in $L^2(P)$ and also $\{D_t(F_1^{(n)} F_2^{(n)})\}_{n=1}^\infty$ converges in $L^2(P \times \lambda)$. Hence we can conclude by Theorem 3.3. □

See also Problem 3.1.

A version of the chain rule can be formulated as follows, see also [169].

Theorem 3.5. Chain rule. *Let $G \in \mathbb{D}_{1,2}$ and $g \in C^1(\mathbb{R})$ with bounded derivative. Then $g(G) \in \mathbb{D}_{1,2}$ and*

$$D_t g(G) = g'(G) D_t G. \tag{3.12}$$

Here $g'(x) = \frac{d}{dx} g(x)$.

Proof The result can be derived as a corollary to a forthcoming general result. See Theorem 6.3 and Corollary 6.4. □

Remark 3.6. Another chain rule requiring only the Lipschitz continuity of φ can be found in [169, Proposition 1.2.4].

3.2.2 Malliavin Derivative and Conditional Expectation

We now present some preliminary results on conditional expectations.

Definition 3.7. *Let G be a Borel set in $[0, T]$. We define $\mathcal{F}_G$ to be the completed σ-algebra generated by all random variables of the form*

$$F = \int_0^T \chi_A(t) dW(t),$$

for all Borel sets $A \subseteq G$.

Thus if $G = [0, t]$, for any $t \in [0, T]$ fixed, we have that $\mathcal{F}_{[0,t]} = \mathcal{F}_t$. Note that if G_1, G_2 are Borel sets in $[0, T]$, then $\mathcal{F}_{G_1} \cap \mathcal{F}_{G_2} = \mathcal{F}_{G_1 \cap G_2}$.

Lemma 3.8. *For any $g \in L^2([0, T])$ we have*

$$E\left[\int_0^T g(t)dW(t) | \mathcal{F}_G\right] = \int_0^T \chi_G(t)g(t)dW(t).$$

Proof By definition of conditional expectation, it is sufficient to verify that the random variable

$$\int_0^T \chi_G(t)g(t)dW(t) \qquad \text{is } \mathcal{F}_G\text{-measurable} \qquad (3.13)$$

and that

$$E\left[F\int_0^T g(t)dW(t)\right] = E\left[F\int_0^T \chi_G(t)g(t)dW(t)\right] \qquad (3.14)$$

for all bounded $\mathcal{F}_G$-measurable random variables F.

To prove (3.13) we may assume that g is continuous, because the continuous functions are dense in $L^2([0, T])$. If g is continuous, then

$$\int_0^T \chi_G(t)g(t)dW(t) = \lim_{\Delta t_i \to 0} \sum_{i=0}^n g(t_i) \int_{t_i}^{t_{i+1}} \chi_G(t)dW(t),$$

where the limit is in $L^2(P)$ for the vanishing mesh Δt_i of the partitions $0 = t_0 < \ldots < t_n = T$. Since each term in the sum is $\mathcal{F}_G$-measurable, the sum is also $\mathcal{F}_G$-measurable. Then by taking a subsequence converging P-a.s. we conclude that the limit represents an $\mathcal{F}_G$-measurable random variable.

To prove (3.14) we may assume $F = \int_0^T \chi_A(t)dW(t)$ for some $A \subseteq G$. Then by the Itô isometry we have

$$E\left[F\int_0^T g(t)dW(t)\right] = E\left[\int_0^T \chi_A(t)g(t)dt\right],$$

and also

$$E\left[F\int_0^T \chi_G(t)g(t)dW(t)\right] = E\left[\int_0^T \chi_A(t)\chi_G(t)g(t)dt\right] = E\left[\int_0^T \chi_A(t)g(t)dt\right].$$

Then the proof can be completed by a density argument. $\square$

Lemma 3.9. *Let $G \subseteq [0,T]$ be a Borel set and $v = v(t)$, $t \in [0,T]$, be a stochastic process such that*

(1)   for all t, $v(t)$ is measurable with respect to $\mathcal{F}_t \cap \mathcal{F}_G$

(2) $E\big[\int_0^T v^2(t)dt\big] < \infty$.

Then

$$\int_G v(t)dW(t) \qquad \text{is } \mathcal{F}_G\text{-measurable.}$$

Proof By a standard approximation procedure it is sufficient to consider v to be an elementary process of the form

$$v(t) = \sum_{i=1}^n v_i \chi_{(t_i, t_{i+1}]}(t),$$

where $0 = t_0 < t_1 < \cdots < t_n = T$ and v_i are $\mathcal{F}_{t_i} \cap \mathcal{F}_G$-measurable random variables such that (2) is satisfied. For such v we have

$$\int_G v(t)dW(t) = \sum_{i=1}^n v_i \int_{G \cap (t_i, t_{i+1}]} 1\, dW(t),$$

which is a sum of products of $\mathcal{F}_G$-measurable functions and hence $\mathcal{F}_G$-measurable.   □

Lemma 3.10. *Let $u = u(t)$, $t \in [0,T]$, be an $\mathbb{F}$-adapted stochastic process in $L^2(P \times \lambda)$. Then*

$$E\Big[\int_0^T u(t)dW(t)|\mathcal{F}_G\Big] = \int_G E[u(t)|\mathcal{F}_G]dW(t).$$

Proof Lemma 3.9 guarantees that $\int_G E[u(t)|\mathcal{F}_G]dW(t)$ is $\mathcal{F}_G$-measurable. Then it suffices to verify that

$$E\Big[F\int_0^T u(t)dW(t)\Big] = E\Big[F\int_G E[u(t)|\mathcal{F}_G]dW(t)\Big]$$

for all F of the form $F = \int_A dW(t)$, where $A \subseteq G$ is a Borel set. In this case we obtain by the Itô isometry that

$$E\Big[F\int_0^T u(t)dW(t)\Big] = E\Big[\int_0^T \chi_A(t)u(t)dt\Big] = \int_A E[u(t)]dt$$

and

$$E\Big[F\int_G E[u(t)|\mathcal{F}_G]dW(t)\Big] = E\Big[\int_0^T \chi_A(t)\chi_G(t)E[u(t)|\mathcal{F}_G]dt\Big]$$

$$= \int_0^T \chi_A(t)E\big[E[u(t)|\mathcal{F}_G]\big]dt$$

$$= \int_A E[u(t)]dt.$$

A density argument completes the proof. $\square$

Proposition 3.11. *Let* $f_n \in \widetilde{L}^2([0,T]^n)$, $n = 1,2,\dots$ *Then*

$$E[I_n(f_n)|\mathcal{F}_G] = I_n[f_n\chi_G^{\otimes n}], \tag{3.15}$$

where $(f_n\chi_G^{\otimes n})(t_1,\dots,t_n) = f_n(t_1,\dots,t_n)\chi_G(t_1)\cdots\chi_G(t_n)$.

Proof We proceed by induction on n. For $n = 1$ we have

$$E[I_1(f_1)|\mathcal{F}_G] = E\Big[\int_0^T f_1(t_1)dW(t_1)|\mathcal{F}_G\Big] = \int_0^T f_1(t_1)\chi_G(t_1)dW(t_1) = I_1\big[f_1\chi_G^{\otimes 1}\big]$$

by Lemma 3.10. Assume that (3.15) holds for $n = k$. Then, again by Lemma 3.10, we have

$$E[I_{k+1}(f_{k+1})|\mathcal{F}_G]$$

$$= (k+1)!E\Big[\int_0^T\int_0^{t_{k+1}}\cdots\int_0^{t_2} f_{k+1}(t_1,\dots,t_{k+1})dW(t_1)\cdots dW(t_k)dW(t_{k+1})|\mathcal{F}_G\Big]$$

$$= (k+1)!\int_0^T E\Big[\int_0^{t_{k+1}}\cdots\int_0^{t_2} f_{k+1}(t_1,\dots,t_{k+1})dW(t_1)\cdots dW(t_k)|\mathcal{F}_G\Big]$$

$$\cdot \chi_G(t_{k+1})dW(t_{k+1})$$

$$=\dots= (k+1)!\int_0^T\int_0^{t_{k+1}}\cdots\int_0^{t_2} f_{k+1}(t_1,\dots,t_{k+1})\chi_G(t_1)\cdots\chi_G(t_{k+1})dW(t_1)\cdots dW(t_{k+1})$$

$$= I_{k+1}[f_{k+1}\chi_G^{\otimes(k+1)}],$$

and the proof is complete. $\square$

Proposition 3.12. *If* $F \in \mathbb{D}_{1,2}$, *then* $E[F|\mathcal{F}_G] \in \mathbb{D}_{1,2}$ *and*

$$D_t E[F|\mathcal{F}_G] = E[D_t F|\mathcal{F}_G]\chi_G(t).$$

Proof   First assume that $F = I_n(f_n)$ for some $f_n \in \widetilde{L}^2([0,T]^n)$. By Proposition 3.11 we have

$$
\begin{aligned}
D_t E[F|\mathcal{F}_G] &= D_t E[I_n(f_n)|\mathcal{F}_G] \\
&= D_t I_n(f_n \chi_G^{\otimes n}) \\
&= n I_{n-1}[f_n(\cdot, t)\chi_G^{\otimes(n-1)}(\cdot)\chi_G(t)] \qquad (3.16) \\
&= n I_{n-1}[f_n(\cdot, t)\chi_G^{\otimes(n-1)}(\cdot)]\chi_G(t) \\
&= E[D_t F|\mathcal{F}_G]\chi_G(t).
\end{aligned}
$$

Next, let $F = \sum_{n=0}^{\infty} I_n(f_n)$ belong to $\mathbb{D}_{1,2}$. Let $F_k = \sum_{n=0}^{k} I_n(f_n)$. Then

$$
F_k \to F \quad \text{in} \quad L^2(\Omega) \quad \text{and} \quad D_t F_k \to D_t F \quad \text{in} \quad L^2(P \times \lambda)
$$

as $k \to \infty$. By (3.16) we have

$$
D_t E[F_k|\mathcal{F}_G] = E[D_t F_k|\mathcal{F}_G]\chi_G(t),
$$

for all k, and taking the limit with convergence in $L^2(P \times \lambda)$ of this, as $k \to \infty$, we obtain the result. $\square$

Corollary 3.13. *Let $u = u(s)$, $s \in [0,T]$, be an $\mathbb{F}$-adapted stochastic process and assume that $u(s) \in \mathbb{D}_{1,2}$ for all s. Then*

(i) $D_t u(s)$, $s \in [0,T]$, is $\mathbb{F}$-adapted for all t;
(ii) $D_t u(s) = 0$, for $t > s$.

Proof   By Proposition 3.12 we have that

$$
D_t u(s) = D_t E[u(s)|\mathcal{F}_s] = E[D_t u(s)|\mathcal{F}_s]\chi_{[0,s]}(t) = E[D_t u(s)|\mathcal{F}_s]\chi_{[t,T]}(s),
$$

from which (i) and (ii) follow immediately. $\square$

3.3 Malliavin Derivative and Skorohod Integral

3.3.1 Skorohod Integral as Adjoint Operator to the Malliavin Derivative

The following result shows that the Malliavin derivative is the adjoint operator of the Skorohod integral.

Theorem 3.14. Duality formula. *Let $F \in \mathbb{D}_{1,2}$ be $\mathcal{F}_T$-measurable and let u be a Skorohod integrable stochastic process. Then*

$$
E\left[F \int_0^T u(t)\delta W(t)\right] = E\left[\int_0^T u(t)D_t F\, dt\right]. \qquad (3.17)
$$

Proof Let $F = \sum_{n=0}^{\infty} I_n(f_n)$ and, for all t, $u(t) = \sum_{k=0}^{\infty} I_k(g_k(\cdot, t))$ be the chaos expansions of F and $u(t)$, respectively. Then

$$E\Big[F \int_0^T u(t)\delta W(t)\Big] = E\Big[\sum_{n=0}^{\infty} I_n(f_n) \int_0^T \sum_{k=0}^{\infty} I_k(g_k(\cdot, t))\delta W(t)\Big]$$

$$= E\Big[\sum_{n=0}^{\infty} I_n(f_n) \sum_{k=0}^{\infty} I_{k+1}(\widetilde{g}_k)\Big]$$

$$= E\Big[\sum_{k=0}^{\infty} I_{k+1}(f_{k+1}) I_{k+1}(\widetilde{g}_k)\Big] \qquad (3.18)$$

$$= \sum_{k=0}^{\infty} (k+1)! \int_{[0,T]^{k+1}} f_{k+1}(x)\widetilde{g}_k(x)dx$$

$$= \sum_{k=0}^{\infty} (k+1)! \big(f_{k+1}, \widetilde{g}_k\big)_{L^2([0,T]^{k+1})},$$

where $\widetilde{g}_k$ is the symmetrization of $g_k(x_1, ..., x_n, t)$ as a function of $n + 1$ variables (see (2.1)). On the other side we have

$$E\Big[\int_0^T u(t)D_t F dt\Big] = E\Big[\int_0^T \Big(\sum_{k=0}^{\infty} I_k(g_k(\cdot, t))\Big)\Big(\sum_{n=1}^{\infty} nI_{n-1}(f_n(\cdot, t))\Big)dt\Big]$$

$$= \int_0^T \sum_{k=0}^{\infty} E\big[(k+1)I_k(g_k(\cdot, t))I_k(f_{k+1}(\cdot, t))\big]dt$$

$$\qquad (3.19)$$

$$= \int_0^T \sum_{k=0}^{\infty} (k+1)k! \big(f_{k+1}(\cdot, t), g_k(\cdot, t)\big)_{L^2([0,T]^k)}dt$$

$$= \sum_{k=0}^{\infty} (k+1)! \big(f_{k+1}, g_k\big)_{L^2([0,T]^{k+1})}.$$

Now

$$\big(f_{k+1}, \widetilde{g}_k\big)_{L^2([0,T]^{k+1})} = \int_0^T \big(f_{k+1}(\cdot, t), \widetilde{g}_k(\cdot, t)\big)_{L^2([0,T]^k)}dt$$

$$= \frac{1}{k+1} \sum_{j-1}^{k+1} \int_0^T \big(f_{k+1}(\cdot, t_j), g_k(\cdot, t_j)\big)_{L^2([0,T]^k)}dt_j \qquad (3.20)$$

$$= \int_0^T \big(f_{k+1}(\cdot, t), g_k(\cdot, t)\big)_{L^2([0,T]^k)}dt$$

$$= \big(f_{k+1}, g_k\big)_{L^2([0,T]^{k+1})}.$$

Therefore, by (3.20) combined with (3.18) and (3.19) the result follows. □

3.3.2 An Integration by Parts Formula and Closability of the Skorohod Integral

Theorem 3.15. Integration by parts. *Let $u(t)$, $t \in [0,T]$, be a Skorohod integrable stochastic process and $F \in \mathbb{D}_{1,2}$ such that the product $Fu(t)$, $t \in [0,T]$, is Skorohod integrable. Then*

$$F \int_0^T u(t)\delta W(t) = \int_0^T Fu(t)\delta W(t) + \int_0^T u(t)D_t F dt. \qquad (3.21)$$

Proof First assume that $F \in \mathbb{D}_{1,2}^0$ (see Theorem 3.4). Choose $G \in \mathbb{D}_{1,2}^0$. By Theorem 3.14 and Theorem 3.4 we get

$$E\left[G \int_0^T Fu(t)\delta W(t)\right] = E\left[\int_0^T Fu(t)D_t G dt\right]$$

$$= E\left[GF \int_0^T u(t)\delta W(t)\right] - E\left[G \int_0^T u(t)D_t F dt\right].$$

Since the set of all $G \in \mathbb{D}_{1,2}^0$ is dense in $L^2(P)$, it follows that

$$F \int_0^T u(t)\delta W(t) = \int_0^T Fu(t)\delta W(t) + \int_0^T u(t)D_t F dt \quad P - a.s.$$

Then the result follows for general $F \in \mathbb{D}_{1,2}$ by approximating F by $F^{(n)} \in \mathbb{D}_{1,2}^0$ such that $F^{(n)} \longrightarrow F$ in $L^2(P)$ and $D_t F^{(n)} \longrightarrow D_t F$ in $L^2(P \times \lambda)$, for $n \to \infty$. $\square$

Remark 3.16. The arguments of the proof of Theorem 3.15 actually show that the assumption of the Skorohod integrability of Fu can be replaced by requiring the existence of the integrals

$$F \int_0^T u(t)\delta W(t) \quad \text{and} \quad \int_0^T u(t)D_t F dt$$

in $L^2(P)$.

We can now use the duality formula to prove the following important result.

Theorem 3.17. Closability of the Skorohod integral. *Suppose that $u_n(t)$, $t \in [0,T]$, $n = 1, 2, ...$, is a sequence of Skorohod integrable stochastic processes and that the corresponding sequence of Skorohod integrals*

$$\delta(u_n) := \int_0^T u_n(t)\delta W(t), \quad n = 1, 2, ...$$

converges in $L^2(P)$. Moreover, suppose that

$$\lim_{n\to\infty} u_n = 0 \quad \text{in } L^2(P \times \lambda).$$

Then

$$\lim_{n\to\infty} \delta(u_n) = 0 \quad \text{in } L^2(P).$$

Proof By Theorem 3.14, we have that

$$\left(\delta(u_n), F\right)_{L^2(P)} = \left(u_n, D.F\right)_{L^2(P \times \lambda)} \longrightarrow 0, \quad n \to \infty,$$

for all $F \in \mathbb{D}_{1,2}$. We conclude that $\delta(u_n) \longrightarrow 0$ weakly in $L^2(P)$. Since $\{\delta(u_n)\}_{n=0}^{\infty}$ is convergent in $L^2(P)$, we obtain that $\delta(u_n) \longrightarrow 0$ in $L^2(P)$.

$\square$

3.3.3 A Fundamental Theorem of Calculus

The next result gives a useful connection between differentiation and Skorohod integration.

Theorem 3.18. The fundamental theorem of calculus. *Let* $u = u(s)$, $s \in [0, T]$, *be a stochastic process such that*

$$E\left[\int_0^T u^2(s)ds\right] < \infty \tag{3.22}$$

and assume that, for all $s \in [0, T]$, $u(s) \in \mathbb{D}_{1,2}$ *and that, for all* $t \in [0, T]$, $D_t u \in Dom(\delta)$. *Assume also that*

$$E\left[\int_0^T \left(\delta(D_t u)\right)^2 dt\right] < \infty. \tag{3.23}$$

Then $\int_0^T u(s)\delta W(s)$ *is well-defined and belongs to* $\mathbb{D}_{1,2}$ *and*

$$D_t\left(\int_0^T u(s)\delta W(s)\right) = \int_0^T D_t u(s)\delta W(s) + u(t). \tag{3.24}$$

Proof First assume that

$$u(s) = I_n(f_n(\cdot, s)),$$

where $f_n(t_1, \ldots, t_n, s)$ is symmetric with respect to $t_1, \ldots, t_n$. Then

$$\int_0^T u(s)\delta W(s) = I_{n+1}[\widetilde{f_n}],$$

where

$$\widetilde{f_n}(x_1, \ldots, x_{n+1}) = \frac{1}{n+1}\left[f_n(\cdot, x_1) + \ldots + f_n(\cdot, x_{n+1})\right]$$

is the symmetrization of f_n as a function of all its $n+1$ variables. Hence

$$D_t\left(\int_0^T u(s)\delta W(s)\right) = (n+1)I_n[\widetilde{f}_n(\cdot, t)], \qquad (3.25)$$

where

$$\widetilde{f}_n(\cdot, t) = \frac{1}{n+1}\Big[f_n(t, \cdot, x_1) + \ldots + f_n(t, \cdot, x_n) + f_n(\cdot, t)\Big] \qquad (3.26)$$

(since f_n is symmetric with respect to its first n variables, we may choose t to be the first of them, in the first n terms on the right-hand side). Combining (3.25) with (3.26) we get

$$D_t\left(\int_0^T u(s)\delta W(s)\right) = I_n\Big[f_n(t, \cdot, x_1) + \ldots + f_n(t, \cdot, x_n) + f_n(\cdot, t)\Big] \qquad (3.27)$$

$$= I_n\Big[f_n(t, \cdot, x_1) + \ldots + f_n(t, \cdot, x_n)\Big] + u(t)$$

(the integration in I_n is with respect to $(x_1, \ldots, x_n)$). To compare this with the right-hand side of (3.24) we consider

$$\delta(D_t u) = \int_0^T D_t u(s)\delta W(s)$$

$$= \int_0^T n I_{n-1}[f_n(\cdot, t, s)]\delta W(s)$$

$$= n I_n[\widehat{f}_n(\cdot, t, \cdot)], \qquad (3.28)$$

where

$$\widehat{f}_n(x_1, \ldots, x_{n-1}, t, x_n) = \frac{1}{n}\Big[f_n(t, \cdot, x_1) + \ldots + f_n(t, \cdot, x_n)\Big]$$

is the symmetrization of $f_n(x_1, \ldots, x_{n-1}, t, x_n)$ with respect to $x_1, \ldots, x_n$. Then, from (3.28) we get

$$\int_0^T D_t u(s)\delta W(s) = I_n\Big[f_n(t, \cdot, x_1) + \ldots + f_n(t, \cdot, x_n)\Big]. \qquad (3.29)$$

Comparing (3.27) and (3.29) we obtain (3.24).

Next, consider the general case when

$$u(s) = \sum_{n=0}^{\infty} I_n[f_n(\cdot, s)].$$

Define

$$u_m(s) = \sum_{n=0}^{m} I_n[f_n(\cdot, s)], \qquad m = 1, 2, \ldots.$$

By (3.22) we have $\|u - u_m\|_{L^2(P \times \lambda)}^2 \longrightarrow 0, m \to \infty$.
Then by the above argument we have

$$D_t(\delta(u_m)) = \delta(D_t u_m) + u_m(t), \quad \text{for all } m. \tag{3.30}$$

By (3.28) we see that (3.23) is equivalent to saying that

$$E\left[\int_0^T (\delta(D_t u))^2 dt\right] = \sum_{n=1}^{\infty} n^2 n! \int_0^T \|\widehat{f}_n(\cdot, t, \cdot)\|_{L^2([0,T]^n)}^2 dt$$

$$= \sum_{n=1}^{\infty} n^2 n! \|\widehat{f}_n\|_{L^2([0,T]^{n+1})}^2 < \infty, \tag{3.31}$$

since $D_t u \in Dom(\delta)$. Hence, for $m \to \infty$,

$$\|\delta(D_t u) - \delta(D_t u_m)\|_{L^2(P \times \lambda)}^2 = \sum_{n=m+1}^{\infty} n^2 n! \|\widehat{f}_n\|_{L^2([0,T]^{n+1})}^2 \longrightarrow 0. \tag{3.32}$$

Therefore, by (3.30)

$$D_t(\delta(u_m)) \to \delta(D_t u) + u(t), \quad m \to \infty,$$

in $L^2(P \times \lambda)$. Note that

$$(n+1)\widetilde{f}_n(\cdot, t) = n\widehat{f}_n(\cdot, t, \cdot) + f_n(\cdot, t)$$

and hence

$$(n+1)! \|\widetilde{f}_n\|_{L^2([0,T]^{n+1})}^2 \leq \frac{2n^2 n!}{n+1} \|\widehat{f}_n\|_{L^2([0,T]^{n+1})}^2 + \frac{2n!}{n+1} \|f_n\|_{L^2([0,T]^{n+1})}^2.$$

Therefore,

$$\|\delta(u)\|_{\mathbb{D}_{1,2}}^2 = \sum_{n=0}^{\infty} (n+1)(n+1)! \|\widetilde{f}_n\|_{L^2([0,T]^{n+1})}^2$$

$$\leq \sum_{n=0}^{\infty} \left[2n^2 n! \|\widehat{f}_n\|_{L^2([0,T]^{n+1})}^2 + 2n! \|f_n\|_{L^2([0,T]^{n+1})}^2\right]$$

$$\leq 2\|\delta(D_t u)\|_{L^2(P \times \lambda)}^2 + 2\|u\|_{L^2(P \times \lambda)}^2 < \infty,$$

by (3.31) and (3.22). Then we conclude that $\delta(u)$ is well-defined and belongs to $\mathbb{D}_{1,2}$. By similar computations, we obtain

$$\left\| D_t\left(\int_0^T u(s)\delta W(s) \right) - D_t\left(\int_0^T u_m(s)\delta W(s) \right) \right\|_{L^2(P\times\lambda)}^2$$

$$= \left\| \sum_{n=m+1}^\infty (n+1)I_n(\widetilde{f}_n(\cdot,t)) \right\|_{L^2(P\times\lambda)}^2$$

$$= \int_0^T \sum_{n=m+1}^\infty (n+1)^2 n! \|\widetilde{f}_n(\cdot,t)\|_{L^2([0,T]^n)}^2 dt \tag{3.33}$$

$$\leq 2 \sum_{n=m+1}^\infty \left[n^2 n! \|\widehat{f}_n\|_{L^2([0,T]^{n+1})}^2 + n! \|f_n\|_{L^2([0,T]^{n+1})}^2 \right],$$

which vanishes when $m \to \infty$. Hence given (3.32) and (3.33), we obtain (3.24):

$$D_t(\delta(u)) = \delta(D_t u) + u(t),$$

by letting $m \to \infty$ in (3.30).   □

Corollary 3.19. *Let u be as in Theorem 3.18 and assume in addition that $u(s)$, $s \in [0,T]$, is $\mathbb{F}$-adapted. Then*

$$D_t\left(\int_0^T u(s)dW(s) \right) = \int_t^T D_t u(s)dW(s) + u(t). \tag{3.34}$$

Proof This is an immediate consequence of Theorem 3.18 and Corollary 3.13.

□

3.4 Exercises

Problem 3.1. Let ξ, ζ be orthonormal functions in $L^2([0,T])$. Using the properties of Hermite polynomials compute directly the following:

(a) $I_1(\xi)I_2(\zeta^{\otimes 2})$
(b) $I_3(\xi \hat{\otimes} \zeta^{\otimes 2})$
(c) $D_t I_3(\xi \hat{\otimes} \zeta^{\otimes 2})$ [*Hint.* Use (1.14), (3.5)–(3.9)].

Using the chain rule compute:

(d) $D_t(I_1(\xi)I_2(\zeta^{\otimes 2}))$.

Compare the results in (c) and (d).

Problem 3.2. (*) Find the Malliavin derivative $D_t F$ of the following random variables:

(a) $F = W(T)$.

(b) $F = \int\limits_0^T s^2 dW(s)$.

(c) $F = \int\limits_0^T \int\limits_0^{t_2} \cos(t_1 + t_2) dW(t_1) dW(t_2)$.

(d) $F = 3W(s_0)W^2(t_0) + \log(1 + W^2(s_0))$, for given $s_0, t_0 \in [0, T]$.

(e) $F = \int\limits_0^T W(t_0)\delta W(t)$, for a given $t_0 \in [0, T]$. [*Hint.* Use Problem 2.4 (b).]

Problem 3.3. (*)

(a) Find the Malliavin derivative $D_t F$, when

$$F = e^G \quad \text{with} \quad G = \int\limits_0^T g(s)dW(s), \quad g \in L^2([0, T]),$$

by using that $F = \sum_{n=0}^\infty I_n[f_n]$, with

$$f_n(t_1, \ldots, t_n) = \frac{1}{n!} \exp\left\{\frac{1}{2}\|g\|_{L^2([0,T])}^2\right\} g(t_1) \cdots g(t_n)$$

(see Problem 1.1 and Problem 1.3 (d)).

(b) Verify that the result in (a) can be expressed in terms of the chain rule:
$D_t e^G = e^G D_t G$.

(c) Find the Malliavin derivative of $F = e^G$ with $G = W(t_0)$, for a given $t_0 \in [0, T]$.

Problem 3.4. Use the integration by parts formula (Theorem 3.15) to compute the Skorohod integrals

$$\int_0^T F\delta W(t),$$

for the random variables F given in Problem 3.2 and in Problem 3.3.

Problem 3.5. Use the integration by parts formula to compute the Skorohod integrals in Problem 2.4.

Problem 3.6. Let $u = u(t)$, $t \in [0, T]$, be a stochastic process such that

$$E\left[\int_0^T u^2(t)dt\right] < \infty.$$

Suppose that there exists a constant K (which can depend on u) such that

$$\left|E\left[\int_0^T D_t F u(t)dt\right]\right| \leq K\|F\|_{L^2(P)}, \quad \text{for all } F \in \mathbb{D}_{1,2}.$$

Show that u is Skorohod integrable.

4

Integral Representations and the Clark–Ocone Formula

In this chapter we present explicit stochastic integral representations for random variables in terms of the Malliavin derivative. The central result is the celebrated Clark–Ocone formula. See [46, 47, 98, 173]. We also discuss some generalization of this formula that turns to be central in the application to hedging in mathematical finance. Another application of the Clark–Ocone formula appears in the sensitivity analysis. This is also presented in the last section of this chapter.

4.1 The Clark–Ocone Formula

Theorem 4.1. The Clark–Ocone formula. *Let $F \in \mathbb{D}_{1,2}$ be $\mathcal{F}_T$-measurable. Then*

$$F = E[F] + \int_0^T E[D_t F | \mathcal{F}_t] dW(t). \tag{4.1}$$

Remark 4.2. This theorem gives a representation of the random variable F in terms of Itô stochastic integrals. With respect to this, the present result appears as a version of the Itô integral representation theorem for random variables (see [179] and Problem 1.4). However, the achievement of this result is deeply different from Itô's theorem. In fact this result gives an unexpected link between Sobolev space differential calculus and Itô calculus and it provides an *explicit* representation of the integrand in terms of the Malliavin derivative. This is the core of the Clark–Ocone formula. The fact that the integrand can be expressed in explicit terms turns out to be of fundamental importance in many fields of application. In the forthcoming sections we discuss the applications of this formula in mathematical finance.

Before coming to the proof of this major result we would like to address the attention to the fact that this formula can only be applied to random variables belonging to $\mathbb{D}_{1,2}$. Much has been done in the recent literature to extend the applicability of this result beyond $\mathbb{D}_{1,2}$. Chap. 6 will present a generalization of the Clark–Ocone formula to the random variables in $L^2(P)$ in the framework of white noise analysis.

We can also refer the reader to the following closely related works in the framework of classical stochastic analysis: [59, 60, 70]. Here, for any arbitrary random variable in $L^2(P)$, the integrand in the Itô integral representation theorem is *explicitly* given in terms of the *non-anticipating stochastic derivative*. These results are then extended to cover integration with respect to general martingales and random fields.

Remark 4.3. The $\mathbb{F}$-adapted process $\varphi = E[D_t F | \mathcal{F}_t]$, $t \in [0, T]$, admits a predictable modification. This can be shown, for example, using the arguments suggested in Problem 2.1.

Let us now detail the proof of the Clark–Ocone formula.

Proof Write $F = \sum\limits_{n=0}^{\infty} I_n(f_n)$ with $f_n \in \tilde{L}^2([0,T]^n)$, $n = 1, 2, \ldots$. Then by Proposition 3.11 and Definition 2.2

$$\int_0^T E[D_t F | \mathcal{F}_t] dW(t) = \int_0^T E\Big[\sum_{n=1}^{\infty} n I_{n-1}(f_n(\cdot, t)) | \mathcal{F}_t\Big] dW(t)$$

$$= \int_0^T \sum_{n=1}^{\infty} n E[I_{n-1}(f_n(\cdot, t)) | \mathcal{F}_t] dW(t)$$

$$= \int_0^T \sum_{n=1}^{\infty} n I_{n-1}[f_n(\cdot, t) \cdot \chi_{[0,t]}^{\otimes(n-1)}(\cdot)] dW(t)$$

$$= \int_0^T \sum_{n=1}^{\infty} n(n-1)! J_{n-1}[f_n(\cdot, t) \chi_{[0,t]}^{\otimes(n-1)}] dW(t)$$

$$= \sum_{n=1}^{\infty} n! J_n[f_n(\cdot)] = \sum_{n=1}^{\infty} I_n[f_n]$$

$$= \sum_{n=0}^{\infty} I_n[f_n] - I_0[f_0] = F - E[F]. \qquad \square$$

The following result is a particular case of Theorem 3.14. The proof is here presented as an application of the Clark-Ocone formula.

Corollary 4.4. Duality formula. *Let $F \in \mathbb{D}_{1,2}$ be $\mathcal{F}_T$-measurable and let u be an $\mathbb{F}$-adapted process with*

$$E\Big[\int_0^T u^2(t)dt\Big] < \infty.$$

Then

$$E\Big[F\int_0^T u(t)dW(t)\Big] = E\Big[\int_0^T u(t)D_tF dt\Big].$$

Proof By the Clark–Ocone theorem and the Itô isometry we have

$$E\Big[F\int_0^T u(t)dW(t)\Big] = E\Big[\Big(E[F] + \int_0^T E[D_tF|\mathcal{F}_t]dW(t)\Big)\int_0^T u(t)dW(t)\Big]$$

$$= E\Big[\int_0^T u(t)E[D_tF|\mathcal{F}_t]dt\Big]$$

$$= E\Big[\int_0^T u(t)D_tF dt\Big]. \qquad \Box$$

4.2 The Clark–Ocone Formula under Change of Measure

We proceed to prove the *Clark–Ocone formula under change of measure*. This formula expresses an $\mathcal{F}_T$-measurable random variable F as a stochastic integral with respect to a process of the form

$$\widetilde{W}(t) = \int_0^t u(s)ds + W(t), \qquad 0 \le t \le T, \qquad (4.2)$$

where $u(s)$, $s \in [0,T]$, is a given $\mathbb{F}$-adapted stochastic process satisfying the Novikov condition, that is,

$$E\Big[\exp\Big\{\frac{1}{2}\int_0^T u^2(s)ds\Big\}\Big] < \infty.$$

By the Girsanov theorem (see Problem 4.1) the process $\widetilde{W}(t) = \widetilde{W}(\omega,t)$, $\omega \in \Omega$, $t \in [0,T]$, is a Wiener process (with respect to the filtration $\mathbb{F}$) under the new probability measure Q defined on $(\Omega, \mathcal{F}_T)$ by

$$Q(d\omega) = Z(T,\omega)P(d\omega), \qquad (4.3)$$

where

$$Z(t) = \exp\Big\{-\int_0^t u(s)dW(s) - \frac{1}{2}\int_0^t u^2(s)ds\Big\}, \qquad 0 \le t \le T. \qquad (4.4)$$

We let E_Q denote the expectation with respect to Q, while $E_P = E$ denotes the expectation with respect to P. The following result was first proved in [174].

Theorem 4.5. The Clark–Ocone formula under change of measure.
Let $F \in \mathbb{D}_{1,2}$ be $\mathcal{F}_T$-measurable. Suppose that

$$E_Q[|F|] < \infty \tag{4.5}$$

$$E_Q\left[\int_0^T |D_t F|^2 dt\right] < \infty. \tag{4.6}$$

Moreover, assume that $u(s) \in \mathbb{D}_{1,2}$ for a.a.s, $Z(T)F \in \mathbb{D}_{1,2}$ and

$$E_Q\left[|F| \int_0^T \left(\int_0^T D_t u(s) dW(s) + \int_0^T u(s) D_t u(s) ds\right)^2 dt\right] < \infty. \tag{4.7}$$

Then

$$F = E_Q[F] + \int_0^T E_Q\left[\left(D_t F - F \int_t^T D_t u(s) d\widetilde{W}(s)\right)\big|\mathcal{F}_t\right] d\widetilde{W}(t). \tag{4.8}$$

Remark 4.6. Note that we cannot obtain an integral representation with respect to $\widetilde{W}$ simply by applying the Clark–Ocone formula to our new Wiener process $\widetilde{W}(t)$, $t \in [0, T]$, because F is only assumed to be $\mathcal{F}_T$-measurable, not $\widetilde{\mathcal{F}}_T$-measurable, where $\widetilde{\mathcal{F}}_T$ is the σ-algebra generated by $\widetilde{W}(t)$, $t \in [0, T]$. In general, in fact, we have $\widetilde{\mathcal{F}}_T \subseteq \mathcal{F}_T$ and usually $\widetilde{\mathcal{F}}_T \neq \mathcal{F}_T$.

For a generalization of this result to $F \in L^2(P)$ see Theorem 6.41.

The proof of this results exploits several properties of the Malliavin derivative and the Skorohod integral presented in the previous chapter.

Lemma 4.7. The Bayes rule. Let μ and ν be two probability measures on a measurable space $(\Omega, \mathcal{G})$ such that $\nu(d\omega) = f(\omega)\mu(d\omega)$ for some $f \in L^1(\mu)$. Further, let X be a random variable on $(\Omega, \mathcal{G})$ such that $X \in L^1(\nu)$. Let $\mathcal{H} \subset \mathcal{G}$ be a σ-algebra. Then

$$E_\nu[X|\mathcal{H}] \cdot E_\mu[f|\mathcal{H}] = E_\mu[fX|\mathcal{H}]. \tag{4.9}$$

Proof See e.g. [179, Lemma 8.6.2]. □

Corollary 4.8. Let Q and Z be as in (4.3) and (4.4) respectively. Suppose $G \in L^1(Q)$. Then

$$E_Q[G|\mathcal{F}_t] = \frac{E[Z(T)G|\mathcal{F}_t]}{Z(t)}. \tag{4.10}$$

Lemma 4.9. Let F and u be as in Theorem 4.5 where Q and Z are defined in (4.3) and (4.4). Then

$$D_t(Z(T)F) = Z(T)\left[D_t F - F\left(u(t) + \int_t^T D_t u(s) d\widetilde{W}(s)\right)\right]. \tag{4.11}$$

Proof By the chain rule and Corollary 3.19, we have

$$D_t Z(T) = Z(T)\Big[- D_t \int_0^T u(s)dW(s) - \frac{1}{2}D_t \int_0^T u^2(s)ds\Big]$$

$$= Z(T)\Big[- \int_t^T D_t u(s)dW(s) - u(t) - \int_0^T u(s)D_t u(s)ds\Big]$$

$$= Z(T)\Big[- \int_t^T D_t u(s)d\widetilde{W}(s) - u(t)\Big]. \qquad \square$$

Proof of Theorem 4.5.   Suppose that (4.5)–(4.7) hold and put

$$Y(t) = E_Q[F|\mathcal{F}_t]$$

and

$$\Lambda(t) = Z^{-1}(t) = \exp\Big\{ \int_0^t u(s)dW(s) + \frac{1}{2}\int_0^t u^2(s)ds\Big\}.$$

Note that

$$\Lambda(t) = \exp\Big\{ \int_0^t u(s)d\widetilde{W}(s) - \frac{1}{2}\int_0^t u^2(s)ds\Big\}. \tag{4.12}$$

By Corollary 4.8, Theorem 4.1, and Proposition 3.12 we can write

$$Y_t = \Lambda(t)E[Z(T)F|\mathcal{F}_t]$$

$$= \Lambda(t)\Big[E[E[Z(T)F|\mathcal{F}_t]] + \int_0^T E[D_s E[Z(T)F|\mathcal{F}_t]|\mathcal{F}_s]dW(s)\Big]$$

$$= \Lambda(t)\Big[E[Z(T)F] + \int_0^t E[D_s(Z(T)F)|\mathcal{F}_s]dW(s)\Big]$$

$$=: \Lambda(t)U(t).$$

By (4.12) and the Itô formula we have

$$d\Lambda(t) = \Lambda(t)u(t)d\widetilde{W}(t). \tag{4.13}$$

Combining (4.2), (4.13), and (4.11) we get

$$dY(t) = \Lambda(t)E[D_t(Z(T)F)|\mathcal{F}_t]dW(t) + \Lambda(t)u(t)U(t)d\widetilde{W}(t)$$
$$+ \Lambda(t)u(t)E[D_t(Z(T)F)|\mathcal{F}_t]dW(t)d\widetilde{W}(t)$$
$$= \Lambda(t)E[D_t(Z(T)F)|\mathcal{F}_t]d\widetilde{W}(t) + u(t)Y(t)d\widetilde{W}(t)$$
$$= \Lambda(t)\Big(E[Z(T)D_tF|\mathcal{F}_t] - E[Z(T)Fu(t)|\mathcal{F}_t]$$
$$- E[Z(T)F\int_t^T D_tu(s)d\widetilde{W}(s)|\mathcal{F}_t]\Big)d\widetilde{W}(t) + u(t)Y(t)d\widetilde{W}(t).$$

Hence

$$dY(t) = \Big(E_Q[D_tF|\mathcal{F}_t] - E_Q[Fu(t)|\mathcal{F}_t] - E_Q\Big[F\int_t^T D_tu(s)d\widetilde{W}(s)|\mathcal{F}_t\Big]\Big)d\widetilde{W}(t)$$
$$+ u(t)E_Q[F|\mathcal{F}_t]d\widetilde{W}(t)$$
$$= E_Q\Big[D_tF - F\int_t^T D_tu(s)d\widetilde{W}(s)|\mathcal{F}_t\Big]d\widetilde{W}(t).$$

Since

$$Y(T) = E_Q[F|\mathcal{F}_T] = F$$

and

$$Y(0) = E_Q[F|\mathcal{F}_0] = E_Q[F],$$

we see that Theorem 4.5 follows from (4.2). The conditions (4.5)–(4.7) are needed to make all the above operations valid. We omit the details. □

4.3 Application to Finance: Portfolio Selection

We end this section by explaining how the generalized Clark–Ocone theorem can be applied in portfolio analysis. Suppose we have two possibility of investment in

(a) A risk-less asset (e.g., a bond), with price dynamics

$$\begin{cases} dS_0(t) = \rho(t)S_0(t)dt, \\ S_0(0) = 1. \end{cases} \tag{4.14}$$

(b) A risky asset (e.g., a stock), with price dynamics

$$\begin{cases} dS_1(t) = \mu(t)S_1(t)dt + \sigma(t)S_1(t)dW(t) \\ S_1(0) > 0. \end{cases} \tag{4.15}$$

Here $\rho(t) = \rho(t,\omega)$, $\mu(t) = \mu(t,\omega)$, and $\sigma(t) = \sigma(t,\omega)$, $\omega \in \Omega$, $t \in [0,T]$, are $\mathbb{F}$-adapted processes satisfying the condition

$$E\left[\int_0^T \{|\rho(t)| + |\mu(t)| + \sigma^2(t)\}dt\right] < \infty.$$

Moreover, let $\sigma(t) \neq 0$, $t \in [0,T]$.

Let $\theta_0(t) = \theta_0(t,\omega)$, $\theta_1(t) = \theta_1(t,\omega)$, $\omega \in \Omega$, $t \in [0,T]$, denote the number of units invested at time t in investments (a) and (b), respectively. Then the corresponding *value* at time t, $V^\theta(t) = V^\theta(t,\omega)$, $\omega \in \Omega$, $t \in [0,T]$, of this *portfolio* $\theta(t) := (\theta_0(t), \theta_1(t))$, $t \in [0,T]$, is given by

$$V^\theta(t) = \theta_0(t)S_0(t) + \theta_1(t)S_1(t). \tag{4.16}$$

The portfolio $\theta(t)$ is called *self-financing* if

$$dV^\theta(t) = \theta_0(t)dS_0(t) + \theta_1(t)dS_1(t). \tag{4.17}$$

Assume from now on that $\theta(t)$, $t \in [0,T]$, is self-financing. Then by substituting

$$\theta_0(t) = \frac{V^\theta(t) - \theta_1(t)S_1(t)}{S_0(t)} \tag{4.18}$$

from (4.16) in (4.17) and using (4.14) we get

$$dV^\theta(t) = \rho(t)(V^\theta(t) - \theta_1(t)S_1(t))dt + \theta_1(t)dS_1(t).$$

Then by (4.15) this can be written

$$dV^\theta(t) = [\rho(t)V^\theta(t) + (\mu(t) - \rho(t))\theta_1(t)S_1(t)]dt + \sigma(t)\theta_1(t)S_1(t)dW(t). \tag{4.19}$$

Suppose now that we are required to find a portfolio $\theta(t)$, $t \in [0,T]$, which leads to a given value

$$V^\theta(T) = F, \quad P - \text{a.s.} \tag{4.20}$$

at a given fixed future time T, where the given F is $\mathcal{F}_T$-measurable. Such a portfolio is called *replicating* (or *hedging*) *portfolio*. Then the following questions arise:

- *What initial fortune $V^\theta(0)$ is needed to achieve this, and what portfolio $\theta(t)$, $t \in [0,T]$, should we use?*
- *Are $V^\theta(0)$ and $\theta(t)$, $t \in [0,T]$, unique?*

This type of questions appears in option pricing. For example, in the classical Black–Scholes model we can consider F given by the *European call option*:

$$F = (S_1(T) - K)^+,$$

where K is the *exercise price* at *maturity* T. Then $V^\theta(0)$ represents the *price of the option*. The study of the above questions led to the celebrated Black–Scholes formula [39]. See also Example 4.11 hereafter.

Because of the relation (4.18) we see that we might as well consider $(V^\theta(t), \theta_1(t))$, $t \in [0, T]$, to be the unknown $\mathbb{F}$-adapted processes. Then (4.19)–(4.20) constitute what is known as a *backward stochastic differential equation* (BSDE), namely the *final* value $V^\theta(T)$ is given and one seeks the value of $(V^\theta(t), \theta_1(t))$ for $0 \le t < T$. Note that since V^θ is $\mathbb{F}$-adapted, we have that $V^\theta(0)$ is $\mathcal{F}_0$-measurable and therefore a *constant*. The general theory of BSDE gives that (under reasonable conditions on F, ρ, μ, and σ) (4.19)–(4.20) has a *unique* solution of $\mathbb{F}$-adapted processes $(V^\theta(t), \theta_1(t))$, $t \in [0, T]$. See, for example, [187].

However, in many situations, this general theory says little about how to find this solution explicitly. This is where the generalized Clark–Ocone theorem enters the scene.

Define

$$u(t) = \frac{\mu(t) - \rho(t)}{\sigma(t)}, \quad t \in [0, T], \tag{4.21}$$

and put

$$\widetilde{W} = \widetilde{W}(t) = \int_0^t u(s)ds + W(t), \quad t \in [0, T], \tag{4.22}$$

as in (4.2) (recall that the Novikov condition on $u(t)$, $t \in [0, T]$, is assumed to hold). Then $\widetilde{W}$ is a Wiener process with respect to the measure Q defined by (4.3)–(4.4). In terms of $\widetilde{W}$, (4.19) gets the form

$$dV^\theta(t) = \left[\rho(t)V^\theta(t) + (\mu(t) - \rho(t))\theta_1(t)S_1(t)\right]dt + \sigma(t)\theta_1(t)S_1(t)d\widetilde{W}(t)$$
$$-\sigma(t)\theta_1(t)S_1(t)\sigma^{-1}(t)(\mu(t) - \rho(t))dt,$$

that is,

$$dV^\theta(t) = \rho(t)V^\theta(t)dt + \sigma(t)\theta_1(t)S_1(t)d\widetilde{W}(t). \tag{4.23}$$

Define

$$U^\theta(t) = e^{-\int_0^t \rho(s)ds}V^\theta(t).$$

Then, substituting in (4.23), we get

$$dU^\theta(t) = e^{-\int_0^t \rho(s)ds}\sigma(t)\theta_1(t)S_1(t)d\widetilde{W}(t)$$

or

$$e^{-\int_0^T \rho(s)ds}V^\theta(T) = V^\theta(0) + \int_0^T e^{-\int_0^t \rho(s)ds}\sigma(t)\theta_1(t)S_1(t)d\widetilde{W}(t).$$

By the generalized Clark–Ocone theorem applied to

$$G := e^{-\int_0^T \rho(s)ds}F \tag{4.24}$$

we get

$$G = E_Q[G] + \int_0^T E_Q\Big[(D_t G - G \int_t^T D_t u(s) d\widetilde{W}(s)) | \mathcal{F}_t\Big] d\widetilde{W}(t). \qquad (4.25)$$

By uniqueness we conclude from (4.24) and (4.25) that

$$V^\theta(0) = E_Q[G] \qquad (4.26)$$

and the required risky investment at time t is

$$\theta_1(t) = e^{\int_0^t \rho(s) ds} \sigma^{-1}(t) S_1^{-1}(t) E_Q\Big[(D_t G - G \int_t^T D_t u(s) d\widetilde{W}(s)) | \mathcal{F}_t\Big]. \qquad (4.27)$$

Example 4.10. Suppose $\rho(t) = \rho$, $\mu(t) = \mu$, and $\sigma(t) = \sigma \neq 0$, $t \in [0, T]$, are constants. Then

$$u(t) = u = \frac{\mu - \rho}{\sigma}$$

is also constant and hence $D_t u = 0$. Therefore, by (4.27)

$$\theta_1(t) = e^{\rho(t-T)} \sigma^{-1} S_1^{-1}(t) E_Q[D_t F | \mathcal{F}_t].$$

For example, if the payoff function is

$$F = \exp\{\alpha W(T)\} \qquad (\alpha \neq 0 \text{ constant}),$$

then by the chain rule (see Problem 3.3) we get

$$\theta_1(t) = e^{\rho(t-T)} \sigma^{-1} S_1^{-1}(t) E_Q\Big[\alpha \exp\{\alpha W(T)\} | \mathcal{F}_t\Big]$$

$$= e^{\rho(t-T)} \alpha \sigma^{-1} S_1^{-1}(t) Z^{-1}(t) E\Big[Z(T) \exp\{\alpha W(T)\} | \mathcal{F}_t\Big]. \qquad (4.28)$$

Note that

$$Z(T) \exp\{\alpha W(T)\} = M(T) \exp\Big\{\frac{1}{2}(\alpha - u)^2 T\Big\},$$

where $M(t) := \exp\{(\alpha - u) W(t) - \frac{1}{2}(\alpha - u)^2 t\}$ is a martingale. This gives

$$\theta_1(t) = e^{\rho(t-T)} \alpha \sigma^{-1} S_1^{-1}(t) Z^{-1}(t) M(t) \exp\Big\{\frac{1}{2}(\alpha - u)^2 T\Big\}$$

$$= e^{\rho(t-T)} \alpha \sigma^{-1} \exp\Big\{(\alpha - \sigma) W(t) + (\frac{1}{2}\sigma^2 + \frac{1}{2}u^2 - \mu) t + \frac{1}{2}(\alpha - u)^2 (T - t)\Big\}.$$

Example 4.11. **The Black–Scholes formula.** Finally, let us illustrate the aforementioned method by using it to prove the celebrated Black–Scholes formula [39]. As in Example 4.10 let us assume that $\rho(t) = \rho$, $\mu(t) = \mu$, and $\sigma(t) = \sigma \neq 0$, $t \in [0, T]$, are constants. Then

$$u = \frac{\mu - \rho}{\sigma}$$

is constant and hence $D_t u = 0$. Hence

$$\theta_1(t) = e^{\rho(t-T)}\sigma^{-1}S_1^{-1}(t)E_Q[D_t F \mid \mathcal{F}_t] \tag{4.29}$$

as in Example 4.10. However, in this case F represents the payoff at a fixed time T of a European call option, which gives the owner the right to buy the stock with value $S_1(T)$ at a fixed exercise price K, say. Thus, if $S_1(T) > K$ the owner of the option gets the profit $S_1(T) - K$, and if $S_1(T) \leq K$ the owner does not exercise the option and the profit is 0. Hence in this case

$$F = (S_1(T) - K)^+. \tag{4.30}$$

Thus, we may write

$$F = f(S_1(T)), \tag{4.31}$$

where

$$f(x) = (x - K)^+. \tag{4.32}$$

The function f is not differentiable at $x = K$, so we cannot use the chain rule directly to evaluate $D_t F$ from (4.31). However, we can approximate f by C^1 functions f_n with the property that

$$f_n(x) = f(x) \qquad \text{for} \quad |x - K| \geq \frac{1}{n}$$

and

$$0 \leq f_n'(x) \leq 1 \qquad \text{for all } x.$$

Putting

$$F_n = f_n(S_1(T))$$

and applying Theorem 3.3, we can see that

$$\begin{aligned}
D_t F &= \lim_{n \to \infty} D_t F_n \\
&= \chi_{[K,\infty]}(S_1(T))D_t S_1(T) \\
&= \chi_{[K,\infty]}(S_1(T))S_1(T)\sigma.
\end{aligned}$$

Hence by (4.29)

$$\theta_1(t) = e^{\rho(t-T)}S_1^{-1}(t)E_Q\left[S_1(T)\chi_{[K,\infty]}(S_1(T))|\mathcal{F}_t\right].$$

By the Markov property of $S_1(t)$ this is the same as

$$\theta_1(t) = e^{\rho(t-T)}S_1^{-1}(t)E_Q^y\left[S_1(T-t)\chi_{[K,\infty]}(S_1(T-t))\right]_{|y=S_1(t)},$$

where E_Q^y is the expectation when $S_1(0) = y$. Since

$$\begin{aligned}
dS_1(t) &= \mu S_1(t)dt + \sigma S_1(t)dW(t) \\
&= (\mu - \sigma u)S_1(t)dt + \sigma S_1(t)d\widetilde{W}(t) \\
&= \rho S_1(t)dt + \sigma S_1(t)d\widetilde{W}(t),
\end{aligned}$$

we have

$$S_1(t) = S_1(0)\exp\left\{(\rho - \frac{1}{2}\sigma^2)t + \sigma\widetilde{W}(t)\right\}$$

and hence

$$\theta_1(t) = e^{\rho(t-T)}S_1^{-1}(t)E^y[Y(T-t)\chi_{[K,\infty]}(Y(T-t))]_{|y=S_1(t)}, \tag{4.33}$$

where

$$Y(t) = S_1(0)\exp\left\{(\rho - \frac{1}{2}\sigma^2)t + \sigma W(t)\right\}.$$

Since the distribution of $W(t)$ is well-known, we can express the solution (4.33) explicitly in terms of quantities involving $S_1(t)$ and the normal distribution function.

In this model, $\theta_1(t)$ represents the number of units we must invest in the risky investment at times $t \leq T$ in order to be guaranteed to get the payoff $F = (S_1(T)-K)^+$ (P−a.s.) at time T. The constant $V^\theta(0)$ represents the corresponding initial fortune needed to achieve this. Thus $V^\theta(0)$ is the (unique) initial fortune, which makes it possible to establish a self-financing portfolio with the same payoff at time T as the option gives. Hence $V^\theta(0)$ deserves to be called *the right price* for such an option. By (4.26) this is given by

$$V^\theta(0) = E_Q[e^{-\rho T}F] \tag{4.34}$$
$$= e^{-\rho T}E_Q[(S_1(T) - K)^+]$$
$$= e^{-\rho T}E[(Y(T) - K)^+], \tag{4.35}$$

which again can be expressed explicitly by the normal distribution function.

Remark 4.12. In the *Markovian* case, that is, when the price $S_1(t)$, $t \in [0,T]$, is given by a stochastic differential equation of the form

$$dS_1(t) = \mu(S_1(t))S_1(t)dt + \sigma(S_1(t))S_1(t)dW(t),$$

where $\mu : \mathbb{R} \to \mathbb{R}$ and $\sigma : \mathbb{R} \to \mathbb{R}$ are given functions, then there is a well-known alternative method for finding the option price $V^\theta(0)$ and the corresponding replicating portfolio $\theta_1(t)$, $t \in [0, T]$. One assumes that the value process has the form

$$V^\theta(t) = f(t, S_1(t))$$

for some function $f : \mathbb{R}^2 \to \mathbb{R}$ and deduces a deterministic partial differential equation, which determines f. See Kolmogorov backward partial differential equations and the Feynman–Kac formula in, for example, [129]. See also Sect. 4.4. Then θ_1 is given by

$$\theta_1(t) = \left[\frac{\partial f(t, x)}{\partial x} \right]_{x=S_1(t)} \sigma(S_1(t)).$$

For example, this method can be applied to the cases of Example 4.10 and, with some attention, Example 4.11. However, the method does not work in the non-Markovian case. The method based on the Clark–Ocone formula has the advantage that it does not depend on a Markovian setup.

4.4 Application to Sensitivity Analysis and Computation of the "Greeks" in Finance

In 1999, Fourniée et al. [81] found a remarkable numerically tractable method for computing the derivatives of expectations of functions of Itô diffusions with respect to some of the parameters involved. Their method (but not the final results) used tools from the Malliavin calculus that we have presented. In general we could call results of this type *sensitivity results*. In the setting of mathematical finance this becomes a method for computing the so-called "greeks." These are quantities representing the market sensitivities of financial derivatives to the variation of the model parameters. They are important tools in risk management and hedging. Note that the name greeks was given because these quantities are often denoted by Greek letters. If V_t, $t \in [0, T]$, is the payoff process of some derivative, we have, for example, the following:

- "Delta" measures the sensitivity to changes in the initial price x of the underlying asset: $\Delta = \frac{\partial V}{\partial x}$ (important for hedging purposes)
- "Gamma" measures the rate of change in the delta: $\Gamma = \frac{\partial^2 V}{\partial x^2}$
- "Rho" measures the sensitivity to the applicable interest rate r: $\rho = \frac{\partial V}{\partial r}$
- "Theta" measures the sensitivity to the amount of time to expiration date: $\Theta = -\frac{\partial V}{\partial T}$
- "Vega" (indicated by ν) measures the sensitivity to volatility σ: $\nu = \frac{\partial V}{\partial \sigma}$

Remark 4.13. Though very interesting quantities to be considered, in many cases, the greeks cannot be expressed in closed form and require numerical methods for the computation. Qualitatively, being V computed as an expectation, the greeks are basically derivatives of expectations. From [89] we see that one of the most flexible methods is the application of Monte Carlo simulation on top of a finite difference approximation of the derivatives. However, this contains intrinsically two kinds of errors: one on the approximation of the derivatives and the other on the numerical computation of the expectations. In particular, most of the inefficiency is revealed when dealing with discontinuous payoffs. Other methods are in use to overcome a generally poor convergence rate [10, 24, 27]. An efficient method was introduced starting from [52], where it was suggested to take the differential of the payoff process inside the expectation. Then [43] introduced the method of differentiation of the density function, moving in this way the differentiation from the payoff function to the density function and they introduced the so-called *likelihood ratio*, for example,

$$\Delta = \frac{\partial}{\partial x} E\big[\varphi(X^x(T))\big] = E\Big[\varphi(X^x(T))\frac{\partial}{\partial x}\log p(X^x(T))\Big].$$

This method (so-called *density method*) is very efficient, but has, however, the disadvantage of requiring an *explicit* expression of the density function.

Here we present only the study of the greek delta, which is connected with the so-called Δ-*hedging*. For simplicity we only consider one-dimensional processes. For more information we refer to [81] and [80]. See also [161].

Consider again a market model of the form

$$\text{risk free asset} \qquad \begin{cases} dS_0(t) = \rho(t)S_0(t)dt \\ S_0(0) = 1 \end{cases} \qquad (4.36)$$

$$\text{risky asset} \qquad \begin{cases} dS_1(t) = S_1(t)\big[\mu(t)dt + \sigma(t)dW(t)\big] \\ S_1(0) = x > 0. \end{cases} \qquad (4.37)$$

Assume now that $\rho(t) = \rho$ is constant and the coefficients μ and σ are *Markovian*, that is, (with abuse of notation) $\mu(t) = \mu(S_1(t))$ and $\sigma(t) = \sigma(S_1(t)) \neq 0$, $0 \le t \le T$.

Suppose we want to replicate an $\mathcal{F}_T$-measurable Markovian payoff

$$F = \varphi(S_1(T)),$$

where $\varphi : \mathbb{R} \longrightarrow \mathbb{R}$ is, for convenience, considered bounded. To this end, let us try to find a self-financing portfolio $\theta(t) = (\theta_0(t), \theta_1(t))$, $0 \le t \le T$, and a function $f(t, x)$, $0 \le t \le T$, $x > 0$, such that the value process $V^\theta(t)$, $0 \le t \le T$, given by

$$V^\theta(t) = \theta_0(t)S_0(t) + \theta_1(t)S_1(t)$$

is of the form

$$V^\theta(t) = f(t, S_1(t)), \quad t \in [0, T].$$

In particular, $f(T, x) = \varphi(x)$, $x > 0$. By the Itô formula we have

$$dV(t) = \frac{\partial f}{\partial t}(t, S_1(t))dt + \frac{\partial f}{\partial x}(t, S_1(t))dS_1(t) + \frac{1}{2}\frac{\partial^2 f}{\partial x^2}(t, S_1(t))\sigma^2(S_1(t))S_1^2(t)dt. \tag{4.38}$$

Since θ is self-financing, we have

$$dV^\theta(t) = \theta_0(t)S_0(t)\rho dt + \theta_1(t)dS_1(t). \tag{4.39}$$

Comparing (4.38) and (4.39) we get

$$\theta_0(t)S_0(t)\rho + \theta_1(t)S_1(t)\mu(S_1(t))$$
$$= \frac{\partial f}{\partial t}(t, S_1(t)) + \frac{\partial f}{\partial x}(t, S_1(t))S_1(t)\mu(S_1(t)) + \frac{1}{2}\frac{\partial^2 f}{\partial x^2}(t, S_1(t))\sigma^2(S_1(t))S_1^2(t) \tag{4.40}$$

and

$$\theta_1(t)\sigma(S_1(t))S_1(t) = \frac{\partial f}{\partial x}(t, S_1(t))\sigma(S_1(t))S_1(t). \tag{4.41}$$

Equation (4.41) holds if and only if

$$\theta_1(t) = \frac{\partial f}{\partial x}(t, S_1(t)) \qquad \text{(the "Δ-hedge")}. \tag{4.42}$$

Substituted into (4.40), this gives

$$[f(t, S_1(t)) - S_1(t)\frac{\partial f}{\partial x}(t, S_1(t))]\rho = \frac{\partial f}{\partial t}(t, S_1(t)) + \frac{1}{2}\frac{\partial^2 f}{\partial x^2}(t, S_1(t))\sigma^2(S_1(t))S_1^2(t),$$

that is, $f(t, x)$ must satisfy the famous Black–Scholes equation

$$\begin{cases} \frac{\partial f}{\partial t}(t, x) - \rho f(t, x) + \rho x \frac{\partial f}{\partial x}(t, x) + \frac{1}{2}\sigma^2(x)x^2 \frac{\partial^2 f}{\partial x^2}(t, x) = 0, \ t < T, \\ f(T, x) = \varphi(x). \end{cases} \tag{4.43}$$

By the Feynman–Kac formula (see, e.g., [129]) the solution of this equation is

$$f(t, S_1(t)) = E^x\left[e^{-\rho(T-t)}\varphi(X(T-t))\right]_{|x=S_1(t)} = e^{-\rho(T-t)}E^x\left[\varphi(X(T-t))\right]_{|x=S_1(t)},$$

where $X(t) = X^x(t)$, $0 \le t \le T$, is the solution of the stochastic differential equation

$$dX(t) = X(t)[\rho dt + \sigma(X(t))dW(t)]; \qquad X(0) = x > 0$$

(here above we have used the standard notation $E^x[\varphi(X(T))] = E[\varphi(X^x(T))]$). Therefore, to compute the "Δ-hedge" $\theta_1(t)$, $t \in [0, T]$, we need to compute

$$\frac{\partial f}{\partial x}(t, x) = e^{-\rho(T-t)}\frac{\partial}{\partial x}E^x[\varphi(X(T-t))]$$
$$= e^{-\rho(T-t)}\frac{\partial}{\partial x}E[\varphi(X^x(T-t))]. \tag{4.44}$$

To do this numerically may be problematic if φ is not smooth. Note that in the applications φ may even be discontinuous, as is the case with binary options.

However, as shown in [81], one can use Malliavin calculus to transform the expression (4.44) into a form that is more suitable for numerical computations (see, e.g., [16, 90]). We now present this approach.

Consider a general Itô diffusion $X^x(t)$, $t \geq 0$, given by

$$dX^x(t) = b(X^x(t))dt + \sigma(X^x(t))dW(t), \qquad X^x(0) = x \in \mathbb{R},$$

where $b : \mathbb{R} \longrightarrow \mathbb{R}$ and $\sigma : \mathbb{R} \longrightarrow \mathbb{R}$ are given functions in $C^1(\mathbb{R})$ and $\sigma(x) \neq 0$ for all $x \in \mathbb{R}$.

The *first variation process* $Y(t) := \frac{\partial}{\partial x}X^x(t)$, $t \geq 0$, is given by

$$dY(t) = b'(X^x(t))Y(t)dt + \sigma'(X^x(t))Y(t)dW(t), \qquad Y(0) = 1,$$

that is,

$$Y(t) = \exp\left\{ \int_0^t \left[b'(X^x(u)) - \frac{1}{2}(\sigma'(X^x(u)))^2\right]du + \int_0^t \sigma'(X^x(u))dW(u)\right\}.$$
(4.45)

See [138] for a rigorous justification of (4.45). Fix $T > 0$ and define

$$g(x) := E^x[\varphi(X(T))] = E[\varphi(X^x(T))].$$

Then we have the following result (see [81]).

Theorem 4.14. *Let $a(t)$, $t \in [0,T]$, be a continuous deterministic function such that*

$$\int_0^T a(t)dt = 1.$$

Then

$$g'(x) = E^x\left[\varphi(X(T)) \int_0^T a(t)\sigma^{-1}(X(t))Y(t)dW(t)\right]. \qquad (4.46)$$

The random variable

$$\pi^\Delta = \int_0^T a(t)\sigma^{-1}(X(t))Y(t)dW(t)$$

is a so-called Malliavin weight.

Some versions of (4.46) are known as Bismut-Elworthy-Li formulae, see e.g. [78].

Example 4.15. Suppose

$$dX(t) = \rho X(t)dt + \sigma_0 X(t)dW(t),$$

with ρ and $\sigma_0 \neq 0$ constants. Choose

$$a(t) = \frac{1}{T}, \qquad t \in [0,T].$$

Then
$$g'(x) = E^x \left[\varphi(X(T)) \frac{W(T)}{x\sigma_0 T} \right].$$

To see this, we can observe that in this case we have $b(x) = \rho x$, $\sigma(x) = \sigma_0 x$ and hence
$$dY(t) = \rho Y(t)dt + \sigma_0 Y(t)dW(t), \qquad Y(0) = 1,$$

and therefore
$$\sigma^{-1}(X(t))Y(t) = \sigma_0^{-1} x^{-1} Y^{-1}(t)Y(t).$$

The proof of Theorem 4.14 is split into two lemmata.

Lemma 4.16.

$$D_s X(t) = Y(t)Y^{-1}(s)\sigma(X(s))\chi_{[0,t]}(s). \tag{4.47}$$

Proof Since

$$X(t) = x + \int_0^t b(X(u))du + \int_0^t \sigma(X(u))dW(u),$$

then we have, by Theorem 3.18, for $t \geq s$,

$$Z(t) := D_s X(t) = \int_s^t b'(X(u))D_s X(u)du + \int_s^t \sigma'(X(u))D_s X(u)dW(u) + \sigma(X(s)).$$

Therefore

$$\begin{cases} dZ(t) = b'(X(t))Z(t)dt + \sigma'(X(t))Z(t)dW(t), & t \geq s, \\ Z(s) = \sigma(X(s)). \end{cases}$$

The solution of this equation is

$$Z(t) = \sigma(X(s)) \exp\left\{ \int_s^t [b'(X(u)) - \frac{1}{2}(\sigma'(X(u)))^2]du + \int_s^t \sigma'(X(u))dW(u) \right\}, \ t \geq s.$$

Comparing with (4.45) we obtain

$$Z(t) = \sigma(X(s))Y(t)Y^{-1}(s), \qquad t \geq s. \qquad \square$$

Lemma 4.17. *Let $a(t)$, $t \in [0,T]$, be a deterministic function such that*

$$\int_0^T a(t)dt = 1.$$

Then

$$Y(T) = \int_0^T D_s X(T)a(s)\sigma^{-1}(X(s))Y(s)ds. \tag{4.48}$$

Proof By Lemma 4.16 we have, with $t = T$,

$$Y(T) = Z(T)Y(s)\sigma(X(s))^{-1}, \qquad s \in [0, T].$$

Hence

$$Y(T) = \int_0^T Y(T)a(s)ds = \int_0^T D_s X(T)a(s)\sigma^{-1}(X(s))Y(s)ds. \qquad \square$$

Proof of Theorem 4.14. First assume that φ is smooth with bounded derivative. Then we have

$$g'(x) = E\big[\varphi'(X^x(T))\frac{d}{dx}X^x(T)\big] = E\big[\varphi'(X^x(T))Y(T)\big]$$

$$= E^x\big[\int_0^T \varphi'(X(T))D_s X(T)a(s)\sigma^{-1}X(s))Y(s)ds\big]$$

$$= E^x\big[\int_0^T D_s(\varphi(X(T))a(s)\sigma^{-1}X(s))Y(s)ds\big]$$

$$= E^x\big[\varphi(X(T))\int_0^T a(s)\sigma^{-1}(X(s))Y(s)dW(s)\big],$$

where we have used Lemma 4.17, the chain rule, and Corollary 4.4. This completes the proof when φ is smooth.

In the general case we approximate φ pointwise boundedly a.e. with respect to Lebesgue measure on $[0, T]$ by smooth functions φ_m, each with bounded derivative. Define

$$g_m(x) = E^x\big[\varphi_m(X(T))\big].$$

Then by the above

$$g'_m(x) = E^x\big[\varphi_m(X(T))\Lambda\big], \qquad m = 1, 2, ...,$$

where

$$\Lambda = \int_0^T a(s)\sigma^{-1}(X(s))Y(s)dW(s).$$

Hence

$$\lim_{m \to \infty} g'_m(x) = E^x\big[\varphi(X(T))\Lambda\big] =: h(x)$$

pointwise boundedly in x. Thus, for $m \to \infty$, we have

$$g_m(x) = g_m(0) + \int_0^x g'_m(t)dt \longrightarrow g(0) + \int_0^x h(t)dt.$$

Hence

$$g(x) = \lim_{m \to \infty} g_m(x) = g(0) + \int_0^x h(t)dt,$$

which implies that g is differentiable and $g'(x) = h(x)$. $\square$

Remark 4.18. Theorem 4.14 is an interesting result for several reasons. Note that, on the one side, the application of formula (4.46) does neither need the differentiability of the payoff function nor know the density function of the diffusion (typical of the density method, see Remark 4.13). On the other side we need to know the diffusion. Note also that the weighting function a in (4.46) is independent of the payoff and it is not unique. Some studies of how to characterize and choose the weights can be found, for example, in [21].

4.5 Exercises

Problem 4.1. (*) Recall the *Girsanov theorem* (see, e.g., [179, Theorem 8.26]). Let $Y(t) \in \mathbb{R}^n$ be an Itô process of the form

$$dY(t) = \beta(t)dt + \gamma(t)dW(t), \qquad t \leq T,$$

where $\beta(t) \in \mathbb{R}^n$, $\gamma(t) \in \mathbb{R}^{n \times m}$, $t \in [0, T]$, are $\mathbb{F}$-adapted and $W(t)$, $t \in [0, T]$, is an m-dimensional Wiener process. Suppose there exist $\mathbb{F}$-adapted processes $u(t) \in \mathbb{R}^m$ and $\alpha(t) \in \mathbb{R}^n$, $t \in [0, T]$, such that

$$\gamma(t)u(t) = \beta(t) - \alpha(t)$$

and such that the Novikov condition

$$E\left[\exp\left\{\frac{1}{2}\int_0^T u^2(s)ds\right\}\right] < \infty$$

holds. Put

$$Z(t) = \exp\left\{-\int_0^t u(s)dW(s) - \frac{1}{2}\int_0^t u^2(s)ds\right\}, \qquad t \leq T,$$

and define a measure Q on $\mathcal{F}_T$ by

$$dQ = Z(T)dP.$$

Then

$$\widetilde{W}(t) := \int_0^t u(s)ds + W(t), \qquad 0 \leq t \leq T$$

is a Wiener process with respect to Q, and in terms of $\widetilde{W}$ the process Y has the stochastic integral representation

$$dY(t) = \alpha(t)dt + \gamma(t)d\widetilde{W}(t).$$

(a) Show that $\widetilde{W}$ is an $\mathbb{F}$-martingale with respect to Q. [*Hint.* Apply Itô formula to $Y(t) := Z(t)\widetilde{W}(t)$.]

(b) Suppose $X(t) = at + W(t) \in \mathbb{R}$, $t \leq T$, where $a \in \mathbb{R}$ is a constant. Find a probability measure Q on $\mathcal{F}_T$ such that X is a Wiener process with respect to Q.

(c) Let $a, b, c \neq 0$ be real constants and define

$$dY(t) = bY(t)dt + cY(t)dW(t).$$

Find a probability measure Q and a Wiener process $\widetilde{W}$ with respect to Q such that

$$dY(t) = aY(t)dt + cY(t)d\widetilde{W}(t).$$

Problem 4.2. (*) Verify the Clark–Ocone formula

$$F = E[F] + \int\limits_0^T E[D_t F | \mathcal{F}_t] dW(t)$$

for the following $\mathcal{F}_T$-measurable random variables F:

(a) $F = W(T)$,

(b) $F = \int\limits_0^T W(s)ds$,

(c) $F = W^2(T)$,

(d) $F = W^3(T)$,

(e) $F = \exp W(T)$,

(f) $F = (W(T) + T)\exp\big\{-W(T) - \frac{1}{2}T\big\}$.

Problem 4.3. (*) Let $\widetilde{W}(t) = \int\limits_0^t u(s)ds + W(t)$ and Q be as in Exercise 4.1. Use the generalized Clark–Ocone formula to find the $\mathbb{F}$-adapted process $\widetilde{\varphi}$, such that

$$F = E_Q[F] + \int\limits_0^T \widetilde{\varphi}(t)d\widetilde{W}(t)$$

in the following cases:

(a) $F = W^2(T)$ and $u(t)$, $t \in [0, T]$, is deterministic.

(b) $F = \exp\big\{\int\limits_0^T \lambda(t)dW(t)\big\}$ and the processes $\lambda(t)$ and $u(t)$, $t \in [0, T]$, are deterministic.

(c) F is like in (b) and $u(t) = W(t)$, $t \in [0, T]$.

Problem 4.4. (*) Suppose we have a market with two investments of type (4.14) and (4.15). Find the initial fortune $V^\theta(0)$ and the number of units $\theta_1(t)$, which must be invested at time t in the risky investment to produce the terminal value $V^\theta(T) = F = W(T)$ when $\rho(t) = \rho > 0$ is constant and the price $S_1(t)$, $t \in [0, T]$, of the risky investment is given by:

(a) $dS_1(t) = \mu S_1(t)dt + \sigma S_1(t)dW(t)$; μ, σ constants ($\sigma \neq 0$). This is the case of the geometric Brownian motion.

(b) $dS_1(t) = cdW(t)$; $c \neq 0$ constant.

(c) $dS_1(t) = \mu S_1(t)dt + cdW(t)$; μ, c constants. This is the case of the Ornstein–Uhlenbeck process. [*Hint.* $S_1(t) = e^{\mu t}S_1(0) + c\int_0^T e^{\mu(t-s)}dW(s)$.]

5

White Noise, the Wick Product, and Stochastic Integration

This chapter gives an introduction to the white noise analysis and its relation to the analysis discussed in the first two chapters. This is a useful alternative approach for several reasons. First, it allows us to represent the Malliavin derivative as a natural directional derivative (or *stochastic gradient*, to be more precise). Second, it makes it possible to obtain an extension of the Clark–Ocone formula from $\mathbb{D}_{1,2}$ to $L^2(P)$. Moreover, it provides a natural platform for the *Wick product*, which is closely related to Skorohod integration (see (5.28)). For example, we shall see that the Wick calculus can be used to simplify the computation of these integrals considerably.

The Wick product was introduced by C.G. Wick in 1950 [226] as a renormalization technique in quantum physics. In stochastic analysis this concept, or rather a relative of it, was introduced by Hida and Ikeda in 1967 [101]. In 1989, Meyer and Yan [165] extended the construction to cover Wick products of stochastic distributions (Hida distributions), including the white noise.

The Wick product has turned out to be a very useful tool in stochastic analysis in general. For example, it can be used to facilitate both the theory and the explicit calculations in stochastic integration and stochastic differential equations.

General references for this chapter are [88, 99, 102, 103, 107, 135, 150, 151, 152, 177, 231].

5.1 White Noise Probability Space

We start with the construction of the *white noise probability space*. Let $\mathcal{S} = \mathcal{S}(\mathbb{R}^d)$ be *the Schwartz space of rapidly decreasing smooth $C^\infty(\mathbb{R}^d)$ real functions* on $\mathbb{R}^d$. The space $\mathcal{S} = \mathcal{S}(\mathbb{R}^d)$ is a Fréchet space with respect to the family of seminorms:

$$\|f\|_{K,\alpha} := \sup_{x \in \mathbb{R}^d} \left\{ (1 + |x|^K) |\partial^\alpha f(x)| \right\},$$

G.Di Nunno et al., *Malliavin Calculus for Lévy Processes with Applications to Finance*,
© Springer-Verlag Berlin Heidelberg 2009

where $K = 0, 1, ...$, $\alpha = (\alpha_1, ..., \alpha_d)$ is a multi-index with $\alpha_j = 0, 1, ...$ $(j = 1, ..., d)$ and

$$\partial^\alpha f := \frac{\partial^{|\alpha|}}{\partial x_1^{\alpha_1} \cdots \partial x_d^{\alpha_d}} f$$

for $|\alpha| = \alpha_1 + ... + \alpha_d$, see, for example, [79, 208] as general references for the properties of these spaces.

Let $\mathcal{S}' = \mathcal{S}'(\mathbb{R}^d)$ be its dual, the space of *tempered distributions*. Let $\mathcal{B}$ denote the family of all Borel subsets of $\mathcal{S}'(\mathbb{R}^d)$ equipped with the weak* topology. If $\omega \in \mathcal{S}'$ and $\phi \in \mathcal{S}$ we let

$$\omega(\phi) = \langle \omega, \phi \rangle \tag{5.1}$$

denote the action of ω on ϕ. For example, if ω is a measure m on $\mathbb{R}^d$ then

$$\langle \omega, \phi \rangle = \int_{\mathbb{R}^d} \phi(x) dm(x)$$

and, in particular, if this measure m is concentrated on $x_0 \in \mathbb{R}^d$, then

$$\langle \omega, \phi \rangle = \phi(x_0)$$

is the evaluation of ϕ at $x_0 \in \mathbb{R}^d$.

From now on we consider the one-dimensional case, that is, $d = 1$. We fix the sample space to be $\Omega = \mathcal{S}'(\mathbb{R})$ and $\mathcal{F} = \mathcal{B}$. By the Bochner–Minlos–Sazonov theorem (see, e.g., [87]), there exists a probability measure P on Ω such that

$$\int_\Omega e^{i\langle \omega, \phi \rangle} P(d\omega) = e^{-\frac{1}{2}\|\phi\|^2}, \quad \phi \in \mathcal{S},$$

where

$$\|\phi\|^2 = \|\phi\|_{L^2(\mathbb{R})}^2 = \int_\mathbb{R} |\phi(x)|^2 dx.$$

The measure P is called the *white noise probability measure* and $(\Omega, \mathcal{F}, P) = (\mathcal{S}', \mathcal{B}, P)$ is called *the white noise probability space*.

Definition 5.1. *The (smoothed) white noise process is the measurable map*

$$w : \mathcal{S} \times \mathcal{S}' \to \mathbb{R}$$

given by

$$w(\phi, \omega) = w_\phi(\omega) = \langle \omega, \phi \rangle, \quad \phi \in \mathcal{S}, \ \omega \in \mathcal{S}'. \tag{5.2}$$

From w_ϕ we can construct a Wiener process $W(t)$, $t \in \mathbb{R}$, as follows:

Step 1 The isometry

$$E[w_\phi^2] = \|\phi\|^2, \quad \phi \in \mathcal{S}, \tag{5.3}$$

holds true where, according to our notation, the left-hand side is

$$E[w_\phi^2] = \int_{\mathcal{S}'} \langle \omega, \phi \rangle^2 P(d\omega).$$

Step 2 Use Step 1 to define the value $\langle \omega, \psi \rangle$ for arbitrary $\psi \in L^2(\mathbb{R})$, as $\langle \omega, \psi \rangle := \lim \langle \omega, \phi_n \rangle$, where $\phi_n \in \mathcal{S}$, $n = 1, 2, ...$, and $\phi_n \to \psi$ in $L^2(\mathbb{R})$.

Step 3 Use Step 2 to define

$$\widetilde{W}(t, \omega) := \langle \omega, \chi_{[0,t]} \rangle, \quad t \in \mathbb{R},$$

by choosing

$$\psi(s) = \chi_{[0,t]}(s) = \begin{cases} 1 \text{ if } s \in [0,t) \text{ (or } s \in [t,0), \text{ if } t < 0), \\ 0 \qquad\qquad \text{otherwise} , \end{cases}$$

which belongs to $L^2(\mathbb{R})$ for all $t \in \mathbb{R}$.

Step 4 By means of the Kolmogorov lemma (see [153]), we can prove that $\widetilde{W}(t)$, $t \in \mathbb{R}$, has a *continuous version* $W(t)$, $t \in \mathbb{R}$, that is, for all t, $P\{\widetilde{W}(t) = W(t)\} = 1$. This continuous process $W(t)$, $t \in \mathbb{R}$, is a Wiener process.

Note that when the Wiener process $W(t, \omega)$, $t \in \mathbb{R}$, $\omega \in \Omega$ with $\Omega := \mathcal{S}'(\mathbb{R})$ is constructed this way, then each ω is interpreted as a tempered distribution.

From the aforementioned Step 2, it follows that the smoothed white noise w_ϕ can be extended to all (deterministic) $\phi \in L^2(\mathbb{R})$ and that the relation between smoothed white noise w_ϕ and the Wiener process $W(t)$, $t \in \mathbb{R}$, is

$$w_\phi(\omega) = \int_{\mathbb{R}} \phi(t) dW(t, \omega), \quad \omega \in \Omega, \qquad \phi \in L^2(\mathbb{R}), \tag{5.4}$$

where the integral on the right-hand side is the Wiener–Itô integral (see Problem 5.1). Note that the isometry (5.3) is then the *Itô isometry*.

5.2 The Wiener–Itô Chaos Expansion Revisited

As in Chap. 1 let the *Hermite polynomials* $h_n(x)$ be defined by

$$h_n(x) = (-1)^n e^{\frac{1}{2}x^2} \frac{d^n}{dx^n}(e^{-\frac{1}{2}x^2}), \quad n = 0, 1, 2, \ldots,$$

cf. (1.13). We recall that, for example,

$$h_0(x) = 1, h_1(x) = x, h_2(x) = x^2 - 1, h_3(x) = x^3 - 3x$$
$$h_4(x) = x^4 - 6x^2 + 3, h_5(x) = x^5 - 10x^3 + 15x, \ldots$$

Let e_k be the kth *Hermite function* defined by

$$e_k(x) := \pi^{-\frac{1}{4}}((k-1)!)^{-\frac{1}{2}} e^{-\frac{1}{2}x^2} h_{k-1}(\sqrt{2}x), \quad k = 1, 2, \ldots. \tag{5.5}$$

Then $\{e_k\}_{k \geq 1}$ constitutes an orthonormal basis for $L^2(\mathbb{R})$ and $e_k \in \mathcal{S}(\mathbb{R})$ for all k; see, for example, [222].
    Define

$$\theta_k(\omega) := \langle \omega, e_k \rangle = w_{e_k}(\omega) = \int_{\mathbb{R}} e_k(x) dW(x, \omega), \quad \omega \in \Omega. \tag{5.6}$$

Let $\mathcal{J}$ denote the set of all finite multi-indices $\alpha = (\alpha_1, \alpha_2, \ldots, \alpha_m)$, $m = 1, 2, \ldots$, of non-negative integers α_i. If $\alpha = (\alpha_1, \ldots, \alpha_m) \in \mathcal{J}$, $\alpha \neq 0$, we put

$$H_\alpha(\omega) := \prod_{j=1}^{m} h_{\alpha_j}(\theta_j(\omega)), \quad \omega \in \Omega. \tag{5.7}$$

By a result of Itô [120] we have that

$$I_m(e^{\hat{\otimes}\alpha}) = \prod_{j=1}^{m} h_{\alpha_j}(\theta_j) = H_\alpha, \tag{5.8}$$

cf. (1.11). We set $H_0 := 1$. Here and in the sequel the functions $e_1, e_2, \ldots$ are defined in (5.5) and $\otimes$ and $\hat{\otimes}$ denote the tensor product and the symmetrized tensor product, respectively. For example, if f and g are real functions on $\mathbb{R}$ then

$$(f \otimes g)(x_1, x_2) = f(x_1)g(x_2)$$

and

$$(f\hat{\otimes}g)(x_1, x_2) = \frac{1}{2}\Big[f(x_1)g(x_2) + f(x_2)g(x_1)\Big].$$

Hereafter we let

$$\epsilon^{(k)} = (0, 0, ..., 1, 0, ..., 0), \tag{5.9}$$

with 1 on kth position. Then

$$H_{\epsilon^{(k)}}(\omega) = h_1(\theta_k(\omega)) = \langle \omega, e_k \rangle,$$

or, if $\alpha = (3, 0, 2)$, then

$$H_{3,0,2} = h_3(\theta_1)h_0(\theta_2)h_2(\theta_3) = (\theta_1^3 - 3\theta_1)(\theta_3^2 - 1).$$

The family $\{H_\alpha\}_{\alpha \in \mathcal{J}}$ is an orthogonal basis for the Hilbert space $L^2(P)$. In fact, we have the following result (see [107, Theorem 2.2.4]).

Theorem 5.2. *The Wiener–Itô chaos expansion theorem. The family* $\{H_\alpha\}_{\alpha \in \mathcal{J}}$ *constitutes an orthogonal basis of* $L^2(P)$*. More precisely, for all* $\mathcal{F}_T$*-measurable* $X \in L^2(P)$ *there exist (uniquely determined) numbers* $c_\alpha \in \mathbb{R}$ *such that*

$$X = \sum_{\alpha \in \mathcal{J}} c_\alpha H_\alpha \quad \in L^2(P). \tag{5.10}$$

Moreover, we have

$$\|X\|^2_{L^2(P)} = \sum_{\alpha \in \mathcal{J}} \alpha! c_\alpha^2, \tag{5.11}$$

where $\alpha! = \alpha_1! \alpha_2! \cdots \alpha_m!$ *for* $\alpha = (\alpha_1, \alpha_2, \ldots, \alpha_m)$.

Example 5.3. To find the chaos expansion of the Wiener process $W(t)$ at time t, we proceed as follows:

$$\begin{aligned}
W(t) &= \int_{\mathbb{R}} \chi_{[0,t]}(s) dW(s) \\
&= \int_{\mathbb{R}} \sum_k (\chi_{[0,t]}, e_k)_{L^2(\mathbb{R})} e_k(s) dW(s) \\
&= \sum_k \left(\int_0^t e_k(y) dy \right) \int_{\mathbb{R}} e_k(s) dW(s) \\
&= \sum_k \left(\int_0^t e_k(y) dy \right) H_{\epsilon^{(k)}}.
\end{aligned}$$

Let us compare Theorem 5.2 with the equivalent formulation of this theorem in terms of *iterated Itô integrals* (see Chap. 1). In fact, if $\psi(t_1, t_2, \ldots, t_n)$ is a real symmetric function in its n variables $t_1, \ldots, t_n$ and $\psi \in L^2(\mathbb{R}^n)$, that is,

$$\|\psi\|_{L^2(\mathbb{R}^n)} := \left[\int_{\mathbb{R}^n} |\psi(t_1, t_2, \ldots, t_n)|^2 dt_1 dt_2 \cdots dt_n \right]^{1/2} < \infty,$$

then its *n-tuple Itô integral* is defined by

$$\begin{aligned}
I_n(\psi) &:= \int_{\mathbb{R}^n} \psi dW^{\otimes n} \\
&:= n! \int_{-\infty}^{\infty} \int_{-\infty}^{t_n} \int_{-\infty}^{t_{n-1}} \cdots \int_{-\infty}^{t_2} \psi(t_1, t_2, \ldots, t_n) dW(t_1) dW(t_2) \cdots dW(t_n),
\end{aligned}$$

where the integral on the right-hand side consists of n iterated Itô integrals. Note that the integrand at each step is adapted to the filtration $\mathbb{F}$. Applying the Itô isometry n times we see that

$$E\left[\left(\int_{\mathbb{R}^n} \psi dW^{\otimes n}\right)^2\right] = n!\|\psi\|^2_{L^2(\mathbb{R}^n)}, \quad n = 1, 2, \dots.$$

For $n = 0$ we adopt the convention that

$$I_0(\psi) := \int_{\mathbb{R}^0} \psi dW^{\otimes 0} = \psi = \|\psi\|_{L^2(\mathbb{R}^0)},$$

for ψ constant. Let $\tilde{L}^2(\mathbb{R}^n)$ denote the set of symmetric real functions on $\mathbb{R}^n$, which are square integrable with respect to Lebesque measure (see Chap. 1). Then we have the following result, cf. Theorem 1.10.

Theorem 5.4. The Wiener–Itô chaos expansion theorem. *For all $\mathcal{F}_T$-measurable $X \in L^2(P)$ there exist (uniquely determined) functions $f_n \in \tilde{L}^2(\mathbb{R}^n)$ such that*

$$X = \sum_{n=0}^{\infty} \int_{\mathbb{R}^n} f_n dW^{\otimes n} = \sum_{n=0}^{\infty} I_n(f_n) \quad \in L^2(P). \tag{5.12}$$

Moreover, we have the isometry

$$\|X\|^2_{L^2(P)} = \sum_{n=0}^{\infty} n!\|f_n\|^2_{L^2(\mathbb{R}^n)}. \tag{5.13}$$

Remark 5.5. The connection between these two expansions in Theorem 5.2 and Theorem 5.4 is given by

$$f_n = \sum_{\alpha \in \mathcal{J}:|\alpha|=n} c_\alpha e_1^{\otimes \alpha_1} \hat{\otimes} e_2^{\otimes \alpha_2} \hat{\otimes} \cdots \hat{\otimes} e_m^{\otimes \alpha_m}, \quad n = 0, 1, 2, \dots, \tag{5.14}$$

where $|\alpha| = \alpha_1 + \cdots + \alpha_m$ for $\alpha = (\alpha_1, \dots, \alpha_m) \in \mathcal{J}$, $m = 1, 2, \dots$. Recall that the functions $e_1, e_2, \dots$ are defined in (5.5) and $\otimes$ and $\hat{\otimes}$ denote the tensor product and the symmetrized tensor product, respectively.

Note that since $H_\alpha = I_m(e^{\hat{\otimes}\alpha})$, for $\alpha \in \mathcal{J}$, $|\alpha| = m$, we get that

$$m!\|e^{\hat{\otimes}\alpha}\|^2_{L^2(\mathbb{R}^m)} = \alpha!, \tag{5.15}$$

by combining (5.11) and (5.13) for $X = X_\alpha$.

Analogous to the test functions $\mathcal{S}(\mathbb{R})$ and the tempered distributions $\mathcal{S}'(\mathbb{R})$ on the real line $\mathbb{R}$, there is a useful space of *stochastic test functions* $(\mathcal{S})$ and a space of *stochastic distributions* $(\mathcal{S})^*$ on the white noise probability space. In the following we use the notation

$$(2\mathbb{N})^\alpha = \prod_{j=1}^{m} (2j)^{\alpha_j}, \quad \text{for} \quad \alpha = (\alpha_1, \dots, \alpha_m) \in \mathcal{J}. \tag{5.16}$$

Definition 5.6. *(a) Let $k \in \mathbb{R}$. We say that $f = \sum\limits_{\alpha \in \mathcal{J}} a_\alpha H_\alpha \in L^2(P)$ belongs to the Hida test function Hilbert space $(\mathcal{S})_k$ if*

$$\|f\|_k^2 := \sum_{\alpha \in \mathcal{J}} \alpha! a_\alpha^2 (2\mathbb{N})^{\alpha k} < \infty. \tag{5.17}$$

We define the Hida test function space $(\mathcal{S})$ as the space

$$(\mathcal{S}) = \bigcap_{k \in \mathbb{R}} (\mathcal{S})_k$$

equipped with the projective topology, that is, $f_n \longrightarrow f$, $n \to \infty$, in $(\mathcal{S})$ if and only if $\|f_n - f\|_k \longrightarrow 0$, $n \to \infty$, for all k.
(b) Let $q \in \mathbb{R}$. We say that the formal sum $F = \sum\limits_{\alpha \in \mathcal{J}} b_\alpha H_\alpha$ belongs to the Hida distribution Hilbert space $(\mathcal{S})_{-q}$ if

$$\|F\|_{-q}^2 := \sum_{\alpha \in \mathcal{J}} \alpha! c_\alpha^2 (2\mathbb{N})^{-\alpha q} < \infty. \tag{5.18}$$

We define the Hida distribution space $(\mathcal{S})^*$ as the space

$$(\mathcal{S})^* = \bigcup_{q \in \mathbb{R}} (\mathcal{S})_{-q}$$

equipped with the inductive topology, that is, $F_n \longrightarrow F$, $n \to \infty$, in $(\mathcal{S})^*$ if and only if there exists q such that $\|F_n - F\|_{-q} \longrightarrow 0$, $n \to \infty$.
(c) If $F = \sum_{\alpha \in \mathcal{J}} b_\alpha H_\alpha \in (\mathcal{S})^*$, we define the generalized expectation $E[F]$ of F by

$$E[F] = b_0. \tag{5.19}$$

(Note that if $F \in L^2(P)$, then the generalized expectation coincides with the usual expectation, since $E[H_\alpha] = 0$ for all $\alpha \neq 0$).

Note that $(\mathcal{S})^*$ is the dual of $(\mathcal{S})$. The action of $F = \sum\limits_{\alpha} b_\alpha H_\alpha \in (\mathcal{S})^*$ on $f = \sum\limits_{\alpha} a_\alpha H_\alpha \in (\mathcal{S})$ is given by

$$\langle F, f \rangle = \sum_{\alpha} \alpha! a_\alpha b_\alpha.$$

We have the inclusions

$$(\mathcal{S}) \subset (\mathcal{S})_k \subset L^2(P) \subset (\mathcal{S})_{-q} \subset (\mathcal{S})^*, \qquad \text{for all} \quad k, q.$$

Example 5.7. (a) The smoothed white noise w_ϕ belongs to $(\mathcal{S})$ if $\phi \in \mathcal{S}(\mathbb{R})$. In fact, if $\phi = \sum\limits_{j=1}^{\infty} c_j e_j$ we have

$$w_\phi = \sum_{j=1}^{\infty} c_j H_{\epsilon(j)}, \tag{5.20}$$

and so using (5.17) we can see that $w_\phi \in (\mathcal{S})$ if and only if

$$\sum_{j=1}^{\infty} c_j^2 (2j)^k < \infty$$

for all k, which holds because $\phi \in \mathcal{S}(\mathbb{R})$; see, for example, [205].

(b) The *singular* (also *pointwise*) *white noise* $\overset{\bullet}{W}(t)$, $t \in \mathbb{R}$, is defined as follows:

$$\overset{\bullet}{W}(t) := \sum_k e_k(t) H_{\epsilon(k)}. \tag{5.21}$$

Using (5.18) one can verify that $\overset{\bullet}{W}(t) \in (\mathcal{S})^*$ for all t, as follows. We have

$$\|\overset{\bullet}{W}(t)\|_{-q}^2 = \sum_{k=0}^{\infty} e_k^2(t) \epsilon^{(k)}! \left((2\mathbb{N})^{\epsilon^{(k)}} \right)^{-q} = \sum_{k=0}^{\infty} e_k^2(t)(2k)^{-q} < \infty, \qquad q \geq 2,$$

because

$$\sup_{t \in \mathbb{R}} |e_k(t)| = \mathcal{O}\left(k^{-1/12}\right),$$

see [105]. Similarly we see that by (5.3) we get

$$\frac{d}{dt} W(t) = \frac{d}{dt} \sum_k \left(\int_0^t e_k(y) dy \right) H_{\epsilon(k)} = \overset{\bullet}{W}(t), \tag{5.22}$$

where the derivative is taken in $(\mathcal{S})^*$ (see Problem 5.2). This justifies the name *white noise* for $\overset{\bullet}{W}$.

5.3 The Wick Product and the Hermite Transform

In addition to a canonical vector space structure, the spaces $(\mathcal{S})$ and $(\mathcal{S})^*$ also have a natural multiplication given by the Wick product.

Definition 5.8. *If $X = \sum_\alpha a_\alpha H_\alpha \in (\mathcal{S})^*, Y = \sum_\beta b_\beta H_\beta \in (\mathcal{S})^*$ then the Wick product $X \diamond Y$ of X and Y is defined by*

$$X \diamond Y := \sum_{\alpha,\beta} a_\alpha b_\beta H_{\alpha+\beta} = \sum_\gamma \left(\sum_{\alpha+\beta=\gamma} a_\alpha b_\beta \right) H_\gamma. \tag{5.23}$$

Using (5.18) and (5.17) one can now verify the following:

$$X, Y \in (\mathcal{S})^* \Rightarrow X \diamond Y \in (\mathcal{S})^*. \tag{5.24}$$

$$X, Y \in (\mathcal{S}) \Rightarrow X \diamond Y \in (\mathcal{S}). \tag{5.25}$$

See [107, Lemma 2.4.4]. Note, however, that $X, Y \in L^2(P) \not\Rightarrow X \diamond Y \in L^2(P)$ in general. See [107, Example 2.4.8].

Example 5.9. (1) The Wick square of the singular white noise is

$$(\overset{\bullet}{W})^{\diamond 2}(t) = \sum_{k,m=1}^{\infty} e_k(t)e_m(t)H_{\epsilon(k)+\epsilon(m)}.$$

One can show that

$$(\overset{\bullet}{W})^{\diamond 2}(t) \in (\mathcal{S})^*, \quad t \in \mathbb{R}, \tag{5.26}$$

see Problem 5.4.

(2) The Wick square of the smoothed white noise is

$$(w_\phi)^{\diamond 2} = \sum_{k,m=1}^{\infty} c_k c_m H_{\epsilon(k)+\epsilon(m)} \quad if \quad \phi = \sum_{k=1}^{\infty} c_k e_k \in L^2(\mathbb{R}).$$

Since

$$H_{\epsilon(k)+\epsilon(m)} = \begin{cases} H_{\epsilon(k)} \cdot H_{\epsilon(m)} & if \ k \neq m \\ H_{\epsilon(k)}^2 - 1 & if \ k = m, \end{cases}$$

we see that

$$(w_\phi)^{\diamond 2} = w_\phi^2 - \sum_{k=1}^{\infty} c_k^2 = w_\phi^2 - \|\phi\|^2. \tag{5.27}$$

Note, in particular, that $(w_\phi)^{\diamond 2}$ is not positive. In fact, $E[(w_\phi)^{\diamond 2}] = 0$ by (5.21) and the fact that $E[H_\alpha] = 0$ for $\alpha \neq 0$ (see Theorem 5.2).

Before proceeding forward, we list some reasons that the Wick product is natural to use in stochastic calculus:

(a) First, note that if (at least) one of the factors X, Y is deterministic, then

$$X \diamond Y = X \cdot Y.$$

Therefore, the two types of products, the Wick product and the ordinary (ω-pointwise) product, coincide in the deterministic calculus. So when one extends a deterministic model to a stochastic model by introducing noise, it is not obvious which interpretation to choose for the products involved. The choice should be based on additional modeling and mathematical considerations.

(b) The Wick product is the only product that is defined for singular white noise $\overset{\bullet}{W}$. Pointwise product $X \cdot Y$ does not make sense in $(\mathcal{S})^*$!

(c) The Wick product has been used for 50 years already in quantum physics as a renormalization procedure.

(d) There is a fundamental relation between Itô/Skorohod integrals and Wick products, given by

$$\int_{\mathbb{R}} Y(t)\delta W(t) = \int_{\mathbb{R}} Y(t) \diamond \dot{W}(t)dt \qquad (5.28)$$

(see [151], [23]). Here the integral on the right is interpreted as a Bochner integral with values in $(\mathcal{S})^*$. See Theorem 5.20 later.

(e) A big class of strong solutions to stochastic differential equations can be explicitly solved by using the Wick product. See [145].

5.3.1 Some Basic Properties of the Wick Product

We list below some useful properties of the Wick product. Some are easy to prove, others harder. For complete proofs see [107].

For arbitrary $X, Y, Z \in (\mathcal{S})^*$ we have

$$X \diamond Y = Y \diamond X \qquad \text{(commutative law)}, \qquad (5.29)$$

$$X \diamond (Y \diamond Z) = (X \diamond Y) \diamond Z \qquad \text{(associative law)}, \qquad (5.30)$$

$$X \diamond (Y + Z) = (X \diamond Y) + (X \diamond Z) \qquad \text{(distributive law)}. \qquad (5.31)$$

In view of the above we can define the *Wick powers*

$$X^{\diamond n} = X \diamond X \diamond \cdots \diamond X \quad (n \text{ times}) \quad \text{for} \quad X \in (\mathcal{S})^*, \quad n = 1, 2, \dots.$$

We put $X^{\diamond 0} = 1$. Similarly, the *Wick exponential* of $X \in (\mathcal{S})^*$ is defined by

$$\exp^{\diamond} X = \sum_{n=0}^{\infty} \frac{1}{n!} X^{\diamond n}, \qquad (5.32)$$

if convergent in $(\mathcal{S})^*$. Thus the Wick algebra obeys the same rules as the ordinary algebra. For example,

$$(X + Y)^{\diamond 2} = X^{\diamond 2} + 2X \diamond Y + Y^{\diamond 2} \qquad (5.33)$$

(no Itô formula!) and

$$\exp^{\diamond}(X + Y) = \exp^{\diamond}(X) \diamond \exp^{\diamond}(Y). \qquad (5.34)$$

Note, however, that combinations of ordinary products and Wick products require caution. For example, in general we have

$$X \cdot (Y \diamond Z) \neq (X \cdot Y) \diamond Z.$$

Note that since $E[H_\alpha] = 0$ for all $\alpha \neq 0$, we have that if $X = \sum_{\alpha \in \mathcal{J}} c_\alpha H_\alpha \in L^2(P)$, then

$$E[X] = c_0.$$

From this we deduce the remarkable property of the Wick product:

$$E[X \diamond Y] = E[X] \cdot E[Y], \tag{5.35}$$

whenever X, Y and $X \diamond Y$ are P-integrable. Note that it is *not* required that X and Y are independent!

By induction it follows from (5.35) that

$$E[\exp^\diamond X] = \exp E[X]. \tag{5.36}$$

From Example 5.9 (2) we deduce that

$$w_\phi \diamond w_\psi = w_\phi \cdot w_\psi - \frac{1}{2} \int_{\mathbb{R}} \phi(t)\psi(t)dt, \ \phi, \psi \in L^2(\mathbb{R}).$$

In particular,

$$W^{\diamond 2}(t) = W^2(t) - t, \qquad t \geq 0. \tag{5.37}$$

Moreover, if $\operatorname{supp}\phi \cap \operatorname{supp}\psi = \emptyset$, then

$$w_\phi \diamond w_\psi = w_\phi \cdot w_\psi. \tag{5.38}$$

Hence if $0 \leq t_1 \leq t_2 \leq t_3 \leq t_4$, then

$$(W(t_4) - W(t_3)) \diamond (W(t_2) - W(t_1)) = (W(t_4) - W(t_3)) \cdot (W(t_2) - W(t_1)). \tag{5.39}$$

More generally, it can be proved that if F is $\mathcal{F}_t$-measurable and $h > 0$, then

$$F \diamond (W(t + h) - W(t)) = F \cdot (W(t + h) - W(t)). \tag{5.40}$$

For a proof see, for example, [107, Exercise 2.22].

5.3.2 Hermite Transform and Characterization Theorem for $(\mathcal{S})_{-1}$

There is a useful tool in white noise analysis, called *Hermite transform* or the *$\mathcal{H}$-transform*, which transforms elements $X \in (\mathcal{S})_{-1}$ into deterministic functions $\mathcal{H}X(z_1, z_2, ...)$ of complex variables $z_j \in \mathbb{C}$, $j = 1, 2, ...$, with values in $\mathbb{C}$. This concept was introduced in [150] and relies on the expansion of elements along the basis $\{H_\alpha\}_{\alpha \in \mathcal{J}}$ in the *Kondratiev distribution space* $(\mathcal{S})_{-1} \supseteq (\mathcal{S})^*$. The construction of $(\mathcal{S})_{-1}$ is similar to that of $(\mathcal{S})^*$ in Definition 5.6.

The *Kondratiev distribution space* $(\mathcal{S})_{-1}$ is the inductive limit of the Hilbert spaces

$$(S)_{-1,q}, q \geq 0. \tag{5.41}$$

For $q \geq 0, (S)_{-1,q}$ consists of all (formal) chaos expansions

$$F = \sum_{\alpha \in J} c_\alpha H_\alpha$$

such that

$$\|F\|_{-1,q}^2 := \sum_{\alpha \in J} c_\alpha^2 (2\mathbb{N})^{-\alpha q} < \infty. \tag{5.42}$$

The space $(S)_{-1}$ can be regarded as the topological dual of the *Kondratiev test function space* $(S)_1$, which is the projective limit of Hilbert spaces

$$(S)_{1,q}, q \geq 0, \tag{5.43}$$

with norms $\| \cdot \|_{1,q}, q \geq 0$, given by

$$\|f\|_{1,q}^2 := \sum_{\alpha \in J} a_\alpha^2 (2\mathbb{N})^{\alpha q} < \infty, \tag{5.44}$$

if f has chaos expansion

$$f = \sum_{\alpha \in J} a_\alpha H_\alpha.$$

It follows from the definition of $(S)_1$ and $(S)_{-1}$ that

$$(S)_1 \hookrightarrow (S) \hookrightarrow L^2(P) \hookrightarrow (S)^* \hookrightarrow (S)_{-1} \tag{5.45}$$

in the sense of continuous inclusions.

The *Hermite transform* of $X = \sum_\alpha c_\alpha H_\alpha \in (\mathcal{S})_{-1}$, denoted by $\mathcal{H}X$ or $\widetilde{X}$, is defined by

$$\mathcal{H}X(z) := \widetilde{X}(z) := \sum_\alpha c_\alpha z^\alpha \in \mathbb{C}, \tag{5.46}$$

where $z = (z_1, z_2, ...) \in \mathbb{C}^{\mathbb{N}}$, that is, in the space of $\mathbb{C}$-valued sequences, and where $z^\alpha = z_1^{\alpha_1} z_2^{\alpha_2} ....$ We have that $\mathcal{H}X(z)$ is absolutely convergent on the infinite dimensional neighborhood

$$\mathbb{K}_q(R) := \left\{ (z_1, z_2, ...) \in \mathbb{C}^{\mathbb{N}} : \sum_{\alpha \neq 0} |z^\alpha|^2 (2\mathbb{N})^{q\alpha} < R^2 \right\} \tag{5.47}$$

for some $0 < q \leq R < \infty$. The latter follows from the fact that

$$\left|\widetilde{X}(z)\right| \leq \sum_\alpha |c_\alpha| \, |z|^\alpha$$

$$\leq \left(\sum_\alpha |c_\alpha|^2 \, \alpha! (2\mathbb{N})^{-\alpha q}\right)^{\frac{1}{2}} \left(\sum_\alpha |z^\alpha|^2 \, (2\mathbb{N})^{\alpha q}\right)^{\frac{1}{2}}$$

$$\leq \left(\sum_\alpha |c_\alpha|^2 \, (2\mathbb{N})^{-\alpha q}\right)^{\frac{1}{2}} \left(\sum_\alpha |z^\alpha|^2 \, (2\mathbb{N})^{\alpha q}\right)^{\frac{1}{2}}$$

$$= \|X\|_{-1,q} \left(\sum_\alpha |z^\alpha|^2 \, (2\mathbb{N})^{\alpha q}\right)^{\frac{1}{2}}.$$

Similarly, one observes that $\mathcal{H}X(z)$ is convergent for all $z = (z_1, z_2, ...) \in \mathbb{C}_c^{\mathbb{N}}$ (space of all finite sequences in $\mathbb{C}^{\mathbb{N}}$).

As an example, consider the Hermite transform of the singular white noise $\overset{\bullet}{W}(t)$. Since $\overset{\bullet}{W}(t) = \sum_{k \geq 1} e_k(t) H_{\epsilon^{(k)}}$ we see that

$$\mathcal{H}(\overset{\bullet}{W}(t))(z) = \sum_{k \geq 1} e_k(t) z_k. \tag{5.48}$$

Using the definition of the Wick product (Definition 5.8) extended to $(S)_{-1}$ we see that the Hermite transform converts the Wick product into an ordinary (complex) product, that is,

$$\mathcal{H}(X \diamond Y)(z) = \mathcal{H}(X)(z) \cdot \mathcal{H}(Y)(z) \tag{5.49}$$

for all $z = (z_1, z_2, ...) \in \mathbb{C}^{\mathbb{N}}$ such that $\mathcal{H}(X)(z)$ and $\mathcal{H}(Y)(z)$ exist.

An important property of the Hermite transform $\mathcal{H}$ is that it can be used to characterize distributions in $(S)_{-1}$. In view of appplications to come, we resort to the following characterization theorem for Kondratiev distributions ([107, Theorem 2.6.11]).

Theorem 5.10. Characterization theorem for $(S)_{-1}$.
(i) Let $X = \sum_\alpha c_\alpha H_\alpha \in (S)_{-1}$. Then there exists some $q, M_q < \infty$ such that

$$\left|\widetilde{X}(z)\right| \leq \sum_\alpha |c_\alpha| \, |z|^\alpha \leq M_q \left(\sum_\alpha |c_\alpha|^2 \, (2\mathbb{N})^{\alpha q}\right)^{\frac{1}{2}}$$

for all $z \in \mathbb{C}_c^{\mathbb{N}}$. In particular, $\widetilde{X}$ is a bounded (analytic) function on $\mathbb{K}_q(R)$ for all $R < \infty$.
(ii) Consider the power series

$$f(z) = \sum_\alpha b_\alpha z^\alpha$$

for coefficients $b_\alpha \in \mathbb{R}$ and $z = (z_1, z_2, ...) \in \mathbb{C}_c^{\mathbb{N}}$. Assume that

$$\sum_{\alpha} |b_\alpha| \, |z|^\alpha < \infty$$

for all $z \in \mathbb{K}_q(R)$ for some $q < \infty$ and $R > 0$. Further require that

$$\sup_{z \in \mathbb{K}_q(R)} |f(z)| < \infty.$$

Then there exists a unique distribution X in $(S)_{-1}$ such that

$$\widetilde{X}(z) = f(z)$$

for all $z \in \mathbb{C}_c^{\mathbb{N}}$. Moreover, X has the representation

$$X = \sum_{\alpha} b_\alpha H_\alpha.$$

Proof The first statement is an immediate consequence of the Cauchy–Schwartz inequality. As for the proof of (ii) the reader is referred to [107, Theorem 2.6.11]. □

Remark 5.11. Using Theorem 5.10 one can show the existence of "Wick versions" of analytic functions: Let $X \in (\mathcal{S})_{-1}$. Suppose that $f : U \longrightarrow \mathbb{C}$ is an analytic function in the neighborhood $U \subseteq \mathbb{C}$ of $\zeta_0 := E[X]$, that is, f has the power series expansion

$$f(z) = \sum_{k \geq 0} a_k (z - \zeta_0)^k.$$

Assume that the Taylor coefficients a_k are real-valued. Then the *Wick version* of f applied to X, denoted by $f^\diamond(X)$, is the unique element $Y \in (\mathcal{S})_{-1}$ such that

$$\widetilde{Y}(z) = f(\widetilde{X}(z))$$

on $\mathbb{K}_q(R)$ for some $q < \infty$ and $R > 0$. If we choose, for example, $f(z) = \exp(z)$, we obtain the Wick version of the exponential function, see (5.32).

The Hermite transform also serves as a useful tool to describe the topology of $(\mathcal{S})_{-1}$. In particular, the convergence of sequences of Hida distributions can be characterized as follows.

Theorem 5.12. *A sequence $X_n, n \geq 1$, converges to a X in $(\mathcal{S})_{-1}$ if and only if there exist $q, M < \infty$, $R > 0$ such that*

$$\sup_{z \in \mathbb{K}_q(R)} \left| \widetilde{X}_n(z) \right| \leq M$$

for all $n \geq 1$ and

$$\widetilde{X}_n(z) \longrightarrow \widetilde{X}(z)$$

as n tends to infinity for all $z \in \mathbb{K}_q(R)$.

Proof See [107, Theorem 2.8.1] $\square$.

Combining Remark 5.11 and Theorem 5.12 we find that $\sum_{k=0}^{n} \frac{1}{k!} X^{\Diamond k}$ converges in $(\mathcal{S})_{-1}$ to the Wick version of the exponential function (5.32).

Remark 5.13. The Hermite transform is closely related to the so-called $\mathcal{S}$-transform. See [102]. In [193] a characterization of the Hida test function and distribution space in terms of the $\mathcal{S}$-transform can be found.

Using Theorem 5.10, Remark 5.11, and Theorem 5.12 we can derive the following chain rule in $(\mathcal{S})_{-1}$; see [107].

Proposition 5.14. Wick chain rule. *Assume that*

$$X(\cdot) : \mathbb{R} \longrightarrow (\mathcal{S})_{-1}$$

is continuously differentiable and let $f : \mathbb{C} \longrightarrow \mathbb{C}$ *be entire (i.e., analytic on* $\mathbb{C}$*), such that* $f(\mathbb{R}) \subseteq \mathbb{R}$. *Then the following chain rule*

$$\frac{d}{dt} f^{\Diamond}(X(t)) = (f')^{\Diamond}(X(t)) \diamond \frac{d}{dt} X(t)$$

holds true in $(\mathcal{S})_{-1}$.

5.3.3 The Spaces $\mathcal{G}$ and $\mathcal{G}^*$

We now introduce another pair of dual spaces, $\mathcal{G}$ and $\mathcal{G}^*$, which is sometimes useful, see [194, 2] and the references therein for more information.

Definition 5.15. *(a)* Let $\lambda \in \mathbb{R}$. Then the space $\mathcal{G}_\lambda$ consists of all formal expansions

$$X = \sum_{n=0}^{\infty} \int_{\mathbb{R}^n} f_n dW^{\otimes n} \tag{5.50}$$

such that

$$\|X\|_{\mathcal{G}_\lambda} := \left(\sum_{n=0}^{\infty} n! e^{2\lambda n} \|f_n\|_{L^2(\mathbb{R}^n)}^2 \right)^{1/2} < \infty. \tag{5.51}$$

For each $\lambda \in \mathbb{R}$, the space $\mathcal{G}_\lambda$ is a Hilbert space with inner product

$$(X, Y)_{\mathcal{G}_\lambda} = \sum_{n=0}^{\infty} n! e^{2\lambda n} (f_n, g_n)_{L^2(\mathbb{R}^n)} \tag{5.52}$$

for every

$$X = \sum_{n=0}^{\infty} \int_{\mathbb{R}^n} f_n dW^{\otimes n}, \qquad Y = \sum_{n=0}^{\infty} \int_{\mathbb{R}^n} g_n dW^{\otimes n}.$$

Note that $\lambda_1 \leq \lambda_2$ implies $\mathcal{G}_{\lambda_2} \subseteq \mathcal{G}_{\lambda_1}$. Define

$$\mathcal{G} = \bigcap_{\lambda \in \mathbb{R}} \mathcal{G}_\lambda = \bigcap_{\lambda > 0} \mathcal{G}_\lambda, \tag{5.53}$$

with projective limit topology.

(b) $\mathcal{G}^*$ is defined to be the dual of $\mathcal{G}$. Hence

$$\mathcal{G}^* = \bigcup_{\lambda \in \mathbb{R}} \mathcal{G}_\lambda = \bigcup_{\lambda < 0} \mathcal{G}_\lambda, \tag{5.54}$$

with inductive limit topology.

Remark 5.16. Note that an element $Y \in \mathcal{G}^*$ can be represented as a formal sum

$$Y = \sum_{n=0}^{\infty} \int_{\mathbb{R}^n} g_n dW^{\otimes n}, \tag{5.55}$$

where $g_n \in \widetilde{L}^2(\mathbb{R}^n)$ and $\|Y\|_{\mathcal{G}_\lambda} < \infty$ for *some* $\lambda \in \mathbb{R}$, while an $X \in \mathcal{G}$ satisfies $\|X\|_{\mathcal{G}_\lambda} < \infty$ for *all* $\lambda \in \mathbb{R}$.

If $X \in \mathcal{G}$ and $Y \in \mathcal{G}^*$ have the representations (5.50) and (5.55), respectively, then the action of Y on X, $\langle Y, X \rangle$, is given by

$$\langle Y, X \rangle = \sum_{n=0}^{\infty} n!(f_n, g_n)_{L^2(\mathbb{R}^n)}. \tag{5.56}$$

One can show that

$$(\mathcal{S}) \subset \mathcal{G} \subset L^2(P) \subset \mathcal{G}^* \subset (\mathcal{S})^*. \tag{5.57}$$

The space $\mathcal{G}^*$ is not big enough to contain the singular white noise $\overset{\bullet}{W}$. However, in some cases, it does contain the solution of stochastic differential equations and thus allows to deduce some useful properties of the solution itself. A test function space related to $\mathcal{G}$ was introduced in [203] to construct strong solutions to stochastic differential equations with discontinuous coefficients. We also mention that the pair $(\mathcal{G}, \mathcal{G}^*)$ can be completely characterized by using Bargmann–Segal spaces and the $\mathcal{S}$-transform, which is closely related to the $\mathcal{H}$-transform. See [95].

Like $(\mathcal{S})$ and $(\mathcal{S})^*$, the spaces $\mathcal{G}$ and $\mathcal{G}^*$ are closed under Wick product [194, Theorem 2.7]:

$$X_1, X_2 \in \mathcal{G} \Rightarrow X_1 \diamond X_2 \in \mathcal{G}. \tag{5.58}$$

$$Y_1, Y_2 \in \mathcal{G}^* \Rightarrow Y_1 \diamond Y_2 \in \mathcal{G}^*. \tag{5.59}$$

Finally, we note that, since

$$H_\alpha = \int_{\mathbb{R}^n} e^{\hat{\otimes}\alpha} dW^{\otimes n} = I_n(e^{\hat{\otimes}\alpha}),$$

with $\alpha \in \mathcal{J}$, $|\alpha| = n$, we get

$$\|H_\alpha\|_{\mathcal{G}_\lambda}^2 = n! e^{2\lambda n} \|e^{\hat{\otimes}\alpha}\|_{L^2(\mathbb{R}^n)}^2 = \alpha! e^{2\lambda n},$$

by (5.15). Therefore, for $F = \sum_{\alpha \in \mathcal{J}} c_\alpha H_\alpha \in \mathcal{G}^*$, we have

$$\|F\|_{\mathcal{G}_\lambda}^2 = \sum_{\alpha \in \mathcal{J}} c_\alpha^2 \alpha! e^{2\lambda|\alpha|}, \tag{5.60}$$

for some $\lambda \in \mathbb{R}$.

5.3.4 The Wick Product in Terms of Iterated Itô Integrals

The definition we have given of the Wick product is based on Theorem 5.2, because only this setting is general enough to include the singular white noise. However, it is useful to know how the Wick product is expressed in terms of chaos expansion for elements of $L^2(P)$ (see Theorem 5.4) or, more generally, for elements of $\mathcal{G}^*$. Then we have the following result.

Theorem 5.17. *Suppose* $X = \sum_{n=0}^{\infty} I_n(f_n) \in \mathcal{G}^*$ *and* $Y = \sum_{m=0}^{\infty} I_m(g_m) \in \mathcal{G}^*$. *Then the Wick product of* X *and* Y *can be expressed by*

$$X \diamond Y = \sum_{n,m=0}^{\infty} I_{n+m}(f_n \hat{\otimes} g_m) = \sum_{k=0}^{\infty} \left(\sum_{n+m=k} I_k(f_n \hat{\otimes} g_m) \right). \tag{5.61}$$

For example, integration by parts gives that for the deterministic functions $f, g \in L^2(\mathbb{R})$, we have

$$\left(\int_{\mathbb{R}} f(x) dW(x) \right) \diamond \left(\int_{\mathbb{R}} g(y) dW(y) \right) = \int_{\mathbb{R}^2} (f \hat{\otimes} g)(x,y) dW^{\otimes 2}$$

$$= \int_{\mathbb{R}} \int_{-\infty}^{y} (f(x)g(y) + f(y)g(x)) dW(x) dW(y)$$

$$= \int_{\mathbb{R}} g(y) \int_{-\infty}^{y} f(x) dW(x) dW(y) + \int_{\mathbb{R}} f(y) \int_{-\infty}^{y} g(x) dW(x) dW(y) \tag{5.62}$$

$$= \left(\int_{\mathbb{R}} f(x) dW(x) \right) \left(\int_{\mathbb{R}} y(y) dW(y) \right) - \int_{\mathbb{R}} f(t)g(t) dt \,.$$

Note that from (5.61) we have that

$$\left(\int_{\mathbb{R}} g(t) dW(t) \right)^{\diamond n} = I_n(g^{\otimes n}), \qquad g \in L^2(\mathbb{R}). \tag{5.63}$$

Combining this with (1.15) we get

$$\theta^{\diamond n} = \|g\|^n h_n\left(\frac{\theta}{\|g\|}\right), \qquad (5.64)$$

with $\theta = \int_{\mathbb{R}} g dW$. In particular,

$$W^{\diamond n}(t) = t^{n/2} h_n\left(\frac{W(t)}{\sqrt{t}}\right), \qquad n = 0, 1, 2, \ldots. \qquad (5.65)$$

Moreover, combining (5.64) with the *generating formula* for Hermite polynomials

$$\exp\left\{tx - \frac{t^2}{2}\right\} = \sum_{n=0}^{\infty} \frac{t^n}{n!} h_n(x) \qquad (5.66)$$

(see Exercise 1.1), we get

$$\exp\left\{\int_{\mathbb{R}} g(t) dW(t) - \frac{1}{2}\|g\|^2\right\} = \sum_{n=0}^{\infty} \frac{\|g\|^n}{n!} h_n\left(\frac{\theta}{\|g\|}\right)$$

$$= \sum_{n=0}^{\infty} \frac{1}{n!} \theta^{\diamond n} = \exp^{\diamond} \theta.$$

Hence

$$\exp^{\diamond}\left\{\int_{\mathbb{R}} g dW\right\} = \exp\left\{\int_{\mathbb{R}} g dW - \frac{1}{2}\|g\|^2\right\}, \qquad g \in L^2(\mathbb{R}). \qquad (5.67)$$

In particular,

$$\exp^{\diamond}\{W(t)\} = \exp\left\{W(t) - \frac{1}{2}t\right\}, \qquad t \geq 0. \qquad (5.68)$$

5.3.5 Wick Products and Skorohod Integration

We now prove the fundamental relation (5.28) between Wick products and Skorohod integration.

Definition 5.18. *A function* $Y : \mathbb{R} \longrightarrow (\mathcal{S})^*$ *(also called an* $(\mathcal{S})^*$*-valued process) is* $(\mathcal{S})^*$*-integrable if*

$$< Y(t), f > \in L^1(\mathbb{R}), \qquad \text{for all} \quad f \in (\mathcal{S}).$$

Then the $(\mathcal{S})^*$*-integral of* Y*, denoted by* $\int_{\mathbb{R}} Y(t) dt$*, is the (unique) element in* $(\mathcal{S})^*$ *such that*

$$< \int_{\mathbb{R}} Y(t) dt, f > = \int_{\mathbb{R}} < Y(t), f > dt, \qquad \text{for all} \quad f \in (\mathcal{S}). \qquad (5.69)$$

Remark 5.19. The fact that (5.69) does indeed define $\int_{\mathbb{R}} Y(t)dt$ as an element of $(\mathcal{S})^*$ is a consequence of [102, Proposition 8.1].

Note that if Y is $(\mathcal{S})^*$-integrable, then so is $Y\chi_{(a,b]}$, for all $a, b \in \mathbb{R}$, and we put

$$\int_a^b Y(t)dt := \int_{\mathbb{R}} Y(t)\chi_{(a,b]}(t)dt.$$

Theorem 5.20. *Assume that $Y(t)$, $t \in \mathbb{R}$, is a Skorohod integrable stochastic process (recall Definition 2.2). Then $Y(t) \diamond \overset{\bullet}{W}(t)$, $t \in \mathbb{R}$, is integrable in $(\mathcal{S})^*$ and*

$$\int_{\mathbb{R}} Y(t)\delta W(t) = \int_{\mathbb{R}} Y(t) \diamond \overset{\bullet}{W}(t)dt. \tag{5.70}$$

Proof Recall $L^2(P) \subseteq (\mathcal{S})^*$. We may assume that, for any t, $Y(t)$ has the representation

$$Y(t) = \sum_{\alpha \in \mathcal{J}} c_\alpha(t)H_\alpha = \sum_{n=0}^{\infty} I_n\big(f_n(\cdot, t)\big).$$

Then the right-hand side of (5.70) can be written as

$$\int_{\mathbb{R}} \Big(\sum_{\alpha \in \mathcal{J}} c_\alpha(t)H_\alpha \Big) \diamond \Big(\sum_k e_k(t)H_{\epsilon^{(k)}} \Big)dt = \sum_{\alpha,k}(c_\alpha, e_k)_{L^2(\mathbb{R})}H_{\alpha+\epsilon^{(k)}}. \tag{5.71}$$

For the left-hand side we obtain, using (5.8) and (5.14),

$$
\begin{aligned}
\int_{\mathbb{R}} Y(t)\delta W(t) &= \int_{\mathbb{R}} \sum_{n=0}^{\infty} I_n\big(f_n(\cdot, t)\big)\delta W(t) \\
&= \sum_{n=0}^{\infty} \int_{\mathbb{R}} I_n\Big(\sum_{|\alpha|=n} c_\alpha(t)e^{\hat{\otimes}\alpha} \Big)\delta W(t) \\
&= \sum_{n=0}^{\infty} \int_{\mathbb{R}} I_n\Big(\sum_{|\alpha|=n} \sum_{k=1}^{\infty} (c_\alpha, e_k)_{L^2(\mathbb{R})} e_k(t)e^{\hat{\otimes}\alpha} \Big)\delta W(t) \\
&= \sum_{n=0}^{\infty} I_{n+1}\Big(\sum_{|\alpha|=n} \sum_{k=1}^{\infty} (c_\alpha, e_k)_{L^2(\mathbb{R})} \big(e^{\hat{\otimes}\alpha}\hat{\otimes}e_k\big) \Big) \\
&= \sum_{n=0}^{\infty} \sum_{|\alpha|=n} \sum_{k=1}^{\infty} (c_\alpha, e_k)_{L^2(\mathbb{R})} I_{n+1}\big(e^{\hat{\otimes}(\alpha+\epsilon^{(k)})}\big) \\
&= \sum_{\alpha,k} (c_\alpha, e_k)_{L^2(\mathbb{R})} H_{\alpha+\epsilon^{(k)}}.
\end{aligned}
\tag{5.72}
$$

Comparing (5.71) and (5.72), we get (5.70). $\square$

Corollary 5.21. *If $Y(t) = \sum_{i=1}^{n} c_i \chi_{(t_i, t_{i+1}]}(t)$, $t \in \mathbb{R}$, with $c_i \in (\mathcal{S})^*$ for $i = 1, ..., n$ and $t_1 < \cdots < t_n$. Then we have*

$$\int_{\mathbb{R}} Y(t) \diamond \overset{\bullet}{W}(t) dt = \sum_{i=1}^{n} c_i \diamond \big(W(t_{i+1}) - W(t_i) \big).$$

In view of the aforementioned theorem, the following terminology is natural.

Definition 5.22. *Suppose Y is an $(\mathcal{S})^*$-valued process such that*

$$\int_{\mathbb{R}} Y(t) \diamond \overset{\bullet}{W}(t) dt \in (\mathcal{S})^*,$$

then we call this integral the generalized Skorohod integral *of Y.*

Combining the aforementioned properties with the fundamental relation (5.28) for Skorohod integration, we get a powerful calculation technique for stochastic integration. First of all, note that, by (5.28),

$$\int_0^T \overset{\bullet}{W}(t) dt = W(T). \tag{5.73}$$

(See Problem 5.3). Moreover, using (5.30) one can deduce that

$$\int_0^T X \diamond Y(t) \diamond \overset{\bullet}{W}(t) dt = X \diamond \int_0^T Y(t) \diamond \overset{\bullet}{W}(t) dt, \tag{5.74}$$

if X does not depend on t. Compare this with the fact that for Skorohod integrals we generally have

$$\int_0^T X \cdot Y(t) \delta W(t) \neq X \cdot \int_0^T Y(t) \delta W(t), \tag{5.75}$$

even if X does not depend on t.

To illustrate the use of Wick calculus, let us again consider Example 2.5:

$$\int_0^T W(t) [W(T) - W(t)] \delta W(t) = \int_0^T W(t) \diamond (W(T) - W(t)) \diamond \overset{\bullet}{W}(t) dt$$

$$= \int_0^T W(t) \diamond W(T) \diamond \overset{\bullet}{W}(t) dt - \int_0^T W^{\diamond 2}(t) \diamond \overset{\bullet}{W}(t) dt$$

$$= W(T) \diamond \int_0^T W(t) \diamond \overset{\bullet}{W}(t) dt - \frac{1}{3} W^{\diamond 3}(T) = \frac{1}{6} W^{\diamond 3}(T)$$

$$= \frac{1}{6} [W^3(T) - 3TW(T)],$$

where we have correspondingly used (5.39), (5.31), (5.74), and (5.65).

We proceed to establish some useful properties of generalized Skorohod integrals.

Lemma 5.23. *Suppose $f \in (\mathcal{S})$ and $G(t) \in (\mathcal{S})_{-q}$ for all $t \in \mathbb{R}$, for some $q \in \mathbb{N}$. Put*

$$\hat{q} = q + \frac{1}{\log 2}.$$

Then

$$\int_{\mathbb{R}} | < G(t) \diamond \dot{W}(t), f > | dt \le \|f\|_{\hat{q}} \left(\int_{\mathbb{R}} \|G(t)\|_{-q}^2 dt \right)^{1/2}.$$

Proof Suppose $G(t) = \sum_{\alpha \in \mathcal{J}} a_\alpha(t) H_\alpha$, $f = \sum_{\beta \in \mathcal{J}} b_\beta H_\beta$. Then

$$< G(t) \diamond \dot{W}(t), f >= < \sum_{\alpha,k} a_\alpha(t) e_k(t) H_{\alpha+\epsilon^{(k)}}, \sum_{\beta \in \mathcal{J}} b_\beta H_\beta >$$

$$= \sum_{\alpha,k} a_\alpha(t) e_k(t) b_{\alpha+\epsilon^{(k)}} (\alpha + \epsilon^{(k)})!.$$

Hence

$$\int_{\mathbb{R}} | < G(t) \diamond \dot{W}(t), f > | dt \le \sum_{\alpha,k} |b_{\alpha+\epsilon^{(k)}}| \alpha! (\alpha_k + 1) \int_{\mathbb{R}} |a_\alpha(t) e_k(t)| dt$$

$$\le \sum_{\alpha,k} |b_{\alpha+\epsilon^{(k)}}| \alpha! (\alpha_k + 1) \left(\int_{\mathbb{R}} a_\alpha^2(t) dt \right)^{1/2}$$

$$\le \left(\sum_{\alpha,k} b_{\alpha+\epsilon^{(k)}}^2 (\alpha + \epsilon^{(k)})! (2\mathbb{N})^{\hat{q}(\alpha+\epsilon^{(k)})} \right)^{1/2}$$

$$\cdot \left(\sum_{\alpha,k} \left(\int_{\mathbb{R}} a_\alpha^2(t) dt \right) \alpha! (\alpha_k + 1)(2\mathbb{N})^{-\hat{q}(\alpha+\epsilon^{(k)})} \right)^{1/2}$$

$$\le \|f\|_{\hat{q}} \left(\sum_{\alpha,k} \left(\int_{\mathbb{R}} a_\alpha^2(t) dt \right) \alpha! (\alpha_k+1)(2k)^{-\frac{\alpha_k}{\log 2}} (2\mathbb{N})^{-q\alpha} \right)^{1/2}$$

$$\le \|f\|_{\hat{q}} \left(\int_{\mathbb{R}} \|G(t)\|_{-q}^2 dt \right)^{1/2}. \quad \sqcup$$

Using this result we obtain the following theorem.

Theorem 5.24. *(1) Suppose $G : \mathbb{R} \longrightarrow (\mathcal{S})_{-q}$ satisfies*

$$\int_{\mathbb{R}} \|G(t)\|_{-q}^2 dt < \infty, \quad \text{for some } q \in \mathbb{N}.$$

Then

$$\int_{\mathbb{R}} G(t) \diamond \dot{W}(t) dt \quad \text{exists in } (\mathcal{S})^*.$$

(2) Suppose $F(t)$, $F_n(t)$, $n = 1, 2, ...$, are elements of $(\mathcal{S})_{-q}$ for all $t \in \mathbb{R}$ and

$$\int_{\mathbb{R}} \|F_n(t) - F(t)\|^2_{-q} dt \longrightarrow 0, \qquad n \to \infty.$$

Then

$$\int_{\mathbb{R}} F_n(t) \diamond \dot{W}(t) dt \longrightarrow \int_{\mathbb{R}} F(t) \diamond \dot{W}(t) dt, \quad n \to \infty,$$

in the weak*-topology on $(\mathcal{S})^*$.

Proof (1) The proof follows from Lemma 5.23 and Definition 5.18.
(2) By Lemma 5.23 we have

$$|<\int_{\mathbb{R}} (F_n(t) - F(t)) \diamond \dot{W}(t) dt, f >| \leq \int_{\mathbb{R}} |< (F_n(t) - F(t)) \diamond \dot{W}(t), f >| dt$$

$$\leq \|f\|_{\hat{q}} \int_{\mathbb{R}} \|F_n(t) - F(t)\|^2_{-q} dt \longrightarrow 0, \quad n \to \infty. \quad \square$$

5.4 Exercises

Problem 5.1. Prove equation (5.4). [*Hint.* First consider step functions ϕ of the form $\phi(t) = \sum_i e_i \chi_{(a_i, a_{i+1}]}(t), \quad t \in \mathbb{R}$.]

Problem 5.2. Prove equation (5.22), that is, that

$$\frac{d}{dt} W(t) = \dot{W}(t),$$

where the derivative exists in $(\mathcal{S})^*$.

Problem 5.3. Prove that the $(\mathcal{S})^*$-valued process $\dot{W}$ is $(\mathcal{S})^*$-integrable (cf. Definition 5.18) and that

$$\int_0^T \dot{W}(t) dt = W(T), \qquad T > 0.$$

Problem 5.4. Prove (5.26), that is, that

$$(\dot{W})^{\diamond 2}(t) \in (\mathcal{S})^*, \qquad t \in \mathbb{R}.$$

Problem 5.5. (*) Use the identity (5.28) and Wick calculus to compute the following Skorohod integrals:

(a) $\int\limits_0^T W(T) \delta W(t)$;

(b) $\int\limits_0^T \int\limits_0^T g(s) dW(s) \delta W(t)$, for the deterministic function $g \in L^2([0, T])$;

(c) $\int\limits_0^T W^2(t_0)\delta W(t)$, where $t_0 \in [0, T]$ is fixed;

(d) $\int\limits_0^T \exp(W(T))\delta W(t)$.

Compare with your calculations in Problem 2.4!

Problem 5.6. Show that $(\mathcal{S})$ and $(\mathcal{S})^*$ are vector spaces.

Problem 5.7. (a) Let $f \in L^2(\mathbb{R})$ be deterministic. Prove that

$$\frac{d}{dt}\int_{-\infty}^t f(s)dW(s) = f(t)\overset{\bullet}{W}(t) \quad \text{in} \quad (\mathcal{S}^*).$$

(b) Let u be a Skorohod integrable process. Prove that

$$\frac{d}{dt}\int_{-\infty}^t u(s)\delta W(s) = u(t) \diamond \overset{\bullet}{W}(t) \quad \text{in} \quad (\mathcal{S}^*).$$

Problem 5.8. (a) Solve the stochastic differential equation

$$dX(t) = X(t)[\mu dt + \sigma dW(t)], \qquad X(0) = x > 0, \qquad (5.76)$$

where the parameters μ, σ, and the initial value x are constants, via the following guidelines. First, rewrite the equation in the form

$$\frac{d}{dt}X(t) = X(t) \diamond [\mu + \sigma\overset{\bullet}{W}(t)], \qquad X(0) = x > 0, \qquad (5.77)$$

and regard it as an ordinary differential equation in the $(\mathcal{S})^*$-valued function $X(t)$, $t > 0$. Then use the Wick calculus in $(\mathcal{S})^*$.
(b) Equation (5.77) also makes sense if the initial value $X(0)$ is not constant, but a given element in $(\mathcal{S})^*$ (e.g., $X(0) \in L^2(P)$). What would the solution of (5.77) be in this case?

Problem 5.9. (a) Proceed as in Problem 5.8, but this time with the stochastic differential equation

$$dX(t) = \alpha dt + \beta X(t)dW(t), \qquad X(0) = x \in \mathbb{R},$$

which can be written as the following differential equation in $(\mathcal{S})^*$:

$$\frac{dX(t)}{dt} = \alpha + \beta X(t) \diamond \overset{\bullet}{W}(t), \qquad X(0) = x \in \mathbb{R}.$$

(b) What can be said about the solution if we only know that $X(0) \in (\mathcal{S})^*$?

The Hida–Malliavin Derivative
on the Space $\Omega = \mathcal{S}'(\mathbb{R})$

6.1 A New Definition of the Stochastic Gradient and a Generalized Chain Rule

In Chap. 3 we have introduced the Malliavin derivative in terms of chaos expansions. However, the original definition was tailored for the Wiener space where the derivative is given as a differential operator. See Appendix. Although the Wiener space is a natural space to work on when dealing with the Brownian motion, this original approach has the disadvantage that the directional derivative in the direction $\gamma = \gamma(t)$, $t \in [0, T]$, where

$$\gamma(t) = \int_0^t g(s)ds, \qquad g \in L^2([0, T]),$$

is represented by a gradient with respect to g, cf. (A.12) and (A.13).

If, however, we follow the suggestion of Hida [99] and work on the space $\Omega = \mathcal{S}'(\mathbb{R})$ instead (as in Chap. 5), then we obtain a complete agreement between the directional derivative and the corresponding *stochastic gradient* (or *the Hida–Malliavin derivative*). We now explain this in more detail.

From now on we assume that the Brownian motion $W(t)$, $t \in \mathbb{R}$, is constructed on the space $(\Omega, \mathcal{B}, P)$ with $\Omega = \mathcal{S}'(\mathbb{R})$ as in Chap. 5.

Definition 6.1. *(1) Let $F \in L^2(P)$ and let $\gamma \in L^2(\mathbb{R})$ be deterministic. Then the directional derivative of F in $(\mathcal{S})^*$ (respectively, in $L^2(P)$) in the direction γ is defined by*

$$D_\gamma F(\omega) = \lim_{\varepsilon \to 0} \frac{1}{\varepsilon} \big[F(\omega + \varepsilon\gamma) - F(\omega) \big] \tag{6.1}$$

whenever the limit exists in $(\mathcal{S})^*$ (respectively, in $L^2(P)$).
(2) Suppose there exists a function $\psi : \mathbb{R} \longrightarrow (\mathcal{S})^*$ (respectively, $\psi : \mathbb{R} \longrightarrow L^2(P)$) such that

G.Di Nunno et al., *Malliavin Calculus for Lévy Processes with Applications to Finance*,
© Springer-Verlag Berlin Heidelberg 2009

$$\int_{\mathbb{R}} \psi(t)\gamma(t)dt \quad \text{converges in } (\mathcal{S})^* \text{ (respectively, in } L^2(P)) \text{ and}$$

$$D_\gamma F = \int_{\mathbb{R}} \psi(t)\gamma(t)dt, \quad \text{for all } \gamma \in L^2(\mathbb{R}), \tag{6.2}$$

then we say that F is Hida–Malliavin differentiable in $(\mathcal{S})^*$ (respectively, in $L^2(P)$) and we write

$$\psi(t) = D_t F, \quad t \in \mathbb{R}.$$

We call $D_t F$ the Hida–Malliavin derivative in $(\mathcal{S})^*$ (respectively, in $L^2(P)$) or the stochastic gradient of F at t.

When it is clear from the context if $D_t F$ is a Malliavin derivative in $(\mathcal{S})^*$ or in $L^2(P)$, we will not write further specifications than "Malliavin derivative" and "Malliavin differentiable."

Example 6.2. (1) Suppose $F(\omega) = <\omega, f> = \int_{\mathbb{R}} f(t)dW(t)$, $f \in L^2(\mathbb{R})$. Then

$$D_\gamma F = \frac{1}{\varepsilon}\big[<\omega + \varepsilon\gamma, f> - <\omega, f> \big] = <\gamma, f> = \int_{\mathbb{R}} f(t)\gamma(t)dt.$$

Therefore, F is Hida–Malliavin differentiable and

$$D_t \Big(\int_{\mathbb{R}} f(t)dW(t) \Big) = f(t), \quad t - a.a.$$

(2) Let $F \in L^2(P)$ be Hida–Malliavin differentiable in $L^2(P)$ for a.a. t. Suppose that $\varphi \in C^1(\mathbb{R})$ and $\varphi'(F)D_t F \in L^2(P \times \lambda)$. Then if $\gamma \in L^2(\mathbb{R})$ we have

$$\begin{aligned}
D_\gamma\big(\varphi(F)\big) &= \lim_{\varepsilon \to 0} \frac{1}{\varepsilon}\big[\varphi(F(\omega + \varepsilon\gamma)) - \varphi(F(\omega))\big] \\
&= \lim_{\varepsilon \to 0} \frac{1}{\varepsilon}\big[\varphi(F(\omega) + \varepsilon D_\gamma F) - \varphi(F(\omega))\big] \\
&= \frac{1}{\varepsilon}\varphi'(F(\omega))\varepsilon D_\gamma F = \varphi'(F)D_\gamma F \\
&= \int_{\mathbb{R}} \varphi'(F)D_t F \gamma(t)dt.
\end{aligned}$$

This proves that $\varphi(F)$ is also Hida–Malliavin differentiable and we have the *chain rule*

$$D_t\big(\varphi(F)\big) = \varphi'(F)D_t F, \tag{6.3}$$

see also Lemma A.12.

More generally, the same proof gives the following extension of Theorem 3.4 and justifies Theorem 3.5.

Theorem 6.3. Chain rule. *Let $F_1, ..., F_m \in L^2(P)$ be Hida–Malliavin differentiable in $L^2(P)$. Suppose that $\varphi \in C^1(\mathbb{R}^m)$, $D_t F_i \in L^2(P)$, for all $t \in \mathbb{R}$, and $\frac{\partial \varphi}{\partial x_i}(F)D.F_i \in L^2(P \times \lambda)$ for $i = 1, ..., m$, where $F = (F_1, ..., F_m)$. Then $\varphi(F)$ is Hida–Malliavin differentiable and*

$$D_t \varphi(F) = \sum_{i=1}^{m} \frac{\partial \varphi}{\partial x_i}(F) D_t F_i. \tag{6.4}$$

It follows from the above that if $h_n(x)$, $x \in \mathbb{R}$, is the Hermite polynomial of order n and $f \in L^2(\mathbb{R})$, then

$$D_t\big(h_n(< \omega, f >)\big) = n h_{n-1}\big(< \omega, f > \big) f(t).$$

Therefore, we obtain that if $f_n \in \widetilde{L}^2(\mathbb{R}^n)$ then

$$D_t\big(I_n(f_n)\big) = n I_{n-1}\big(f_n(\cdot, t)\big). \tag{6.5}$$

We conclude that the Hida–Malliavin derivative defined above coincides with the Malliavin derivative defined in Chap. 3 on the space $\mathbb{D}_{1,2}$. See also Appendix. To simplify the terminology, we also use the term Malliavin derivative in this setting.

Corollary 6.4. *Let $F = (F_1, ..., F_m)$ with $F_i \in \mathbb{D}_{1,2}$ for all $i = 1, ..., m$. Let $\varphi \in C^1(\mathbb{R}^m)$, with $\left|\frac{\partial \varphi}{\partial x_i}\right|$ bounded for all $i = 1, ..., m$. Then $\varphi(F) \in \mathbb{D}_{1,2}$ and*

$$D_t \varphi(F) = \sum_{i=1}^{m} \frac{\partial \varphi}{\partial x_i}(F) D_t F_i. \tag{6.6}$$

It is useful to note how the Malliavin derivative can be expressed in terms of the Wiener–Itô chaos expansion (see Theorem 5.2). To this aim observe that from (5.7) and the chain rule (6.3) we have

$$D_t H_\alpha = \sum_{k=1}^{m} \prod_{j \neq k} h_{\alpha_j}(\theta_j) \alpha_k h_{\alpha_k - 1}(\theta_k) e_k(t) = \sum_{k=1}^{m} \alpha_k e_k(t) H_{\alpha - \epsilon^{(k)}}. \tag{6.7}$$

More generally, we have the following definition.

Definition 6.5. *If $F = \sum_{\alpha \in \mathcal{J}} c_\alpha H_\alpha \in (\mathcal{S})^*$, we define $D_t F$ as the Malliavin derivative of F at t in $(\mathcal{S})^*$ the following expansion:*

$$D_t F = \sum_{\alpha \in \mathcal{J}} \sum_{k=1}^{\infty} c_\alpha \alpha_k e_k(t) H_{\alpha - \epsilon^{(k)}} \tag{6.8}$$

whenever this sum converges in $(\mathcal{S})^*$. We denote $Dom(D_t)$ the set of all $F \in (\mathcal{S})^*$ for which the above series converges in $(\mathcal{S})^*$.

Remark 6.6. Note that $L^2(P) \subseteq Dom(D_t) \subseteq (\mathcal{S})^*$. See Problem 6.1.

Lemma 6.7. *Suppose $F \in Dom(D_t)$ for a.a. t and $q \in \mathbb{N}$. Then*

(1)

$$\int_{\mathbb{R}} \|D_t F\|_{-\hat{q}}^2 dt \leq \|F\|_{-q}^2, \quad \hat{q} \geq 2q + \frac{1}{\log 2}.$$

(2) Suppose $F_N \in Dom(D_t)$ for a.a. t and for all $N = 1, 2, \dots$. Suppose $F_N \longrightarrow F$, $N \to \infty$, in $(\mathcal{S})^*$. Then there exists a subsequence F_{N_k}, $k = 1, 2, \dots$, such that

$$D_t F_{N_k} \longrightarrow D_t F, \quad k \to \infty, \text{ in } (\mathcal{S}^*), \text{ for a.a. } t \in \mathbb{R}.$$

Proof (1) Suppose $F = \sum_{\alpha \in \mathcal{J}} c_\alpha H_\alpha$. Then by (6.8) we have

$$D_t F = \sum_{\alpha \in \mathcal{J}} \sum_{k=1}^{\infty} c_\alpha \alpha_k e_k(t) H_{\alpha - \epsilon^{(k)}} = \sum_{\beta \in \mathcal{J}} \sum_{k=1}^{\infty} c_{\beta + \epsilon^{(k)}} (\beta_k + 1) e_k(t) H_\beta = \sum_{\beta \in \mathcal{J}} g_\beta(t) H_\beta,$$

where

$$g_\beta(t) = \sum_{k=1}^{\infty} c_{\beta + \epsilon^{(k)}} (\beta_k + 1) e_k(t).$$

Since $F \in (\mathcal{S})^*$, there exists $q \in \mathbb{N}$ such that

$$\|F\|_{-q}^2 := \sum_{\alpha \in \mathcal{J}} c_\alpha^2 \alpha! (2\mathbb{N})^{-q\alpha} < \infty.$$

Note that

$$\int_{\mathbb{R}} g_\beta^2(t) dt = \int_{\mathbb{R}} \left(\sum_{k=1}^{\infty} c_{\beta + \epsilon^{(k)}} (\beta_k + 1) e_k(t) \right)^2 dt = \sum_{k=1}^{\infty} c_{\beta + \epsilon^{(k)}}^2 (\beta_k + 1)^2.$$

Therefore, using $(x + 1) x e^{-x} \leq 1$ for all $x \geq 0$, we get

$$\int_{\mathbb{R}} \|D_t F\|_{-\hat{q}}^2 dt = \int_{\mathbb{R}} \sum_{\beta \in \mathcal{J}} g_\beta^2(t) \beta! (2\mathbb{N})^{-\hat{q}\beta} dt$$

$$= \sum_{\beta \in \mathcal{J}} \sum_{k=1}^{\infty} c_{\beta + \epsilon^{(k)}}^2 (\beta_k + 1)^2 \beta! (2\mathbb{N})^{-\hat{q}\beta}$$

$$= \sum_{\beta \in \mathcal{J}} \sum_{k=1}^{\infty} c_{\beta + \epsilon^{(k)}}^2 (\beta_k + \epsilon^{(k)})! \beta_k (\beta_k + 1)(2k)^{-\frac{\beta_k}{\log 2}} (2\mathbb{N})^{-2q\beta_k}$$

$$\leq \sum_{\beta \in \mathcal{J}} \sum_{k=1}^{\infty} c_{\beta + \epsilon^{(k)}}^2 (\beta_k + \epsilon^{(k)})! (2\mathbb{N})^{-q(\beta + \epsilon^{(k)})}$$

$$\leq \sum_{\alpha \in \mathcal{J}} c_\alpha^2 \alpha! (2\mathbb{N})^{-q\alpha} = \|F\|_{-q}^2,$$

which gives (1).

(2) For the given F and F_N, $N = 1, 2, ...$, we have

$$\int_{\mathbb{R}} \|D_t(F_N - F)\|_{-\hat{q}}^2 dt \leq \|F_N - F\|_{-q}^2 \longrightarrow 0, \quad N \to \infty,$$

by (1), if $q \in \mathbb{N}$ is large enough. Hence there exists a subsequence F_{N_k}, $k = 1, 2, ...$, such that

$$\|D_t(F_{N_k} - F)\|_{-q}^2 \longrightarrow 0, \quad k \to \infty, \text{ for a.a. } t \in \mathbb{R}. \quad \square$$

6.2 Calculus of the Hida–Malliavin Derivative and Skorohod Integral

6.2.1 Wick Product vs. Ordinary Product

The Malliavin derivative can be used to obtain an explicit relation beween the Wick product and the ordinary product in the case when one of the factors is a first order chaos element. In fact the following result holds true.

Theorem 6.8. *Suppose $g \in L^2(\mathbb{R})$ and $F \in \mathbb{D}_{1,2}$. Then*

$$F \diamond \int_{\mathbb{R}} g(t)dW(t) = F \int_{\mathbb{R}} g(t)dW(t) - \int_{\mathbb{R}} g(t)D_t F dt. \qquad (6.9)$$

Proof To ease the notation, let $\| \cdot \| = \| \cdot \|_{L^2(\mathbb{R})}$ and $(\cdot, \cdot) = (\cdot, \cdot)_{L^2(\mathbb{R})}$. For $y \in \mathbb{R}$ we define (5.67),

$$G_y := \exp^{\diamond} \left\{ y \int_{\mathbb{R}} g(t)dW(t) \right\} = \exp \left\{ y \int_{\mathbb{R}} g(t)dW(t) - \frac{1}{2} y^2 \|g\|^2 \right\}.$$

Choose $F = \exp^{\diamond} \left\{ \int_{\mathbb{R}} f(t)dW(t) \right\} = \exp \left\{ \int_{\mathbb{R}} f(t)dW(t) - \frac{1}{2}\|f\|^2 \right\}$, where $f \in L^2(\mathbb{R})$. Then

$$F \diamond G_y = \exp^{\diamond} \left\{ \int_{\mathbb{R}} f(t)dW(t) \right\} \diamond \exp^{\diamond} \left\{ y \int_{\mathbb{R}} g(t)dW(t) \right\}$$

$$= \exp^{\diamond} \left\{ \int_{\mathbb{R}} [f(t) + yg(t)]dW(t) \right\}$$

$$= \exp \left\{ \int_{\mathbb{R}} [f(t) + yg(t)]dW(t) - \frac{1}{2}\|f + yg\|^2 \right\}$$

$$= \exp^{\diamond} \left\{ \int_{\mathbb{R}} f(t)dW(t) \right\} \exp^{\diamond} \left\{ \int_{\mathbb{R}} yg(t)dW(t) \right\} \exp \left\{ -y(f, g) \right\}$$

$$= F G_y \exp \left\{ -y(f, g) \right\}.$$

Now differentiating with respect to y, we get

$$\frac{d}{dy}(F \diamond G_y) = F \diamond \left(G_y \diamond \int_{\mathbb{R}} g(t)dW(t)\right) \tag{6.10}$$

and

$$\frac{d}{dy}\left(F G_y \exp\{-y(f,g)\}\right)$$

$$= F G_y \left[\int_{\mathbb{R}} g(t)dW(t) \exp\{-y(f,g)\} - (f,g)\exp\{-y(f,g)\}\right]. \tag{6.11}$$

Comparing (6.10) and (6.11) we get

$$F \diamond \left(G_y \diamond \int_{\mathbb{R}} g(t)dW(t)\right) = F G_y \exp\{-y(f,g)\}\left[\int_{\mathbb{R}} g(t)dW(t) - (f,g)\right].$$

In particular, putting $y = 0$ we get

$$F \diamond \int_{\mathbb{R}} g(t)dW(t) = F \int_{\mathbb{R}} g(t)dW(t) - F \int_{\mathbb{R}} f(t)g(t)dt$$

$$= F \int_{\mathbb{R}} g(t)dW(t) - \int_{\mathbb{R}} g(t)D_t F dt.$$

This proves the result if $F = \exp^{\diamond}\left\{\int_{\mathbb{R}} f(t)dW(t)\right\}$ for some $f \in L^2(\mathbb{R})$. Since linear combinations of such F's are dense in $\mathbb{D}_{1,2}$, the result follows by an approximation argument. $\square$

Example 6.9. Choose $F = \int_{\mathbb{R}} f(t)dW(t)$ with $f \in L^2(\mathbb{R})$. Then (6.9) gives

$$\left(\int_{\mathbb{R}} f(t)dW(t)\right) \diamond \left(\int_{\mathbb{R}} g(t)dW(t)\right) = \left(\int_{\mathbb{R}} f(t)dW(t)\right)\left(\int_{\mathbb{R}} g(t)dW(t)\right)$$

$$- \int_{\mathbb{R}} f(t)g(t)dt, \tag{6.12}$$

which is in agreement with (5.62).

Remark 6.10. A general formula for the relation between Wick products and ordinary products can be found in [109].

6.2.2 Closability of the Hida–Malliavin Derivative

We now apply Theorem 6.8 to obtain the following useful result (see also Lemma A.15).

Theorem 6.11. Integration by parts. *Let* $G, X \in \mathbb{D}_{1,2}$ *and* $\gamma \in L^2(\mathbb{R})$. *Then*

$$E[X D_\gamma G] = E[X G <\omega, \gamma>] - E\left[G \int_{\mathbb{R}} \gamma(t)D_t X dt\right] \tag{6.13}$$

where, as before, $<\omega, \gamma> = \int_{\mathbb{R}} \gamma(t)dW(t)$.

Proof By the Girsanov theorem in [107, Corollary 2.10.4 (b)], we have

$$E[XG(\omega + \varepsilon\gamma)] = E[G(\omega) \cdot (\exp^\diamond\{< \omega, \varepsilon\gamma >\} \diamond X)]$$

for all $\varepsilon > 0$. Hence, by Theorem 6.8,

$$E[XD_\gamma G] = E\Big[X \lim_{\varepsilon \to 0} \frac{G(\omega + \varepsilon\gamma) - G(\omega)}{\varepsilon}\Big]$$

$$= E\Big[\lim_{\varepsilon \to 0} \frac{1}{\varepsilon} G(\omega) \big((\exp^\diamond\{< \omega, \varepsilon\gamma >\} - 1) \diamond X\big)\Big]$$

$$= E[G(\omega) (< \omega, \gamma > \diamond X)]$$

$$= E\Big[G \big(< \omega, \gamma > X - \int_{\mathbb{R}} \gamma(t)D_t X \, dt\big)\Big]$$

$$= E[GX < \omega, \gamma >] - E\Big[G \int_{\mathbb{R}} \gamma(t)D_t X \, dt\Big]. \qquad \square$$

From this we obtain the following important result.

Theorem 6.12. Closability of the Hida–Malliavin derivative. *Suppose* $G, G_N \in \mathbb{D}_{1,2}$ *for* $N = 1, 2, \ldots$ *and that* $\lim_{N\to\infty} G_N = G$ *in* $L^2(P)$ *and that the sequence* DG_N, $N = 1, 2, \ldots$ *converges in* $L^2(P \times \lambda)$ *to some limit. Then* $\lim_{N\to\infty} DG_N = DG$ *in* $L^2(P \times \lambda)$.

Proof We may assume that $G = 0$. For all $X \in \mathbb{D}_{1,2}$ and all N we have

$$E\Big[XD_\gamma G_N\Big] = E\Big[XG_N < \omega, \gamma >\Big] - E\Big[G_N \int_{\mathbb{R}} \gamma(t)D_t X \, dt\Big],$$

by Theorem 6.11. Hence,

$$E\Big[X \int_{\mathbb{R}} D_t G_N \gamma(t) dt\Big] = E[XD_\gamma G_N] \longrightarrow 0, \quad N \to \infty.$$

Since this holds for all $X \in \mathbb{D}_{1,2}$ and all $\gamma \in L^2(\mathbb{R})$, we conclude that

$$\lim_{N\to\infty} DG_N = 0 \qquad \text{weakly in } L^2(P \times \lambda).$$

By assumption the sequence DG_N, $N = 1, 2, \ldots$, converges (strongly) in $L^2(P \times \lambda)$. Hence, $\lim_{N\to\infty} DG_N = 0$ (strongly) in $L^2(P \times \lambda)$. $\square$

6.2.3 Wick Chain Rule

We now apply this to obtain some versions of the Wick chain rule for Malliavin derivatives.

Theorem 6.13. Wick chain rule.

(1) Let $F, G \in \mathbb{D}_{1,2}$ and $F \diamond G \in \mathbb{D}_{1,2}$. Then

$$D_t(F \diamond G) = F \diamond D_t G + D_t F \diamond G, \quad t \in \mathbb{R}.$$

(2) Let $F \in \mathbb{D}_{1,2}$ and $F^{\diamond n} \in \mathbb{D}_{1,2}$. Then

$$D_t(F^{\diamond n}) = n F^{\diamond(n-1)} \diamond D_t F \qquad (n = 1, 2, ...).$$

(3) Let $F \in \mathbb{D}_{1,2}$ be Malliavin differentiable and assume that

$$\exp^{\diamond} F = \sum_{n=0}^{\infty} \frac{1}{n!} F^{\diamond n} \in \mathbb{D}_{1,2}.$$

Then

$$D_t \exp^{\diamond} F = \exp^{\diamond} F \diamond D_t F.$$

Proof (1) By Theorem 6.12, it suffices to prove this in the case when

$$F = \exp^{\diamond}(<\omega, f>) = \exp(<\omega, f> - \frac{1}{2}\|f\|^2_{L^2(\mathbb{R})})$$

and

$$G = \exp^{\diamond}(<\omega, g>) = \exp(<\omega, g> - \frac{1}{2}\|g\|^2_{L^2(\mathbb{R})}),$$

where $f, g \in L^2(\mathbb{R})$ are deterministic, $<\omega, f> = \int_{\mathbb{R}} f(s) dW(s)$, and $<\omega, g> = \int_{\mathbb{R}} g(s) dW(s)$. In this case we have, by Theorem 6.3,

$$
\begin{aligned}
D_t(F \diamond G) &= D_t\big(\exp^{\diamond}(<\omega, f>) \diamond \exp^{\diamond}(<\omega, g>)\big) \\
&= D_t \exp^{\diamond}(<\omega, f+g>) = D_t\big(\exp\{<\omega, f+g> - \frac{1}{2}\|f+g\|^2_{L^2(\mathbb{R})}\}\big) \\
&= \exp\{<\omega, f+g> - \frac{1}{2}\|f+g\|^2_{L^2(\mathbb{R})}\}(f(t) + g(t)) \\
&= \exp^{\diamond}(<\omega, f+g>)(f(t) + g(t)).
\end{aligned}
$$

On the other hand, again by Theorem 6.3,

$$
\begin{aligned}
D_t F \diamond G + F \diamond D_t G &= D_t\big(\exp\{<\omega, f> - \frac{1}{2}\|f\|^2_{L^2(\mathbb{R})}\}\big) \diamond G \\
&\quad + F \diamond D_t\big(\exp\{<\omega, g> - \frac{1}{2}\|g\|^2_{L^2(\mathbb{R})}\}\big) \\
&= \exp\{<\omega, f> - \frac{1}{2}\|f\|^2_{L^2(\mathbb{R})}\}f(t) \diamond G \\
&\quad + F \diamond \exp\{<\omega, g> - \frac{1}{2}\|g\|^2_{L^2(\mathbb{R})}\}g(t) \\
&= (F \diamond G)(f(t) + g(t)) \\
&= \exp^{\diamond}(<\omega, f+g>)(f(t) + g(t)).
\end{aligned}
$$

This ends the proof of (1).

(2) This part of the proof is left as an exercise, see Problem 6.2.

(3) By (2) we have

$$D_t\Big(\sum_{n=0}^{N}\frac{1}{n!}F^{\diamond n}\Big) = \sum_{n=1}^{N-1}\frac{1}{(n-1)!}F^{\diamond(n-1)}\diamond D_tF \longrightarrow \exp^\diamond F \diamond D_tF, \qquad N\to\infty.$$

So if we put

$$H_N = \sum_{n=0}^{N}\frac{1}{n!}F^{\diamond n}, \quad H = \sum_{n=0}^{\infty}\frac{1}{n!}F^{\diamond n},$$

then

$$\lim_{N\to\infty} H_N = H \qquad \text{in } L^2(P)$$

and

$$\lim_{N\to\infty} DH_N = \exp^\diamond G \diamond DF \qquad \text{in } L^2(P\times\lambda).$$

Hence $D_tH = \exp^\diamond G \diamond D_tF$, by Theorem 6.12. □

Example 6.14. Consider $F = I_n(f)$ and $G = I_m(g)$, with $f \in \widetilde{L}^2([0,T]^n)$ and $g \in \widetilde{L}^2([0,T]^m)$. Then $F\diamond G = I_{n+m}(f\hat\otimes g)$. So $F\diamond G \in \mathbb{D}_{1,2}$.

6.2.4 Integration by Parts, Duality Formula, and Skorohod Isometry

We end this section by establishing some fundamental properties of Skorohod integrals.

Theorem 6.15. Integration by parts. *Suppose $u(t)$, $0 \le t \le T$, is Skorohod integrable and $F \in \mathbb{D}_{1,2}$. Suppose that*

$$E\Big[\int_0^T u^2(t)\big(D_tF\big)^2 dt\Big] < \infty. \tag{6.14}$$

Then $F\,u(t)$, $0 \le t \le T$, is Skorohod integrable and

$$\int_0^T F\,u(t)\delta W(t) = F\int_0^T u(t)\delta W(t) - \int_0^T u(t)D_tF\,dt. \tag{6.15}$$

Proof First assume that $u(t)$, $0 \le t \le T$, is a simple function, that is, it has the form of a finite linear combination of the form $u(t) = \sum_i a_i\chi_{(t_i,t_{i+1}]}(t)$, $0 \le t \le T$, where $a_i \in \mathbb{D}_{1,2}$ for all i. Then by applying Theorem 6.8 twice we get

$$\int_0^T F u(t)\delta W(t) = \sum_i (Fa_i) \diamond \Delta W(t_i)$$

$$= \sum_i Fa_i \Delta W(t_i) - \sum_i \int_{t_i}^{t_i+1} D_t(Fa_i)dt$$

$$= F\sum_i a_i \Delta W(t_i) - \sum_i \int_{t_i}^{t_i+1} D_t(Fa_i)dt$$

$$= F\Big(\sum_i a_i \diamond \Delta W(t_i) + \sum_i \int_{t_i}^{t_i+1} D_t a_i dt\Big) - \sum_i \int_{t_i}^{t_i+1} D_t(Fa_i)dt$$

$$= F\int_0^T u(t)\delta W(t) - \int_0^T u(t)D_t F dt.$$

Now approximate the general u by a sequence u_m of simple functions in $Dom\,(\delta) \subseteq L^2(P\times\lambda)$ converging to u in $L^2(P\times\lambda)$. We omit the details.   □

Since Skorohod integrals have expectation null, we obtain the following immediate consequence of Theorem 6.15. This result is an extension of Corollary 4.4.

Corollary 6.16. Duality formula. *Let F and $u(t)$, $0 \le t \le T$, be as in Theorem 6.15. Then*

$$E\Big[F\int_0^T u(t)\delta W(t)\Big] = E\Big[\int_0^T u(t)D_t F dt\Big]. \tag{6.16}$$

The following result gives an expression for the $L^2(P)$-norm of a Skorohod integral in terms of Malliavin derivatives. This is often more useful than the isometry given in (2.4).

Theorem 6.17. Isometry for Skorohod integrals. *Let u be a measurable process such that $u(s) \in \mathbb{D}_{1,2}$ for a.a. s and*

$$E\Big[\int_0^T u^2(t)dt + \int_0^T \int_0^T |D_t u(s)D_s u(t)|dsdt\Big] < \infty.$$

Then u is Skorohod integrable and

$$E\Big[\Big(\int_0^T u(s)\delta W(s)\Big)^2\Big] = E\Big[\int_0^T u^2(t)dt + \int_0^T \int_0^T D_t u(s)D_s u(t)dsdt\Big]. \tag{6.17}$$

Proof First assume that u satisfies, in addition to the hypothesis given in the statement, also the conditions of Theorem 3.18 and Theorem 6.15. Then, with $F = \int_0^T u(s)\delta W(s)$, we have

$$E\left[\left(\int_0^T u(t)\delta W(t)\right)^2\right] = E\left[F\int_0^T u(t)\delta W(t)\right]$$

$$= E\left[\int_0^T u(t)D_t F\, dt\right]$$

$$= E\left[\int_0^T u(t)\left[u(t) + \int_0^T D_t u(s)\delta W(s)\right]dt\right]$$

$$= E\left[\int_0^T u^2(t)\, dt + \int_0^T\int_0^T D_t u(s)D_s u(t)\, ds\, dt\right],$$

where we have used Theorem 6.15 and Theorem 3.18, correspondingly. The general case follows by an approximation argument. $\square$

Remark 6.18. Let u be as in Theorem 6.17 and assume that, in addition, u is $\mathbb{F}$-adapted. Then, by Corollary 3.13 (2)

$$D_t u(s) = 0, \qquad t > s,$$

and therefore

$$D_t u(s)D_s u(t) = 0, \qquad \text{a.a. } (s,t) \in [0,T]\times[0,T].$$

Thus the isometry (6.17) reduces to

$$E\left[\left(\int_0^T u(t)\delta W(t)\right)^2\right] = E\left[\int_0^T u^2(t)\, dt\right].$$

This, as expected, is the classical isometry for Itô stochastic integrals. Recall in fact that, in the case of $\mathbb{F}$-adapted integrands, the Skorohod integral coincides with the Itô integral, see Theorem 2.9.

6.3 Conditional Expectation on $(\mathcal{S})^*$

It is natural to try to extend the concept of conditional expectation to $(\mathcal{S})^*$ as follows.

Definition 6.19. *If $F = \sum_{\alpha\in\mathcal{J}} c_\alpha H_\alpha \in (\mathcal{S})^*$ and $t > 0$, we define the (generalized) conditional expectation $E[F|\mathcal{F}_t]$ of F with respect to $\mathcal{F}_t$ by*

$$E[F|\mathcal{F}_t] := \sum_{\alpha\in\mathcal{J}} c_\alpha E[H_\alpha|\mathcal{F}_t] \tag{6.18}$$

when convergent in $(\mathcal{S})^*$.

Note that if $F = I_n(e^{\hat{\otimes} \alpha})$ (cf. (5.14)) then

$$E[F|\mathcal{F}_t] = I_n[e^{\hat{\otimes}\alpha}(x)\chi_{[0,t]}(\max_i x_i)]$$

$$= I_n\Big[\sum_{|\beta|=n}\Big(\int_0^t e^{\hat{\otimes}\alpha}(y)e^{\hat{\otimes}\beta}(y)dy\Big)e^{\hat{\otimes}\beta}(x)\Big]$$

$$= I_n\Big[\sum_{|\beta|=n}a_{\alpha,\beta}e^{\hat{\otimes}\beta}(x)\Big]$$

$$= \sum_{|\beta|=n}a_{\alpha,\beta}H_\beta,$$

where

$$a_{\alpha,\beta} = \int_0^t e^{\hat{\otimes}\alpha}(y)e^{\hat{\otimes}\beta}(y)dy.$$

In particular,

$$\sum_{|\beta|=n}a_{\alpha,\beta}^2 \le 1.$$

Therefore,

$$\sum_{|\beta|=n}\beta!a_{\alpha,\beta}^2 = \|E[F|\mathcal{F}_t]\|_{L^2(P)}^2 = \alpha! \tag{6.19}$$

However, it is not clear from this if $E[F|\mathcal{F}_t] \in (\mathcal{S})^*$ for all $F \in (\mathcal{S})^*$. But we do have the following result.

Lemma 6.20. *Suppose that F, G, $E[F|\mathcal{F}_t]$, and $E[G|\mathcal{F}_t]$ belong to $(\mathcal{S})^*$, then*

$$E[F \diamond G|\mathcal{F}_t] = E[F|\mathcal{F}_t] \diamond E[G|\mathcal{F}_t]. \tag{6.20}$$

Proof We may assume without loss of generality that $F = I_n(f_n)$ and $G = I_m(g_m)$ for some $f_n \in \widetilde{L}^2(\mathbb{R}^n)$ and $g_m \in \widetilde{L}^2(\mathbb{R}^m)$. Then

$$E[F \diamond G|\mathcal{F}_t] = E[I_n(f_n) \diamond I_m(g_m)|\mathcal{F}_t]$$

$$= E[I_{n+m}(f_n\hat{\otimes}g_m)|\mathcal{F}_t]$$

$$= I_{n+m}(f_n\hat{\otimes}g_m(x_1, ..., x_n, y_1, ..., y_m)\chi_{[0,t]}(\max_{i,j}\{x_i, y_j\}))$$

$$= I_n(f_n(x_1, ..., x_n)\chi_{[0,t]}(\max_i x_i)) \diamond I_m(g_m(y_1, ..., y_m)\chi_{[0,t]}(\max_j y_j))$$

$$= E[F|\mathcal{F}_t] \diamond E[G|\mathcal{F}_t]. \qquad \square$$

Corollary 6.21. *Let F, G be as in Lemma 6.20 and assume in addition that $F, G \in L^1(P)$. Then*

$$E[F \diamond G] = E[F] \cdot E[G]. \tag{6.21}$$

Proof Set $t = 0$ in Lemma 6.20. $\square$

Corollary 6.22. *Let $F \in (\mathcal{S})^*$ and suppose that*

$$\exp^\diamond F := \sum_{n=0}^{\infty} \frac{1}{n!} F^{\diamond n} \in (\mathcal{S})^*$$

and also $E[F|\mathcal{F}_t] \in (\mathcal{S})^*$ and $\exp^\diamond E[F|\mathcal{F}_t] \in (\mathcal{S})^*$. Then

$$E[\exp^\diamond F|\mathcal{F}_t] = \exp^\diamond\{E[F|\mathcal{F}_t]\}. \tag{6.22}$$

In particular, if, in addition, $F \in L^1(P)$, we have

$$E[\exp^\diamond F] = \exp\{E[F]\}. \tag{6.23}$$

Proof By Lemma 6.20 we have

$$E[\exp^\diamond F|\mathcal{F}_t] = E[\sum_{n=0}^{\infty} \frac{1}{n!} F^{\diamond n}|\mathcal{F}_t]$$

$$= \sum_{n=0}^{\infty} \frac{1}{n!} E[F|\mathcal{F}_t]^{\diamond n} = \exp^\diamond\{E[F|\mathcal{F}_t]\}. \qquad \square$$

The conditional expectation of white noise has a particularly simple form.

Lemma 6.23.

$$E[\dot{W}(s)|\mathcal{F}_t] = \dot{W}(s)\chi_{[0,t]}(s). \tag{6.24}$$

Proof Consider

$$E[\dot{W}(s)|\mathcal{F}_t] = E[\sum_{i=1}^{\infty} e_i(s)H_{\epsilon(i)}|\mathcal{F}_t]$$

$$= \sum_{i=1}^{\infty} e_i(s)E[I_1(e_i)|\mathcal{F}_t] = \sum_{i=1}^{\infty} e_i(s)I_1(e_i\chi_{[0,t]})$$

$$= \sum_{i=1}^{\infty} e_i(s)I_1\Big(\sum_{j=1}^{\infty}(e_i\chi_{[0,t]}, e_j)_{L^2(\mathbb{R})}e_j\Big)$$

$$= \sum_{j=1}^{\infty}\Big[\sum_{i=1}^{\infty}(e_i\chi_{[0,t]}, e_j)_{L^2(\mathbb{R})}e_i(s)\Big]I_1(e_j)$$

$$= \sum_{j=1}^{\infty}\Big[\sum_{i=1}^{\infty}(e_j\chi_{[0,t]}, e_i)_{L^2(\mathbb{R})}e_i(s)\Big]I_1(e_j)$$

$$= \sum_{j=1}^{\infty} e_j(s)\chi_{[0,t]}(s)I_1(e_j) = \dot{W}(s)\chi_{[0,t]}(s). \qquad \square$$

We apply this to prove the following useful result.

Theorem 6.24. *Suppose $Y(s)$, $s \in \mathbb{R}$, is Skorohod integrable and $E[Y(s)|\mathcal{F}_t] \in (\mathcal{S})^*$ for all $s \in \mathbb{R}$. Then*

$$E[\int_{\mathbb{R}} Y(s)\delta W(s)|\mathcal{F}_t] = \int_0^t E[Y(s)|\mathcal{F}_t]\delta W(s). \tag{6.25}$$

Proof By Theorem 5.20, Lemma 6.20, and Lemma 6.23 we have

$$E[\int_{\mathbb{R}} Y(s)\delta W(s)|\mathcal{F}_t] = E[\int_{\mathbb{R}} Y(s) \diamond \dot{W}(s)ds|\mathcal{F}_t]$$

$$= \int_{\mathbb{R}} E[Y(s)|\mathcal{F}_t] \diamond E[\dot{W}(s)|\mathcal{F}_t]ds$$

$$= \int_{\mathbb{R}} E[Y(s)|\mathcal{F}_t] \diamond \dot{W}(s)\chi_{[0,t]}(s)ds$$

$$= \int_0^t E[Y(s)|\mathcal{F}_t]\delta W(s). \qquad \square$$

Corollary 6.25. *Let $Y(s)$, $s \in \mathbb{R}$, be as in Theorem 6.24. Then*

$$E[\int_t^{\infty} Y(s)\delta W(s)|\mathcal{F}_t] = 0. \tag{6.26}$$

6.4 Conditional Expectation on $\mathcal{G}^*$

If we restrict ourselves to $\mathcal{G}^*$ (see Sect. 5.3.3), the conditional expectation operator is easier to handle. First, it can be defined as follows.

Definition 6.26. *Let $F = \sum_{n=0}^{\infty} I_n(f_n) \in \mathcal{G}^*$. Then*

$$E[F|\mathcal{F}_t] = \sum_{n=0}^{\infty} I_n(f_n\chi_{[0,t]^n}). \tag{6.27}$$

Note that

$$\|E[F|\mathcal{F}_t]\|_{\mathcal{G}_\lambda} \leq \|F\|_{\mathcal{G}_\lambda} \qquad \lambda \in \mathbb{R}. \tag{6.28}$$

In particular,

$$E[F|\mathcal{F}_t] \in \mathcal{G}^*, \quad \text{for all } F \in \mathcal{G}^*. \tag{6.29}$$

Therefore, the results stated above for $(\mathcal{S})^*$ have simpler formulations when restricted to $\mathcal{G}^*$.

Lemma 6.27. *Suppose $F, G \in \mathcal{G}^*$. Then*

$$E[F \diamond G|\mathcal{F}_t] = E[F|\mathcal{F}_t] \diamond E[G|\mathcal{F}_t].$$

Lemma 6.28. *Suppose $F \in \mathcal{G}^*$ and $\exp^\diamond F \in \mathcal{G}^*$. Then*

$$E[\exp^\diamond F|\mathcal{F}_t] = \exp^\diamond E[F|\mathcal{F}_t].$$

From now on we use the following notation. Let $e_1, e_2, ...$ be the Hermite functions and set

$$X_i = X_i(\omega) = <\omega, e_i> := \int_{\mathbb{R}} e_i(s)dW(s), \qquad i = 1, 2, ..., \qquad (6.30)$$

and

$$X_i^{(t)} = X_i^{(t)}(\omega) := \int_0^t e_i(s)dW(s), \qquad i = 1, 2, ..., \qquad (6.31)$$

and

$$X = (X_1, X_2, ...), \qquad X^{(t)} = (X_1^{(t)}, X_2^{(t)}, ...). \qquad (6.32)$$

Then

$$X^{\diamond\alpha} = X_1^{\diamond\alpha_1} \diamond \cdots \diamond X_m^{\diamond\alpha_m} = H_\alpha$$

if $\alpha = (\alpha_1, ..., \alpha_m) \in \mathcal{J}$.

Note that we can restate (6.27) in Definition 6.26 as follows: *let* $F = \sum_{\alpha \in \mathcal{J}} c_\alpha X^{\diamond\alpha} \in \mathcal{G}^*$. *Then*

$$E[F|\mathcal{F}_t] = \sum_{\alpha \in \mathcal{J}} c_\alpha (X^{(t)})^{\diamond\alpha}.$$

In fact, we have $E[X_i|\mathcal{F}_t] = X_i^{(t)}$.

Definition 6.29. *Let* $T > 0$ *be constant. We say that* $F \in \mathcal{G}^*$ *is* $\mathcal{F}_T$-*measurable if*

$$E[F|\mathcal{F}_T] = F.$$

Then the above arguments yield the following result.

Corollary 6.30. *The element* $F \in \mathcal{G}^*$ *is* $\mathcal{F}_T$-*measurable if and only if* F *can be written as*

$$F = \sum_{\alpha \in \mathcal{J}} c_\alpha (X^{(T)})^{\diamond\alpha}$$

(convergence in $\mathcal{G}^*$*) for some real numbers* c_α, $\alpha \in \mathcal{J}$.

6.5 A Generalized Clark–Ocone Theorem

A drawback with the classical Clark–Ocone theorem (Theorem 4.1) is that it requires $F \in \mathbb{D}_{1,2}$. For several reasons it is of interest to extend the Clark–Ocone theorem to $F \in L^2(P)$. Mathematically this is natural because the Itô representation theorem is indeed valid for all $F \in L^2(P)$. Moreover, in financial applications it is important to be able to deal with claims that are not necessarily in $\mathbb{D}_{1,2}$, such as binary options of the form

$$F = \chi_{[K,\infty)}(W(T)), \qquad (K > 0 \text{ constant}).$$

To obtain such a generalization, we apply the white noise machinery developed in the previous sections. This was first done in the paper [2] and our presentation is based on that paper. We first establish some auxiliary results.

Lemma 6.31. *Let* $P(x) = \sum_\alpha c_\alpha x^\alpha$ *be a polynomial in* $x = (x_1, ..., x_n) \in \mathbb{R}^n$. *Let* $X = (X_1, ..., X_n)$ *be as in* (6.27)–(6.29). *Then*

$$D_t P(X) = \sum_{i=1}^n \frac{\partial P}{\partial x_i}(X_1, ..., X_n) e_i(t) = \sum_\alpha c_\alpha \sum_i \alpha_i X^{\alpha - \epsilon^{(i)}} e_i(t) \qquad (6.33)$$

and

$$D_t P^\diamond(X) = \sum_{i=1}^n \left(\frac{\partial P}{\partial x_i}\right)^\diamond (X_1, ..., X_n) e_i(t) = \sum_\alpha c_\alpha \sum_i \alpha_i X^{\diamond(\alpha - \epsilon^{(i)})} e_i(t). \qquad (6.34)$$

Proof These statements (6.33) and (6.34) are just reformulations of the chain rules in Theorem 6.3 and Theorem 6.13.   □

Lemma 6.32. Chain rule. *Let* $P(x) = \sum_\alpha c_\alpha x^\alpha$ *be a polynomial in* $x = (x_1, ..., x_n) \in \mathbb{R}^n$. *Consider* $X^{(t)} = (X_1^{(t)}, ..., X_n^{(t)})$, $t \geq 0$, *with* $X_i^{(t)}$, $i = 1, ..., n$, *as defined in* (6.31). *Then*

$$\frac{d}{dt} P^\diamond(X^{(t)}) = \sum_{j=i}^n \left(\frac{\partial P}{\partial x_j}\right)^\diamond (X^{(t)}) \diamond e_j(t) \dot{W}(t) \quad in \ (\mathcal{S})^*. \qquad (6.35)$$

Proof By Problem 5.7 we have

$$\frac{d}{dt} \int_0^t e_j(s) dW(s) = e_j(t) \dot{W}(t).$$

Hence, the result follows by the Wick chain rule (Theorem 6.13).   □

We can now prove a preliminary version of the extended Clark–Ocone formula.

Lemma 6.33. The Clark–Ocone formula for polynomials. *Let* $F \in \mathcal{G}^*$ *be* $\mathcal{F}_T$-*measurable and assume that*

$$F = P^\diamond(X)$$

for some polynomial $P(x) = \sum_\alpha c_\alpha x^\alpha$, *where* $X = (X_1, ..., X_n)$ *is as in* (6.30) *and* (6.32). *Then* $F = P^\diamond(X^{(T)})$ *and*

$$F = E[F] + \int_0^T E[D_t F | \mathcal{F}_t] dW(t).$$

Proof Since F is $\mathcal{F}_T$-measurable, we have

$$F = E[F | \mathcal{F}_T] = P^\diamond(E[X | \mathcal{F}_T]) = P^\diamond(X^{(T)}).$$

Moreover, we have

$$
\int_0^T E[D_t F | \mathcal{F}_t] dW(t) = \int_0^T E\Big[\sum_{j=1}^n \Big(\frac{\partial P}{\partial x_j}\Big)^\diamond (X) e_j(t) | \mathcal{F}_t\Big] dW(t)
$$

$$
= \int_0^T \sum_{j=i}^n \Big(\frac{\partial P}{\partial x_j}\Big)^\diamond (X^{(t)}) e_j(t) \diamond \dot{W}(t) dt
$$

$$
= \int_0^T \frac{d}{dt} P^\diamond(X^{(t)}) dt = P^\diamond(X^{(T)}) - P^\diamond(X^{(0)})
$$

$$
= F - P^\diamond(0) = F - E[F],
$$

using Lemma 6.31, Lemma 6.20, and Lemma 6.32.   □

We proceed to extend this result to $L^2(P)$. To this aim, the following lemma is crucial.

Lemma 6.34. *Let $F \in \mathcal{G}^*$. Then we have*

(i)   The estimate

$$
\int_{\mathbb{R}} \|D_t F\|_{\mathcal{G}_{-q-1}}^2 dt \leq e^{2q} \|F\|_{\mathcal{G}_{-q}}^2 \tag{6.36}
$$

holds for all $q \in \mathbb{N}$,

(ii) $D_t F \in \mathcal{G}^*$, for a.a. $t \geq 0$,

(iii) Suppose $F_N \in \mathcal{G}^*$, $N = 1, 2, ...$, and

$$
F_N \longrightarrow F, \quad N \to \infty, \quad in \ \mathcal{G}^*.
$$

Then there exists a subsequence F_{N_k}, $k = 1, 2, ...$, such that

$$
D_t F_{N_k} \longrightarrow D_t F, \quad k \to \infty, \quad in \ \mathcal{G}^*, \ for \ a.a. \ t \geq 0
$$

and

$$
E[D_t F_{N_k} | \mathcal{F}_t] \longrightarrow E[D_t F | \mathcal{F}_t], \quad k \to \infty, \ in \ \mathcal{G}^* \ for \ a.a. \ t.
$$

Proof (i)   Let $F = \sum_{\alpha \in \mathcal{J}} c_\alpha H_\alpha$. Then by (6.7) we have

$$
D_t F = \sum_{\alpha \in \mathcal{J}} c_\alpha \sum_{i=1}^\infty \alpha_i e_i(t) H_{\alpha - \epsilon^{(i)}}
$$

$$
= \sum_{\beta \in \mathcal{J}} \Big(\sum_{i=1}^\infty c_{\beta + \epsilon^{(i)}} (\beta_i + 1) e_i(t) \Big) H_\beta
$$

$$
= \sum_{\beta \in \mathcal{J}} g_\beta(t) H_\beta,
$$

where

$$g_\beta(t) = \sum_{i=1}^{\infty} c_{\beta+\epsilon^{(i)}}(\beta_i + 1)e_i(t).$$

Since $F \in \mathcal{G}^*$, there exists $q < \infty$ such that

$$\|F\|_{\mathcal{G}_{-q}}^2 := \sum_{\alpha \in \mathcal{J}} c_\alpha^2 \alpha! e^{-2q|\alpha|} = \sum_{n=0}^{\infty} \sum_{|\alpha|=n} c_\alpha^2 \alpha! e^{-2qn} < \infty,$$

see (5.60). We can now prove that

$$\|D_F\|_{\mathcal{G}_{-q-1}}^2 := \sum_{\beta \in \mathcal{J}} g_\beta^2(t)\beta! e^{-2(q+1)|\beta|} < \infty, \qquad \text{for a.a. } t.$$

In fact, note that

$$\int_{\mathbb{R}} g_\beta^2(t)dt = \int_{\mathbb{R}} \Big(\sum_{i=1}^{\infty} c_{\beta+\epsilon^{(i)}}(\beta_i + 1)e_i(t) \Big)^2 dt = \sum_{i=1}^{\infty} c_{\beta+\epsilon^{(i)}}^2 (\beta_i + 1)^2.$$

Therefore, using that $(x+1)e^{-x} \le 1$ for all $x \ge 0$,

$$\sum_{\beta \in \mathcal{J}} \Big(\int_{\mathbb{R}} g_\beta^2(t)dt \Big)\beta! e^{-2(q+1)|\beta|} = \sum_{\beta \in \mathcal{J}} \Big(\sum_{i=1}^{\infty} c_{\beta+\epsilon^{(i)}}^2 (\beta_i + 1)^2 \Big)\beta! e^{-2(q+1)|\beta|}$$

$$\le \sum_{\beta \in \mathcal{J}} (|\beta| + 1)e^{-2(q+1)|\beta|} \sum_{i=1}^{\infty} c_{\beta+\epsilon^{(i)}}^2 (\beta + \epsilon^{(i)})!$$

$$\le \sum_{\beta \in \mathcal{J}} e^{-(2q+1)|\beta|} \sum_{\alpha \in \mathcal{J}:\, |\alpha|=|\beta|+1} c_\alpha^2 \alpha!$$

$$= \sum_{n=0}^{\infty} \sum_{|\alpha|=n+1} c_\alpha^2 \alpha! e^{-(2q+1)n}$$

$$\le e^{2q} \|F\|_{\mathcal{G}_{-q}}^2 < \infty.$$

This proves

$$\int_{\mathbb{R}} \|D_t F\|_{\mathcal{G}_{-q-1}}^2 dt \le e^{2q} \|F\|_{\mathcal{G}_{-q}}^2.$$

(ii)   From the estimate in (i), we have that

$$D_t F \in \mathcal{G}_{-q-1} \subset \mathcal{G}^*, \qquad \text{for a.a. } t.$$

(iii)   We may assume that $F = 0$. By (i) we get that

$$\|D_t F_N\|_{\mathcal{G}_{-q-1}} \longrightarrow 0, \qquad N \to \infty, \qquad \text{in } L^2(\mathbb{R}).$$

So there exists a subsequence $\|D_t F_{N_k}\|_{\mathcal{G}_{-q-1}}$, $k = 1, 2, ...,$ such that

$$D_t F_{N_k} \longrightarrow 0, \qquad k \to \infty, \qquad \text{in } \mathcal{G}^*, \text{ a.a. } t.$$

This proves the first part of (iii). The same argument combined with (6.28) yields the second part of (iii).   □

The following theorem was first proved in [2].

Theorem 6.35. The Clark–Ocone theorem for $L^2(P)$. *Let $F \in L^2(P)$ be $\mathcal{F}_T$-measurable. Then the process*

$$(t, \omega) \longrightarrow E[D_t F | \mathcal{F}_t](\omega)$$

is an element of $L^2(P \times \lambda)$ and

$$F = E[F] + \int_0^T E[D_t F | \mathcal{F}_t] dW(t), \tag{6.37}$$

where $D_t F$ is the Malliavin derivative of F in $\mathcal{G}^*$.

Proof Let $F = \sum_{\alpha \in \mathcal{J}} c_\alpha H_\alpha$ be the chaos expansion of F and put

$$F_N = \sum_{\alpha \in \mathcal{J}_N} c_\alpha H_\alpha.$$

Here $\mathcal{J}_N = \{\alpha \in \mathcal{J} : |\alpha| \le N, l(\alpha) \le N\}$, where $l(\alpha) = \max\{i : \alpha_i \ne 0\}$ is the length of α. Then, by Lemma 6.33 we have

$$F_N = E[F_N] + \int_0^T E[D_t F_N | \mathcal{F}_t] dW(t),$$

for all N. By the Itô representation theorem we know that there exists a unique $\mathbb{F}$-adapted process $u = u(t)$, $t \in [0, T]$, such that

$$E\left[\int_0^T u^2(t) dt\right] < \infty$$

and

$$F = E[F] + \int_0^T u(t) dW(t). \tag{6.38}$$

Since $F_N \longrightarrow F$, $N \to \infty$, in $L^2(P)$, we get

$$E\left[\int_0^T \left(E[D_t F_N | \mathcal{F}_t] - u(t)\right)^2 dt\right] = E\left[(F_N - F - E[F_N] + E[F])^2\right] \longrightarrow 0, \quad N \to \infty.$$

Therefore,

$$E[D_t F_N | \mathcal{F}_t] \longrightarrow u, \quad N \to \infty, \quad \text{in } L^2(P \times \lambda). \tag{6.39}$$

On the other hand, by Lemma 6.34 (iii),

$$E[D_t F_{N_k} | \mathcal{F}_t] \longrightarrow E[D_t F | \mathcal{F}_t], \quad k \to \infty, \quad \text{in } \mathcal{G}^* \tag{6.40}$$

for some subsequence F_{N_k}, $k = 1, 2, \dots$. Combining (6.39) and (6.40) we conclude that

$$u(t) = E[D_t F | \mathcal{F}_t] \quad P - a.e.$$

for a.a. t. Hence (6.37) follows from (6.38). $\quad\square$

We now extend this even further to a Clark–Ocone theorem for $\mathcal{G}^*$. To this end, we first prove some auxiliary results.

Lemma 6.36. *Let* $G \in \mathcal{G}_{-q} \subset (\mathcal{S})^*$ *for some* $q \in \mathbb{N}$ *and put*

$$\hat{q} = \frac{2q}{\log 2}.$$

Let $f \in (\mathcal{S})$. *Then*

$$| < G, f > | \leq \|G\|_{\mathcal{G}_{-q}} \|f\|_{\hat{q}},$$

where $\| \cdot \|_{\hat{q}}$ *is the* $(\mathcal{S})_{\hat{q}}$*-norm defined in* (5.17), *and* $< G, f >$ *is the action of* $G \in (\mathcal{S})^*$ *on* $f \in (\mathcal{S})$.

Proof Suppose $G = \sum_{\alpha \in \mathcal{J}} a_\alpha H_\alpha$ and $f = \sum_{\beta \in \mathcal{J}} b_\beta H_\beta$. Then

$$| < G, f > | = | \sum_{\alpha \in \mathcal{J}} a_\alpha b_\alpha \alpha! |$$

$$= | \sum_{m=0}^{\infty} \sum_{|\alpha|=m} a_\alpha b_\alpha \alpha! |$$

$$\leq \sum_{m=0}^{\infty} \Big(\sum_{|\alpha|=m} a_\alpha^2 \alpha! \Big)^{1/2} \Big(\sum_{|\alpha|=m} b_\alpha^2 \alpha! \Big)^{1/2}$$

$$\leq \Big(\sum_{m=0}^{\infty} \big(\sum_{|\alpha|=m} a_\alpha^2 \alpha! \big) e^{-2qm} \Big)^{1/2} \Big(\sum_{m=0}^{\infty} \big(\sum_{|\alpha|=m} b_\alpha^2 \alpha! \big) e^{2qm} \Big)^{1/2}$$

$$= \|G\|_{\mathcal{G}_{-q}} \Big(\sum_{\alpha \in \mathcal{J}} b_\alpha^2 \alpha! (2\mathbb{N})^{\hat{q}\alpha} \Big)^{1/2}$$

$$= \|G\|_{\mathcal{G}_{-q}} \|f\|_{\hat{q}}. \qquad \square$$

Lemma 6.37. *Let* $f \in (\mathcal{S})$ *and let* $G(t) \in \mathcal{G}^*$ *for all* $t \in \mathbb{R}$. *Choose* $q \in \mathbb{N}$ *and put*

$$\hat{q} = \frac{2q}{\log 2}.$$

Then

$$\int_{\mathbb{R}} | < G(t) \diamond \dot{W}(t), f > | \, dt \leq \|f\|_{2\hat{q}} \Big(\int_{\mathbb{R}} \|G(t)\|_{\mathcal{G}_{-q}}^2 \, dt \Big)^{1/2},$$

where $\|f\|_{2\hat{q}}$ *is the* $(\mathcal{S})_{2\hat{q}}$*-norm defined in* (5.17).

Proof Suppose $G(t) = \sum_{\alpha \in \mathcal{J}} a_\alpha(t) H_\alpha$ and $f = \sum_{\beta \in \mathcal{J}} b_\beta H_\beta$. Then

$$< G(t) \diamond \dot{W}(t), f > = < \sum_{\alpha,k} a_\alpha(t) e_k(t) H_{\alpha+\epsilon^{(k)}}, \sum_{\beta \in \mathcal{J}} b_\beta H_\beta >$$

$$= \sum_{\alpha,k} a_\alpha(t) e_k(t) b_{\alpha+\epsilon^{(k)}} (\alpha + \epsilon^{(k)})!$$

Hence

$$
\int_{\mathbb{R}} | < G(t) \diamond \dot{W}(t), f > | dt \leq \sum_{\alpha,k} |b_{\alpha+\epsilon^{(k)}}| \alpha! (\alpha_k + 1) \int_{\mathbb{R}} |a_\alpha(t) e_k(t)| dt
$$

$$
\leq \sum_{\alpha,k} |b_{\alpha+\epsilon^{(k)}}| \left(\int_{\mathbb{R}} a_\alpha^2(t) dt \right)^{1/2} \alpha! (\alpha_k + 1)
$$

$$
\leq \left(\sum_{\alpha,k} b_{\alpha+\epsilon^{(k)}}^2 (\alpha + \epsilon^{(k)})! (2\mathbb{N})^{2\hat{q}(\alpha+\epsilon^{(k)})} \right)^{1/2}
$$

$$
\cdot \left(\sum_{\alpha,k} \left(\int_{\mathbb{R}} a_\alpha^2(t) dt \right) \alpha! (\alpha_k+1)(2\mathbb{N})^{-2\hat{q}(\alpha+\epsilon^{(k)})} \right)^{1/2}
$$

$$
\leq \|f\|_{2\hat{q}} \left(\sum_{\alpha,k} \left(\int_{\mathbb{R}} a_\alpha^2(t) dt \right) \alpha! (\alpha_k+1) e^{-4q(\alpha_k+1)} \right)^{1/2}
$$

$$
\leq \|f\|_{2\hat{q}} \left(\int_{\mathbb{R}} \|G(t)\|_{\mathcal{G}_{-q}}^2 dt \right)^{1/2}. \qquad \square
$$

Lemma 6.38. *Suppose F, F_N, $N=1,2,...$, are elements of $\mathcal{G}^*$ and $F_N \longrightarrow F$, $N \to \infty$, in $\mathcal{G}^*$. Then*

$$
\lim_{N \to \infty} \int_0^T E[D_t F_N | \mathcal{F}_t] \diamond \dot{W}(t) dt = \int_0^T E[D_t F | \mathcal{F}_t] \diamond \dot{W}(t) dt,
$$

where the convergence is in $(\mathcal{S})^*$.

Proof Note that since

$$
\int_{\mathbb{R}} \|E[D_t F | \mathcal{F}_t]\|_{\mathcal{G}_{-q-1}}^2 dt \leq \int_{\mathbb{R}} \|D_t F\|_{\mathcal{G}_{-q-1}}^2 dt
$$

$$
\leq e^{-2q} \|F\|_{\mathcal{G}_{-q}}^2
$$

by (6.36), we have that

$$
\int_{\mathbb{R}} | < E[D_t F | \mathcal{F}_t] \diamond \dot{W}(t), f > | dt < \infty
$$

by Lemma 6.37. Therefore, $E[D_t F | \mathcal{F}_t] \diamond \dot{W}(t)$ is integrable in $(\mathcal{S})^*$ and similarly with $E[D_t F_N | \mathcal{F}_t] \diamond \dot{W}(t)$ for all N (see Definition 5.18). Moreover, by Lemma 6.37 and (6.36),

$$
| < \int_0^T E[D_t F_N | \mathcal{F}_t] \diamond \dot{W}(t) dt - \int_0^T E[D_t F | \mathcal{F}_t] \diamond \dot{W}(t) dt, f > |
$$

$$
\leq \|f\|_{2\hat{q}} \left(\int_{\mathbb{R}} \|E[D_t (F_N - F) | \mathcal{F}_t]\|_{\mathcal{G}_{-q}}^2 dt \right)^{1/2}
$$

$$
\leq \|f\|_{2\hat{q}} e^q \|F_N - F\|_{\mathcal{G}_{-q+1}} \longrightarrow 0, \qquad N \to \infty,
$$

if $q \in \mathbb{N}$ is large enough. Since this holds for all $f \in (\mathcal{S})$, the result follows. $\square$

We now have all the ingredients for the proof of the second main theorem of this section, which was first proved in [2].

Theorem 6.39. The Clark–Ocone theorem for $\mathcal{G}^*$. *Let* $F \in \mathcal{G}^*$ *be* $\mathcal{F}_T$-*measurable. Then* $D_t F \in \mathcal{G}^*$ *and* $E[D_t F | \mathcal{F}_t] \in \mathcal{G}^*$ *for a.a.* t, $E[D_t F | \mathcal{F}_t] \diamond \overset{\bullet}{W}(t)$, $t \in [0, T]$, *is integrable in* $(\mathcal{S})^*$ *and*

$$F = E[F] + \int_0^T E[D_t F | \mathcal{F}_t] \diamond \overset{\bullet}{W}(t) dt,$$

where $E[F] = E[F | \mathcal{F}_0]$ *is the generalized expectation of* F *(see Definition 5.6) (a).*

Proof If $F = \sum_{\alpha \in \mathcal{J}} a_\alpha H_\alpha$, define $F_N = \sum_{\alpha \in \mathcal{J}_N} a_\alpha H_\alpha$, as in the proof of Theorem 6.35. Then by Lemma 6.33 we have

$$F_N = E[F_N] + \int_0^T E[D_t F_N | \mathcal{F}_t] \diamond \overset{\bullet}{W}(t) dt,$$

for all N. Since $F_N \longrightarrow F$, $N \to \infty$, in $\mathcal{G}^*$ we have

$$F = \lim_{N \to \infty} F_N$$

$$= E[F] + \lim_{N \to \infty} \int_0^T E[D_t F_N | \mathcal{F}_t] \diamond \overset{\bullet}{W}(t) dt$$

$$= E[F] + \int_0^T E[D_t F | \mathcal{F}_t] \diamond \overset{\bullet}{W}(t) dt,$$

by Lemma 6.38. □

Example 6.40. **The digital option.** Let us return to the payoff

$$F = \chi_{[k,\infty)}(W(T))$$

of a digital option, mentioned in the beginning of this section. By combining the Clark–Ocone theorem for $L^2(P)$ (see Theorem 6.35) with results from Chap. 7 about the Donsker delta function, we obtain the representation

$$\chi_{[K,\infty)}(W(T)) = P\{W(T) \geq K\} + \int_0^T \frac{1}{2\pi(T-t)} \exp\left(-\frac{(K-W(t))^2}{2(T-t)}\right) dW(t).$$
(6.41)

See Problem 7.1 for details. See also Corollary 7.13.

To end this section we present the analog of Theorem 4.5 in the white noise setting. Using white noise techniques the Clark–Ocone formula under change of measure as presented in Theorem 4.5 can be extended to square integrable random variables that are not necessarily Malliavin differentiable. As indicated earlier, where we derived a generalized Clark–Ocone theorem (Theorem 6.35) under the original measure, such a representation is especially convenient when it comes to finding portfolio strategies of payoff functions that are not Malliavin differentiable.

Theorem 6.41. The generalized Clark–Ocone formula under change of measure. *Let* $Q(d\omega) = Z(\omega, T)P(d\omega)$ *be the change of measure as given in (4.3) such that* $Z(t)$, $0 \le t \le T$, *is a P-martingale. Require that* $F \in L^2(P) \cap L^2(Q)$ *is a* $\mathcal{F}_T$-*measurable random variable. Denote by* D_t *the (Hida–) Malliavin derivative on the space of generalized random variables* $\mathcal{G}^*$. *Further, suppose that* $u(t)$, $t \in [0, T]$, *is a square integrable* $\mathbb{F}$-*adapted measurable process satisfying the following conditions:*

(1) $\int_0^T D_t(u(s))dW(s)$ *exists in the sense of Definition 5.22 and is contained in* $\mathcal{G}^*$ *for all* $0 \le t \le T$

(2) $Z(T)F \in L^2(P)$

Then

$$F = E_Q[F] + \int_0^T E_Q\Big[D_t F - F \int_t^T D_t u(s) d\widetilde{W}(s) \big| \mathcal{F}_t\Big] d\widetilde{W}(t),$$

where $\widetilde{W}(t) = W(t) + \int_0^t u(s)ds$ *is a Brownian motion under measure* Q.

Proof The proof is based on white noise techniques as developed in Sect. 6.5 and can be found in [184]. □

6.6 Exercises

Problem 6.1. Prove $L^2(P) \subseteq Dom(D_t) \subseteq (\mathcal{S})^*$. See Remark 6.6.

Problem 6.2. The Wick chain rule. Let F be Malliavin differentiable and let $n \in \mathbb{N}$. Show that

$$D_t(F^{\diamond n}) = nF^{\diamond(n-1)} \diamond D_t F.$$

Problem 6.3. Verify that if $g \in L^2(\mathbb{R})$, $n \in \mathbb{N}$, then

$$W^n(T) \diamond \int_0^T g(t)\delta W(t) = W^n(T) \int_0^T g(t)\delta W(t) - nW^{n-1}(T) \int_0^T g(t)dt.$$

Problem 6.4. (*) Show that the singular noise $\overset{\bullet}{W}$ does not belong to the domain of the Hida–Malliavin derivative as given in (6.8).

Problem 6.5. Generalized Bayes formula. Let $Q(d\omega) = Z(T)P(d\omega)$, where $Z(t), 0 \le t \le T$, is the Doleans–Dade exponential in (4.4). Further, let $G \in \mathcal{G}^*$ and assume that $Z(T)G$ belongs to $\mathcal{G}^*$. Show that the following *generalized Bayes formula* holds:

$$E_Q[G|\mathcal{F}_t] = \frac{E_Q[Z(T)G|\mathcal{F}_t]}{Z(t)}.$$

Problem 6.6. Consider the digital option with payoff given by

$$F = \chi_{[K,\infty)}(W(T)).$$

Use Theorem 6.41 and relation 6.41 to derive a replicating hedging strategy for this option in the case of the Black–Scholes market of Example 4.10. [*Hint.* Employ the arguments of Sect. 4.3 and Problem 6.5.]

7

The Donsker Delta Function and Applications

In this chapter we use white noise calculus to define the Donsker delta function of a Brownian motion. The Donsker delta function of a Brownian motion can be considered the time derivative of local time of a Brownian motion on a distribution space. We aim at employing this concept to determine explicit formulae for replicating portfolios in a Black–Scholes market for a class of contingent claims.

7.1 Motivation: An Application of the Donsker Delta Function to Hedging

As a motivation we start by considering the following problem from mathematical finance. Fix $T > 0$ and consider the following financial market with two securities:

- A risk-free asset (e.g., a bank account), where the price $S_0(t)$, $t \in [0, T]$, per unit at time t is given by the differential equation

$$dS_0(t) = \rho(t)S_0(t)dt, \ S_0(0) - 1. \tag{7.1}$$

- A risky asset (e.g. a stock), where the price $S_1(t)$, $t \in [0, T]$, per unit at time t is given by the stochastic differential equation

$$dS_1(t) = \mu(t)S_1(t)dt + \sigma(t)S_1(t)dW(t), \ S_1(0) > 0 \text{ constant.} \tag{7.2}$$

Here $\rho(t), \mu(t)$, and $\sigma(t)$, $t \in [0, T]$, are given deterministic functions with the property that

$$\int_0^T \left(|\rho(s)| + |\mu(s)| + \sigma^2(s) \right) ds < \infty.$$

For simplicity, we assume that σ is bounded away from zero. This market was discussed in Sect. 4.3. For convenience of the reader we repeat the arguments here.

Since this simple generalization of the Black–Scholes market is complete, it is well known that any claim F can be hedged, that is, there exists a *replicating (self-financing) portfolio* for F.

The problem we study is as follows:

- How do we find explicitly such a replicating portfolio for F?

We want to describe this more in detail: Let $(\theta_0(t), \theta_1(t))$, $t \in [0, T]$, be a *portfolio*. Then the *value* $V(t)$ of this portfolio at time t is defined by

$$V(t) = \theta_0(t)S_0(t) + \theta_1(t)S_1(t). \tag{7.3}$$

The portfolio is called *self-financing* if

$$dV(t) = \theta_0(t)dS_0(t) + \theta_1(t)dS_1(t). \tag{7.4}$$

This means that no external funds are added to the portfolio and that no funds are extracted from the portfolio as time evolves. We consider only self-financing portfolios. From (7.3) we get

$$\theta_0(t) = \frac{V(t) - \theta_1(t)S_1(t)}{S_0(t)}. \tag{7.5}$$

Substituting this in (7.4) and using (7.1) and (7.2), we get

$$dV(t) = \left(V(t) - \theta_1(t)S_1(t)\right)\frac{dS_0(t)}{S_0(t)} + \theta_1(t)dS_1(t)$$
$$= \rho(t)V(t)dt + \theta_1(t)S_1(t)\left((\mu(t) - \rho(t))\,dt + \sigma(t)dW(t)\right).$$

Since $\sigma(t) \neq 0$ for a.a. t, this can be written

$$dV(t) = \rho(t)V(t)dt + \sigma(t)\theta_1(t)S_1(t)(\alpha(t)dt + dW(t)), \tag{7.6}$$

where

$$\alpha(t) = \frac{\mu(t) - \rho(t)}{\sigma(t)}. \tag{7.7}$$

Multiplying (7.6) by the integrating factor $e^{-\int_0^t \rho(s)ds}$, we get

$$d\left(e^{-\int_0^t \rho(s)ds}V(t)\right) = e^{-\int_0^t \rho(s)ds}\sigma(t)\theta_1(t)S_1(t)(\alpha(t)dt + dW(t)).$$

Hence

$$e^{-\int_0^t \rho(s)ds}V(T) = V(0) + \int_0^T e^{-\int_0^t \rho(s)ds}\sigma(t)\theta_1(t)S_1(t)(\alpha(t)dt + dW(t)). \tag{7.8}$$

Now suppose that F is a given European contingent T-claim, that is, F is a given $\mathcal{F}_T$-measurable, lower bounded random variable. To *hedge* such a claim means to find a constant $V(0)$ and a self-financing portfolio $(\theta_0(t), \theta_1(t))$,

$t \in [0, T]$, such that the corresponding value process $V(t)$, $t \in [0, T]$, starts up with value $V(0)$ for $t = 0$ and ends up with the value

$$V(T) = F \quad P\text{-a.s.} \tag{7.9}$$

at time T. The amount $V(0)$ is then the market value of F at time 0. We also require that the process $V(t)$, $0 \le t \le T$, is (ω, t)-a.s. lower bounded. By (7.8) combined with (7.5) we see that it suffices to find $V(0)$ and a process $u(t)$ such that

$$e^{-\int_0^T \rho(s)ds} F = V(0) + \int_0^T u(t)(\alpha(t)dt + dW(t)) \tag{7.10}$$

and

$$P\left(\int_0^T u^2(s)ds < \infty\right) = P\left(\int_0^T |u(s)\alpha(s)|\, ds < \infty\right) = 1,$$

such that $\int_0^t u(s, \omega)(\alpha(s)ds + dW(s))$, $0 \le t \le T$, is lower bounded. If such a process is found, we put

$$\theta_1(t) = e^{\int_0^t \rho(s)ds} \sigma(t)^{-1} S_1(t)^{-1} u(t)$$

and solve for $\theta_0(t)$ using (7.5). It is well known and easy to see by the Girsanov theorem that if α is such that

$$\int_0^T \alpha^2(s)ds < \infty \quad P\text{-a.s.,}$$

and satisfies the Novikov condition, then $V(0)$ is unique and given by

$$V(0) = E_Q\left[e^{-\int_0^T \rho(s)ds} F\right] \tag{7.11}$$

(provided this quantity is finite), where E_Q denotes the expectation with respect to the measure Q defined on $\mathcal{F}_T$ by

$$dQ = \exp\left\{-\int_0^T \alpha(s)dW(s) - \frac{1}{2}\int_0^T \alpha^2(s)ds\right\}dP, \tag{7.12}$$

so that

$$\widetilde{W}(t) := \int_0^t \alpha(s)ds + W(t) \tag{7.13}$$

is a Brownian motion with respect to Q. To find $u(t)$, there are several known methods:

(a) If the claim F is of *Markovian type*, that is,

$$F = h(S_1(T)),$$

for some (deterministic) function $h : \mathbb{R} \longrightarrow \mathbb{R}$, then $u(t)$ can (in principle) be found by solving a deterministic boundary value problem for a parabolic partial differential equation. See Sect. 4.4 for details. See also, for example, [39, 73, 164].

(b) For some not necessarily Markovian type claims F one can (in principle) apply the generalized Clark–Ocone theorem see Theorem 4.5 to express $u(t)$ as

$$u(t) = E_Q\left[D_t F \mid \mathcal{F}_t\right] \qquad (7.14)$$

for the Malliavin derivative D_t. See Sect. 4.3, in the case of a European call option. The problems with this formula are the following:

- It is in general difficult to compute conditional expectations
- The classical Malliavin derivative $D_t F$ exists only under additional restrictions on F, that is, $F \in \mathbb{D}_{1,2}$, see Problem 7.3. In the previous chapter we solved this difficulty by extending the Clark–Ocone formula to $L^2(P)$ and even $\mathcal{G}^*$ (see Theorem 6.35 and Theorem 6.39).

The purpose of this chapter is to give an alternative approach based on white noise calculus and the Donsker delta function. In particular, we deal with claims of the type

$$F = f(Y(T)),$$

where

$$Y(t) = \int_0^t \psi(s)ds + \int_0^t \phi(s)dW(s), \quad 0 \le t \le T.$$

See Theorem 7.23 and Corollary 7.24 together with the remarks following the corollary. Although the question of hedging posed at the beginning of the chapter could also be solved by method (a) and method (b), it is conceivable that the white noise approach can cover some cases that are not well adapted to Methods (a) and (b). Moreover, it may give new insights. See (7.36) and the corresponding remark. The following presentation is based on [2, 3].

7.2 The Donsker Delta Function

The Donsker delta function is a generalized white noise functional, which has been treated in several papers within white noise analysis (see, e.g., [100, 140, 139] and also [102] and the references therein). For completeness we give an independent presentation here.

Definition 7.1. Let $Y : \Omega \longrightarrow \mathbb{R}$ be a random variable, which also belongs to the Hida distribution space $(\mathcal{S})^*$. Then a continuous function

$$\delta_Y(\cdot) : \mathbb{R} \longrightarrow (\mathcal{S})^*$$

is called Donsker delta function of Y if it has the property that

$$\int_{\mathbb{R}} g(y)\delta_Y(y)dy = g(Y) \ a.e. \qquad (7.15)$$

for all (measurable) $g : \mathbb{R} \longrightarrow \mathbb{R}$ such that the integral on the left-hand side converges.

Proposition 7.2. *Suppose Y is a normally distributed random variable with mean m and variance v > 0 (on our white noise space). Then δ_Y is unique and is given by the expression*

$$\delta_Y(y) = \frac{1}{\sqrt{2\pi v}} \exp^{\diamond} \left\{ -\frac{(y-Y)^{\diamond 2}}{2v} \right\} \in (\mathcal{S})^*. \tag{7.16}$$

Proof Let $G_Y(y)$ denote the right-hand side of (7.16). It follows from the characterization theorem for $(\mathcal{S})_{-1}$ (see Theorem 5.10 and Theorem 5.12) that $G_Y(y) \in (\mathcal{S})^*$ for all y and that $y \longmapsto G_Y(y)$ is continuous for $y \in \mathbb{R}$. We verify that G_Y satisfies (7.15), that is, that

$$\int_{\mathbb{R}} g(y)G_Y(y)dy = g(Y) \text{ a.e.} \tag{7.17}$$

First, let us assume that g has the form

$$g(y) = e^{\lambda y} \text{ for some } \lambda \in \mathbb{C}. \tag{7.18}$$

Then by taking the Hermite transform of the left-hand side of (7.17), we get

$$\mathcal{H}\left(\int_{\mathbb{R}} g(y)G_Y(y)dy \right) = \int_{\mathbb{R}} g(y)\mathcal{H}\left(G_Y(y)\right)dy$$

$$= \int_{\mathbb{R}} e^{\lambda y} \frac{1}{\sqrt{2\pi v}} \exp\left\{ -\frac{\left(y - \widetilde{Y}\right)^2}{2v} \right\}dy, \quad (7.19)$$

where $\widetilde{Y} = \widetilde{Y}(z)$ is the Hermite transform of Y at $z = (z_1, z_2, ...) \in \mathbb{C}^{\mathbb{N}}$. See Sect. 5.3 for an introduction to Hermite transforms. The expression (7.19) may be regarded as the expectation of $e^{\lambda Z}$, where Z is a normally distributed random variable with mean $\widetilde{Y}$ and variance v. Now set $Z = Y - m - \widetilde{Y}$ is such a random variable. Hence (7.19) can be written as $E\left[e^{\lambda(Y-m-\widetilde{Y})}\right]$, which by the well known formula for the characteristic function of a normal random variable is equal to $\exp\left\{ \lambda\widetilde{Y} + \frac{1}{2}\lambda^2 v \right\}$. We conclude that

$$\mathcal{H}\left(\int_{\mathbb{R}} g(y)G_Y(y)dy \right) = \exp\left\{ \lambda\widetilde{Y} + \frac{1}{2}\lambda^2 v \right\}$$

$$= \mathcal{H}\left(\exp^{\diamond}\left\{ \lambda Y + \frac{1}{2}\lambda^2 v \right\} \right)$$

$$= \mathcal{H}\left(\exp\left\{ \lambda Y \right\} \right) = \mathcal{H}(g(Y)).$$

This proves that (7.17) holds for functions g given by (7.18). Therefore, (7.17) also holds for linear combinations of such functions. By a well known density argument, (7.17) holds for all g such that the integral in (7.15) converges. Uniqueness follows easily from (7.15). □

Lemma 7.3. *Let* $\psi : [0,T] \longrightarrow \mathbb{R}$, $\phi : [0,T] \longrightarrow \mathbb{R}$ *be deterministic functions such that* $\int_0^T |\psi(s)| \, ds < \infty$ *and* $\|\phi\|_{L^2([0,T])}^2 = \int_0^T \phi^2(s) ds < \infty$. *Define*

$$Y(t) = \int_0^t \psi(s)ds + \int_0^t \phi(s)dW(s), \quad 0 \le t \le T. \tag{7.20}$$

Then

$$\exp^\diamond \left\{ -\frac{(y - Y(T))^{\diamond 2}}{2\|\phi\|_{L^2([0,T])}^2} \right\}$$

$$= \exp\left\{ -\frac{y^2}{2\|\phi\|_{L^2([0,T])}^2} \right\} + \int_0^T \exp^\diamond \left\{ -\frac{(y - Y(t))^{\diamond 2}}{2\|\phi\|_{L^2([0,T])}^2} \right\}$$

$$\diamond \frac{y - Y(t)}{\|\phi\|_{L^2([0,T])}^2} \diamond \left(\Psi(t) + \phi(t)\overset{\bullet}{W}(t) \right) dt, \tag{7.21}$$

where $\overset{\bullet}{W}$ *denotes the white noise of* W *(5.21).*

Proof This follows directly from some fundamental results on Wick calculus (see Problem 5.7) plus the chain rule in $(\mathcal{S})^*$ (see Proposition 5.14). □

We are coming to the first main result in this section:

Theorem 7.4. *Let* $\phi : [0,T] \longrightarrow \mathbb{R}$, $\alpha : [0,T] \longrightarrow \mathbb{R}$ *be deterministic functions such that*

$$0 < \|\phi\|_{L^2([0,T])}^2 = \int_0^T \phi^2(s)ds < \infty \quad \text{and} \quad 0 \le \int_0^T \alpha^2(s)ds < \infty. \tag{7.22}$$

Define

$$Y(t) = \int_0^t \phi(s)dW(s) + \int_0^t \phi(s)\alpha(s)ds, 0 \le t \le T. \tag{7.23}$$

Let $f : \mathbb{R} \longrightarrow \mathbb{R}$ *be bounded. Then*

$$f(Y(T)) = V(0) + \int_0^T u(t) \diamond \left(\alpha(t) + \overset{\bullet}{W}(t) \right) dt, \tag{7.24}$$

where

$$V(0) = \int_\mathbb{R} \frac{f(y)}{\sqrt{2\pi}\|\phi\|_{L^2([0,T])}} \exp\left\{ -\frac{y^2}{2\|\phi\|_{L^2([0,T])}^2} \right\} dy \tag{7.25}$$

and

$$u(t)$$
$$= \phi(t) \cdot \int_\mathbb{R} \frac{f(y)}{\sqrt{2\pi}\|\phi\|_{L^2([0,T])}} \exp^\diamond \left\{ -\frac{(y - Y(t))^{\diamond 2}}{2\|\phi\|_{L^2([0,T])}^2} \right\} \diamond \frac{y - Y(t)}{\|\phi\|_{L^2([0,T])}^2} dy. \tag{7.26}$$

Proof The proof is a consequence of Proposition 7.2 and Lemma 7.3 for $\psi(s) = \phi(s)\alpha(s)$.   □

In view of applications of Theorem 7.4 we need the following two results:

Proposition 7.5. *Let $\phi \in L^2(\mathbb{R})$ with $\|\phi\|_{L^2(\mathbb{R})} = 1$. Suppose that*

$$X = \sum_{k=0}^{\infty} a_k w(\phi)^{\diamond k} \in (\mathcal{S})^*,$$

where $w(\phi) = \int_{\mathbb{R}} \phi(s)dW(s)$ (see (5.2)). Define $f(z) = \sum_{k=0}^{\infty} a_k z^k$ for $z \in \mathbb{C}$. Assume that $y \longmapsto f(x+iy)$ is integrable with respect to the measure $e^{-y^2/2}dy$ on $\mathbb{R}$ for all $x \in \mathbb{R}$. Set

$$F(x) = \int_{\mathbb{R}} f(x+iy)e^{-y^2/2} \frac{dy}{\sqrt{2\pi}}.$$

Further require that $V := F(w(\phi)) \in L^2(P)$. Then $X = V$ P-a.e., that is,

$$X = \int_{\mathbb{R}} f(x+iy)e^{-y^2/2} \frac{dy}{\sqrt{2\pi}} \bigg|_{x=w(\phi)} \qquad P\text{-a.e.} \qquad (7.27)$$

Proof For a proof of this result the reader is referred to Sect. 4.1 of [88].   □

Proposition 7.6. *Let $p(x) = ax^2 + bx + c$, where a, b, c are real constants. Let ψ be as in (7.20) and suppose that $2|a|\|\psi\|_{L^2(\mathbb{R})}^2 < 1$. Define*

$$Y = \int_{\mathbb{R}} \psi(s)dW(s).$$

Then

$$\exp^{\diamond}\left\{aY^{\diamond 2} + bY + c\right\}$$

$$= K_{\psi}^{-1} \exp\left\{K_{\psi}^{-2}\left(aY^2 + bY + c + \frac{(4ac - b^2)\|\psi\|_{L^2(\mathbb{R})}^2}{2}\right)\right\}, \quad (7.28)$$

where the constant K_{ψ} is defined by

$$K_{\psi} = \sqrt{1 + 2a\|\psi\|_{L^2(\mathbb{R})}^2}. \qquad (7.29)$$

Proof We want to expand the Wick product along a base with $\phi := \psi/\|\psi\|_{L^2(\mathbb{R})}$ as its first base element. Note that $Y = \|\psi\|_{L^2(\mathbb{R})} w(\phi)$. Adopting the notation of Proposition 7.5 let us consider

$$f(z) := \exp\left\{a\|\psi\|_{L^2(\mathbb{R})}^2 z^2 + b\|\psi\|_{L^2(\mathbb{R})} z + c\right\}.$$

Choose $x \in \mathbb{R}$. Then

$$
F(x) := \int_{\mathbb{R}} f(x + iy) \exp\left\{-\frac{y^2}{2}\right\} \frac{dy}{\sqrt{2\pi}}
$$

$$
= \int_{\mathbb{R}} \exp\left\{a \|\psi\|_{L^2(\mathbb{R})}^2 \left(x^2 - y^2 + 2ixy\right) + b \|\psi\|_{L^2(\mathbb{R})} (x + iy) + c\right\}
$$

$$
\times \exp\left\{\frac{-y^2}{2}\right\} \frac{dy}{\sqrt{2\pi}}
$$

$$
= \int_{\mathbb{R}} \exp\left\{a \|\psi\|_{L^2(\mathbb{R})}^2 x^2 + b \|\psi\|_{L^2(\mathbb{R})} x + c\right\}
$$

$$
\exp\left\{i\left(2xa \|\psi\|_{L^2(\mathbb{R})} + b\right) \|\psi\|_{L^2(\mathbb{R})} y - \left(\frac{1}{2} + a \|\psi\|_{L^2(\mathbb{R})}^2\right) y^2\right\} \frac{dy}{\sqrt{2\pi}}
$$

$$
= \exp\left\{a \|\psi\|_{L^2(\mathbb{R})}^2 x^2 + b \|\psi\|_{L^2(\mathbb{R})} x + c\right\}
$$

$$
\cdot \frac{1}{\sqrt{1 + 2a \|\psi\|_{L^2(\mathbb{R})}^2}} \exp\left\{-\frac{\left(2xa \|\psi\|_{L^2(\mathbb{R})} + b\right)^2 \|\psi\|_{L^2(\mathbb{R})}^2}{2 + 4a \|\psi\|_{L^2(\mathbb{R})}^2}\right\}.
$$

In the aforementioned calculation, we employed the well-known formula

$$
\frac{1}{\sqrt{2\pi}} \int_{\mathbb{R}} \exp\left\{i\alpha t - \beta^2 t^2\right\} dt = \frac{1}{\sqrt{2\beta}} \exp\left\{-\frac{\alpha^2}{4\beta}\right\}.
$$

The latter implies that

$$
V := F(w(\phi)) = \exp\left\{a \|\psi\|_{L^2(\mathbb{R})}^2 w(\phi)^2 + b \|\psi\|_{L^2(\mathbb{R})} w(\phi) + c\right\}
$$

$$
\cdot \frac{1}{\sqrt{1 + 2a \|\psi\|_{L^2(\mathbb{R})}^2}} \exp\left\{-\frac{\left(2aw(\phi) \|\psi\|_{L^2(\mathbb{R})} + b\right)^2 \|\psi\|_{L^2(\mathbb{R})}^2}{2 + 4a \|\psi\|_{L^2(\mathbb{R})}^2}\right\}
$$

$$
= \frac{1}{\sqrt{1 + 2a \|\psi\|_{L^2(\mathbb{R})}^2}} \exp\left\{a \|\psi\|_{L^2(\mathbb{R})}^2 w(\phi)^2 + b \|\psi\|_{L^2(\mathbb{R})} w(\phi) + c\right\}
$$

$$
\cdot \exp\left\{-\frac{2a^2 \|\psi\|_{L^2(\mathbb{R})}^2 w(\phi)^2 + 2ab \|\psi\|_{L^2(\mathbb{R})}^2 w(\phi) + \frac{1}{2}b^2 \|\psi\|_{L^2(\mathbb{R})}^2}{1 + 2a \|\psi\|_{L^2(\mathbb{R})}^2}\right\}.
$$

So

$$
V = \frac{1}{\sqrt{1 + 2a \|\psi\|_{L^2(\mathbb{R})}^2}}
$$

$$
\cdot \exp\left\{\frac{a \|\psi\|_{L^2(\mathbb{R})}^2 w(\phi)^2 + 2a^2 \|\psi\|_{L^2(\mathbb{R})}^4 w(\phi)^2 + b \|\psi\|_{L^2(\mathbb{R})} w(\phi)}{1 + 2a \|\psi\|_{L^2(\mathbb{R})}^2}\right\}
$$

$$\cdot \exp \left\{ \frac{2ab \, \|\psi\|^3_{L^2(\mathbb{R})} \, w(\phi) + c + 2ac \, \|\psi\|^2_{L^2(\mathbb{R})}}{1 + 2a \, \|\psi\|^2_{L^2(\mathbb{R})}} \right\}$$

$$\cdot \exp \left\{ - \frac{2a^2 \, \|\psi\|^2_{L^2(\mathbb{R})} \, w(\phi)^2 - 2ab \, \|\psi\|^2_{L^2(\mathbb{R})} \, w(\phi) - \frac{1}{2} b^2 \, \|\psi\|^2_{L^2(\mathbb{R})}}{1 + 2a \, \|\psi\|^2_{L^2(\mathbb{R})}} \right\}.$$

Hence

$$V = \frac{1}{\sqrt{1 + 2a \, \|\psi\|^2_{L^2(\mathbb{R})}}} \exp \left\{ \frac{\left(a + 2a^2 \, \|\psi\|^2_{L^2(\mathbb{R})} - 2a^2\right) \|\psi\|^2_{L^2(\mathbb{R})} \, w(\phi)^2}{1 + 2a \, \|\psi\|^2_{L^2(\mathbb{R})}} \right\}$$

$$\cdot \exp \left\{ \frac{\left(b + 2ab \, \|\psi\|^2_{L^2(\mathbb{R})} - 2ab \, \|\psi\|_{L^2(\mathbb{R})}\right) \|\psi\|_{L^2(\mathbb{R})} \, w(\phi)}{1 + 2a \, \|\psi\|^2_{L^2(\mathbb{R})}} \right\}$$

$$\cdot \exp \left\{ \frac{c + (2ac - \frac{1}{2} b^2) \, \|\psi\|^2_{L^2(\mathbb{R})}}{1 + 2a \, \|\psi\|^2_{L^2(\mathbb{R})}} \right\}.$$

Since $w(\phi)$ is Gaussian, we see that $V \in L^2(P)$. Thus, relation (7.28) follows from Proposition 7.5. $\square$

Corollary 7.7. *Retain the conditions of the previous proposition. Then*

$$\exp^\diamond \left\{ a \, (y - Y)^{\diamond 2} \right\} = K_\psi^{-1} \exp \left\{ a K_\psi^{-2} \, (y - Y)^2 \right\}. \tag{7.30}$$

Proof Just note that $b^2 - 4ac = 0$ in this case. $\square$

Corollary 7.8. *Let $\phi(t), Y(t)$ be as in Theorem 7.4. Assume that*

$$\|\phi\|^2_{L^2([t,T])} = \int_t^T \phi(s) ds > 0$$

for $t < T$. Then

$$\frac{1}{\|\phi\|_{L^2([0,T])}} \exp^\diamond \left\{ - \frac{(y - Y(t))^{\diamond 2}}{2 \, \|\phi\|^2_{L^2([0,T])}} \right\} = \frac{1}{\|\phi\|_{L^2([t,T])}} \exp \left\{ - \frac{(y - Y(t))^2}{2 \, \|\phi\|^2_{L^2([t,T])}} \right\}.$$

$$\tag{7.31}$$

Proof Set $\psi(s) = \phi(s) \chi_{[0,t]}$ in Corollary 7.7 and compute a and K_ψ. $\square$

Lemma 7.9. *Suppose that $\phi(t)$, $Y(t)$, and $\|\phi\|_{L^2([t,T])}$ satisfy the conditions of Corollary 7.8. Then*

$$\frac{1}{\|\phi\|_{L^2([0,T])}} \exp^{\diamond}\left\{-\frac{(y-Y(t))^{\diamond 2}}{2\|\phi\|^2_{L^2([0,T])}}\right\} \diamond \frac{(y-Y(t))}{\|\phi\|^2_{L^2([0,T])}}$$

$$= \frac{1}{\|\phi\|_{L^2([t,T])}} \exp\left\{-\frac{(y-Y(t))^2}{2\|\phi\|^2_{L^2([t,T])}}\right\} \cdot \frac{(y-Y(t))}{\|\phi\|^2_{L^2([t,T])}}. \quad (7.32)$$

Proof Differentiation of both sides of (7.31) with respect to y (see Proposition 5.14) yields the result. $\square$

The next result provides a more explicit (and familiar) representation than the one given in Theorem 7.4.

Theorem 7.10. *Require $\phi(t), Y(t)$ to be as in Theorem 7.4. Further assume that*

$$\|\phi\|_{L^2([t,T])} > 0 \quad (7.33)$$

for all $t < T$. Let $f : \mathbb{R} \longrightarrow \mathbb{R}$ be bounded. Then

$$f(Y(T)) = V(0) + \int_0^T g(t)\,(\alpha(t)dt + dW(t)),$$

where

$$V(0) = \int_{\mathbb{R}} \frac{f(y)}{\sqrt{2\pi}\,\|\phi\|_{L^2([0,T])}} \exp\left\{-\frac{y^2}{2\|\phi\|^2_{L^2([0,T])}}\right\}dy \quad (7.34)$$

and

$$g(t) = \phi(t)\int_{\mathbb{R}} \frac{f(y)}{\sqrt{2\pi}\,\|\phi\|_{L^2([t,T])}} \exp\left\{-\frac{(y-Y(T))^2}{2\|\phi\|^2_{L^2([t,T])}}\right\}\frac{y-Y(T)}{\|\phi\|^2_{L^2([t,T])}}dy. \quad (7.35)$$

Proof We apply Theorem 7.4, and therefore we consider the process

$$u(t) := \phi(t)\int_{\mathbb{R}} \frac{f(y)}{\sqrt{2\pi}\,\|\phi\|_{L^2([0,T])}} \exp^{\diamond}\left\{-\frac{(y-Y(t))^{\diamond 2}}{2\|\phi\|^2_{L^2([0,T])}}\right\} \diamond \frac{y-Y(t)}{\|\phi\|^2_{L^2([0,T])}}dy.$$

It follows from Lemma 7.9 that $u(t) = g(t)$. Thus

$$E\left[\int_0^T u^2(t)dt\right] = E\left[\int_0^T g^2(t)dt\right] < \infty.$$

So Theorem 7.10 is a consequence of Theorem 7.4, since $\int_0^T g(t) \diamond \overset{\bullet}{W}_t dt = \int_0^T g(t)dB_t$ in the L^2-case (see Theorem 5.20). $\square$

Remark 7.11. The conclusion of Theorem 7.10 still holds without the condition (7.33), if we interpret $g(t)$ as zero when $\|\phi\|_{L^2([t,T])} = 0$.

Remark 7.12. Although the expression (7.35) clearly exhibits computational tractability compared to the Wick version (7.26), it should be mentioned that (7.26) may convey some insight, which is not evident from (7.35). For example, we might ask for the limiting behavior as $t \longrightarrow T$ of the replicating portfolio $g(t)$ in (7.35). If $\phi(t)$ is continuous at $t = T$, then (7.26) implies

$$\lim_{t \longrightarrow T} g(t) = \lim_{t \longrightarrow T} u(t)$$

$$= \phi(T) \int_{\mathbb{R}} \frac{f(y)}{\sqrt{2\pi} \, \|\phi\|_{L^2([0,T])}} \exp^{\diamond} \left\{ -\frac{(y-Y(T))^{\diamond 2}}{2\, \|\phi\|_{L^2([0,T])}^2} \right\} \diamond \frac{y-Y(T)}{\|\phi\|_{L^2([0,T])}^2} dy. \tag{7.36}$$

This limit clearly exists in $(\mathcal{S})^*$.

Corollary 7.13. *The digital payoff* $F = \chi_{[K,\infty)}\left(Y(T)\right)$ *takes the representation*

$$\chi_{[K,\infty)}\left(Y(T)\right) = V(0) + \int_0^T g(t) \left(\alpha(t)dt + dW(t)\right),$$

where

$$V(0) = \int_K^\infty \frac{1}{\sqrt{2\pi} \, \|\phi\|_{L^2([0,T])}} \exp \left\{ -\frac{y^2}{2\, \|\phi\|_{L^2([0,T])}^2} \right\} dy \tag{7.37}$$

and

$$u(t) = \frac{\phi(t)}{\sqrt{2\pi} \, \|\phi\|_{L^2([t,T])}} \cdot \exp \left\{ -\frac{(K-Y(t))^2}{2\, \|\phi\|_{L^2([t,T])}^2} \right\}. \tag{7.38}$$

Proof Relation (7.37) follows from (7.35) for $f(y) = \chi_{[K,\infty)}(y)$ by performing integration with respect to y. $\quad\square$

Remark 7.14. To be precise, the hedging strategy with respect to a contingent T-claim of type $F = h(Z(T))$ with

$$Z(t) = \int_0^t \phi(s)dW(s), \quad t \in [0,T],$$

can be determined as follows: set $Y(t) = Z(t) + \int_0^t \alpha(s)\phi(s)ds$ and let

$$f(x) := e^{-\int_0^T \rho(s)ds} h\left(x - \int_0^T \alpha(s)\phi(s)ds \right).$$

Then $e^{-\int_0^T \rho(s)ds} F = f(Y(T))$ and $V(0)$ and $u(t)$ in (7.10) can be explicitly computed from the expressions given in Theorem 7.10.

7.3 The Multidimensional Case

In this section, we generalize the results of the previous section to the multidimensional case. Let $W(t) = (W_1(t), ..., W_n(t))^T$ be a n-dimensional Brownian motion and $\dot{W}(t) = (\dot{W}_1(t), ..., \dot{W}_n(t))$ be the n-dimensional white noise.

Definition 7.15. *Let* $Y = (Y_1, ..., Y_n) : \Omega \longrightarrow \mathbb{R}^n$ *be a random variable such that* $Y_i \in (\mathcal{S})^*$, $i = 1, ..., n$. *Then a continuous function*

$$\delta_Y(\cdot) : \mathbb{R}^n \longrightarrow (\mathcal{S})^*$$

is called the Donsker delta function *of* Y *if it satisfies*

$$\int_{\mathbb{R}^n} g(y)\delta_Y(y)dy = g(Y) \quad a.e. \tag{7.39}$$

for all measurable functions $g : \mathbb{R}^n \longrightarrow \mathbb{R}$, *provided the integral on left-hand side exists. Here, and in the sequel,* $dy = dy_1...dy_n$ *denotes the n-dimensional Lebesgue measure.*

Proposition 7.16. *Let* $Y : \Omega \longrightarrow \mathbb{R}^n$ *be a normally distributed random variable with mean* $m = E[Y]$ *and covariance matrix* $C = (c_{ij})_{1 \le i,j \le n}$. *Suppose that* C *is invertible with inverse* $A = (a_{ij})_{1 \le i,j \le n}$. *Then* $\delta_Y(y)$ *is uniquely determined and given by the expression*

$$\delta_Y(y) = (2\pi)^{-n/2} \sqrt{|A|} \exp^\diamond \left\{ -\frac{1}{2} \sum_{i,j=1}^n a_{ij} (y_i - Y_i) \diamond (y_j - Y_j) \right\}, \tag{7.40}$$

where $|A|$ *is the determinant of* A.

Proof Let us denote the right-hand side of (7.40) by $G_Y(y)$. We want to verify that G_Y fulfills (7.39), that is,

$$\int_{\mathbb{R}^n} g(y)G_Y(y)dy = g(Y) \quad \text{a.e.} \tag{7.41}$$

To this end let us first assume that g has the form

$$g(y) = e^{\lambda \cdot y} = e^{\lambda_1 y_1 + ... + \lambda_n y_n} \tag{7.42}$$

for some $\lambda = (\lambda_1, ..., \lambda_n) \in \mathbb{C}^n$. Then applying the $\mathcal{H}$-transform to the left-hand side of (7.41), we obtain that

$$\mathcal{H} \left(\int_{\mathbb{R}^n} g(y)G_Y(y)dy \right)$$

$$= \int_{\mathbb{R}^n} e^{\lambda \cdot y} \mathcal{H} (G_Y(y)) \, dy$$

$$= \int_{\mathbb{R}^n} e^{\lambda \cdot y} (2\pi)^{-n/2} \sqrt{|A|} \exp \left\{ -\frac{1}{2} \sum_{i,j=1}^n a_{ij} \left(y_i - \tilde{Y}_i \right) \cdot \left(y_j - \tilde{Y}_j \right) \right\} dy,$$

$$\tag{7.43}$$

where $\widetilde{Y}(z) = (\widetilde{Y}_1(z), ..., \widetilde{Y}_n(z)) := (\mathcal{H}(Y_1)(z), ..., \mathcal{H}(Y_n)(z))$ is the Hermite transform of Y at $z = (z_1, z_2, ...) \in \mathbb{C}^{\mathbb{N}}$. Relation (7.43) may be regarded as the expectation of $e^{\lambda \cdot Z}$, where Z is a normally distributed random variable with mean $\widetilde{Y}$ and covariance matrix $C = A^{-1}$. Now $Z := Y - m + \widetilde{Y}$ is such a random variable. Hence (7.43) can be written as $E\left[e^{\lambda \cdot (Y - m + \widetilde{Y})}\right]$, which by the well-known formula for the characteristic function of a normal random variable is equal to $\exp\left\{\lambda \cdot \widetilde{Y} + \frac{1}{2}\sum_{i,j=1}^{n} c_{ij}\lambda_i\lambda_j\right\}$. By the properties of the $\mathcal{H}$-transform, we conclude that

$$\mathcal{H}\left(\int_{\mathbb{R}^n} g(y)G_Y(y)dy\right) = \exp\left\{\lambda \cdot \widetilde{Y} + \frac{1}{2}\sum_{i,j=1}^{n} c_{ij}\lambda_i\lambda_j\right\}$$

$$= \mathcal{H}\left(\exp^{\diamond}\left\{\lambda \cdot Y + \frac{1}{2}\sum_{i,j=1}^{n} c_{ij}\lambda_i\lambda_j\right\}\right)$$

$$= \mathcal{H}\left(e^{\lambda \cdot Y}\right) = \mathcal{H}\left(g(Y)\right).$$

The latter shows that (7.41) holds for all functions g given by (7.42). Hence using, for example, the Fourier transform, we see that (7.41) is valid for all g in the sense of (7.39). It remains to prove uniqueness: If $H_1 : \mathbb{R}^n \longrightarrow (\mathcal{S})^*$ and $H_2 : \mathbb{R}^n \longrightarrow (\mathcal{S})^*$ are two continuous functions such that

$$\int_{\mathbb{R}^n} g(y)H_i(y)dy = g(Y) \text{ a.e. for } i = 1, 2 \tag{7.44}$$

for all g such that the integral converges, then, in particular, (7.44) must hold for all continuous functions with compact support. But then we clearly must have

$$H_1(y) = H_2(y) \text{ for a.a. } y \in \mathbb{R}^n$$

and hence for all $y \in \mathbb{R}^n$ by continuity. $\square$

In the sequel, let $\psi : [0, T] \longrightarrow \mathbb{R}^n, \phi : [0, T] \longrightarrow \mathbb{R}^{n \times n}$ be deterministic functions such that

$$\int_0^T |\psi(s)|\, ds < \infty \text{ and } \|\phi\|^2 := \sum_{i,j=1}^{n} \int_0^T \phi_{ij}^2(s)ds < \infty. \tag{7.45}$$

Define

$$Y(t) = \int_0^t \phi(s)dW(s) + \int_0^t \psi(s)ds$$

$$= \int_0^t \left(\phi(s)\dot{W}(s) + \psi(s)\right) ds, 0 \leq t \leq T$$

$$m = E\left[Y(T)\right] = \int_0^T \psi(s)ds \in \mathbb{R}^n$$

and, for $1 \le i, j \le n$,

$$c_{ij} = E\left[(Y_i(T) - m_i)(Y_j(T) - m_j)\right] = \int_0^T \left(\phi \cdot \phi^T\right)_{ij}(s)ds.$$

Require that the matrix $C = (c_{ij})_{1 \le i,j \le n}$ is invertible and set

$$A = (a_{ij})_{1 \le i,j \le n} = C^{-1}.$$

Further, define

$$H(t) = H(t, y)$$

$$= \exp^\diamond \left\{ -\frac{1}{2} \sum_{i,j=1}^n a_{ij} (y_i - Y_i(t)) \diamond (y_j - Y_j(t)) \right\}$$

$$= \exp^\diamond \left\{ -\frac{1}{2} (y - Y(t))^T \diamond A (y - Y(t)) \right\}, \quad 0 \le t \le T. \quad (7.46)$$

Lemma 7.17. *The value $H(T)$ in (7.46) takes the form*

$$H(T) = H(0) + \int_0^T H(t)$$

$$\diamond \left(\frac{1}{2} \sum_{i,j=1}^n a_{ij} \left((y_i - Y_i(t)) \diamond \left(\phi_j(t)\dot{W}(t) + \psi_j(t) \right) \right. \right.$$

$$\left. \left. + (y_j - Y_j(t)) \diamond \left(\phi_i(t)\dot{W}(t) + \psi_i(t) \right) \right) \right) dt, \quad (7.47)$$

where ϕ_j is the jth row of the matrix ϕ.

Proof By the Wick chain rule (see Theorem 6.13) we get

$$H(T) = H(0) + \int_0^T \frac{dH}{dt} dt$$

$$= H(0) + \int_0^T H(t)$$

$$\diamond \left(\frac{d}{dt} \left(-\frac{1}{2} \sum_{i,j=1}^n a_{ij} (y_i - Y_i(t)) \diamond (y_j - Y_j(t)) \right) \right) dt$$

$$= H(0) + \int_0^T H(t)$$

$$\diamond \left(\frac{1}{2} \sum_{i,j=1}^n a_{ij} \left((y_i - Y_i(t)) \diamond \frac{d}{dt} Y_j(t) + (y_j - Y_j(t)) \diamond \frac{d}{dt} Y_i(t) \right) \right) dt$$

$$= H(0) + \int_0^T H(t)$$

$$\diamond \left(\frac{1}{2} \sum_{i,j=1}^n a_{ij} \left((y_i - Y_i(t)) \diamond \left(\phi_j(t)\dot{W}(t) + \psi_j(t) \right) \right. \right.$$

$$\left. \left. + (y_j - Y_j(t)) \diamond \left(\phi_i(t)\dot{W}(t) + \psi_i(t) \right) \right) \right) dt. \quad \Box$$

We can now prove the main result of this section.

Theorem 7.18. *Suppose that* $\alpha : [0,T] \longrightarrow \mathbb{R}^n$ *is a deterministic function such that*

$$\|\alpha\|_{L^2([0,T];\mathbb{R}^n)}^2 = \int_0^T |\alpha|^2(s)ds < \infty. \tag{7.48}$$

Let $\phi : [0,T] \longrightarrow \mathbb{R}^{n \times n}$ *be as in (7.45) and define*

$$Y(t) = \int_0^t \phi(s)dW(s) + \int_0^t \phi(s)\alpha(s)ds, 0 \le t \le T. \tag{7.49}$$

Further let $f : \mathbb{R}^n \longrightarrow \mathbb{R}$ *be bounded. Then*

$$f(Y(T)) = V(0) + \int_0^T u(t) \diamond \left(\alpha(t) + \dot{W}(t) \right) dt, \tag{7.50}$$

where

$$V(0) = (2\pi)^{-n/2} \sqrt{|A|} \int_{\mathbb{R}^n} f(y) \exp\left\{ -\frac{1}{2} y^T A y \right\} dy \tag{7.51}$$

and

$$u(t) = (2\pi)^{-n/2} \sqrt{|A|} \int_{\mathbb{R}^n} (f(y)$$

$$\exp^\diamond \left\{ -\frac{1}{2} (y - Y(t))^T \diamond A (y - Y(t)) \right\}$$

$$\diamond (y - Y(t))^T A\phi(t) \Big) dy. \tag{7.52}$$

Proof Let us apply Proposition 7.16 and Lemma 7.17 for $\psi(t) = \phi(t)\alpha(t)$ and $a = \int_0^T \phi(t)\alpha(t)dt$. Then we obtain that

$$f(Y(T)) = \int_{\mathbb{R}^n} f(y)\delta_{Y(T)}(y)dy = (2\pi)^{-n/2} \sqrt{|A|} \int_{\mathbb{R}^n} \Big[f(y)$$

$$\exp^\diamond \left\{ -\frac{1}{2} \sum_{i,j=1}^n a_{ij} (y_i - Y_i(T)) \diamond (y_j - Y_j(T)) \right\} \Big] dy$$

$$= (2\pi)^{-n/2} \sqrt{|A|} \int_{\mathbb{R}^n} f(y) H(T,y) dy$$

$$= (2\pi)^{-n/2} \sqrt{|A|} \int_{\mathbb{R}^n} f(y) H(0,y) dy$$

$$+ (2\pi)^{-n/2} \sqrt{|A|} \int_{\mathbb{R}^n} f(y) \Big[\int_0^T H(t,y)$$

$$\diamond \Big(\frac{1}{2} \sum_{i,j=1}^n a_{ij} \Big[(y_i - Y_i(t)) \diamond \Big(\phi_j(t) \dot{W}(t) + \psi_j(t) \Big)$$

$$+ (y_j - Y_j(t)) \diamond \Big(\phi_i(t) \dot{W}(t) + \psi_i(t) \Big) \Big] \Big) dt \Big] dy.$$

So

$$f(Y(T)) = (2\pi)^{-n/2} \sqrt{|A|} \int_{\mathbb{R}^n} f(y) H(0,y) dy$$

$$+ (2\pi)^{-n/2} \sqrt{|A|} \int_0^T \Big(\int_{\mathbb{R}^n} f(y) H(t,y) \diamond (y - Y(t))^T A\phi(t) dy \Big)$$

$$\diamond \Big(\alpha(t) + \dot{W}(t) \Big) dt,$$

which by (7.46) is the same as (7.50)–(7.52).   □

Lemma 7.19. *Suppose that $\phi(t)$ and $Y(t)$ are as in (7.45) and (7.49). Let $0 \leq t \leq T$ and define the $n \times n$ matrix*

$$C_{[t,T]} = \int_t^T \phi(s)\phi^T(s) ds. \tag{7.53}$$

Further require that

$$|C_{[t,T]}| > 0 \tag{7.54}$$

and set

$$A_{[t,T]} = C_{[t,T]}^{-1}. \tag{7.55}$$

Then

$$\sqrt{|A_{[0,T]}|} \exp^{\diamond} \Big\{ -\frac{1}{2} (y - Y(t))^T \diamond A_{[0,T]} (y - Y(t)) \Big\}$$

$$= \sqrt{|A_{[t,T]}|} \exp \Big\{ -\frac{1}{2} (y - Y(t))^T \cdot A_{[t,T]} (y - Y(t)) \Big\}. \tag{7.56}$$

Proof The proof follows the line of reasoning in the proof of Corollary 7.8. Therefore we omit the details.   □

Corollary 7.20. *Choose* $\phi(t)$, $Y(t)$, *and* $\left|A_{[t,T]}\right|$ *to be as in Lemma 7.19.*
Then

$$
\sqrt{\left|A_{[0,T]}\right|} \exp^{\diamond}\left\{ -\frac{1}{2}\left(y - Y(t)\right)^{T} \diamond A_{[0,T]}\left(y - Y(t)\right) \right\}
$$
$$
\diamond \left(y - Y(t)\right)^{T} A_{[0,T]}\phi(t)
$$
$$
= \sqrt{\left|A_{[t,T]}\right|} \exp\left\{ -\frac{1}{2}\left(y - Y(t)\right)^{T} \cdot A_{[t,T]}\left(y - Y(t)\right) \right\}
$$
$$
\cdot \left(y - Y(t)\right)^{T} A_{[t,T]}\phi(t). \tag{7.57}
$$

Proof Differentiate (7.56) with respect to $y_1, ..., y_n$. $\square$

We can now give an explicit representation of $f(Y(T))$ without Wick product.

Theorem 7.21. *Keep the notation and conditions of Lemma 7.19. Let* f :
$\mathbb{R}^n \longrightarrow \mathbb{R}$ *be bounded. Then*

$$
f(Y(T)) = V(0) + \int_0^T u(t)\left(\alpha(t)dt + dW(t)\right), \tag{7.58}
$$

where $V(0)$ *is as in (7.51) and*

$$
u(t) = (2\pi)^{-n/2} \sqrt{\left|A_{[t,T]}\right|} \int_{\mathbb{R}^n} \Big[f(y)
$$
$$
\exp\left\{ -\frac{1}{2}\left(y - Y(t)\right)^{T} \cdot A_{[t,T]}\left(y - Y(t)\right) \right\}
$$
$$
\cdot \left(y - Y(t)\right)^{T} A_{[t,T]}\phi(t) \Big]\, dy. \tag{7.59}
$$

Example 7.22. The general results in Theorem 7.18 and Theorem 7.21 can be employed to analyze the replicating portfolios for some exotic options. For example, one can study portfolios of *path dependent* options like a *knock-out option* of the form

$$
F = \chi_{[K,\infty)}\left(\max_{0 \le t \le T} Z(t)\right), \tag{7.60}
$$

where

$$
Z(t) = \int_0^t \phi(s)dW(s), \quad t \in [0,T],
$$

is a one-dimensional process. The idea is the following. Let $0 = t_0 < t_1 < ... < t_n = T$ be an equidistant partition of $[0,T]$, and define

$$
\phi(t) = \begin{pmatrix} \chi_{[0,t_1]}(t) & 0 & \cdots & 0 \\ \chi_{[0,t_2]}(t) & 0 & \cdots & 0 \\ \vdots & \vdots & \ddots & \vdots \\ \chi_{[0,t_n]}(t) & 0 & \cdots & 0 \end{pmatrix} \in \mathbb{R}^{n \times n}.
$$

Then

$$\phi\phi^T = \begin{pmatrix} \chi_{[0,t_1]}(t) \; \chi_{[0,t_1]}(t) \; \cdots \; \chi_{[0,t_1]}(t) \\ \chi_{[0,t_1]}(t) \; \chi_{[0,t_2]}(t) \; \cdots \; \chi_{[0,t_2]}(t) \\ \vdots \quad\quad \vdots \quad\; \ddots \quad \vdots \\ \chi_{[0,t_1]}(t) \; \chi_{[0,t_2]}(t) \; \cdots \; \chi_{[0,t_n]}(t) \end{pmatrix}.$$

Thus

$$C = \int_0^T \phi\phi^T(t)dt = \begin{pmatrix} t_1 \; t_1 \; \cdots \; t_1 \\ t_1 \; t_2 \; \cdots \; t_2 \\ \vdots \; \vdots \; \ddots \; \vdots \\ t_1 \; t_2 \; \cdots \; t_n \end{pmatrix}.$$

Since

$$|C| = t_1(t_2 - t_1)(t_3 - t_2) \cdot \dots \cdot (t_n - t_{n-1}) = (\Delta t)^n,$$

where

$$\Delta t = t_i - t_{i-1} = \frac{T}{n} \neq 0, \quad 1 \leq i \leq n,$$

the matrix C is invertible. Hence Theorem 7.18 can be applied to

$$Y(t) := \int_0^t \phi(s)dW(s) + \int_0^t \phi(s)\alpha(s)ds$$

$$= \begin{pmatrix} W_1(t \wedge t_1) + \int_0^{t \wedge t_1} \alpha_1(s)ds \\ \vdots \\ W_1(t \wedge t_n) + \int_0^{t \wedge t_n} \alpha_1(s)ds \end{pmatrix},$$

where $\alpha = (\alpha_1, ..., \alpha_n) : [0, T] \longrightarrow \mathbb{R}^n$ is as in (7.48). In this case, we observe that

$$A = C^{-1} = \frac{n}{T} \begin{pmatrix} 2 & -1 & 0 & \cdots & \cdots & 0 \\ -1 & 2 & -1 & \cdots & \cdots & 0 \\ 0 & -1 & 2 & -1 & \cdots & 0 \\ \vdots & \vdots & \ddots & \ddots & \ddots & \vdots \\ 0 & 0 & \cdots & -1 & 2 & -1 \\ 0 & 0 & \cdots & & -1 & 1 \end{pmatrix}.$$

Now let $f : \mathbb{R}^n \longrightarrow \mathbb{R}$ be bounded. Then

$$f(Y(T))$$

$$= f\left(W_1(t_1) + \int_0^{t_1} \alpha_1(s)ds, ..., W_1(t_n) + \int_0^{t_n} \alpha_1(s)ds\right).$$

In particular, if $\alpha_1 = 0$, we get the following representation by Theorem 7.18:

$$f(W_1(t_1), ..., W_1(t_n)) = V(0) + \int_0^T u_1(t)dW_1(t),$$

where

$$u_1(t) = \left(\frac{n}{2\pi T}\right)^{n/2} \int_{\mathbf{R}^n} f(y_1, ..., y_n)$$

$$\cdot \exp^\diamond \left\{ -\frac{n}{2T} \left(\sum_{i=1}^{n-1} (y_i - W_1(t \wedge t_i))^{\diamond 2} + (y_n - W_1(t))^{\diamond 2} \right. \right.$$

$$\left. -2 \sum_{i=1}^{n-1} (y_i - W_1(t \wedge t_i)) \diamond (y_{i+1} - W_1(t \wedge t_{i+1})) \right) \right\}$$

$$\diamond \frac{n}{T} \left(2 \sum_{i=1}^{n-1} (y_i - W_1(t \wedge t_i)) \chi_{[0,t_i]}(t) + (y_n - W_1(t)) \right.$$

$$-\sum_{i=1}^{n-1} (y_i - W_1(t \wedge t_i)) \chi_{[0,t_{i+1}]}(t)$$

$$\left. -\sum_{i=1}^{n-1} (y_{i+1} - W_1(t \wedge t_{i+1})) \chi_{[0,t_i]}(t) \right) dy. \tag{7.61}$$

Thus, we see that if F is the knock-out option

$$\chi_{[K,\infty)} \left(\max_{0 \le t \le T} W_1(t) \right),$$

then we obtain an approximation of the corresponding replicating portfolio $u(t)$ by choosing n large and $f(y_1, ..., y_n) = \max\{y_i : 1 \le i \le n\}$ in (7.61). With some extra work, one can obtain a similar representation without Wick products by using Theorem 7.21.

7.4 Exercises

Problem 7.1. In this problem we combine the generalized Clark–Ocone theorem (see Theorem 6.35) with results about the Donsker delta function to give an alternative proof of Corollary 7.13. More precisely, we prove the representation (6.41) for the digital payoff

$$F = \chi_{[K,\infty)}(W(T)).$$

We proceed as follows. By Theorem 6.35 we can write

$$F = E[F] + \int_0^T E[D_t F | \mathcal{F}_t] dW(t).$$

To find $D_t F$, we first represent F in terms of the Donsker delta function

$$\delta_{W(T)}(y) = \frac{1}{\sqrt{2\pi T}} \exp^\diamond \left\{ -\frac{(y - W(T))^{\diamond 2}}{2T} \right\},$$

and we get

$$\chi_{[K,\infty)}(W(T)) = \int_{\mathbb{R}} \chi_{[K,\infty)}(y) \frac{1}{\sqrt{2\pi T}} \exp^{\diamond} \left\{ -\frac{(y - W(T))^{\diamond 2}}{2T} \right\} dy.$$

(a) Use the Wick chain rule (see Theorem 6.13) to deduce that

$$D_t \chi_{[K,\infty)}(W(T)) = \int_K^{\infty} \frac{1}{\sqrt{2\pi T}} \exp^{\diamond} \left\{ -\frac{(y - W(T))^{\diamond 2}}{2T} \right\} \diamond \frac{y - W(T)}{T} dy.$$

(b) Use (a) and Lemma 6.27 and Lemma 6.28 to prove that

$$E\left[D_t \chi_{[K,\infty)}(W(T)) | \mathcal{F}_t\right] = \int_K^{\infty} \frac{1}{\sqrt{2\pi T}} \exp^{\diamond} \left\{ -\frac{(y - W(t))^{\diamond 2}}{2T} \right\} \diamond \frac{y - W(t)}{T} dy.$$

(c) Note that

$$\frac{d}{dy} \exp^{\diamond} \left\{ -\frac{(y - W(t))^{\diamond 2}}{2T} \right\} = \exp^{\diamond} \left\{ -\frac{(y - W(t))^{\diamond 2}}{2T} \right\} \diamond \frac{y - W(t)}{T}$$

and combine this with (b) to obtain

$$E\left[D_t \chi_{[K,\infty)}(W(T)) | \mathcal{F}_t\right] = \frac{1}{\sqrt{2\pi T}} \exp^{\diamond} \left\{ -\frac{(K - W(t))^{\diamond 2}}{2T} \right\}.$$

(d) Finally, combine (c) with Corollary 7.8 to obtain the representation (6.41).

Problem 7.2. (*) Let Y be normally distributed with mean m and variance v. Define

$$Z := e^Y.$$

Show that the Donsker delta function of Z is given by

$$\delta_Z(z) = \frac{1}{\sqrt{2\pi v}} \frac{1}{z} \exp^{\diamond} \left\{ -\frac{(\log z - \log Z)^{\diamond 2}}{2v} \right\} \chi_{[0,\infty)}(z).$$

Problem 7.3. Define $F = \chi_{[K,\infty)}(W(T))$, where $K \in \mathbb{R}$ is constant. Show that $F \notin \mathbb{D}_{1,2}$.

8

The Forward Integral and Applications

8.1 A Motivating Example

The following example is based on [191]. Suppose we have a financial market on $[0, T]$ ($T > 0$) with two investment possibilities:

(1) A *risk free* asset with unit price $S_0(t)$ at time t given by

$$dS_0(t) = \rho S_0(t)dt, \qquad S_0(0) = 1,$$

(2) A *risky* asset with unit price $S_1(t)$ at time t given by

$$dS_1(t) = S_1(t)\big[\mu dt + \sigma dW(t)\big], \qquad S_1(0) > 0.$$

Here ρ, μ, and $\sigma > 0$ are given constants.

If we represent a portfolio in this market by the fraction $\pi(t)$ of the total wealth $X(t)$ invested in the risky asset at time t, then the dynamics of the wealth process $X_\pi(t) = X(t)$, $t \geq 0$, of a self-financing portfolio π is

$$
\begin{aligned}
dX(t) &= (1 - \pi(t))X(t)\rho dt + \pi(t)X(t)[\mu dt + \sigma dW(t)] \\
&= X(t)[(\rho + (\mu - \rho)\pi(t))dt + \pi(t)\sigma dW(t)], \qquad X(0) = x > 0.
\end{aligned}
\tag{8.1}
$$

Let $\mathcal{A}_{\mathbb{F}}$ be the set of all $\mathbb{F}$-adapted portfolios π such that

$$\int_0^T \pi^2(s)ds < \infty \quad P\text{-a.s.}$$

If $\pi \in \mathcal{A}_{\mathbb{F}}$, then the solution $X(t) = X^x(t)$, $t \in [0, T]$, of (8.1) is

$$X(t) = x \exp\left\{ \int_0^t \sigma\pi(s)dW(s) + \int_0^t \big[\rho + (\mu - \rho)\pi(s) - \frac{1}{2}\sigma^2\pi^2(s)\big]ds\right\}. \tag{8.2}$$

In this classical setting it is well-known that the portfolio $\pi_{\mathbb{F}}^*$ that maximizes

$$E\Big[\log X_\pi^x(T)\Big] \qquad (T > 0) \tag{8.3}$$

over all $\pi \in \mathcal{A}_{\mathbb{F}}$ is given by

$$\pi_{\mathbb{F}}^*(t) = \frac{\mu - \rho}{\sigma^2}, \qquad t \in [0, T], \tag{8.4}$$

and the corresponding value function is

$$V_{\mathbb{F}}(x) = \sup_{\pi \in \mathcal{A}_{\mathbb{F}}} E\Big[\log X_{\pi}^x(T)\Big] = E\Big[\log X_{\pi_{\mathbb{F}}^*}^x(T)\Big]$$

$$= \log x + \Big(\rho + \frac{(\mu - \rho)^2}{2\sigma^2}\Big)T, \quad x > 0. \tag{8.5}$$

We now ask, what happens if the trader has, at any time t, more information than $\mathcal{F}_t$? More precisely, suppose $\mathbb{G} = \{\mathcal{G}_t, t \geq 0\}$ is a filtration such that

$$\mathcal{F}_t \subseteq \mathcal{G}_t, \quad t \in [0, T] \ (T > 0),$$

representing the information available to the trader. Then if we allow the trader's portfolios to be $\mathbb{G}$-adapted, then what would the maximal value of (8.3) be?

In [191] the special case

$$\mathcal{G}_t := \mathcal{F}_t \vee \sigma(W(T_0)) \quad (T_0 > T) \tag{8.6}$$

is considered. This means that the trader knows the value of $W(T_0)$ (which is equivalent to knowing $S_1(T_0)$) in addition to the basic information $\mathcal{F}_t$ at any time $t \in [0, T]$. We call such a trader an *insider* (or *informed trader*). In this case we have the following result, first proved in [122].

Theorem 8.1. *The process $W(t)$, $t \geq 0$, is a semimartingale with respect to $\mathbb{G}$ (under the measure P). Its semimartingale decomposition is*

$$W(t) = \hat{W}(t) + \int_0^t \frac{W(T_0) - W(s)}{T_0 - s} ds, \tag{8.7}$$

where $\hat{W}(t)$, $t \geq 0$, is a Brownian motion with respect to $\mathbb{G}$.

Using this we can interpret the integrals in (8.1) and (8.2) as integrals with respect to a semimartingale as integrator, that is, we can write

$$\int_0^t \varphi(s) dW(s) = \int_0^t \varphi(s) d\hat{W}(s) + \int_0^t \varphi(s) \frac{W(T_0) - W(s)}{T_0 - s} ds$$

for $\varphi(s) = X(s)\pi(s)$ and $\varphi(s) = \pi(s)$, $s \geq 0$, in (8.1) and (8.2), respectively.

Let $\pi_{\mathbb{G}}^*$ denote the optimal portfolio among all insider's portfolios $\mathcal{A}_{\mathbb{G}}$, that is, all $\mathbb{G}$-adapted processes π such that

$$\int_0^T \pi^2(s) ds < \infty \quad P\text{-a.s.}$$

Let $V_{\mathbb{G}}$ be the value function

$$V_{\mathbb{G}}(x) = \sup_{\pi \in \mathcal{A}_{\mathbb{G}}} E\Big[\log X_\pi^x(T)\Big] = E\Big[\log X_{\pi_{\mathbb{G}}^*}^x(T)\Big], \quad x > 0. \tag{8.8}$$

In this framework, [191] proved that

$$\pi_{\mathbb{G}}^*(t) = \frac{\mu - \rho}{\sigma^2} + \frac{W(T_0) - W(t)}{\sigma(T_0 - t)}, \quad t \in [0, T], \tag{8.9}$$

and

$$V_{\mathbb{G}}(x) = V_{\mathbb{F}}(x) + \frac{1}{2\sigma^2} \log\Big(\frac{T_0}{T_0 - T}\Big), \quad x > 0. \tag{8.10}$$

We could regard the difference

$$V_{\mathbb{G}}(x) - V_{\mathbb{F}}(x) = \frac{1}{2\sigma^2} \log\Big(\frac{T_0}{T_0 - T}\Big)$$

as the *value* of the additional information $S_1(T_0)$ the insider has at her disposal.

This example opens a few questions that we would like to address:

(1) What happens if we consider a more general insider filtration $\mathbb{G} = \{\mathcal{G}_t, t \geq 0\}$, that is, with the only requirement that

$$\mathcal{F}_t \subseteq \mathcal{G}_t, \quad \text{for all } t \in [0, T]?$$

(2) What if we consider a more general utility function $U(x)$, $x \geq 0$, that is, if the optimization problem is to maximize

$$E\Big[U(X_\pi^x(T))\Big]?$$

(3) What if we allow the financial market model for the prices to have jumps?

Question (3) will be discussed in Part II of the book. To handle questions (1) and (2) we need to find a mathematical interpretation of the integrals

$$\text{``} \int_0^t X(s)\pi(s)dW(s) \text{''} \quad \text{and} \quad \text{``} \int_0^t \pi(s)dW(s) \text{''}$$

in (8.1) and (8.2), respectively. Note that in both cases the integrands are *not* adapted with respect to the filtration generated by the integrator, actually they embody some *anticipating* information that the insider has and uses. We also emphasize that, in general, the process W need not be a semimartingale with respect to the general (larger) filtration $\mathbb{G}$. Here is a simple example that illustrates this.

Example 8.2. Define $\mathcal{G}_t := \mathcal{F}_T$, for all $t \in [0, T]$. Suppose that W is a semi-martingale with respect to $\mathbb{G}$, with semimartingale decomposition

$$W(t) = A(t) + M(t), \quad t \in [0, T], \quad \text{with } W(0) = 0,$$

where M is a $\mathbb{G}$-martingale and A is a $\mathbb{G}$-adapted process with bounded variation and $A(0) = 0$. Then

$$M(t) = E[M(T)|\mathcal{G}_t] = E[M(T)|\mathcal{F}_T] = M(T), \quad t \in [0, T].$$

Hence $M(t) = M(0)$, $t \in [0, T]$. Thus

$$W(t) = A(t) + M(0), \quad t \in [0, T],$$

which is a process of bounded variation. This is absurd.

The same argument applies to show that if

$$\mathcal{G}_t := \mathcal{F}_{t + \delta(t)}, \quad \text{for some } \delta(t) > 0, \quad t \in [0, T], \tag{8.11}$$

then W is not a semimartingale with respect to $\mathbb{G}$, (see Problem 8.1).

This shows that to be able to handle the questions above related to insider trading, we must go beyond the semimartingale context. It turns out that the *forward integral*, to be introduced in the next section, provides the natural framework for modeling insider trading.

More about insider trading can be found in the forthcoming Sect. 8.6 and in Chap. 16.

8.2 The Forward Integral

The forward integral with respect to Brownian motion was first defined in the seminal paper [209] and further studied in [210, 211]. This integral was introduced in the modeling of insider trading in [33] and then applied extensively in questions related to insider trading and stochastic control with advanced information (see, e.g., [63, 65, 66, 111, 133, 134, 180, 182]. See also [50] and for the use of Malliavin calculus in insider trading see, for example, [116, 117, 147].

Definition 8.3. *We say that a stochastic process* $\varphi = \varphi(t)$, $t \in [0, T]$, *is forward integrable (in the weak sense) over the interval* $[0, T]$ *with respect to* W *if there exists a process* $I = I(t)$, $t \in [0, T]$, *such that*

$$\sup_{t \in [0, T]} \left| \int_0^t \varphi(s) \frac{W(s + \varepsilon) - W(s)}{\varepsilon} ds - I(t) \right| \longrightarrow 0, \quad \varepsilon \to 0^+, \tag{8.12}$$

in probability. In this case we write

$$I(t) := \int_0^t \varphi(s) d^- W(s), \quad t \in [0, T],$$

and call $I(t)$ *the* forward integral *of* φ *with respect to* W *on* $[0, t]$.

The following results give a more intuitive interpretation of the forward integral as a limit of Riemann sums, see [33, 134].

Lemma 8.4. *Suppose φ is càglàd and forward integrable. Then*

$$\int_0^T \varphi(s) d^- W(s) = \lim_{\Delta t \to 0} \sum_{j=1}^{J_n} \varphi(t_{j-1})(W(t_j) - W(t_{j-1}))$$

with convergence in probability. Here the limit is taken over the partitions $0 = t_0 < t_1 < ... < t_{J_n} = T$ of $t \in [0, T]$ with $\Delta t := max_{j=1,...,J_n}(t_j - t_{j-1}) \longrightarrow 0$, $n \to \infty$.

Proof We may assume that φ is a simple stochastic process, that is,

$$\varphi(t) = \sum_{j=1}^{J_n} \varphi(t_{j-1}) \chi_{(t_{j-1},t_j]}(t), \quad t \in [0, T],$$

because any càglàd stochastic process on $[0, T]$ can be approximated pointwise in ω and uniformly in t by such simple processes. For such φ we have

$$\int_0^T \varphi(s) d^- W(s) = \lim_{\varepsilon \to 0^+} \int_0^T \varphi(s) \frac{W(s + \varepsilon) - W(s)}{\varepsilon} ds$$

$$= \sum_{j=1}^{J_n} \varphi(t_{j-1}) \lim_{\varepsilon \to 0^+} \int_{t_{j-1}}^{t_j} \frac{W(s + \varepsilon) - W(s)}{\varepsilon} ds$$

$$= \sum_{j=1}^{J_n} \varphi(t_{j-1}) \lim_{\varepsilon \to 0^+} \frac{1}{\varepsilon} \int_{t_{j-1}}^{t_j} \int_s^{s+\varepsilon} dW(u) ds$$

$$= \sum_{j=1}^{J_n} \varphi(t_{j-1}) \lim_{\varepsilon \to 0^+} \frac{1}{\varepsilon} \int_{t_{j-1}}^{t_j} \int_{u-\varepsilon}^u ds dW(u)$$

$$= \sum_{j=1}^{J_n} \varphi(t_{j-1})(W(t_j) - W(t_{j-1})),$$

by application of the Fubini theorem. $\square$

Remark 8.5. From the previous lemma we can see that, if the integrand φ is $\mathbb{F}$-adapted, then the Riemann sums are also an approximation to the Itô integral of φ with respect to the Brownian motion. Hence in this case the forward integral and the Itô integral coincide. In this sense we can regard the forward integral as an extension of the Itô integral to a nonanticipating setting. A formal proof of this statement is presented here as a consequence of the forthcoming Lemma 8.9, see Corollary 8.10.

Remark 8.6. The previous result gives the motivation for using forward integrals in the modeling of insider trading in mathematical finance. Suppose the price of a risky asset at time t is represented by $W(t)$ and that a trader buys the stochastic amount α of units of this asset at some random time τ_1. He keeps all these units up to a random time τ_2: $\tau_1 < \tau_2 < T$, when he sells them. The profit obtained by this transaction would be

$$\alpha W(\tau_2) - \alpha W(\tau_1).$$

This is in fact the result obtained with the forward integration of the portfolio

$$\varphi(t) = \alpha \chi_{(\tau_1, \tau_2]}(t), \quad t \in [0, T],$$

with respect to the Brownian motion, that is,

$$\int_0^T \varphi(t) d^- W(t) = \alpha W(\tau_2) - \alpha W(\tau_1).$$

In the sequel we give some useful properties of the forward integral. The following result is an immediate consequence of the definition.

Lemma 8.7. *Suppose φ is a forward integrable stochastic process and G a random variable. Then the product $G\varphi$ is forward integrable stochastic process and*

$$\int_0^T G\varphi(t) d^- W(t) = G \int_0^T \varphi(t) d^- W(t).$$

Remark 8.8. Note that this result does not hold for the Skorohod integral, see Example 2.4.

The next result shows that the forward integral is an extension of the integral with respect to a semimartingale.

Lemma 8.9. *Let $\mathbb{G} := \{\mathcal{G}_t, t \in [0, T]\}$ $(T > 0)$ be a given filtration. Suppose that*

(1) W is a semimartingale with respect to the filtration $\mathbb{G}$
(2) φ is $\mathbb{G}$-predictable and the integral

$$\int_0^T \varphi(t) dW(t),$$

with respect to W, exists.

Then φ is forward integrable and

$$\int_0^T \varphi(t) d^- W(t) = \int_0^T \varphi(t) dW(t).$$

Proof Since W is a semimartingale with respect to $\mathbb{G}$ and φ is $\mathbb{G}$-predictable, we have by the theory of Itô integration with respect to a general (continuous) semimartingale (see e.g. [204]) that

$$\int_0^T \varphi(t)d^-W(t) = \lim_{\varepsilon \to 0^+} \int_0^T \varphi(t)\frac{W(t+\varepsilon) - W(t)}{\varepsilon}dt$$

$$= \lim_{\varepsilon \to 0^+} \int_0^T \frac{1}{\varepsilon}\int_{s-\varepsilon}^s \varphi(t)dt\,dW(s)$$

$$= \int_0^T \varphi(t)dW(t),$$

since

$$\varphi_\varepsilon(s) := \frac{1}{\varepsilon}\int_{s-\varepsilon}^s \varphi(t)dt, \quad s \in [0,T],$$

converges to φ in probability uniformly in s.   □

Corollary 8.10. The forward integral as an extension of the Itô integral. *Let φ be $\mathbb{F}$-predictable and Itô integrable. Then φ is forward integrable and*

$$\int_0^T \varphi(t)d^-W(t) = \int_0^T \varphi(t)dW(t).$$

Proof It is enough to apply Lemma 8.9 with $\mathbb{G} = \mathbb{F}$.   □

8.3 Itô Formula for Forward Integrals

We now turn to the Itô formula for forward integrals. In this connection it is convenient to introduce a notation that is analogous to the classical notation for Itô processes.

Definition 8.11. *A forward process (with respect to W) is a stochastic process of the form*

$$X(t) = x + \int_0^t u(s)ds + \int_0^t v(s)d^-W(s), \quad t \in [0,T], \qquad (8.13)$$

(x constant), where

$$\int_0^T |u(s)|ds < \infty, \quad P\text{-a.s.}$$

and v is a forward integrable stochastic process. A shorthand notation for (8.13) is that

$$d^-X(t) = u(t)dt + v(t)d^-W(t). \qquad (8.14)$$

The next result was first proved in [211].

Theorem 8.12. The one-dimensional Itô formula for forward integrals. *Let*

$$d^-X(t) = u(t)dt + v(t)d^-W(t)$$

be a forward process. Let $f \in C^{1,2}([0,T] \times \mathbb{R})$ *and define*

$$Y(t) = f(t, X(t)), \quad t \in [0, T].$$

Then $Y(t)$, $t \in [0, T]$, *is a forward process and*

$$d^-Y(t) = \frac{\partial f}{\partial t}(t, X(t))dt + \frac{\partial f}{\partial x}(t, X(t))d^-X(t) + \frac{1}{2}\frac{\partial^2 f}{\partial x^2}(t, X(t))v^2(t)dt. \quad (8.15)$$

Sketch of proof. We only sketch the proof and refer to [211] for details. Assume that $f(t, x) = f(x)$, for $t \in [0, T]$ and $x \in \mathbb{R}$. Let $0 = t_0 < t_1 < \ldots < t_N = T$ be a partition of $[0, t]$ with $\Delta t := \max_{j=1,\ldots N}(t_j - t_{j-1})$. Then by Taylor expansion we have

$$f(X(t)) - f(X(0)) = \sum_{j=1}^{N} f(X(t_j)) - f(X(t_{j-1}))$$

$$= \sum_{j=1}^{N} f'(X(t_{j-1}))(X(t_j) - X(t_{j-1})) \quad (8.16)$$

$$+ \frac{1}{2}\sum_{j=1}^{N} f''(\bar{X}_j)(X(t_j) - X(t_{j-1}))^2,$$

for some point $\bar{X}_j \in [X(t_{j-1}), X(t_j)]$. By Lemma 8.7 we have

$$\sum_{j=1}^{N} f'(X(t_{j-1}))(X(t_j) - X(t_{j-1}))$$

$$= \sum_{j=1}^{N} f'(X(t_{j-1}))\left(\int_{t_{j-1}}^{t_j} u(s)ds + \int_{t_{j-1}}^{t_j} v(s)d^-W(s)\right)$$

$$= \sum_{j=1}^{N}\left(\int_{t_{j-1}}^{t_j} f'(X(t_{j-1}))u(s)ds + \int_{t_{j-1}}^{t_j} f'(X(t_{j-1}))v(s)d^-W(s)\right)$$

$$= \int_{0}^{T}\left(\sum_{j=1}^{N} f'(X(t_{j-1}))\chi_{(t_{j-1},t_j]}(s)\right)u(s)ds \quad (8.17)$$

$$+ \int_{0}^{T}\left(\sum_{j=1}^{N} f'(X(t_{j-1}))\chi_{(t_{j-1},t_j]}(s)\right)v(s)d^-W(s)$$

$$\longrightarrow \int_{0}^{T} f'(X(s))u(s)ds + \int_{0}^{T} f'(X(s))v(s)d^-W(s)$$

$$= \int_{0}^{T} f'(X(s))d^-X(s), \quad \Delta t \to 0,$$

with convergence in probability. As in the classical case one can prove also that

$$\sum_{j=1}^{N} f''(\bar{X}_j)(X(t_j) - X(t_{j-1}))^2 \longrightarrow \int_0^T f''(X(s))v^2(s)ds, \quad \Delta t \to 0 \quad (8.18)$$

(in probability). Combining (8.16)–(8.18), we can obtain the result.   □

Remark 8.13. Note the similarity with the classical Itô formula. This similarity extends to the multidimensional case. We state the result without proof.

Theorem 8.14. The multidimensional Itô formula for forward integrals. *Let*

$$d^- X_i(t) = u_i(t)dt + \sum_{j=1}^{m} v_{ij}(t)d^- W_j(t) \quad (i = 1, ..., n)$$

be n forward processes driven by m independent Brownian motions $W_1, ..., W_m$. *Let* $f \in C^{1,2}([0, T] \times \mathbb{R}^n)$ *and define*

$$Y(t) := f(t, X(t)), \quad t \in [0, T].$$

Then Y is a forward process and

$$d^- Y(t) = \frac{\partial f}{\partial t}(t, X(t))dt + \sum_{i=1}^{n} \frac{\partial f}{\partial x_i}(t, X(t))d^- X_i(t)$$

$$+ \frac{1}{2} \sum_{i=1}^{n} \sum_{k=1}^{n} \frac{\partial^2 f}{\partial x_i \partial x_k}(t, X(t))(vv^T)_{ik}(t)dt. \quad (8.19)$$

As in the classical case we can use the Itô formula for forward integrals to solve forward stochastic differential equations. We illustrate this by an example.

Example 8.15. Let $\mu(t), t \in [0, T]$, and $\sigma(t), t \in [0, T]$, be measurable processes such that

(1) σ is forward integrable
(2) $\int_0^T (|\mu(t)| + \sigma^2(t))dt < \infty$ *P*-a.e.

Then the unique solution $Y(t), t \in [0, T]$, of the forward stochastic differential equation,

$$d^- Y(t) = \mu(t)Y(t)dt + \sigma(t)Y(t)d^- W(t), \quad t \in (0, T]$$

$$Y(0) = F, \quad (8.20)$$

where F is an $\mathcal{F}_T$-measurable random variable, is

$$Y(t) = F \exp \left\{ \int_0^t (\mu(s) - \frac{1}{2}\sigma^2(s))ds + \int_0^t \sigma(s)d^-W(s) \right\}.$$

To see this, we apply the Itô formula for forward integrals with $f(x) = e^x$, $x \in \mathbb{R}$, and

$$dX(t) = \left(\mu(t) - \frac{1}{2}\sigma^2(t) \right) dt + \sigma(t)d^-W(t).$$

This is the same procedure as in the classical case. To see that Y is the only solution of (8.20), suppose that $\tilde{Y}$ is another solution. Put

$$Z(t) := Y^{-1}(t)\tilde{Y}(t), \quad t \in [0, T].$$

Then by the two-dimensional Itô formula for forward processes, we have

$$\begin{aligned}
d^-Z(t) &= \tilde{Y}(t)d^-Y^{-1}(t) + Y^{-1}(t)d^-\tilde{Y}(t) + d^-Y^{-1}(t)d^-\tilde{Y}(t) \\
&= Y^{-1}(t)\tilde{Y}(t)\big[(-\mu(t) + \sigma^2(t))dt - \sigma(t)d^-W(t) \\
&\quad + \mu(t)dt + \sigma(t)d^-W(t) + (-\sigma(t))\sigma(t)dt\big] = 0.
\end{aligned}$$

Since $Y(0) = \tilde{Y}(0) = F$, it follows that $Z(t) = 1$, $t \in [0, T]$, which proves uniqueness.

8.4 Relation Between the Forward Integral and the Skorohod Integral

We now turn to the relation between the forward integral and the Skorohod integral.

Definition 8.16. *Let φ be a measurable process. We say that φ is forward integrable in the strong sense if the limit*

$$\lim_{\varepsilon \to 0^+} \int_0^T \varphi(t) \frac{W(t + \varepsilon) - W(t)}{\varepsilon} dt$$

exists in $L^2(P)$.

Definition 8.17. *The class $\mathbb{D}_0$ consists of all measurable processes φ such that*

(1) the trajectories $\varphi(\cdot, \omega) : t \longrightarrow \varphi(t, \omega)$ are càglàd P-a.e.
(2) the random variables $\varphi(t) \in \mathbb{D}_{1,2}$ for all $t \in [0, T]$
(3) the trajectories $t \longrightarrow D_s\varphi(t)(\omega)$ are càglàd s-a.e., P-a.e.
(4) the limit $D_{t^+}\varphi(t) := \lim_{s \to t^+} D_s\varphi(t)$ exists with convergence in $L^2(P)$
(5) φ is Skorohod integrable.

The first version of the following result was proved in [209, Proposition 2.3].

Theorem 8.18. *Suppose $\varphi \in \mathbb{D}_0$. Then φ is forward integrable in the strong sense and*

$$\int_0^T \varphi(t)d^-W(t) = \int_0^T \varphi(t)\delta W(t) + \int_0^T D_{t+}\varphi(t)dt.$$

Proof For all $\varepsilon > 0$ we have

$$\lim_{\varepsilon \to 0^+} \int_0^T \varphi(t)\frac{W(t+\varepsilon) - W(t)}{\varepsilon}dt = \lim_{\varepsilon \to 0^+} \int_0^T \varphi(t)\frac{1}{\varepsilon}\int_t^{t+\varepsilon} 1\, dW(s)dt$$

$$= \lim_{\varepsilon \to 0^+} \left[\int_0^T \varphi(t) \diamond \frac{1}{\varepsilon}\int_t^{t+\varepsilon} 1\, dW(s)dt + \int_0^T \frac{1}{\varepsilon}\int_t^{t+\varepsilon} D_s\varphi(t)dsdt \right]$$

$$= \lim_{\varepsilon \to 0^+} \int_0^T \left(\frac{1}{\varepsilon}\int_{s-\varepsilon}^s \varphi(t)dt \right) \diamond \dot{W}(s)ds + \int_0^T D_{t+}\varphi(t)dt,$$

with convergence in $L^2(P)$ (see Chaps. 5 and 6 for the definition of Wick product, white noise, and the related results). Define

$$\varphi_\varepsilon(s) := \frac{1}{\varepsilon}\int_{s-\varepsilon}^s \varphi(t)dt, \quad s \in [0, T] \quad (\varepsilon > 0).$$

Then

$$\varphi_\varepsilon(s) \longrightarrow \varphi(s), \quad \varepsilon \to 0^+, \quad P\text{-a.s. for all } s.$$

Moreover,

$$E\left[\int_0^T \varphi_\varepsilon^2(s)ds\right] \le E\left[\int_0^T \varphi^2(s)ds\right], \quad \text{for all } \varepsilon > 0.$$

Therefore,

$$\varphi_\varepsilon \longrightarrow \varphi, \quad \varepsilon \to 0^+, \quad \text{in } L^2(P \times \lambda).$$

Moreover, by Definition 8.17, we have

$$E\left[\int_0^T \int_0^T (D_u\varphi_\varepsilon(s) - D_u\varphi(s))^2 dsdu\right]$$

$$\le E\left[\int_0^T \int_0^T \frac{1}{\varepsilon}\int_{s-\varepsilon}^s (D_u\varphi(t) - D_u\varphi(s))^2 dtdsdu\right] \longrightarrow 0, \quad \varepsilon \to 0^+.$$

Applying Theorem 6.17, we obtain

$$\lim_{\varepsilon \to 0^+} \int_0^T \varphi_\varepsilon(s) \diamond \dot{W}(s)ds = \lim_{\varepsilon \to 0^+} \int_0^T \varphi_\varepsilon(s)\delta W(s) = \int_0^T \varphi(s)\delta W(s),$$

with convergence in $L^2(P)$. This completes the proof. $\square$

Corollary 8.19. *Let $\varphi \in \mathbb{D}_0$. Then*

$$E\left[\int_0^T \varphi(t) d^- W(t)\right] = E\left[\int_0^T D_{t^+}\varphi(t) dt\right].$$

Proof Recall that the expectation of the Skorohod integral is 0, cf. (2.6). Then the result follows directly from Theorem 8.18.   □

8.5 Itô Formula for Skorohod Integrals

We now combine the Itô formula for forward integrals with Theorem 8.14 to obtain a version of the Itô formula for Skorohod integrals. See [170] and also [5, 9].

Theorem 8.20. *Let $X = X(t)$, $t \in [0,T]$, be a Skorohod process, that is, a process of the form*

$$X(t) = x + \int_0^t \alpha(s) ds + \int_0^t \beta(s) \delta W(s), \qquad (8.21)$$

with $x \in \mathbb{R}$ constant and the stochastic processes α and β satisfy: $\beta \in \mathbb{D}_0$ and $\int_0^t |\alpha(s)| ds < \infty$ P-a.s. for all $t \in [0,T]$ $(T > 0)$. A short-hand notation of (8.21) is

$$\delta X(t) = \alpha(t) dt + \beta(t) \delta W(t), \qquad X(0) = x. \qquad (8.22)$$

Let $f \in C^{1,2}([0,T] \times \mathbb{R})$ and define

$$Y(t) := f(t, X(t)), \quad t \geq 0.$$

Suppose $D_{t^+}X(t)$, $t \in [0,T]$, exists in $L^2(P \times \lambda)$ and that the stochastic process $\beta(t)\frac{\partial f}{\partial x}(t, X(t))$, $t \geq 0$, is Skorohod integrable. Then $Y = Y(t)$, $t \in [0,T]$, is also a Skorohod process and

$$\delta Y(t) = \left[\frac{\partial f}{\partial t}(t, X(t)) + \frac{1}{2}\frac{\partial^2 f}{\partial x^2}(t, X(t))\beta^2(t) + \frac{\partial^2 f}{\partial x^2}(t, X(t))\beta(t)D_{t^+}X(t)\right] dt$$
$$\qquad (8.23)$$
$$+ \frac{\partial f}{\partial x}(t, X(t))\delta X(t).$$

Proof Using Theorem 8.14, we can write

$$X(t) = x + \int_0^t [\alpha(s) - D_{s^+}\beta(s)] ds + \int_0^t \beta(s) d^- W(s).$$

By the Itô formula for forward integrals we have

$$d^-Y(t) = \frac{\partial f}{\partial t}(t, X(t))dt + \frac{\partial f}{\partial x}(t, X(t))\big[(\alpha(t) - D_{t+}\beta(t))dt + \beta(t)d^-W(t)\big]$$
$$+ \frac{1}{2}\frac{\partial^2 f}{\partial x^2}(t, X(t))\beta^2(t)dt.$$

Hence, again by Theorem 8.14,

$$\delta Y(t) = \frac{\partial f}{\partial t}(t, X(t))dt + \frac{\partial f}{\partial x}(t, X(t))\big[(\alpha(t) - D_{t+}\beta(t))dt + \frac{1}{2}\frac{\partial^2 f}{\partial x^2}(t, X(t))\beta^2(t)dt$$
$$+ \frac{\partial f}{\partial x}(t, X(t))\beta(t)\delta W(t) + D_{t+}\Big(\frac{\partial f}{\partial x}(t, X(t))\beta(t)\Big)dt$$
$$= \frac{\partial f}{\partial t}(t, X(t))dt + \frac{\partial f}{\partial x}(t, X(t))\big[\alpha(t)dt + \beta(t)\delta W(t)\big]$$
$$+ \frac{1}{2}\frac{\partial^2 f}{\partial x^2}(t, X(t))\beta^2(t)dt + \frac{\partial^2 f}{\partial x^2}(t, X(t))\beta(t)D_{t+}X(t)dt. \quad \square$$

Example 8.21. Consider the following Skorohod stochastic differential equation

$$\delta X(t) = \alpha(t)X(t)dt + \beta(t)X(t)\delta W(t),$$
$$X(0) = x > 0. \tag{8.24}$$

Let us try a solution of the form

$$\tilde{X}(t) = x\exp\Big\{\int_0^t \big[\alpha(s) - \frac{1}{2}\beta^2(s)\big]ds + \int_0^t \beta(s)\delta W(s)\Big\}, \quad t \in [0,T]. \tag{8.25}$$

By the Itô formula for Skorohod integrals, we get

$$\delta\tilde{X}(t) = \tilde{X}(t)\delta\tilde{X}(t) + \frac{1}{2}\tilde{X}(t)\beta^2(t)dt + \tilde{X}(t)\beta(t)D_{t+}\tilde{X}(t)dt$$
$$= \alpha(t)\tilde{X}(t)dt + \beta(t)\tilde{X}(t)\delta W(t) + \beta(t)\tilde{X}(t)D_{t+}\tilde{X}(t).$$

Hence $\tilde{X}(t)$, $t \in [0,T]$, is a solution of (8.24) if and only if

$$\beta(t)D_{t+}\tilde{X}(t) = 0 \quad \text{for all } t.$$

If α and β are $\mathbb{F}$-adapted, the aforementioned condition holds. But nothing can be said in general.

If β is *deterministic*, we can solve (8.24) by using white noise analysis, see Problem 5.8. In this case we get that the solution is

$$X(t) = x\exp^\diamond\Big\{\int_0^t \alpha(s)ds + \int_0^t \beta(s)dW(s)\Big\}, \quad t \in [0,T]. \tag{8.26}$$

If, in addition, α is $\mathbb{F}$-adapted, then $\tilde{X}(t) = X(t)$, $t \in [0,T]$.

8.6 Application to Insider Trading Modeling

8.6.1 Markets with No Friction

Let us return to the financial market in Sect. 8.1, but this time with more general coefficients, that is,

$$dS_0(t) = \rho(t)S_0(t)dt, \qquad S_0(0) = 1$$
$$dS_1(t) = S_1(t)\big[\mu(t)dt + \sigma(t)dW(t)\big], \quad S_1(0) > 0,$$

where ρ, μ and $\sigma \neq 0$ are $\mathbb{F}$-adapted stochastic processes such that

$$\int_0^T \Big[|\rho(t)| + |\mu(t)| + \sigma^2(t)\Big] dt < \infty, \quad P\text{-}a.s.$$

Suppose the trader is an *insider*, namely that the trader has access to an information flow represented by a filtration $\mathbb{G} := \{\mathcal{G}_t \subseteq \mathcal{F}, t \in [0,T]\}$ such that

$$\mathcal{F}_t \subseteq \mathcal{G}_t, \quad \text{for all } t \in [0,T].$$

Then the portfolio $\pi := \pi(t)$, $t \in [0,T]$, that the insider could select is allowed to be $\mathbb{G}$-adapted. Motivated by Sect. 8.1, we assume that the wealth process $X(t) = X_\pi(t)$, $t \in [0,T]$, corresponding to π is given by the forward equation

$$d^-X(t) = X(t)\Big[\big(\rho(t) + (\mu(t) - \rho(t))\pi(t)\big)dt + \pi(t)\sigma(t)d^-W(t)\Big] \tag{8.27}$$
$$X(0) = x > 0.$$

By the Itô formula for forward integrals (cf. Theorem 8.8), the solution $X(t) = X^x(t)$, $t \in [0,T]$, of this equation is

$$X(t) = x \exp\Big\{\int_0^t \big[\rho(t) + (\mu(t) - \rho(t))\pi(t) - \frac{1}{2}\sigma^2(t)\pi^2(t)\big]dt + \int_0^t \sigma(t)\pi(t)d^-W(t)\Big\}. \tag{8.28}$$

Remark 8.22. In this model the insider is a "small" investor in the sense that he is a price taker and his actions on the market do not have direct impact on the price dynamics. Equilibrium models and price formation are discussed in [144, 11]. See also [1, 104, 112, 157].

Let

$$U : [0, \infty) \longrightarrow [-\infty, \infty)$$

be a given utility function, which is here assumed to be nondecreasing, concave, lower semi-continuous on $[0, \infty)$ and in $C^1(0, \infty)$. We consider the following problem:

• Find $\pi^* \in \mathcal{A}_\mathbb{G}$ such that

$$\sup_{\pi \in \mathcal{A}_\mathbb{G}} E\big[U(X_\pi^x(T))\big] = E\big[U(X_{\pi^*}^x(T))\big]. \tag{8.29}$$

Here $\mathcal{A}_{\mathbb{G}}$ is the family of *admissible* controls defined as follows.

Definition 8.23. *The set $\mathcal{A}_{\mathbb{G}}$ of admissible portfolios consists of all processes $\pi = \pi(t), t \in [0, T]$, such that*

(1) π is càglàd and $\mathbb{G}$-adapted

(2) $E\left[\int_0^T \left[|\mu(t) - \rho(t)||\pi(t)| + \sigma^2(t)\pi^2(t) \right] dt \right] < \infty$

(3) the product $\pi\sigma$ is càglàd and forward integrable with respect to W

(4) $0 < E\left[U'(X_\pi(T))X_\pi(T) \right] < \infty$. Here $U'(x) = \frac{d}{dx}U(x)$.

Note that the buy–hold–sell portfolios β:

$$\beta(s) = \alpha \chi_{(t,t+h]}(s), \quad s \in [0, T],$$

with $0 \leq t < t + h \leq T$ and α $\mathcal{G}_t$-measurable, belong to $\mathcal{A}_{\mathbb{G}}$.

We say that a portfolio $\pi \in \mathcal{A}_{\mathbb{G}}$ is a *local maximum* for the problem (8.29) if

$$E\left[U(X_{\pi+y\beta}(T)) \right] \leq E\left[U(X_\pi(T)) \right]$$

for all bounded $\beta \in \mathcal{A}_{\mathbb{G}}$ and all $y \in (-\delta, \delta)$ with $\delta > 0$ as in (5) above.

We now use a variational method to find a necessary and sufficient condition for $\pi \in \mathcal{A}_{\mathbb{G}}$ to be a local maximum.

First assume that π is a local maximum. Assume also that, for all $\pi, \beta \in \mathcal{A}_{\mathbb{G}}$ with β bounded, there exists $\delta > 0$ such that $\pi + y\beta \in \mathcal{A}_{\mathbb{G}}$ and for all $y \in (-\delta, \delta)$ the family

$$\left\{ U'(X_{\pi+y\beta}(T))X_{\pi+y\beta}(T) |M_{\pi+y\beta}(T)| \right\}_{y \in (-\delta, \delta)}$$

is uniformly integrable, where

$$M_\pi(t) := \int_0^t \left[\mu(s) - \rho(s) - \sigma^2(s)\pi(s) \right] ds + \int_0^t \sigma(s)dW(s), \quad t \in [0, T].$$

Then for all bounded $\beta \in \mathcal{A}_{\mathbb{G}}$ we have

$$0 = \frac{d}{dy} E\left[U(X_{\pi+y\beta}(T)) \right]_{|y=0}$$

$$= E\left[U'(X_\pi(T))X_\pi(T) \left\{ \int_0^T \beta(s) \left[\mu(s) - \rho(s) - \sigma^2(s)\pi(s) \right] ds \right. \right. \tag{8.30}$$

$$\left. \left. + \int_0^T \beta(s)\sigma(s)d^-W(s) \right\} \right].$$

Now fix t, h: $0 \leq t < t + h \leq T$ and choose

$$\beta(s) = \alpha \chi_{(t,t+h]}(s), \quad s \in [0, T],$$

where α is an arbitrary bounded $\mathcal{G}_t$-measurable random variable. Then (8.30) gives

$$E\Big[U'(X_\pi(T))X_\pi(T)\Big\{\int_t^{t+h}[\mu(s)-\rho(s)-\sigma^2(s)\pi(s)]ds$$

$$+\int_t^{t+h}\sigma(s)d^-W(s)\Big\}\alpha\Big]=0. \tag{8.31}$$

Since this holds for all α, we conclude that

$$E\Big[F_\pi(T)\big(M_\pi(t+h)-M_\pi(t)\big)|\mathcal{G}_t\Big]=0, \tag{8.32}$$

where

$$F_\pi(T)=\frac{U'(X_\pi(T))X_\pi(T)}{E[U'(X_\pi(T))X_\pi(T)]} \tag{8.33}$$

and

$$M_\pi(t):=\int_0^t[\mu(s)-\rho(s)-\sigma^2(s)\pi(s)]ds+\int_0^t\sigma(s)dW(s),\quad t\in[0,T].$$

Define the probability measure Q_π on $\mathcal{F}_T$ by

$$dQ_\pi=F_\pi(T)dP \tag{8.34}$$

and let $E_{Q_\pi}[\cdot]$ denote the expectation with respect to Q_π. Then (8.32) can be written as

$$E_{Q_\pi}\big[M_\pi(t+h)-M_\pi(t)|\mathcal{G}_t\big]=0,$$

which states that $M_\pi(t)$, $t\in[0,T]$, is a $(\mathbb{G},Q_\pi)$-martingale, that is, a martingale with respect to the filtration $\mathbb{G}$ under the measure Q_π.

This argument can be reversed as follows. Suppose M_π is a $(\mathbb{G},Q_\pi)$-martingale. Then

$$E_{Q_\pi}\big[M_\pi(t+h)-M(t)|\mathcal{G}_t\big]=0,$$

for all t,h such that $0\le t<t+h\le T$. Equivalently,

$$E_{Q_\pi}\big[(M_\pi(t+h)-M(t))\alpha\big]=0,$$

for all bounded $\mathcal{G}_t$-measurable α. Thus (8.31) holds. Taking linear combinations we see that (8.30) holds for all càglàd step processes $\beta\in\mathcal{A}_\mathbb{G}$. By our assumption (1) and Lemma 8.4, we get that (8.30) holds for all bounded $\beta\in\mathcal{A}_\mathbb{G}$. Consider the function

$$g(y):y\longrightarrow E\big[U(X_{\pi+y\beta}(T))\big],\quad y\in(-\delta,\delta).$$

For π to give a maximum we need that $g'(0)=0$ and $g''(0)<0$. A sufficient condition for this is

$$xU''(x)+U'(x)\le 0. \tag{8.35}$$

Hence, under the above assumptions, we have the following result, which was first proved in [33].

Theorem 8.24. *A stochastic process* $\pi \in \mathcal{A}_{\mathbb{G}}$ *is a local maximum for the problem* (8.29) *only if the stochastic process*

$$M_\pi(t) = \int_0^t \left[\mu(s) - \rho(s) - \sigma^2(s)\pi(s)\right] ds + \int_0^t \sigma(s) dW(s), \quad t \in [0,T], \quad (8.36)$$

is a $(\mathbb{G}, Q_\pi)$-*martingale. Conversely, suppose* (8.35)–(8.36) *hold. Then* π *is a local maximum.*

Using the Girsanov theorem this result can be rewritten in the following way.

Theorem 8.25. *A stochastic process* $\pi \in \mathcal{A}_{\mathbb{G}}$ *is a local maximum for the problem* (8.29) *only if the stochastic process*

$$\hat{M}_\pi(t) := M_\pi(t) - \int_0^t \frac{d[M_\pi, Z](s)}{Z(s)}, \quad t \in [0,T], \tag{8.37}$$

is a $(\mathbb{G}, P)$-*martingale. Here*

$$Z(t) := E_{Q_\pi}\left[\frac{dP}{dQ_\pi} | \mathcal{G}_t\right] = \left(E[F_\pi(T)|\mathcal{G}_t]\right)^{-1}, \quad t \in [0,T]. \tag{8.38}$$

Conversely, if (8.35) *and* (8.37) *hold, then* π *is a local maximum.*

Proof If $\pi \in \mathcal{A}_{\mathbb{G}}$ is a local maximum, then by Theorem 8.24 the process M_π is a $(\mathbb{G}, Q_\pi)$-martingale. By the Girsanov theorem (see e.g. [204, Theorem III.39]) we get that

$$\hat{M}_\pi(t) := M_\pi(t) - \int_0^t \frac{d[M_\pi, Z](s)}{Z(s)}, \quad t \in [0,T],$$

is a $(\mathbb{G}, P)$-martingale, with

$$Z(t) := E_{Q_\pi}\left[\frac{dP}{dQ_\pi} | \mathcal{G}_t\right] = E\left[(F_\pi(T))^{-1} \frac{F_\pi(T)}{E[F_\pi(T)|\mathcal{G}_t]} | \mathcal{G}_t\right] = \left(E[F_\pi(T)|\mathcal{G}_t]\right)^{-1}, \quad t \in [0,T].$$

Conversely, if $\hat{M}_\pi$ is a $(\mathbb{G}, P)$-martingale, then M_π is a $(\mathbb{G}, Q_\pi)$-martingale and hence π is a local maximum by Theorem 8.24. □

The main feature of Theorem 8.25 is that it gives an explicit connection between a local maximum π and the semimartingale decomposition of W with respect to $\mathbb{G}$. The following result was first proved in [33].

Theorem 8.26. (1) *Suppose* π *is a local maximum for* (8.29). *Then* W *is a semimartingale with respect to* $\mathbb{G}$ *and its semimartingale decomposition is*

$$dW(t) = d\hat{W}(t) + \left[\sigma(t)\pi(t) - \frac{\mu(t) - \rho(t)}{\sigma(t)}\right] dt + \frac{d[M_\pi, Z](t)}{\sigma(t)Z(t)}, \tag{8.39}$$

where $\hat{W}$ *is a* $(\mathbb{G}, P)$-*Brownian motion.*

(2) Conversely, suppose (8.35) holds and W is a semimartingale with respect to $(\mathbb{G}, P)$ *with decomposition*

$$dW(t) = dB(t) + dA(t),$$

where B is a $(\mathbb{G}, P)$-*Brownian motion and A is a* $\mathbb{G}$-*adapted finite variation process. Suppose*

$$dA(t) = \alpha(t)dt \tag{8.40}$$

for some $\mathbb{G}$-*adapted process* α *(i.e.,* $dA(t)$ *is absolutely continuous with respect to dt) and there exists a solution* $\pi \in \mathcal{A}_{\mathbb{G}}$ *of the equation*

$$\sigma(t)\pi(t) + \frac{1}{\sigma(t)Z(t)} \frac{d[M_\pi, Z](t)}{dt} = \alpha(t) + \frac{\mu(t) - \rho(t)}{\sigma(t)}. \tag{8.41}$$

Then π *is a local maximum for (8.29).*

Remark 8.27. Note that since the quadratic variation of $\hat{M}_\pi$ is absolutely continuous, in fact,

$$[M_\pi, M_\pi](t) = \int_0^t \sigma^2(s)ds$$

(see (8.37)), it follows that $d[M_\pi, Z](t)$ is also absolutely continuous with respect to dt. Therefore,

$$\frac{d[M_\pi, Z](t)}{\sigma(t)Z(t)} = \frac{1}{\sigma(t)Z(t)} \frac{d}{dt}[M_\pi, Z](t)\, dt. \tag{8.42}$$

Proof (1) If π is a local maximum, then by Theorem 8.25 the process

$$\sigma^{-1}(t)d\hat{M}_\pi(t) = dW(t) + \sigma^{-1}(t)\Big[\big(\mu(t) - \rho(t) - \sigma^2(t)\pi(t)\big)dt - \frac{d[M_\pi, Z](t)}{Z(t)}\Big],\ t \in [0, T],$$

is a $(\mathbb{G}, P)$-martingale. Since the quadratic variation of the process

$$\sigma^{-1}(t)d\hat{M}_\pi(t), \quad t \in [0, T],$$

is t, $t \in [0, T]$, then we have that

$$d\hat{W}(t) := \sigma^{-1}(t)d\hat{M}_\pi(t), \quad t \in [0, T],$$

is a $(\mathbb{G}, P)$-Brownian motion and (8.39) follows.
(2) Suppose W is a $(\mathbb{G}, P)$-semimartingale with decomposition

$$dW(t) = dB(t) + dA(t)$$

as in (2). Let π be as in (8.41). Then

$$\sigma^{-1}(t)d\hat{M}_\pi(t) = dW(t) + \sigma^{-1}(t)\Big[\big(\mu(t) - \rho(t) - \sigma^2(t)\pi(t)\big)dt - \frac{d[M_\pi, Z](t)}{Z(t)}\Big]$$

$$= dW(t) - dA(t) = dB(t).$$

Hence $\sigma^{-1}(t)d\hat{M}(t)$, $t \in [0, T]$, is a $(\mathbb{G}, P)$-Brownian motion. Therefore, $d\hat{M}_\pi(t)$, $t \in [0, T]$, is a $(\mathbb{G}, P)$-martingale and it follows that π is a local maximum by Theorem 8.25. $\square$

Example 8.28. Let us assume, as in [191], that

$$\mathcal{G}_t = \mathcal{F}_t \vee \sigma(W(T_0)) \quad \text{for some } T_0 > T.$$

Then, by Theorem 8.1, W is a semimartingale with respect to $(\mathbb{G}, P)$, with decomposition

$$dW(t) = dB(t) + \frac{W(T_0) - W(t)}{T_0 - t} dt, \quad t \in [0, T].$$

Suppose $\pi \in \mathcal{A}_{\mathbb{G}}$ is such that (8.41) holds, that is,

$$\sigma(t)\pi(t) + \frac{1}{\sigma(t)Z(t)} \frac{d[M_\pi, Z](t)}{dt} = \frac{W(T_0) - W(t)}{T_0 - t} + \frac{\mu(t) - \rho(t)}{\sigma(t)}. \tag{8.43}$$

Then π is a local maximum for (8.29). In particular, if we choose

$$U(x) = \log x, \quad x > 0,$$

which is the case studied in [191], then we have

$$F_\pi(T) = 1, \quad Q_\pi = P, \quad Z \equiv 1, \quad \text{and } [M_\pi, Z] \equiv 0.$$

This gives

$$\pi(t) = \pi^*(t) = \frac{W(T_0) - W(t)}{\sigma(t)(T_0 - t)} + \frac{\mu(t) - \rho(t)}{\sigma^2(t)}, \quad t \in [0, T], \tag{8.44}$$

which is the solution of the problem (8.29) as given in [191].

8.6.2 Markets with Friction

The market model discussed so far assumes that the prices are not affected by the actions of the insider and that there are no constraints on the sizes and fluctuations of the insider's portfolio. This may be a reasonable assumption in some cases, but clearly not realistic in other situations. For example, let us go back to Example 8.28 where the utility is logarithmic, that is, $U(x) = \log x$, $x \geq 0$, and let us consider $T_0 = T$, then the optimal portfolio π^* for an insider, given by (8.44), will involve more and more frequent and violent fluctuations as the time approaches T. Such a portfolio would necessarily influence the prices on the market.

We now present a market model with friction where the insider is penalized for large volumes and/or rapid fluctuations in the trade. The presentation of this section is based on [111].

Assume that the market consists of the two investment possibilities:

(1) A bond, with price given by

$$dS_0(t) = \rho(t)S_0(t)dt; \quad S_0(0) = 1; \quad 0 \leq t \leq T.$$

(2) A stock, with price given by

$$dS_1(t) = S_1(t) \left[\mu(t)dt + \sigma(t)dW(t)\right] ; \quad 0 \leq t \leq T,$$

where $T > 0$ is constant and $\rho(t)$, $\mu(t)$, and $\sigma(t)$ are given $\mathcal{F}_t$-adapted processes. We assume that

$$E \left[\int_0^T \left\{|\mu(t)| + |\rho(t)| + \sigma^2(t)\right\} dt\right] < \infty$$

$$\sigma(t) \neq 0 \quad \text{for a.a.} \quad (\omega, t) \in \Omega \times [0, T].$$

Let $\mathcal{G}_t \supset \mathcal{F}_t$ be the information filtration available to the insider and let $\pi(t)$ be the portfolio chosen by the insider, measured in terms of the fraction of the total wealth $X(t) = X_\pi(t)$ invested in the stock at time $t \in [0, T]$. Then the corresponding wealth $X(t) = X_\pi(t)$ at time t is modeled by the forward differential equation

$$dX(t) = (1 - \pi(t))X(t)\rho(t)dt + \pi(t)X(t) \left[\mu(t)dt + \sigma(t)d^-W(t)\right]$$
$$= X(t) \left[[\rho(t) + (\mu(t) - \rho(t))\pi(t)]dt + \sigma(t)\pi(t)d^-W(t)\right] . \quad (8.45)$$

For simplicity we assume $X(0) = x = 1$.

We now specify the set $\mathcal{A}_{\mathbb{G},\mathcal{Q}}$ of the admissible portfolios π as follows.

Definition 8.29. *In the sequel we let $\mathcal{A}_{\mathbb{G},\mathcal{Q}}$ denote a linear space of stochastic processes $\pi = \pi(t)$, $t \in [0, T]$, such that (8.46)–(8.49) hold, where*

$\triangleright$ *$\pi(t)$ is $\mathbb{G}$-adapted and the σ-algebra generated by $\{\pi(t) ; \pi \in \mathcal{A}_{\mathbb{G},\mathcal{Q}}\}$*

    is equal to $\mathcal{G}_t$, for all $t \in [0, T]$,   (8.46)

$\triangleright$ *π belongs to the domain of $\mathcal{Q}$ (see later),*   (8.47)

$\triangleright$ *$\sigma(t)\pi(t)$ is forward integrable,*   (8.48)

$\triangleright$ $E \left[\int_0^T |\mathcal{Q}\pi(t)|^2 dt\right] < \infty.$ (8.49)

Here $\mathcal{Q} : \mathcal{A}_{\mathbb{G},\mathcal{Q}} \to \mathcal{A}_{\mathbb{G},\mathcal{Q}}$ is some linear operator measuring the size and/or the fluctuations of the portfolio. For example, we could have

$$\mathcal{Q}\pi(s) = \lambda_1(s)\pi(s), \quad (8.50)$$

where $\lambda_1(s) \geq 0$ is some given weight function. This models the situation where the insider is penalized for large volumes of trade. An alternative choice of $\mathcal{Q}$ would be

$$\mathcal{Q}\pi(s) = \lambda_2(s)\pi'(s), \quad (8.51)$$

for some weight function $\lambda_2(s) \geq 0$. $(\pi'(s) = \dfrac{d}{ds}\pi(s).)$ In this case the insider is penalized for large trade fluctuations. Other choices of $\mathcal{Q}$ are also possible, including combinations of (8.50) and (8.51).

With these definitions we can now specify our problem as follows:

- Find Φ and $\pi^* \in \mathcal{A}_{\mathbb{G},\mathcal{Q}}$ such that

$$\Phi = \sup_{\pi \in \mathcal{A}_{\mathbb{G},\mathcal{Q}}} J(\pi) = J(\pi^*), \tag{8.52}$$

where

$$J(\pi) = E\left[\log(X_\pi(T)) - \frac{1}{2}\int_0^T |\mathcal{Q}\pi(s)|^2 ds\right],$$

$\mathcal{Q} : \mathcal{A}_{\mathbb{G},\mathcal{Q}} \to \mathcal{A}_{\mathbb{G},\mathcal{Q}}$ being a given linear operator as mentioned earlier. We call Φ the *value* of the insider and $\pi^* \in \mathcal{A}_{\mathbb{G},\mathcal{Q}}$ an optimal portfolio (if it exists).

We now proceed to solve this problem above. Using the Itô formula (see Theorem 8.12) we get that the solution of (8.45) is

$$X(t) = \exp\left(\int_0^t \left\{\rho(s) + (\mu(s) - \rho(s))\pi(s) - \frac{1}{2}\sigma^2(s)\pi^2(s)\right\} ds \right.$$
$$\left. + \int_0^t \sigma(s)\pi(s)d^-W(s)\right).$$

Therefore, we get

$$J(\pi) = E\left[\int_0^T \left\{\rho(t) + (\mu(t) - \rho(t))\pi(t) - \frac{1}{2}\sigma^2(t)\pi^2(t)\right\} dt \right.$$
$$\left. + \int_0^T \sigma(t)\pi(t)d^-W(t) - \frac{1}{2}\int_0^T |\mathcal{Q}\pi(t)|^2 dt\right]. \tag{8.53}$$

To maximize $J(\pi)$ we use a perturbation argument as follows. Suppose an optimal insider portfolio $\pi = \pi^*$ exists (in the following we omit the *). Let $\theta \in \mathcal{A}_{\mathbb{G},\mathcal{Q}}$ be another portfolio. Then the function

$$f(y) := J(\pi + y\theta), \quad y \in \mathbb{R},$$

is maximal for $y = 0$ and hence

$$0 = f'(0) = \frac{d}{dy}\left[J(\pi + y\theta)\right]_{y=0}$$

$$= E\left[\int_0^T \left\{(\mu(t) - \rho(t))\theta(t) - \sigma^2(t)\pi(t)\theta(t)\right\} dt \right.$$
$$\left. + \int_0^T \sigma(t)\theta(t)d^-W(t) - \int_0^T \mathcal{Q}\pi(t)\mathcal{Q}\theta(t)dt\right]. \tag{8.54}$$

Let Q^* denote the adjoint operator of Q in the Hilbert space $L^2(P \times \lambda)$, namely,

$$E\left[\int_0^T \alpha(t)(Q\beta)(t)dt\right] = E\left[\int_0^T (Q^*\alpha)(t)\beta(t)dt\right]$$

for all α and β in $\mathcal{A}_{\mathbb{G},Q}$. Then we can rewrite (8.54) as

$$E\left[\int_0^T \{\mu(t) - \rho(t) - \sigma^2(t)\pi(t) - Q^*Q\pi(t)\}\theta(t)dt + \int_0^T \sigma(t)\theta(t)d^-W(t)\right] = 0.$$

(8.55)

Now we apply this to a special choice of θ. Fix $t \in [0,T]$ and $h > 0$ such that $t + h < T$ and choose

$$\theta(s) = \theta_0(t)\chi_{(t,t+h]}(s); \quad s \in [0,T],$$

where $\theta_0(t)$ is $\mathcal{G}_t$-measurable. Then by Lemma 8.7 we have

$$E\left[\int_0^T \sigma(s)\theta(s)d^-W(s)\right] = E\left[\int_t^{t+h} \sigma(s)\theta_0(t)d^-W(s)\right]$$

$$= E\left[\theta_0(t)\int_t^{t+h} \sigma(s)dW(s)\right].$$

Combining this with (8.55) we get

$$E\left[\left(\int_t^{t+h}\{\mu(s) - \rho(s) - \sigma^2(s)\pi(s) - Q^*Q\pi(s)\}ds + \int_t^{t+h}\sigma(s)dW(s)\right)\theta_0(t)\right] = 0.$$

Since this holds for all such $\theta_0(t)$, we conclude that

$$E\left[M(t+h) - M(t)|\mathcal{G}_t\right] = 0,$$

where

$$M(t) := \int_0^t \{\mu(s) - \rho(s) - \sigma^2(s)\pi(s) - E\left[Q^*Q\pi(s)|\mathcal{G}_s\right]\}ds + \int_0^t \sigma(s)dW(s).$$

(8.56)

Since $\sigma \neq 0$ this proves the result stated further.

Theorem 8.30. *Suppose an optimal insider portfolio* $\pi \in \mathcal{A}_{\mathbb{G},Q}$ *for problem* (8.52) *exists. Then*

$$dW(t) = d\hat{W}(t) - \frac{1}{\sigma(t)}\{\mu(t) - \rho(t) - \sigma^2(t)\pi(t) - E\left[Q^*Q\pi(t)|\mathcal{G}_t\right]\}dt,$$

where $\hat{W}(t) := \int_0^t \sigma^{-1}(s)dM(s)$ *is a Brownian motion with respect to* $\mathbb{G}$. *In particular,* $W(t)$, $t \in [0,T]$, *is a semimartingale with respect to* $\mathbb{G}$.

We now use this to find an equation for an optimal portfolio π.

Theorem 8.31. *Assume that there exists a process $\gamma_t(\omega, s)$, $\omega \in \Omega$, $s \in [0, T]$, such that $\gamma_t(s)$ is $\mathcal{G}_t$-measurable for all $s \leq t$ and*

$$t \longmapsto \int_0^t \gamma_t(\omega, s) ds \quad \text{is of finite variation } P\text{-a.s.}$$

and

$$N(t) := W(t) - \int_0^t \gamma_t(s) ds, \quad t \in [0, T], \tag{8.57}$$

is a martingale with respect to $\mathbb{G}$. Assume that $\pi \in \mathcal{A}_{\mathbb{G}, \mathcal{Q}}$ is optimal. Then

$$\sigma^2(t)\pi(t) + E\left[\mathcal{Q}^*\mathcal{Q}\pi(t)|\mathcal{G}_t\right] = \mu(t) - \rho(t) + \sigma(t)\frac{d}{dt}\left(\int_0^t \gamma_t(s) ds\right). \tag{8.58}$$

Proof By comparing (8.56) and (8.57) we get that

$$\sigma(t) dN(t) = dM(t),$$

that is,

$$-\sigma(t)\frac{d}{dt}\left(\int_0^t \gamma_t(s) ds\right) = \mu(t) - \rho(t) - \sigma^2(t)\pi(t) - E\left[\mathcal{Q}^*\mathcal{Q}\pi(t)|\mathcal{G}_t\right]. \quad \square$$

Next we turn to a partial converse of Theorem 8.31.

Theorem 8.32. *Suppose (8.57) holds. Let π be a process solving (8.58). Suppose $\pi \in \mathcal{A}_{\mathbb{G}, \mathcal{Q}}$. Then π is optimal for problem (8.52).*

Proof Substituting

$$dW(t) - dN(t) + \frac{d}{dt}\left(\int_0^t \gamma_t(s) ds\right) dt$$

and

$$\sigma(t)\pi(t) d^- W(t) = \sigma(t)\pi(t) dN(t) + \sigma(t)\pi(t)\frac{d}{dt}\left(\int_0^t \gamma_t(s) ds\right) dt$$

into (8.53), we get

$$J(\pi) = E\left[\int_0^T \left\{\rho(t) + (\mu(t) - \rho(t))\pi(t) - \frac{1}{2}\sigma^2(t)\pi^2(t)\right.\right.$$

$$\left.\left. + \sigma(t)\pi(t)\frac{d}{dt}\left(\int_0^t \gamma_t(s) ds\right) - \frac{1}{2}|\mathcal{Q}\pi(t)|^2\right\} dt\right]. \tag{8.59}$$

This is a concave functional of π, so if we can find $\pi = \pi^* \in \mathcal{A}_{G,\varrho}$ such that

$$\frac{d}{dy}[J(\pi^* + y\theta)]_{y=0} = 0 \quad \text{for all} \quad \theta \in \mathcal{A}_{G,\varrho},$$

then π^* is optimal. By a computation similar to the one leading to (8.55), we get

$$\frac{d}{dy}[J(\pi^* + y\theta)]_{y=0} = E\left[\int_0^T \left\{\mu(t) - \rho(t) - \sigma^2(t)\pi^*(t) \right.\right.$$
$$\left.\left. + \sigma(t)\frac{d}{dt}\int_0^t \gamma_t(s)ds - \mathcal{Q}^*\mathcal{Q}\pi(t)\right\}\theta(t)dt\right].$$

This is 0 if $\pi = \pi^*$ solves (8.58). $\square$

We apply this to some examples and give some further results in the situations presented.

Example 8.33. Choose
$$\mathcal{Q}\pi(t) = \lambda_1(t)\sigma(t)\pi(t), \tag{8.60}$$

where $\lambda_1(t) \geq 0$ is deterministic. Then (8.58) takes the form

$$\sigma^2(t)\pi(t) + \lambda_1^2(t)\sigma^2(t)\pi(t) = \mu(t) - \rho(t) + \sigma(t)\frac{d}{dt}\int_0^t \gamma_t(s)ds$$

or

$$\pi(t) = \pi^*(t) = \frac{\mu(t) - \rho(t) + \sigma(t)\frac{d}{dt}\int_0^t \gamma_t(s)ds}{\sigma^2(t)[1 + \lambda_1^2(t)]}. \tag{8.61}$$

Substituting this into the formula (8.59) for $J(\pi)$ we obtain the following result.

Theorem 8.34. *Suppose (8.57) and (8.60) hold. Let π^* be given by (8.61). If $\pi \in \mathcal{A}_{G,\varrho}$ then π^* is optimal for problem (8.52). Moreover, the insider value is*

$$\Phi = J(\pi^*) \tag{8.62}$$
$$= E\left[\int_0^T \left\{\rho(t) + \frac{1}{2}(1 + \lambda_1^2(t))^{-1}\left(\frac{\mu(t) - \rho(t)}{\sigma(t)} + \frac{d}{dt}\int_0^t \gamma_t(s)ds\right)^2\right\}dt\right].$$

In particular, if we consider the case mentioned in the introduction, where

$$\mathcal{G}_t = \mathcal{F}_t \vee \sigma(W(T_0)) \quad \text{for some } T_0 > T,$$

then, by Theorem 8.1,

$$\gamma_t(s) = \gamma(s) = \frac{W(T_0) - W(s)}{T_0 - s}$$

and (8.61) becomes

$$\pi^*(t) = \sigma^{-2}(t) \left[1 + \lambda_1^2(t)\right]^{-1} \left[\mu(t) - \rho(t) + \frac{\sigma(t)}{T_0 - t} \left(W(T_0) - W(t)\right)\right].$$

The corresponding value is, by (8.62),

$$J(\pi^*) = E\left[\int_0^T \left\{\rho(t) + \frac{1}{2}(1 + \lambda_1^2(t))^{-1} \left(\frac{\mu(t) - \rho(t)}{\sigma(t)} + \frac{W(T_0) - W(t)}{T_0 - t}\right)^2\right\} dt\right].$$

In particular, we see that if $\sigma(t) \geq \sigma_0 > 0$ and

$$\lambda_1(t) = (T_0 - t)^{-\beta} \quad \text{for some constant} \quad \beta > 0, \tag{8.63}$$

then

$$J(\pi^*) \leq C_1 + C_2 \int_0^T (T_0 - t)^{-1+2\beta} dt < \infty,$$

even if $T_0 = T$. Thus if we penalize large investments near $t = T_0$ according to (8.63), the insider gets a finite value even if $T_0 = T$.

Example 8.35. Next we put

$$Q\pi(t) = \pi'(t) \quad \left(= \frac{d}{dt}\pi(t)\right). \tag{8.64}$$

This means that the insider is being penalized for large portfolio fluctuations. Choose $\mathcal{A}_{\mathbb{G},Q}$ to be the set of all continuously differentiable processes π satisfying (8.46)–(8.49) and in addition

$$\pi(0) = \pi(T) = 0 \quad a.s. \tag{8.65}$$

For simplicity assume that

$$\sigma(t) \equiv 1.$$

Then (8.58) gets the form

$$\pi(t) - \pi''(t) = a(t),$$

where

$$a(t) = \mu(t) - \rho(t) + \frac{d}{dt}\left(\int_0^t \gamma_t(s)ds\right).$$

Using the variation of parameter method, we obtain the solution

$$\pi(t) = \int_0^t \sinh(t - s)a(s)ds + K\sinh(t), \tag{8.66}$$

where, as usual, $\sinh(x) = \frac{1}{2}(e^x - e^{-x})$, $x \in \mathbb{R}$, is the hyperbolic sinus function and the constant K is chosen such that $\pi(T) = 0$. In particular, if we again consider the case

$$\mathcal{G}_t = \mathcal{F}_t \vee \sigma(W(T_0)), \quad T_0 > T,$$

so that

$$\gamma_t(s) = \gamma(s) = \frac{W(T_0) - W(s)}{T_0 - s}, \quad 0 \le s \le T.$$

we obtain, by (8.66),

$$\pi(t) = \int_0^t \sinh(t - s) \left[\mu(s) - \rho(s) + \frac{W(T_0) - W(s)}{T_0 - s} \right] ds + K \sinh(t). \quad (8.67)$$

The corresponding value is by (8.59),

$$J(\pi) = E \left[\int_0^T \left\{ \rho(t) + (\mu(t) - \rho(t)) \, \pi(t) - \frac{1}{2} \pi^2(t) \right. \right.$$
$$\left. \left. + \pi(t) \frac{W(T_0) - W(t)}{T_0 - t} - \frac{1}{2} (\pi'(t))^2 \right\} dt \right].$$

Note that if $0 \le t \le T < T_0$ then

$$E \left[\pi(t) \frac{W(T_0) - W(t)}{T_0 - t} \right] \le E \left[\int_0^t \sinh(t - s) \frac{(W(T_0) - W(s)) \, (W(T_0) - W(t))}{(T_0 - s) \, (T_0 - t)} ds \right]$$
$$= \int_0^t \sinh(t - s) \frac{ds}{T_0 - s}.$$

Therefore

$$J(\pi) \le \int_0^T \left(\int_0^t \sinh(t - s) \frac{ds}{T_0 - s} \right) dt \le \int_0^T \frac{\cosh(T - s) - 1}{T - s} ds \quad \text{for all } T_0 > T.$$

Thus, we have proved:

Theorem 8.36. *Suppose $\mathcal{Q}\pi(t) = \pi'(t)$ and $\mathcal{A}_{\mathbb{G},\mathcal{Q}}$ is chosen as in (8.64), (8.65) and assume that $\sigma(t) = 1$. Then the optimal insider portfolio is given by (8.66). In particular, if we choose*

$$\mathcal{G}_t = \mathcal{F}_t \vee \sigma(W(T_0)) \quad \text{with } T_0 > T,$$

then the optimal portfolio π is given by (8.67) and the corresponding insider value $J(\pi)$ is uniformly bounded for $T_0 > T$.

Remark 8.37. Both Example 8.33 and Example 8.35 yield ways to penalize the insider investors so that he would not obtain infinite utility. In Example 8.33, $\lambda_1(t) = (T_0 - t)^{-\beta}$ for some $\beta > 0$. To use this penalization, one needs to know T_0. In Example 8.35, T_0 is not required to be known.

8.7 Exercises

Problem 8.1. Suppose $\mathcal{G}_t = \mathcal{F}_{t+\delta(t)}$, for some $\delta(t) > 0$, $t \in [0, T]$. Show that W is not a semimartingale with respect to $\mathbb{G} = \{\mathcal{G}_t, t \in [0, T]\}$.

Problem 8.2. (*) Compute the following forward integrals:

(a) $\int_0^T W(T)d^-W(t)$

(b) $\int_0^T W(t)[W(T) - W(t)]d^-W(t)$

(c) $\int_0^T \left(\int_0^T g(s)dW(s)\right)d^-W(t)$ for a given (deterministic) function $g \in L^2([0,T])$

(d) $\int_0^T W^2(t_0)d^-W(t)$, where $t_0 \in [0,T]$ is fixed

(e) $\int_0^T \exp\{W(T)\}d^-W(t)$.

Problem 8.3. Solve the following forward stochastic differential equations:

(a) $d^-X(t) = X(t)\mu(t)dt + \beta(t)d^-W(t)$, $t \in [0,T]$,
   $X(0) = F$;

(b) $d^-X(t) = \alpha(t)dt + X(t)\sigma(t)d^-W(t)$, $t \in [0,T]$,
   $X(0) = F$.

Here the coefficients μ, σ, α, and β are given stochastic processes, not necessarily $\mathbb{F}$-adapted, and F is a given $\mathcal{F}_T$-measurable random variable. We assume that σ and β are forward integrable and that

$$\int_0^T \left[|\mu(t)| + |\alpha(t)| + \sigma^2(t) + \beta^2(t)\right]dt < \infty \quad P - a.s.$$

Problem 8.4. Use Theorem 8.18 and the basic rules for forward integration to find the Skorohod integrals in Problem 2.4.

Problem 8.5. (a) Let μ, σ, and α be deterministic functions such that

$$\int_0^T \left[|\mu(t)| + \sigma^2(t) + \alpha^2(t)\right]dt < \infty \quad P\text{-}a.s.$$

and define

$$F = \exp\left\{\int_0^T \alpha(s)dW(s)\right\}.$$

Use Theorem 8.18 to show that the Skorohod stochastic differential equation

$$\begin{cases} \delta X(t) = \mu(t)X(t)dt + \sigma(t)X(t)\delta W(t), & t \in [0,T] \\ X(0) = F \end{cases} \tag{8.68}$$

is equivalent to the forward stochastic differential equation

$$\begin{cases} d^-X(t) = \left[\mu(t)X(t) - \sigma D_{t^+}X(t)\right]dt + \sigma(t)X(t)d^-W(t), & t \in [0,T] \\ X(0) = F. \end{cases}$$

$$\tag{8.69}$$

(b) Using Problem 5.8, the solution of (8.68) is

$$X(t) = F \diamond \exp\left\{ \int_0^t \sigma(s)dW(s) + \int_0^t [\mu(s) - \frac{1}{2}\sigma^2(s)]ds \right\}. \qquad (8.70)$$

With this expression for the stochastic process $X(t)$, $t \in [0, T]$, show that

$$D_{t+}X(t) = \alpha(t)X(t).$$

(c) Substitute this into (8.69) and solve the corresponding forward stochastic differential equation to get

$$X(t) = F \exp\left\{ \int_0^t \sigma(s)dW(s) + \int_0^t [\mu(s) - \frac{1}{2}\sigma^2(s) - \alpha(s)\sigma(s)]ds \right\}. \qquad (8.71)$$

(d) Comparing (8.70) and (8.71), deduce that

$$\exp\left\{ \int_0^T \alpha(s)dW(s) \right\} \diamond \exp\left\{ \int_0^t \sigma(s)dW(s) \right\}$$

$$= \exp\left\{ \int_0^T \alpha(s)dW(s) \right\} \cdot \exp\left\{ \int_0^t \sigma(s)dW(s) - \int_0^t \alpha(s)\sigma(s)ds \right\}, \quad t \in [0, T].$$

Part II

The Discontinuous Case: Pure Jump Lévy Processes

9

A Short Introduction to Lévy Processes

In this chapter we present some basic definitions and results about Lévy processes. We do not aim at being complete in our presentation, but for a smoother reading, we like to include the notions we constantly refer to in this book. We can address the reader to the recent monographs, for example, [8, 32, 115, 213] for a deeper study of Lévy processes. Our presentation follows the survey given in [183, Chap. 1].

9.1 Basics on Lévy Processes

Let $(\Omega, \mathcal{F}, P)$ be a complete probability space.

Definition 9.1. *A one-dimensional Lévy process is a stochastic process* $\eta = \eta(t)$, $t \geq 0$:
$$\eta(t) = \eta(t, \omega), \quad \omega \in \Omega,$$
with the following properties:

(i) $\eta(0) = 0$ *P-a.s.,*

(ii) η *has independent increments, that is, for all* $t > 0$ *and* $h > 0$, *the increment* $\eta(t + h) - \eta(t)$ *is independent of* $\eta(s)$ *for all* $s \leq t$,

(iii) η *has stationary increments, that is, for all* $h > 0$ *the increment* $\eta(t + h) - \eta(t)$ *has the same probability law as* $\eta(h)$,

(iv) *It is stochastically continuous, that is, for every* $t \geq 0$ *and* $\varepsilon > 0$ *then* $\lim_{s \to t} P\{|\eta(t) - \eta(s)| > \varepsilon\} = 0$,

(v) η *has càdlàg paths, that is, the trajectories are right-continuous with left limits.*

A stochastic process η satisfying (i)–(iv) is called a *Lévy process in law*.

The *jump of η at time t* is defined by

$$\Delta\eta(t) := \eta(t) - \eta(t^-).$$

Put $\mathbb{R}_0 := \mathbb{R} \setminus \{0\}$ and let $\mathcal{B}(\mathbb{R}_0)$ be the σ-algebra generated by the family of all Borel subsets $U \subset \mathbb{R}$, such that $\bar{U} \subset \mathbb{R}_0$. If $U \in \mathcal{B}(\mathbb{R}_0)$ with $\bar{U} \subset \mathbb{R}_0$ and $t > 0$, we define

$$N(t, U) := \sum_{0 \le s \le t} \chi_U(\Delta\eta(s)), \tag{9.6}$$

that is, the number of jumps of size $\Delta\eta(s) \in U$ for any s in $0 \le s \le t$. Since the paths of η are càdlàg we can see that $N(t, U) < \infty$ for all $U \in \mathcal{B}(\mathbb{R}_0)$ with $\bar{U} \subset \mathbb{R}_0$; see, e.g. [213]. Moreover, (9.6) defines in a natural way a Poisson random measure N on $\mathcal{B}(0, \infty) \times \mathcal{B}(\mathbb{R}_0)$ given by

$$(a, b] \times U \longmapsto N(b, U) - N(a, U), \qquad 0 < a \le b, \ U \in \mathcal{B}(\mathbb{R}_0),$$

and its standard extension. See e.g. [97], [79]. We call this random measure the *jump measure of η*. Its differential form is denoted by $N(dt, dz)$, $t > 0$, $z \in \mathbb{R}_0$.

The *Lévy measure ν* of η is defined by

$$\nu(U) := E\big[N(1, U)\big], \quad U \in \mathcal{B}(\mathbb{R}_0). \tag{9.7}$$

It is important to note that ν does not need to be a finite measure. It can be possible that

$$\int_{\mathbb{R}_0} \min(1, |z|)\nu(dz) = \infty. \tag{9.8}$$

This is the case when the trajectories of η would appear with many jumps of small size, a situation that is of interest in financial modeling (see, e.g., [17, 48, 75, 216] and references therein).

On the contrary, the Lévy measure always satisfies

$$\int_{\mathbb{R}_0} \min(1, z^2)\nu(dz) < \infty.$$

In fact, a measure ν on $\mathcal{B}(\mathbb{R}_0)$ can be a Lévy measure of some Lévy process η if and only if the condition above holds true. This is due to the following theorem.

Theorem 9.2. The Lévy–Khintchine formula.

(1)   Let η be a Lévy process in law. Then

$$E\big[e^{iu\eta(t)}\big] = e^{i\Psi(u)}, \quad u \in \mathbb{R} \quad (i = \sqrt{-1}), \tag{9.9}$$

with the characteristic exponent

$$\Psi(u) := i\alpha u - \frac{1}{2}\sigma^2 u^2 + \int_{|z|<1} \left(e^{iuz} - 1 - iuz\right)\nu(dz) + \int_{|z|\geq 1} \left(e^{iuz} - 1\right)\nu(dz),$$

$$(9.10)$$

where the parameters $\alpha \in \mathbb{R}$ and $\sigma^2 \geq 0$ are constants and $\nu = \nu(dz)$, $z \in \mathbb{R}_0$, is a σ-finite measure on $\mathcal{B}(\mathbb{R}_0)$ satisfying

$$\int_{\mathbb{R}_0} \min(1, z^2)\nu(dz) < \infty. \qquad (9.11)$$

It follows that ν is the Lévy measure of η.

(2)   Conversely, given the constants $\alpha \in \mathbb{R}$ and $\sigma^2 \geq 0$ and the σ-finite measure ν on $\mathcal{B}(\mathbb{R}_0)$ such that (9.11) holds, then there exists a process η (unique in law) such that (9.9) and (9.10) hold. The process η is a Lévy process in law.

There always exists a càdlàg version of the above Lévy process in law (see, e.g., [213]), which is a Lévy process, cf. Definition 9.1. Using this càdlàg version, we can give the representation (9.7) of the σ-finite measure ν.

We define the *compensated jump measure $\tilde{N}$*, also called the *compensated Poisson random measure*, by

$$\tilde{N}(dt, dz) := N(dt, dz) - \nu(dz)dt. \qquad (9.12)$$

For any t, let $\mathcal{F}_t$ be the σ-algebra generated by the random variables $W(s)$ and $\tilde{N}(ds, dz)$, $z \in \mathbb{R}_0$, $s \leq t$, augmented for all the sets of P-zero probability. Let us equip the given probability space $(\Omega, \mathcal{F}, P)$ with the corresponding filtration

$$\mathbb{F} = \{\mathcal{F}_t, \ t \geq 0\}.$$

A stochastic process $\theta = \theta(t, z)$, $t \geq 0, z \in \mathbb{R}_0$, is called $\mathbb{F}$-adapted if for all $t \geq 0$ and for all $z \in \mathbb{R}_0$, the random variable $\theta(t, z) = \theta(t, z, \omega)$, $\omega \in \Omega$, is $\mathcal{F}_t$-measurable. For any $\mathbb{F}$-adapted process θ such that

$$E\left[\int_0^T \int_{\mathbb{R}_0} \theta^2(t, z)\nu(dz)dt\right] < \infty \quad \text{for some } T > 0, \qquad (9.13)$$

we can see that the process

$$M_n(t) := \int_0^t \int_{|z|\geq \frac{1}{n}} \theta(s, z)\tilde{N}(ds, dz), \quad 0 \leq t \leq T,$$

is a martingale in $L^2(P)$ and its limit

$$M(t) := \lim_{n\to\infty} M_n(t) := \int_0^t \int_{\mathbb{R}_0} \theta(s, z)\tilde{N}(ds, dz), \quad 0 \leq t \leq T, \qquad (9.14)$$

in $L^2(P)$ is also a martingale. Moreover, we have the Itô isometry

$$E\Big[\Big(\int_0^T \int_{\mathbb{R}_0} \theta(t,z)\tilde{N}(dt,dz)\Big)^2\Big] = E\Big[\int_0^T \int_{\mathbb{R}_0} \theta^2(t,z)\nu(dz)dt\Big]. \qquad (9.15)$$

A Wiener process is a special case of a Lévy process. In fact, we have the following general representation theorem (see, e.g., [119, 213].

Theorem 9.3. The Lévy–Itô decomposition theorem. *Let η be a Lévy process. Then $\eta = \eta(t)$, $t \geq 0$, admits the following integral representation*

$$\eta(t) = a_1 t + \sigma W(t) + \int_0^t \int_{|z|<1} z\tilde{N}(ds,dz) + \int_0^t \int_{|z|\geq 1} zN(ds,dz) \qquad (9.16)$$

for some constants $a_1, \sigma \in \mathbb{R}$. Here $W = W(t)$, $t \geq 0$ ($W(0) = 0$), is a standard Wiener process.

In particular, we can see that if the Lévy process has continuous trajectories, then it is of the form

$$\eta(t) = a_1 t + \sigma W(t), \quad t \geq 0.$$

It can be proved that if

$$E\big[|\eta(t)|^p\big] < \infty \quad \text{for some } p \geq 1,$$

then

$$\int_{|z|\geq 1} |z|^p \nu(dz) < \infty,$$

see [213]. In particular, if we assume that

$$E\big[\eta^2(t)\big] < \infty, \qquad t \geq 0, \qquad (9.17)$$

then we have

$$\int_{|z|\geq 1} |z|^2 \nu(dz) < \infty$$

and the representation (9.16) appears as

$$\eta(t) = at + \sigma W(t) + \int_0^t \int_{\mathbb{R}_0} z\tilde{N}(ds,dz), \qquad (9.18)$$

where $a = a_1 + \int_{|z|\geq 1} z\nu(dz)$. A Lévy process of the type above with $\sigma = 0$ is called a *pure jump Lévy process*.

We assume from now on that (9.17) holds and hence that η has the representation (9.18).

Motivated by the representation (9.18), it is natural to consider processes $X = X(t)$, $t \geq 0$, admitting a stochastic integral representation in the form

$$X(t) = x + \int_0^t \alpha(s)ds + \int_0^t \beta(s)dW(s) + \int_0^t \int_{\mathbb{R}_0} \gamma(s,z)\tilde{N}(ds,dz), \qquad (9.19)$$

where $\alpha(t)$, $\beta(t)$, and $\gamma(t, z)$ are predictable processes such that, for all $t > 0$, $z \in \mathbb{R}_0$,

$$\int_0^t \left[|\alpha(s)| + \beta^2(s) + \int_{\mathbb{R}_0} \gamma^2(s, z)\nu(dz) \right] ds < \infty, \qquad P\text{-}a.s. \qquad (9.20)$$

This condition implies that the stochastic integrals are well-defined and local martingales. If we strengthened the condition to

$$E\left[\int_0^t \left[|\alpha(s)| + \beta^2(s) + \int_{\mathbb{R}_0} \gamma^2(s, z)\nu(dz) \right] ds \right] < \infty,$$

for all $t > 0$, then the corresponding stochastic integrals are martingales.

We call such a process an *Itô–Lévy process*. In analogy with the Brownian motion case, we use the short-hand differential notation

$$dX(t) = \alpha(t)dt + \beta(t)dW(t) + \int_{\mathbb{R}_0} \gamma(t, z)\tilde{N}(dt, dz); \qquad X(0) = x \qquad (9.21)$$

for the processes of type (9.19).

Recall that a *predictable process* is a stochastic process measurable with respect to the σ-algebra generated by

$$A \times (s, u] \times B, \quad A \in \mathcal{F}_s, \ 0 \le s < u, \ B \in \mathcal{B}(\mathbb{R}_0).$$

Moreover, any measurable $\mathbb{F}$-adapted and left-continuous (with respect to t) process is predictable.

9.2 The Itô Formula

The following result is fundamental in the stochastic calculus of Lévy processes.

Theorem 9.4. The one-dimensional Itô formula. *Let $X = X(t)$, $t \ge 0$, be the Itô–Lévy process given by (9.19) and let $f : (0, \infty) \times \mathbb{R} \longrightarrow \mathbb{R}$ be a function in $C^{1,2}((0, \infty) \times \mathbb{R})$ and define*

$$Y(t) := f(t, X(t)), \qquad t \ge 0.$$

Then the process $Y = Y(t)$, $t \ge 0$, is also an Itô–Lévy process and its differential form is given by

$$dY(t) = \frac{\partial f}{\partial t}(t, X(t))dt + \frac{\partial f}{\partial x}(t, X(t))\alpha(t)dt + \frac{\partial f}{\partial x}(t, X(t))\beta(t)dW(t)$$

$$+ \frac{1}{2}\frac{\partial^2 f}{\partial x^2}(t, X(t))\beta^2(t)dt + \int_{\mathbb{R}_0} \Big[f(t, X(t) + \gamma(t, z)) - f(t, X(t))$$

$$- \frac{\partial f}{\partial x}(t, X(t))\gamma(t, z)\Big]\nu(dz)dt + \int_{\mathbb{R}_0} \Big[f(t, X(t^-) + \gamma(t, z)) - f(t, X(t^-))\Big]\tilde{N}(dt, dz).$$

$$(9.22)$$

In the multidimensional case we are given a J-dimensional Brownian motion $W(t) = (W_1(t), ..., W_J(t))^T$, $t \geq 0$, and K independent compensated Poisson random measures $\tilde{N}(dt, dz) = (\tilde{N}_1(dt, dz_1), ..., \tilde{N}_K(dt, dz_K))^T$, $t \geq 0$, $z = (z_1, ..., z_K) \in (\mathbb{R}_0)^K$, and n Itô–Lévy processes of the form

$$dX(t) = \alpha(t)dt + \beta(t)dW(t) + \int_{(\mathbb{R}_0)^K} \gamma(t, z)\tilde{N}(dt, dz), \quad t \geq 0,$$

that is,

$$dX_i(t) = \alpha_i(t)dt + \sum_{j=1}^{J} \beta_{ij}(t)dW_j(t) + \sum_{k=1}^{K} \int_{\mathbb{R}_0} \gamma_{ik}(t, z_k)\tilde{N}_k(dt, dz_k), \quad i = 1, ..., n.$$

(9.23)

With this notation we have the following result.

Theorem 9.5. The multidimensional Itô formula. *Let* $X = X(t)$, $t \geq 0$, *be an* n-*dimensional Itô–Lévy process of the form* (9.23). *Let* $f : (0, \infty) \times \mathbb{R}^n \longrightarrow \mathbb{R}$ *be a function in* $C^{1,2}((0, \infty) \times \mathbb{R}^n)$ *and define*

$$Y(t) := f(t, X(t)), \qquad t \geq 0.$$

Then the process $Y = Y(t)$, $t \geq 0$, *is a one-dimensional Itô–Lévy process and its differential form is given by*

$$dY(t) = \frac{\partial f}{\partial t}(t, X(t))dt + \sum_{i=1}^{n} \frac{\partial f}{\partial x_i}(t, X(t))\alpha_i(t)dt$$

$$+ \sum_{i=1}^{n} \sum_{j=1}^{J} \frac{\partial f}{\partial x_i}(t, X(t))\beta_{ij}(t)dW_j(t) + \frac{1}{2}\sum_{i=1}^{n}\sum_{j=1}^{J} \frac{\partial^2 f}{\partial x_i \partial x_j}(t, X(t))(\beta\beta^T)_{ij}(t)dt$$

$$+ \sum_{k=1}^{K} \int_{\mathbb{R}_0} \left[f(t, X(t) + \gamma^{(k)}(t, z)) - f(t, X(t)) - \sum_{i=1}^{n} \frac{\partial f}{\partial x_i} f(t, X(t))\gamma_{ik}(t, z) \right] \nu_k(dz_k)dt$$

$$+ \sum_{k=1}^{K} \int_{\mathbb{R}_0} \left[f(t, X(t^-) + \gamma^{(k)}(t, z)) - f(t, X(t^-)) \right] \tilde{N}_k(dt, dz_k),$$

(9.24)

where $\gamma^{(k)}$ *is the column number* k *of the* $n \times K$ *matrix* $\gamma = [\gamma_{ik}]$.

Example 9.6. **The generalized geometric Lévy process.** Consider the one-dimensional stochastic differential equation for the càdlàg process $Z = Z(t)$, $t \geq 0$:

$$\begin{cases} dZ(t) = Z(t^-)\left[\alpha(t)dt + \beta(t)dW(t) + \int_{\mathbb{R}_0} \gamma(t, z)\tilde{N}(dt, dz)\right], & t > 0, \\ Z(0) = z_0 > 0. \end{cases}$$

(9.25)

Here $\alpha(t)$, $\beta(t)$, and $\gamma(t, z)$, $t \geq 0$, $z \in \mathbb{R}_0$, are given predictable processes with $\gamma(t, z) > -1$, for almost all $(t, z) \in [0, \infty) \times \mathbb{R}_0$ and for all $0 < t < \infty$

$$\int_0^t [|\alpha(s)| + \beta^2(s) + \int_{\mathbb{R}_0} \gamma^2(s,z)\nu(dz)]ds < \infty \quad P\text{-}a.s.,$$

cf. (9.20). We claim that the solution of this equation is

$$Z(t) = z_0 e^{X(t)}, \quad t \geq 0, \tag{9.26}$$

where

$$X(t) = \int_0^t \left[\alpha(s) - \frac{1}{2}\beta^2(s) + \int_{\mathbb{R}_0} \left[\log(1 + \gamma(s,z)) - \gamma(s,z)\right]\nu(dz)\right] ds \tag{9.27}$$

$$+ \int_0^t \beta(s)dW(s) + \int_0^t \int_{\mathbb{R}_0} \log(1 + \gamma(s,z))\tilde{N}(ds,dz).$$

To see this we apply the one-dimensional Itô formula to $Y(t) = f(t, X(t))$, $t \geq 0$, with $f(t,x) = z_0 e^x$ and $X(t)$, as given in (9.27). Then we obtain

$$dY(t) = z_0 e^{X(t)}\left[\left(\alpha(t) - \frac{1}{2}\beta^2(t) + \int_{\mathbb{R}_0} \left[\log(1 + \gamma(t,z)) - \gamma(t,z)\right]\nu(dz)\right)dt + \beta(t)dW(t)\right]$$

$$+ z_0 e^{X(t)}\frac{1}{2}\beta^2(t)dt + \int_{\mathbb{R}_0} z_0 [e^{X(t)+\log(1+\gamma(t,z))} - e^{X(t)} - e^{X(t)}\log(1 + \gamma(t,z))]\nu(dz)dt$$

$$+ \int_{\mathbb{R}_0} z_0 [e^{X(t^-)+\log(1+\gamma(t,z))} - e^{X(t^-)}]\tilde{N}(dt,dz)$$

$$= Y(t^-)\left[\alpha(t)dt + \beta(t)dW(t) + \int_{\mathbb{R}_0} \gamma(t,z)\tilde{N}(dt,dz)\right]$$

as required.

Example 9.7. **The quadratic covariation process.** Let

$$dX_i(t) = \int_{\mathbb{R}_0} \gamma_{i1}(t,z)\tilde{N}_1(dt,dz) + \int_{\mathbb{R}_0} \gamma_{i2}(t,z)\tilde{N}_2(dt,dz), \quad i = 1,2 \tag{9.28}$$

be two pure jump Lévy processes (see (9.23) with $\beta_{ij} = 0$. Define

$$Y(t) = X_1(t)X_2(t), \quad t \geq 0.$$

Then by the two-dimensional Itô formula we have

$$dY(t) = \sum_{k=1,2} \int_{\mathbb{R}_0} [(X_1(t) + \gamma_{1k}(t,z_k))(X_2(t) + \gamma_{2k}(t,z_k)) - X_1(t)X_2(t)$$

$$- \gamma_{1k}(t,z_k)X_2(t) - \gamma_{2k}(t,z_k)X_1(t)]\nu_k(dz_k)dt$$

$$+ \sum_{k=1,2} \int_{\mathbb{R}_0} [(X_1(t^-) + \gamma_{1k}(t,z_k))(X_2(t^-) + \gamma_{2k}(t,z_k)) - X_1(t^-)X_2(t^-)]\tilde{N}_k(dt,dz_k)$$

$$= \sum_{k=1,2} \int_{\mathbb{R}_0} \gamma_{1k}(t,z_k)\gamma_{2k}(t,z_k)\nu_k(dz_k)dt$$

$$+ \sum_{k=1,2} \int_{\mathbb{R}_0} [\gamma_{1k}(t,z_k)X_2(t^-) + \gamma_{2k}(t,z_k)X_1(t^-) + \gamma_{1k}(t,z_k)\gamma_{2k}(t,z_k)]\tilde{N}_k(dt,dz_k)$$

$$= X_1(t^-)dX_2(t) + X_2(t^-)dX_1(t) + \sum_{k=1,2} \int_{\mathbb{R}_0} \gamma_{1k}(t,z_k)\gamma_{2k}(t,z_k)N_k(dt,dz_k).$$

We define the *quadratic covariation process* $[X_1, X_2](t)$, $t \geq 0$, of the processes X_1 and X_2 as

$$d[X_1, X_2](t) := d(X_1(t)X_2(t)) - X_1(t^-)dX_2(t) - X_2(t^-)dX_1(t). \qquad (9.29)$$

Hence, for the processes X_1, X_2 given in (9.28), we have that

$$d[X_1, X_2](t) = \sum_{k=1,2} \int_{\mathbb{R}_0} \gamma_{1k}(t, z_k)\gamma_{2k}(t, z_k)N_k(dt, dz_k). \qquad (9.30)$$

In particular, note that

$$E[X_1(t)X_2(t)] = X_1(0)X_2(0) + \sum_{k=1,2} \int_0^t \int_{\mathbb{R}_0} \gamma_{1k}(s, z_k)\gamma_{2k}(s, z_k)\nu_k(dz_k)ds.$$
$$(9.31)$$

9.3 The Itô Representation Theorem for Pure Jump Lévy Processes

We now proceed to prove the Itô representation theorem for Lévy processes. Since we already know the representation theorem in the continuous case, that is, $\tilde{N} \equiv 0$ (see [179] and Problem 1.4), we concentrate on the pure jump case in this section. We assume that

$$\eta(t) = \int_0^t \int_{\mathbb{R}_0} z\tilde{N}(ds, dz), \quad t \geq 0, \qquad (9.32)$$

that is, $a = \sigma = 0$ in (9.18).

The following representation theorem was first proved by Itô [121]. Here we follow the presentation given in [154]. The proof is based on two lemmata.

Let us consider the filtration $\mathbb{F}$ of the σ-algebras $\mathcal{F}_t$ generated by $\eta(s)$, $s \leq t$ ($t \geq 0$).

Lemma 9.8. *The set of all random variables of the form*

$$\left\{ f(\eta(t_1), ..., \eta(t_n)) : t_i \in [0, T], i = 1, ..., n; f \in C_0^\infty(\mathbb{R}^n), n = 1, 2, ... \right\}$$

is dense in the subspace $L^2(\mathcal{F}_T, P) \subset L^2(P)$ of $\mathcal{F}_T$-measurable square integrable random variables.

Proof The proof follows the same argument as in Lemma 4.3.1 in [179]. See also, for example, [154]. □

Lemma 9.9. *The linear span of all the so-called Wick/Doléans–Dade exponentials*

$$\exp\left\{ \int_0^T \int_{\mathbb{R}_0} h(t)z\chi_{[0,R]}(z)\tilde{N}(dt, dz) - \int_0^T \int_{\mathbb{R}_0} \left[e^{h(t)z\chi_{[0,R]}(z)} \right. \right.$$
$$\left. \left. -1 - h(t)z\chi_{[0,R]}(z) \right]\nu(dz)dt \right\}, \quad h \in C(0, T), R > 0,$$

is dense in $L^2(\mathcal{F}_T, P)$.

Proof The proof follows the same argument as in Lemma 4.3.2 in [179]. See also, for example, [154]. □

We are ready now for the first main result of this section. Note that the representation is in terms of a stochastic integral with respect to $\tilde{N}$ and *not* with respect to η.

Theorem 9.10. The Itô representation theorem. *Let $F \in L^2(P)$ be $\mathcal{F}_T$-measurable. Then there exists a unique predictable process $\Psi = \Psi(t, z)$, $t \geq 0, z \in \mathbb{R}_0$, such that*

$$E\left[\int_0^T \int_{\mathbb{R}_0} \Psi^2(t, z)\nu(dz)dt\right] < \infty$$

for which we have

$$F = E[F] + \int_0^T \int_{\mathbb{R}_0} \Psi(t, z)\tilde{N}(dt, dz). \tag{9.33}$$

Proof First assume that $F = Y(T)$, where

$$Y(t) = \exp\left\{\int_0^t \int_{\mathbb{R}_0} h(s)z\chi_{[0,R]}(z)\tilde{N}(ds, dz) - \int_0^t \int_{\mathbb{R}_0} \left[e^{h(s)z\chi_{[0,R]}(z)}\right.\right.$$
$$\left.\left. -1 - h(s)z\chi_{[0,R]}(z)\right]\nu(dz)ds\right\}, \quad t \in [0, T],$$

for some $h \in C(0, \infty)$, that is, F is a Wick/Doléans–Dade exponential. Then by the Itô formula (cf. Theorem 9.4 and see Problem 9.2)

$$dY(t) = Y(t^-) \int_{\mathbb{R}_0} \left[e^{h(t)z\chi_{[0,R]}(z)} - 1\right]\tilde{N}(dt, dz).$$

Therefore,

$$F = Y(T) = Y(0) + \int_0^T 1dY(t) = 1 + \int_0^T \int_{\mathbb{R}_0} Y(t^-)\left[e^{h(t)z\chi_{[0,R]}(z)} - 1\right]\tilde{N}(dt, dz).$$

So for this F the representation (9.33) holds with

$$\Psi(t, z) = Y(t^-)\left[e^{h(t)z\chi_{[0,R]}(z)} - 1\right].$$

Note that

$$E[Y^2(T)] = 1 + E\left[\int_0^T \int_{\mathbb{R}_0} Y^2(t^-)\left(e^{h(t)z\chi_{[0,R]}(z)} - 1\right)^2\nu(dz)dt\right].$$

If $F \in L^2(\mathcal{F}_T, P)$ (i.e., an $\mathcal{F}_T$-measurable random variable in $L^2(P)$) is arbitrary, we can choose a sequence F_n of linear combinations of Wick/Doléan–Dade exponentials such that $F_n \longrightarrow F$ in $L^2(P)$. See Lemma 9.9. Then we have

$$F_n = E[F_n] + \int_0^T \int_{\mathbb{R}_0} \Psi_n(t, z)\tilde{N}(dt, dz),$$

for all $n = 1, 2, ...$, where

$$E[F_n^2] = (E[F_n])^2 + E\left[\int_0^T \int_{\mathbb{R}_0} \Psi_n^2(t, z)\nu(dz)dt\right] < \infty.$$

Then by the Itô isometry we have the expression

$$E\left[(F_m - F_n)^2\right] = (E[F_m - F_n])^2 + E\left[\int_0^T \int_{\mathbb{R}_0} (\Psi_m(t, z) - \Psi_n(t, z))^2 \nu(dz)dt\right],$$

which vanishes for $m, n \to \infty$. Therefore, Ψ_n, $n = 1, 2, ...$, is a Cauchy sequence in $L^2(P \times \lambda \times \nu)$; hence, it converges to a limit $\Psi \in L^2(P \times \lambda \times \nu)$. This yields (9.33), in fact

$$F = \lim_{n \to \infty} F_n = \lim_{n \to \infty} \left\{ EF_n + \int_0^T \int_{\mathbb{R}_0} \Psi_n(t, z)\tilde{N}(dt, dz) \right\}$$

$$= EF + \int_0^T \int_{\mathbb{R}_0} \Psi(t, z)\tilde{N}(dt, dz).$$

The uniqueness is given by the convergence in L^2-spaces and the Itô isometry.$\quad\Box$

Example 9.11. Choose $F = \eta^2(T)$. To find the representation (9.33) for F we define

$$Y(t) = \eta^2(t) = \left(\int_0^t \int_{\mathbb{R}_0} z\tilde{N}(ds, dz)\right)^2, \quad t \in [0, T].$$

By the Itô formula

$$d\eta^2(t) = \int_{\mathbb{R}_0} \left[(\eta(t) + z)^2 - \eta^2(t) - 2\eta(t)z\right]\nu(dz)dt$$

$$+ \int_{\mathbb{R}_0} \left[(\eta(t^-) + z)^2 - \eta^2(t^-)\right]\tilde{N}(dt, dz)$$

$$= \int_{\mathbb{R}_0} z^2\nu(dz)dt + \int_{\mathbb{R}_0} \left[2\eta(t^-) + z\right]z\tilde{N}(dt, dz)$$

(see Problem 9.2). Hence we get

$$\eta^2(T) = T\int_{\mathbb{R}_0} z^2\nu(dz) + \int_0^T \int_{\mathbb{R}_0} \left[2\eta(t^-) + z\right]z\tilde{N}(dt, dz). \tag{9.34}$$

Remark 9.12. Note that it is *not* possible to write

$$\eta^2(T) = E[\eta^2(T)] + \int_0^T \varphi(t)d\eta(t) \tag{9.35}$$

for some predictable process φ. In fact, the representation (9.35) is equivalent to

$$\eta^2(T) = E[\eta^2(T)] + \int_0^T \int_{\mathbb{R}_0} \varphi(t)z\tilde{N}(dt, dz),$$

which contradicts (9.34), in view of the uniqueness of the Itô stochastic integral representation.

9.4 Application to Finance: Replicability

The fact that the representation (9.35) is not possible has a special interpretation in finance: it means that the *claim* $F = \eta^2(T)$ is not *replicable* in a certain financial market driven by the Lévy process η. We now make this more precise.

Consider a *securities market* with two kinds of investment possibilities:

- A *risk free asset* with a price per unit fixed as

$$S_0(t) = 1, \qquad t \geq 0, \tag{9.36}$$

- n *risky assets* with prices $S_i(t)$, $t \geq 0$ ($i = 1, ..., n$), given by

$$dS_i(t) = S_i(t^-)\left[\sigma_i(t)dW(t) + \int_{\mathbb{R}_0} \gamma_i(t, z)\tilde{N}(dt, dz)\right], \tag{9.37}$$

where $W(t) = \big(W_1(t), ..., W_n(t)\big)^T$, $t \geq 0$, is an n-dimensional Wiener process and $\tilde{N}(dt, dz) = \big(\tilde{N}_1(dt, dz), ..., \tilde{N}_n(dt, dz)\big)^T$, $t \geq 0$, $z \in \mathbb{R}_0$, corresponds to n independent compensated Poisson random measures. The parameters $\sigma_i(t) = \big(\sigma_{i1}(t), ..., \sigma_{in}(t)\big)$, $t \geq 0$ ($i = 1, ..., n$), and $\gamma_i(t, z) = \big(\gamma_{i1}(t, z), ..., \gamma_{in}(t, z)\big)$, $t \geq 0$, $z \in \mathbb{R}_0$ ($i = 1, ..., n$), are predictable processes that satisfy

$$E\left[\sum_{i,j=1}^n \int_0^T \left[\sigma_{ij}^2(t) + \int_{\mathbb{R}_0} \gamma_{ij}^2(t, z)\nu(dz)\right]dt\right] < \infty. \tag{9.38}$$

A random variable $F \in L^2(\mathcal{F}_T, P)$ represents a financial *claim* (or *T-claim*). The claim F is *replicable* if there exists a predictable process $\varphi(t) = \big(\varphi_1(t), ..., \varphi_n(t)\big)^T$, $t \geq 0$, such that

$$\sum_{i=1}^n \int_0^T \varphi_i^2(t)S_i^2(t^-)\left[\sum_{j=1}^n \sigma_{ij}^2(t) + \int_{\mathbb{R}_0} \gamma_{ij}^2(t, z)\nu(dz)\right]dt < \infty \tag{9.39}$$

and

$$F = E[F] + \sum_{i=1}^n \int_0^T \varphi_i(t)dS_i(t). \tag{9.40}$$

If this is the case, then φ is a *replicating portfolio* for the claim F. See Sect. 4.3.

Let $\mathcal{A}$ denote the set of all predictable processes satisfying (9.39). These constitute the set of *admissible portfolios* in this context.

Let us now consider the general Lévy process $\eta(t)$, $t \geq 0$:

$$d\eta(t) = dW(t) + \int_{\mathbb{R}_0} z\tilde{N}(dt, dz).$$

Theorem 9.13. *Let $F \in L^2(\mathcal{F}_T, P)$ be a claim on the market (9.36)–(9.37), with stochastic integral representation of the form*

$$F = E[F] + \int_0^T \alpha(t)dW(t) + \int_0^T \int_{\mathbb{R}_0} \beta(t, z)\tilde{N}(dt, dz). \tag{9.41}$$

Then F is replicable if and only if the integrands $\alpha = (\alpha_1, ..., \alpha_n)$ and $\beta = (\beta_1, ..., \beta_n)$ have the form

$$\alpha_j(t) = \sum_{i=1}^n \varphi_i(t)S_i(t)\sigma_{ij}(t), \qquad t \geq 0, \tag{9.42}$$

$$\beta_j(t, z) = \sum_{i=1}^n \varphi_i(t)S_i(t^-)\gamma_{ij}(t, z), \qquad t \geq 0, z \in \mathbb{R}_0 \qquad (j = 1, ..., n), \tag{9.43}$$

for some process $\varphi \in \mathcal{A}$. In this case φ is a replicating portfolio for F.

Proof First assume that F is replicable with replicating portfolio φ. Then clearly

$$F - E[F] = \sum_{i=1}^n \int_0^T \varphi_i(t)dS_i(t)$$

$$= \int_0^T \sum_{i=1}^n \varphi_i(t)S_i(t)\sigma_i(t)dW(t)$$

$$+ \int_0^T \int_{\mathbb{R}_0} \sum_{i=1}^n \varphi_i(t)S_i(t^-)\gamma_i(t, z)\tilde{N}(dt, dz).$$

In view of the uniqueness, comparing this with the representation (9.41), we obtain the results (9.42)–(9.43). The argument works both ways, so that if (9.42)–(9.43) hold, then the above computation lead to representation (9.40) and the claim F is replicable. $\square$

We refer to [25, 29] for further arguments on the aforementioned result.

Remark 9.14. Note that any $F \in L^2(\mathcal{F}_T, P)$ admits representation in form (9.41) (see, e.g., [71, 121]).

Example 9.15. Returning to Example 9.11 we can see that in a securities market model of the type:

- A risk free asset with price per unit $S_0(t) = 1$, $t \geq 0$,
- One risky asset with price per unit $S_1(t)$, $t \geq 0$, given by

$$dS_1(t) = S_i(t^-) \int_{\mathbb{R}_0} z \tilde{N}(dt, dz), \qquad (9.44)$$

the claim $F = \eta^2(T)$ is *not* replicable since its representation is given by (9.34):

$$\eta^2(T) = T \int_{\mathbb{R}_0} z^2 \nu(dz) + \int_0^T \int_{\mathbb{R}_0} (2\eta(t^-) + z)z \tilde{N}(dt, dz),$$

with $\beta(t, z) = (2\eta(t^-) + z)z$, $t \geq 0$, $z \in \mathbb{R}_0$, which is not of the form (9.43).

Financial markets in which all claims are replicable are called *complete*. Otherwise the market is called *incomplete*. The market model presented here above is an example of an incomplete market.

9.5 Exercises

Problem 9.1. (*) Let $X(t)$, $t \geq 0$, be a one-dimensional Itô–Lévy process:

$$dX(t) = \alpha(t)dt + \beta(t)dW(t) + \int_{\mathbb{R}_0} \gamma(t, z)\tilde{N}(dt, dz).$$

Use the Itô formula to express $dY(t) = df(X(t))$ in the standard form:

$$dY(t) = a(t)dt + b(t)dW(t) + \int_{\mathbb{R}_0} c(t, z)\tilde{N}(dt, dz)$$

in the following cases:

(a) $Y(t) = X^2(t)$, $t \geq 0$,
(b) $Y(t) = \exp\{X(t)\}$, $t \geq 0$,
(c) $Y(t) = \cos X(t)$, $t \geq 0$.

Problem 9.2. (*) Let $h \in L^2([0, T])$ be a càglàd real function. Define

$$X(t) := \int_0^t \int_{\mathbb{R}_0} h(s)z\tilde{N}(ds, dz) - \int_0^t \int_{\mathbb{R}_0} (e^{h(s)z} - 1 - h(s)z)\nu(dz)ds$$

and put

$$Y(t) = \exp\{X(t)\}, \quad t \in [0, T].$$

Show that

$$dY(t) = Y(t^-) \int_{\mathbb{R}_0} (e^{h(t)z} - 1)\tilde{N}(dt, dz).$$

In particular, $Y(t)$, $t \in [0, T]$, is a local martingale.

Problem 9.3. Use the Itô formula to solve the following stochastic differential equations.

(a) The *Lévy–Ornstein–Uhlenbeck process*:

$$dY(t) = \rho(t)Y(t)dt + \int_{\mathbb{R}_0} \gamma(t, z)\tilde{N}(dt, dz), \quad t \in [0, T],$$

where $\rho \in L^1([0, T])$ and γ is a predictable process satisfying (9.20).

(b) A multiplicative noise dynamics:

$$dY(t) = \alpha(t)dt + Y(t^-)\int_{\mathbb{R}_0} \gamma(t, z)\tilde{N}(dt, dz), \quad t \in [0, T],$$

where α, γ are predictable processes satisfying (9.20).

Problem 9.4. (a) **Integration by parts.** Let

$$dX_i(t) = \alpha_i(t)dt + \beta_i(t)dW(t) + \int_{\mathbb{R}_0} \gamma_i(t, z)\tilde{N}(dt, dz) \quad (i = 1, 2),$$

be two Itô–Lévy processes. Use the two-dimensional Itô formula to show that

$$\begin{aligned} d(X_1(t)X_2(t)) = {} & X_1(t^-)dX_2(t) + X_2(t^-)dX_1(t) + \beta_1(t)\beta_2(t)dt \\ & + \int_{\mathbb{R}_0} \gamma_1(t, z)\gamma_2(t, z)N(dt, dz). \end{aligned}$$

(b) **The Itô–Lévy isometry.** In the aforementioned processes, choose $\alpha_i = X_i(0) = 0$ for $i = 1, 2$ and assume that $E[X_i^2(t)] < \infty$ for $i = 1, 2$. Show that

$$E[X_1(t)X_2(t)] = E\left[\int_0^T \left(\beta_1(t)\beta_2(t) + \int_{\mathbb{R}_0} \gamma_1(t, z)\gamma_2(t, z)\nu(dz)\right)dt\right].$$

Problem 9.5. The Girsanov theorem. Let $u = u(t)$, $t \in [0, T]$, and $\theta(t, z) \leq 1$, $t \in [0, T], z \in \mathbb{R}_0$, be predictable processes such that the process

$$\begin{aligned} Z(t) := {} & \exp\left\{-\int_0^t u(s)dW(s) - \frac{1}{2}\int_0^t u^2(s)ds + \int_0^t \int_{\mathbb{R}_0} \log(1-\theta(s, z))\tilde{N}(ds, dz)\right. \\ & \left. + \int_0^t \int_{\mathbb{R}_0} (\log(1 - \theta(s, z)) + \theta(s, z))\nu(dz)ds\right), \quad t \in [0, T], \end{aligned}$$

exists and satisfies $E[Z(T)] = 1$.

(a) Show that

$$dZ(t) = Z(t^-)\left[-u(t)dW(t) - \int_{\mathbb{R}_0} \theta(t, z)\tilde{N}(dt, dz)\right].$$

Thus, Z is a local martingale.

(b) Let

$$dX(t) = \alpha(t)dt + \beta(t)dW(t) + \int_{\mathbb{R}_0} \gamma(t,z)\tilde{N}(dt,dz), \quad t \in [0,T],$$

be an Itô–Lévy process. Find $d\big(Z(t)X(t)\big)$. In particular, note that if u and θ are chosen such that

$$\beta(t)u(t) + \int_{\mathbb{R}_0} \gamma(t,z)\theta(t,z)\nu(dz) = \alpha(t),$$

then $Z(t)X(t)$, $t \in [0,T]$, is a local martingale.

(c) Use the Bayes rule to show that if we define the probability measure Q on $(\Omega, \mathcal{F}_T)$ by

$$dQ = Z(T)dP,$$

then the process X is a local martingale with respect to Q. This is a version of the Girsanov theorem for Lévy processes.

Problem 9.6. (*) Let

$$\eta(t) = \int_0^t \int_{\mathbb{R}_0} z\tilde{N}(dz,dz), \quad t \in [0,T].$$

Find the integrand ψ for the stochastic integral representation

$$F = E[F] + \int_0^T \int_{\mathbb{R}_0} \psi(t,z)\tilde{N}(dt,dz)$$

of the following random variables:

(a) $F = \int_0^T \eta(t)dt$
(b) $F = \eta^3(T)$
(c) $F = \exp\{\eta(T)\}$
(d) $F = \cos\eta(T)$ [*Hint.* If $x \in \mathbb{R}$ and $i = \sqrt{-1}$, then $e^{ix} = \cos x + i\sin x$].

Moreover, in each case above, decide if F is replicable in the following Bachelier–Lévy type market model on the time interval $[0,T]$:

$$\text{risk free asset:} \quad dS_0(t) = 0, \quad S_0(0) = 1$$
$$\text{risky asset:} \quad dS_1(t) = d\eta(t), \quad S_1(0) \in \mathbb{R}.$$

10

The Wiener–Itô Chaos Expansion

In this chapter we show that it is possible to obtain a chaos expansion similar to the one in Chap. 1 in the context of pure jump Lévy processes $\eta = \eta(t)$, $t \geq 0$. However, in this case, the corresponding iterated integrals must be taken with respect to the *compensated Poisson measure* associated with η,

$$\eta(t) = \int_0^t \int_{\mathbb{R}_0} z \tilde{N}(dt, dz), \quad t \geq 0, \tag{10.1}$$

and *not* with respect to η itself. Our method will be basically the same as for Theorem 1.10; therefore, we do not give all the details.

For a study on the construction of chaos expansions and a calculus with respect to η we can refer to e.g. [61, 71].

10.1 Iterated Itô Integrals

Let $L^2((\lambda \times \nu)^n) = L^2(([0, T] \times \mathbb{R}_0)^n)$ be the space of deterministic real functions f such that

$$\|f\|_{L^2((\lambda \times \nu)^n)} = \left(\int_{([0,T] \times \mathbb{R}_0)^n} f^2(t_1, z_1, \ldots, t_n, z_n) dt_1 \nu(dz_1) \cdots dt_n \nu(dz_n) \right)^{1/2} < \infty,$$

where, we recall, $\lambda(dt) = dt$ denotes the Lebesgue measure on $[0, T]$.

The *symmetrization* $\tilde{f}$ of f is defined by

$$\tilde{f}(t_1, z_1, \ldots, t_n, z_n) = \frac{1}{n!} \sum_{\sigma} f(t_{\sigma_1}, z_{\sigma_1}, \ldots, t_{\sigma_n}, z_{\sigma_n}),$$

the sum being taken over all permutations $\sigma = (\sigma_1, \ldots, \sigma_n)$ of $\{1, \ldots, n\}$. Note that the symmetrization is over the n pairs $(t_1, z_1), \ldots, (t_n, z_n)$ and not over the $2n$ variables $t_1, z_1, \ldots, t_n, z_n$. A function $f \in L^2((\lambda \times \nu)^n)$ is called *symmetric* if $f = \tilde{f}$. We denote the space of all symmetric functions in $L^2((\lambda \times \nu)^n)$ by $\tilde{L}^2((\lambda \times \nu)^n)$.

G.Di Nunno et al., *Malliavin Calculus for Lévy Processes with Applications to Finance*,
© Springer-Verlag Berlin Heidelberg 2009

Define

$$G_n := \big\{ (t_1, z_1, ..., t_n, z_n) \ : \ 0 \le t_1 \le ... \le t_n \le T, \ z_i \in \mathbb{R}_0, \ i = 1, ..., n \big\}$$

and let $L^2(G_n)$ be the set of the real functions g on G_n, such that

$$\|g\|_{L^2(G_n)} := \Big(\int_{G_n} g^2(t_1, z_1, ..., t_n, z_n) dt_1 \nu(dz_1) \cdots dt_n \nu(dz_n) \Big)^{1/2} < \infty.$$

Note that, for $f \in \tilde{L}^2((\lambda \times \nu)^n)$, we have $f_{|G_n} \in L^2(G_n)$ and

$$\|f\|_{L^2((\lambda \times \nu)^n)}^2 = n! \|f\|_{L^2(G_n)}^2.$$

Definition 10.1. *For any $g \in L^2(G_n)$, the n-fold iterated integral $J_n(f)$ is the random variable in $L^2(P)$ defined as*

$$J_n(g) := \int_0^T \int_{\mathbb{R}_0} \cdots \int_0^{t_2^-} \int_{\mathbb{R}_0} g(t_1, z_1, ..., t_n, z_n) \tilde{N}(dt_1, dz_1) \cdots \tilde{N}(dt_n, dz_n).$$

We set $J_0(g) = g$ for any $g \in \mathbb{R}$.

If $f \in \tilde{L}^2((\lambda \times \nu)^n)$, we also define

$$I_n(f) := \int_{([0,T] \times \mathbb{R}_0)^n} f(t_1, z_1, ..., t_n, z_n) \tilde{N}^{\otimes n}(dt, dz) := n! J_n(f), \qquad (10.2)$$

where $\tilde{N}^{\otimes n}(dt, dz) = \tilde{N}(dt_1, dz_1) \cdots \tilde{N}(dt_n, dz_n)$. We also call $I_n(f)$ the n-fold iterated integral of f. For any $g \in \tilde{L}^2((\lambda \times \nu)^m)$ and $f \in \tilde{L}^2((\lambda \times \nu)^n)$, we have the following relations

$$E[I_m(g) I_n(f)] = \begin{cases} 0, & n \ne m \\ (g, f)_{L^2((\lambda \times \nu)^n)}, & n = m \end{cases} \quad (m, n = 1, 2, ...), \qquad (10.3)$$

where

$$(g, f)_{L^2((\lambda \times \nu)^n)} := \int_{([0,T] \times \mathbb{R}_0)^n} g(t_1, z_1, ..., t_n, z_n) f(t_1, z_1, ..., t_n, z_n) dt_1 \nu(dz_1) \\ \cdots dt_n \nu(dz_n).$$

10.2 The Wiener–Itô Chaos Expansion

We can now formulate the chaos expansion with respect to the Poisson random measure. See [121].

Theorem 10.2. Wiener–Itô chaos expansion for Poisson random measures. *Let $F \in L^2(P)$ be a $\mathcal{F}_T$-measurable random variable. Then F admits the representation*

$$F = \sum_{n=0}^{\infty} I_n(f_n) \tag{10.4}$$

via a unique sequence of elements $f_n \in \tilde{L}^2((\lambda \times \nu)^n)$, $n = 1, 2, \ldots$. Here we set $I_0(f_0) := f_0$ for the constant values $f_0 \in \mathbb{R}$. Moreover, we have that

$$\|F\|_{L^2(P)}^2 = \sum_{n=0}^{\infty} n! \|f_n\|_{L^2((\lambda \times \nu)^n)}^2. \tag{10.5}$$

Proof The proof follows the lines of the proof of Theorem 1.10; here we sketch only the arguments.

By the Itô representation theorem, there exists a predictable process $\psi_1(t_1, z_1)$, $(t_1, z_1) \in [0, T] \times \mathbb{R}_0$, such that

$$F = E[F] + \int_0^T \int_{\mathbb{R}_0} \Psi_1(t_1, z_1) \tilde{N}(dt_1, dz_1),$$

where

$$\|F\|_{L^2(P)}^2 = \left(E[F]\right)^2 + E\left[\int_0^T \int_{\mathbb{R}_0} \Psi_1^2(t_1, z_1) dt_1 \nu(dz_1)\right] < \infty.$$

Hence for almost all $(t_1, z_1) \in [0, T] \times \mathbb{R}_0$, there exists a predictable process $\Psi_2(t_1, z_1, t_2, z_2)$, $(t_2, z_2) \in [0, t_1] \times \mathbb{R}_0$, such that

$$\Psi(t_1, z_1) = E[\Psi(t_1, z_1)] + \int_0^T \int_{\mathbb{R}_0} \Psi_2(t_1, z_1, t_2, z_2) \tilde{N}(dt_2, dz_2).$$

This gives

$$F = E[F] + \int_0^T \int_{\mathbb{R}_0} E\left[\Psi_1(t_1, z_1)\right] \tilde{N}(dt_1, dz_1)$$

$$+ \int_0^T \int_{\mathbb{R}_0} \int_0^{t_1^-} \int_{\mathbb{R}_0} \Psi_2(t_1, z_1, t_2, z_2) \tilde{N}(dt_2, dz_2) \tilde{N}(dt_1, dz_1).$$

Define

$$g_0 := E[F]$$
$$g_1(t_1, z_1) := E[\Psi_1(t_1, z_1)], \quad (t_1, z_1) \in [0, T] \times \mathbb{R}_0.$$

We repeat the same argument for (t_2, z_2) – almost all the random variables of the integrand process $\Psi_2(t_1, z_1, t_2, z_2)$ and again for the new integrands. This yields

$$F = \sum_{n=0}^{k-1} J(g_n) + \int_{G_k} \Psi_k(t_1, z_1, \ldots, t_k, z_k) \tilde{N}^{\otimes k}(dt, dz).$$

Proceeding as in the proof of Theorem 1.10, we can see that the residual term here above vanishes, that is,

$$\int_{G_k} \Psi_k \tilde{N}^{\otimes k}(dt, dz) \longrightarrow 0, \quad k \to \infty,$$

with the convergence in $L^2(P)$. This gives the chaos expansion

$$F = \sum_{n=0}^{\infty} J_n(g_n)$$

in $L^2(P)$ with $g_n \in L^2(G_n)$, $n = 1, 2, \ldots$. Extend the function g_n on the whole $([0, T] \times \mathbb{R}_0)^n$ by putting $g_n := 0$ on $([0, T] \times \mathbb{R}_0)^n \setminus G_n$ and define $f_n := \tilde{g}_n$. Then

$$I_n(f_n) = n! J_n(f_n) = n! J_n(\tilde{g}_n) = J_n(g_n).$$

Thus we obtain the claim

$$F = \sum_{n=0}^{\infty} I_n(f_n).$$

Moreover, the isometry (10.5) is obtained directly from (10.3).   □

Example 10.3. Let $F = \eta^2(T)$. From (9.34) we have

$$\begin{aligned}
\eta^2(T) &= T \int_{\mathbb{R}_0} z^2 \nu(dz) + \int_0^T \int_{\mathbb{R}_0} \left(2\eta(t^-)z + z^2\right) \tilde{N}(dt, dz) \\
&= T \int_{\mathbb{R}_0} z^2 \nu(dz) + \int_0^T \int_{\mathbb{R}_0} z_1^2 \tilde{N}(dt_1, dz_1) \\
&\quad + \int_0^T \int_{\mathbb{R}_0} \int_0^{t_2^-} \int_{\mathbb{R}_0} 2z_1 z_2 \tilde{N}(dt_2, dz_2) \tilde{N}(dt_1, dz_1) \\
&= \sum_{n=0}^{2} I_n(f_n),
\end{aligned} \tag{10.6}$$

with

$$\begin{aligned}
f_0 &:= T \int_{\mathbb{R}_0} z^2 \nu(dz) \\
f_1(t_1, z_1) &:= z_1^2 \\
f_2(t_1, z_1, t_2, z_2) &= z_1 z_2.
\end{aligned}$$

Example 10.4. Let $F = Y(T)$, where

$$\begin{aligned}
Y(t) = \exp\Big\{ &\int_0^t \int_{\mathbb{R}_0} h(s) z \tilde{N}(ds, dz) \\
&- \int_0^t \int_{\mathbb{R}_0} \left(e^{h(s)z} - 1 - h(s)z\right) \nu(dz) ds \Big\}, \quad t \in [0, T],
\end{aligned}$$

with $h \in L^2([0, T])$ is a càglàd real function. Then by the Itô formula we have

$$dY(t) = Y(t^-) \int_{\mathbb{R}_0} \left(e^{h(t)z} - 1\right) \tilde{N}(dt, dz)$$

and thus

$$Y(T) = 1 + \int_0^T \int_{\mathbb{R}_0} Y(t^-)\left(e^{h(t)z} - 1\right) \tilde{N}(dt, dz).$$

Repeating the argument, we get

$$Y(T) = 1 + \int_0^T \int_{\mathbb{R}_0} \left(1 + \int_0^{t_1^-} \int_{\mathbb{R}_0} Y(t_2^-)\left(e^{h(t_2)z_2} - 1\right) \tilde{N}(dt_2, dz_2)\right)$$
$$\cdot \left(e^{h(t_1)z_1} - 1\right) \tilde{N}(dt_1, dz_1)$$
$$= 1 + \int_0^T \int_{\mathbb{R}_0} \left(e^{h(t_1)z_1} - 1\right) \tilde{N}(dt_1, dz_1)$$
$$+ \int_0^T \int_{\mathbb{R}_0} \int_0^{t_1^-} \int_{\mathbb{R}_0} Y(t_2^-)\left(e^{h(t_2)z_2} - 1\right)\left(e^{h(t_1)z_1} - 1\right) \tilde{N}(dt_2, dz_2) \tilde{N}(dt_1, dz_1).$$

Proceeding by iteration, we obtain

$$Y(T) = \sum_{n=0}^{k-1} I_n(f_n)$$
$$+ \int_0^T \int_{\mathbb{R}_0} \cdots \int_0^{t_2^-} \int_{\mathbb{R}_0} Y(t_1^-) \prod_{i=1}^k \left(e^{h(t_i)z_i} - 1\right) \tilde{N}(dt_1, dz_1) \cdots \tilde{N}(dt_k, dz_k),$$

where

$$f_n(t_1, z_1, ..., t_n, z_n) := \frac{1}{n!} \prod_{i=1}^n \left(e^{h(t_i)z_i} - 1\right)$$
$$= \frac{1}{n!} \left(e^{h(t)z} - 1\right)^{\otimes n}(t_1, z_1, ..., t_n, z_n), \tag{10.7}$$

which leads to the chaos expansion

$$Y(T) = \sum_{n=0}^{\infty} I_n(f_n),$$

with convergence in $L^2(P)$. To prove this we need to verify that

$$E\left[\left(\int_0^T \int_{\mathbb{R}_0} \cdots \int_0^{t_2^-} \int_{\mathbb{R}_0} Y(t_1^-)\left(e^{h(t)z} - 1\right)^{\otimes k} \tilde{N}^{\otimes k}(dt, dz)\right)^2\right] \longrightarrow 0, \quad k \to \infty.$$

This follows from the estimate

$$\int_0^T \int_{\mathbb{R}_0} \cdots \int_0^{t_2} \int_{\mathbb{R}_0} E\big[Y^2(t_1^-)\big]\big|\big(e^{h(t)z}-1\big)^{\otimes k}\big|^2 dt_1\nu(dz_1)\cdots dt_k\nu(dz_k)$$

$$\leq E\big[Y^2(T)\big]\frac{1}{k!}\Big(\int_0^T \int_{\mathbb{R}_0} \big(e^{h(t_1)z_1}-1\big)^2 dt_1\nu(dz_1)\Big)^k \longrightarrow 0, \quad k \to \infty.$$

10.3 Exercises

Problem 10.1. (*) Let

$$\eta(t) = \int_0^t \int_{\mathbb{R}_0} z\tilde{N}(ds, dz), \quad t \in [0, T].$$

Find the chaos expansion of

(a) $F = \eta^3(T)$
(b) $F = \exp \eta(T)$
(c) $F = \int_0^T g(s)d\eta(s)$, where $g \in L^2([0, T])$
(d) $F = \int_0^T g(s)\eta(s)ds$, where $g \in L^2([0, T])$

Problem 10.2. With η as in Problem 10.1, let $\mathbb{F}^\eta = \{\mathcal{F}_t^\eta, \ t \in [0, T]\}$ be the P-augmented filtration of η. Find $F \in L^2(P)$, $\mathcal{F}_T^\eta$-measurable, for which it is *not* possible to write a chaos expansion in terms of $d\eta$, that is,

$$F = E[F] + \sum_{n=1}^{\infty} n! \int_0^T \int_0^{t_n} \cdots \int_0^{t_2} f_n(t_1, ..., t_n)d\eta(t_1) \cdots d\eta(t_n)$$

for a sequence $f_n \in \tilde{L}^2([0, T]^n)$, $n = 1, 2, ...$ [*Hint.* See Example 9.15].

11

Skorohod Integrals

We now use the chaos expansions of Theorem 10.2 to define the *Skorohod integral* with respect to the compensated Poisson random measure $\tilde{N}$. The approach will be similar to the approach used in Chap. 2 for Brownian motion.

11.1 The Skorohod Integral

Definition 11.1. *Let $X = X(t, z)$, $0 \leq t \leq T, z \in \mathbb{R}_0$, be a stochastic process (more precisely a* random field*) such that $X(t, z)$ is an $\mathcal{F}_T$-measurable random variable for all $(t, z) \in [0, T] \times \mathbb{R}_0$ and*

$$E\left[\int_0^T \int_{\mathbb{R}_0} X^2(t, z)\nu(dz)dt\right] < \infty.$$

Then for each (t, z), the random variable $X(t, z)$ has an expansion of the form

$$X(t, z) = \sum_{n=0}^{\infty} I_n(f_n(\cdot, t, z)), \quad where \quad f_n(\cdot, t, z) \in \tilde{L}^2((\lambda \times \nu)^n).$$

Let $\tilde{f}_n(t_1, z_1, ..., t_n, z_n, t_{n+1}, z_{n+1})$ be the symmetrization of $f_n(t_1, z_1, ..., t_n, z_n, t, z)$ as a function of the $n+1$ pairs $(t_1, z_1), ..., (t_n, z_n), (t, z) = (t_{n+1}, z_{n+1})$. Suppose that

$$\sum_{n=0}^{\infty} (n+1)! \|\tilde{f}_n\|_{L^2((\lambda \times \nu)^{n+1})}^2 < \infty. \tag{11.1}$$

Then we say that X is Skorohod integrable and we write $X \in Dom(\delta)$. We define the Skorohod integral *$\delta(X)$ of X with respect to $\tilde{N}$ by*

$$\delta(X) = \int_0^T \int_{\mathbb{R}_0} X(t, z)\tilde{N}(\delta t, dz) := \sum_{n=0}^{\infty} I_{n+1}(\tilde{f}_n). \tag{11.2}$$

G.Di Nunno et al., *Malliavin Calculus for Lévy Processes with Applications to Finance*, © Springer-Verlag Berlin Heidelberg 2009

This definition was first given in [125], see also, for example, [126].

Remark 11.2. (1)   Note that

$$E\Big[\Big(\int_0^T \int_{\mathbb{R}_0} X(t,z)\tilde{N}(\delta t, dz)\Big)^2\Big] = \sum_{n=0}^{\infty}(n+1)!\|\tilde{f}_n\|^2_{L^2((\lambda\times\nu)^{n+1})} < \infty,$$

(11.3)

so $\delta(X) \in L^2(P)$;

(2)   Moreover,

$$E\Big[\int_0^T \int_{\mathbb{R}_0} X(t,z)\tilde{N}(\delta t, dz)\Big] = 0.$$

(11.4)

Next we consider Skorohod integrals with respect to

$$\eta(t) = \int_0^t \int_{\mathbb{R}_0} z\tilde{N}(dt, dz), \quad t \in [0, T].$$

Let $Y = Y(t)$, $t \in [0, T]$, be a measurable stochastic process such that

$$X(t, z) := Y(t)z, \quad (t, z) \in [0, T] \times \mathbb{R}_0,$$

is Skorohod integrable with respect to $\tilde{N}$. Then we define the *Skorohod integral of Y with respect to η* by

$$\int_0^T Y(t)\delta\eta(t) := \int_0^T \int_{\mathbb{R}_0} Y(t)z\tilde{N}(\delta t, dz).$$

(11.5)

11.2 The Skorohod Integral as an Extension of the Itô Integral

Just as in Chap. 2 we can now prove that the Skorohod integral is an extension of the Itô integral.

Theorem 11.3. *(a) Let $X = X(t, z)$, $t \in [0, T], z \in \mathbb{R}_0$, be a predictable process such that*

$$E\Big[\int_0^T \int_{\mathbb{R}_0} X^2(t, z)\nu(dz)dt\Big] < \infty.$$

(11.6)

Then X is both Itô and Skorohod integrable with respect to $\tilde{N}$ and

$$\int_0^T \int_{\mathbb{R}_0} X(t, z)\tilde{N}(\delta t, dz) = \int_0^T \int_{\mathbb{R}_0} X(t, z)\tilde{N}(dt, dz).$$

(11.7)

(b) Let $Y = Y(t)$, $t \in [0, T]$, be a predictable process such that

$$E\left[\int_0^T Y^2(t)dt\right] < \infty. \tag{11.8}$$

Then Y is both Itô and Skorohod integrable with respect to η and

$$\int_0^T Y(t)\delta\eta(t) = \int_0^T Y(t)d\eta(t). \tag{11.9}$$

Proof The proof is similar to the proof of Theorem 2.9 and therefore omitted.□

Example 11.4. What is $\int_0^T \eta(T)\delta\eta(t)$? Since

$$\eta(T) = \int_0^T \int_{\mathbb{R}_0} z\tilde{N}(dt, dz) = I_1(f_1(t_1, z_1)),$$

where $f_1(t_1, z_1) = z_1$, we have

$$\int_0^T \eta(T)\delta\eta(t) = \int_0^T \int_{\mathbb{R}_0} \eta(T)z\tilde{N}(\delta t, dz)$$

$$= \int_0^T \int_{\mathbb{R}_0} I_1(z_1 z)\tilde{N}(\delta t, dz) = I_2(z_1 z_2)$$

$$= 2 \int_0^T \int_{\mathbb{R}_0} \int_0^{t_2^-} \int_{\mathbb{R}_0} z_1 z_2 \tilde{N}(dt_1, dz_1)\tilde{N}(dt_2, dz_2)$$

$$= 2 \int_0^T \int_{\mathbb{R}_0} z_2 \eta(t_2^-)\tilde{N}(dt_2, dz_2) = 2 \int_0^T \eta(t_2^-)d\eta(t_2).$$

By the Itô formula we get

$$2 \int_0^T \eta(t^-)d\eta(t) = \eta^2(T) - \int_0^T \int_{\mathbb{R}_0} z^2 N(dt, dz).$$

Hence,

$$\int_0^T \eta(T)\delta\eta(t) = 2 \int_0^T \eta(t^-)d\eta(t) = \eta^2(T) - \int_0^T \int_{\mathbb{R}_0} z^2 N(dt, dz). \tag{11.10}$$

Additional properties of the Skorohod integral will be proved in Sect. 12.3 (the relation between the Skorohod integral and the Malliavin derivative) and in Sect. 13.3 (the relation between the Skorohod integral and the Wick product).

11.3 Exercises

Problem 11.1. Prove (11.3) and (11.4).

Problem 11.2. (*) Find the following Skorohod integrals:

(a) $\int_0^T \left(\int_0^T g(s)d\eta(s) \right) f(t)\delta\eta(t)$, where f, g are càglàd functions in $L^2([0,T])$

(b) $\int_0^T \left(\int_0^T f(t)d\eta(t) \right) g(s)\delta\eta(s)$, with f, g as in (a)

(c) By comparing (a) and (b) deduce the following *Fubini type formula* for Skorohod integrals

$$\int_0^T \left(\int_0^T g(s)d\eta(s) \right) f(t)\delta\eta(t) = \int_0^T \left(\int_0^T f(t)d\eta(t) \right) g(s)\delta\eta(s) \quad (11.11)$$

Problem 11.3. Let $g \in L^2(\nu) = L^2(\mathbb{R}_0)$ and define

$$X = \int_0^T \int_{\mathbb{R}_0} g(z)\widetilde{N}(dt, dz).$$

Express the following Skorohod integrals in terms of iterated Itô integrals:

(a) $\int_0^T \int_{\mathbb{R}_0} X\widetilde{N}(\delta t, dz)$,

(b) $\int_0^T \int_{\mathbb{R}_0} X^2\widetilde{N}(\delta t, dz)$.

12

The Malliavin Derivative

12.1 Definition and Basic Properties

In the Brownian motion case we saw that there were several ways of defining the Malliavin derivative:

(1) Either as a *stochastic gradient*, using the concept of directional derivatives, either on the Wiener space as in Appendix A (see Definition A.10) or on the space $\Omega = \mathcal{S}'(\mathbb{R}_0)$ as in Chap. 6 (see Definition 6.1).
(2) Or by means of the *chaos expansion* in terms of iterated integrals with respect to Brownian motion (see Lemma A.20).

In the Brownian motion case, those approaches are equivalent and they lead to "essentially" the same differential operator.

We now consider the pure jump martingale case, when

$$\eta(t) := \int_0^t \int_{\mathbb{R}_0} z \tilde{N}(ds, dz), \quad t \in [0, T].$$

In this case, it turns out that the two approaches do not give the same operator and it is necessary to make a choice about which gives the most useful derivative concept. For several reasons we choose the approach based on the chaos expansions (see Theorem 9.15). For example, this is a definition that gives us a Clark–Ocone type theorem for compensated Poisson random measures similar to Theorem 4.1 for the Brownian motion, see Theorem 12.16. For the other approach to the Malliavin calculus we refer to [35, 58].

Definition 12.1. *The* stochastic Sobolev space $\mathbb{D}_{1,2}$ *consists of all* $\mathcal{F}_T$-*measurable random variables* $F \in L^2(P)$ *with chaos expansion*

$$F = \sum_{n=0}^{\infty} I_n(f_n), \quad f_n \in \widetilde{L}^2((\lambda \times \nu)^n),$$

G.Di Nunno et al., *Malliavin Calculus for Lévy Processes with Applications to Finance,*
© Springer-Verlag Berlin Heidelberg 2009

satisfying the convergence criterion

$$\|F\|_{\mathbb{D}_{1,2}}^2 := \sum_{n=1}^{\infty} nn! \|f_n\|_{L^2((\lambda \times \nu)^n)}^2 < \infty. \tag{12.1}$$

Comparing the aforementioned condition with (10.5) we see that $\mathbb{D}_{1,2}$ is strictly contained in the space of all $\mathcal{F}_T$-measurable random variables in $L^2(P)$.

Definition 12.2. *We define the operator D:*

$$L^2(P) \supset \mathbb{D}_{1,2} \ni F \quad \Longrightarrow \quad DF \in L^2(P \times \lambda \times \nu)$$

by

$$D_{t,z}F = \sum_{n=1}^{\infty} n I_{n-1}(f_n(\cdot, t, z)), \qquad F \in \mathbb{D}_{1,2}. \tag{12.2}$$

Here $I_{n-1}(f_n(\cdot, t, z))$ means that the $(n-1)$-fold iterated integral of f_n is regarded as a function of its $(n-1)$ first pairs of variables $(t_1, z_1), ..., (t_{n-1}, z_{n-1})$, while the final pair (t, z) is kept as a parameter. In view of Definition 3.1 for the Brownian motion, it is natural to call $D_{t,z}F$ the Malliavin derivative of F at (t, z).

Note that we indeed have that $DF \in L^2(P \times \lambda \times \nu)$ because

$$\|DF\|_{L^2(\lambda \times \nu \times P)}^2 = \int_0^T \int_{\mathbb{R}_0} E[(D_{t,z}F)^2] \nu(dz) dt$$

$$= \int_0^T \int_{\mathbb{R}_0} \sum_{n=1}^{\infty} n^2 (n-1)! \|f_n(\cdot, t, z)\|_{L^2((\lambda \times \nu)^{n-1})}^2 \nu(dz) dt \tag{12.3}$$

$$= \sum_{n=1}^{\infty} nn! \|f_n\|_{L^2((\lambda \times \nu)^n)}^2 = \|F\|_{\mathbb{D}_{1,2}}^2 < \infty.$$

Example 12.3. Choose $F = \int_0^T \int_{\mathbb{R}_0} f(t, z) \tilde{N}(dt, dz)$, with the deterministic integrand $f \in L^2(\lambda \times \nu)$. Then $F = I_1(f)$ and hence

$$D_{t,z}F = I_0(f(\cdot, t, z)) = f(t, z). \tag{12.4}$$

In particular, if $F = \eta(T) := \int_0^T \int_{\mathbb{R}_0} z \tilde{N}(dt, dz)$, then

$$D_{t,z}\eta(T) = z. \tag{12.5}$$

Example 12.4. Let $F = \eta^2(T)$, then by (10.6) we have

$$\eta^2(T) = I_0(f_0) + I_1(f_1) + I_2(f_2),$$

where

$$f_0 = T \int_{\mathbb{R}_0} z^2 \nu(dz)$$

$$f_1(t_1, z_1) = z_1^2$$

$$f_2(t_1, z_1, t_2, z_2) = z_1 z_2.$$

Hence, by (12.2),

$$D_{t,z} \eta^2(T) = z^2 + 2I_1(f_2(\cdot, t, z))$$

$$= z^2 + 2 \int_0^T \int_{\mathbb{R}_0} z_1 z \tilde{N}(dt_1, dz_1) \tag{12.6}$$

$$= z^2 + 2\eta(T)z.$$

Since $D_{t,z} \eta(T) = z$ (see (12.5)) we conclude that

$$D_{t,z} \eta^2(T) = 2\eta(T) D_{t,z} \eta(T) + \left(D_{t,z} \eta(T)\right)^2$$

$$= \left(\eta(T) + D_{t,z} \eta(T)\right)^2 - \eta^2(T). \tag{12.7}$$

This shows that D does not satisfy the usual chain rule of a differential operator. In fact, it illustrates that D is a *difference* operator and not a differential operator.

Example 12.5. Let $F = Y(T)$, as in Example 10.4. Then by (10.7) we have

$$D_{t,z} F = \sum_{n=1}^{\infty} n I_{n-1}(f_n(\cdot, t, z)) = \sum_{n=1}^{\infty} \frac{n}{n!} (e^{h(t)z} - 1) I_{n-1}(e^{h(t)z} - 1)^{\otimes(n-1)}$$

$$= (e^{h(t)z} - 1) \sum_{n=1}^{\infty} \frac{1}{(n-1)!} I_{n-1}(e^{h(t)z} - 1)^{\otimes(n-1)}$$

$$= F(e^{h(t)z} - 1).$$

Theorem 12.6. Closability of the Malliavin derivative. *Suppose $F \in L^2(P)$ and F_k, $k = 1, 2, ...,$ are in $\mathbb{D}_{1,2}$ and that*

(1) $F_k \longrightarrow F$, $k \to \infty$ in $L^2(P)$
(2) $D_{t,z} F_k$, $k = 1, 2, ...,$ *converges in* $L^2(P \times \lambda \times \nu)$.

Then $F \in \mathbb{D}_{1,2}$ and

$$D_{t,z} F_k \longrightarrow D_{t,z} F, \quad k \to \infty, \quad in \ L^2(P \times \lambda \times \nu).$$

Proof Let $F = \sum_{n=0}^{\infty} I_n(f_n)$ and $F_k = \sum_{n=0}^{\infty} I_n(f_n^{(k)})$, $k = 1, 2, ...$ From (1) we know that

$$f_n^{(k)} \longrightarrow f_n, \quad k \to \infty, \quad in \ L^2((\lambda \times \nu)^n)$$

for all $n = 0, 1, \ldots$. Since (2) holds, we deduce that

$$\sum_{n=0}^{\infty} nn! \|f_n^{(k)} - f_n^{(j)}\|_{L^2((\lambda \times \nu)^n)}^2 = \|D_{t,z}F_k - D_{t,z}F_j\|_{L^2(\lambda \times \nu \times P)}^2 \longrightarrow 0, \quad k, j \to \infty.$$

Hence by the Fatou lemma,

$$\lim_{k \to \infty} \sum_{n=0}^{\infty} nn! \|f_n^{(k)} - f_n\|_{L^2((\lambda \times \nu)^n)}^2$$

$$\le \lim_{k \to \infty} \left(\lim_{j \to \infty} \sum_{n=0}^{\infty} nn! \|f_n^{(k)} - f_n^{(j)}\|_{L^2((\lambda \times \nu)^n)}^2 \right) = 0,$$

which means that $F \in \mathbb{D}_{1,2}$ and

$$D_{t,z}F_k \longrightarrow D_{t,z}F, \quad k \to \infty, \quad \text{in } L^2(P \times \lambda \times \nu). \quad \square$$

12.2 Chain Rules for Malliavin Derivative

As in Example 12.5, let us consider

$$G_1 = \exp \left\{ \int_0^T \int_{\mathbb{R}_0} h_1(s) z \tilde{N}(ds, dz) \right\}, \tag{12.8}$$

with $h_1 \in L^2([0, T])$. Its derivative can be written as

$$D_{t,z}G_1 = G_1(e^{h_1(t)z} - 1). \tag{12.9}$$

Let $\mathbb{D}_{1,2}^{\mathcal{E}}$ denote the set of linear combinations of such exponentials. Now choose $G_2 = \exp\{\int_0^T \int_{\mathbb{R}_0} h_2(t) z \tilde{N}(dt, dz)\} \in \mathbb{D}_{1,2}^{\mathcal{E}}$. Then from the above

$$D_{t,z}(G_1 G_2) = D_{t,z}\left(\exp\left\{ \int_0^T \int_{\mathbb{R}_0} (h_1(t) + h_2(t)) z \tilde{N}(dt, dz) \right\} \right)$$

$$= G_1 G_2 \left(e^{(h_1(t) + h_2(t))z} - 1 \right)$$

$$= \left(G_1 + G_1(e^{h_1(t)z} - 1) \right)\left(G_2 + G_2(e^{h_2(t)z} - 1) \right) - G_1 G_2$$

$$= \left(G_1 + D_{t,z}G_1 \right)\left(G_2 + D_{t,z}G_2 \right) - G_1 G_2$$

$$= G_1 D_{t,z}G_2 + G_2 D_{t,z}G_1 + D_{t,z}G_1 \, D_{t,z}G_2.$$

By linearity this continues to hold if we replace G_1 and G_2 by linear combinations F_1, F_2 of such exponentials. This proves the following result.

Theorem 12.7. Product rule. *Let* $F, G \in \mathbb{D}_{1,2}^{\mathcal{E}}$. *Then* $FG \in \mathbb{D}_{1,2}^{\mathcal{E}}$ *and*

$$D_{t,z}(FG) = F D_{t,z}G + G D_{t,z}F + D_{t,z}F D_{t,z}G. \tag{12.10}$$

By induction it follows that if $F \in \mathbb{D}^{\mathcal{E}}_{1,2}$ then

$$D_{t,z}(F^n) = (F + D_{t,z}F)^n - F^n. \tag{12.11}$$

For a related result see also Lemma 6.1 in [172]. For an extension to the so-called normal martingales, see, for example, Proposition 1 in [196] or Proposition 5 in [199].

More generally we have the following result.

Theorem 12.8. Chain rule. *Let $F \in \mathbb{D}_{1,2}$ and let φ be a real continuous function on $\mathbb{R}$. Suppose $\varphi(F) \in L^2(P)$ and $\varphi(F + D_{t,z}F) \in L^2(P \times \lambda \times \nu)$. Then $\varphi(F) \in \mathbb{D}_{1,2}$ and*

$$D_{t,z}\varphi(F) = \varphi(F + D_{t,z}F) - \varphi(F). \tag{12.12}$$

Proof First assume that φ has compact support and $F \in \mathbb{D}^{\mathcal{E}}_{1,2}$. Then

$$\varphi(F) = \frac{1}{\sqrt{2\pi}} \int_{\mathbb{R}} e^{iyF} \hat{\varphi}(y)dy,$$

where

$$\hat{\varphi}(y) = \frac{1}{\sqrt{2\pi}} \int_{\mathbb{R}} e^{-ixy} \varphi(x)dx$$

is the Fourier transform of φ. By (12.11) and Theorem 12.6 we get that

$$D_{t,z}\varphi(F) = \frac{1}{\sqrt{2\pi}} \int_{\mathbb{R}} \sum_{n=0}^{\infty} \frac{1}{n!} (iy)^n \left((F + D_{t,z}F)^n - F \right) \hat{\varphi}(y)dy$$

$$= \frac{1}{\sqrt{2\pi}} \int_{\mathbb{R}} \left(e^{iy(F+D_{t,z}F)} - e^{iyF} \right) \hat{\varphi}(y)dy$$

$$= \varphi(F + D_{t,z}F) - \varphi(F),$$

so the result holds in this case. For general $F \in \mathbb{D}_{1,2}$ we proceed by approximation. Choose $F_n \in \mathbb{D}^{\mathcal{E}}_{1,2}$, $n = 1, 2, ...$, such that $F_n \longrightarrow F$, $n \to \infty$, in $\mathbb{D}_{1,2}$, see (12.1). Then $\varphi(F_n) \longrightarrow \varphi(F)$ in $L^2(P)$ and $\varphi(F_n + D_{t,z}F_n) - \varphi(F_n) \longrightarrow \varphi(F + D_{t,z}F) - \varphi(F)$ in $\mathbb{D}_{1,2}$. Hence the result holds for all $F \in \mathbb{D}_{1,2}$ in the case of φ with compact support. The extension to the case when $\varphi(F) \in L^2(P)$ and $\varphi(F + D_{t,z}F) \in L^2(P \times \lambda \times \nu)$ follows by a similar limit argument. $\square$

Example 12.9. The chain rule (12.12) is useful for the evaluation of Malliavin derivatives. To illustrate this, consider the following:

(1) The derivative of $\eta^2(T)$ is

$$D_{t,z}\eta^2(T) = \left(\eta(T) + D_{t,z}\eta(T) \right)^2 - \eta^2(T)$$

$$= \left(\eta(T) + z \right)^2 - \eta^2(T) = 2\eta(T)z + z^2,$$

which is what we found in (12.6).

(2) With

$$G = \exp\left(\int_0^T \int_{\mathbb{R}_0} h(t)z\tilde{N}(dt, dz)\right)$$

as in (12.8), the chain rule (12.12) gives

$$D_{t,z}G = \exp\left(\int_0^T \int_{\mathbb{R}_0} h(t)z\tilde{N}(dt, dz) + h(t)z\right) - G$$
$$= G\left(e^{h(t)z} - 1\right),$$

which is (12.9).

(3) Let $F = (\eta(T) - K)^+$ be a *European call payoff*, where $K > 0$ is a constant. Then

$$D_{t,z}F = \left(\eta(T) + z - K\right)^+ - \left(\eta(T) - K\right)^+. \qquad (12.13)$$

12.3 Malliavin Derivative and Skorohod Integral

In this section we explore the relationship between the Malliavin derivative and the Skorohod integral following the same lines as in the Brownian motion case. We also derive useful rules of calculus.

12.3.1 Skorohod Integral as Adjoint Operator to the Malliavin Derivative

For the following result we can also refer to [29, 54, 69, 172].

Theorem 12.10. Duality formula. *Let $X(t, z)$, $t \in [0, T], z \in \mathbb{R}$, be Skorohod integrable and $F \in \mathbb{D}_{1,2}$. Then*

$$E\left[\int_0^T \int_{\mathbb{R}_0} X(t, z)D_{t,z}F\nu(dz)dt\right] = E\left[F \int_0^T \int_{\mathbb{R}_0} X(t, z)\tilde{N}(\delta t, dz)\right].$$
$$(12.14)$$

Proof The proof of this is the same as the proof of the corresponding result in the Brownian motion case. See Theorem 3.14. □

12.3.2 Integration by Parts and Closability of the Skorohod Integral

The following result is basically Theorem 7.1 in [172], here presented in the setting of Poisson random measures.

Theorem 12.11. Integration by parts. *Let $X(t,z)$, $t \in [0,T], z \in \mathbb{R}$, be a Skorohod integrable stochastic process and $F \in \mathbb{D}_{1,2}$ such that the product $X(t,z) \cdot (F + D_{t,z}F)$, $t \in [0,T], z \in \mathbb{R}$, is Skorohod integrable. Then*

$$F \int_0^T \int_{\mathbb{R}_0} X(t,z) \widetilde{N}(\delta t, dz) \tag{12.15}$$

$$= \int_0^T \int_{\mathbb{R}_0} X(t,z)(F + D_{t,z}F) \widetilde{N}(\delta t, dz) + \int_0^T \int_{\mathbb{R}_0} X(t,z) D_{t,z}F \nu(dz)dt.$$

Proof First assume that $F \in \mathbb{D}_{1,2}^{\mathcal{E}}$. Let $G \in \mathbb{D}_{1,2}^{\mathcal{E}}$. Then we obtain by Theorem 12.10 and Theorem 12.7

$$E\left[G \int_0^T \int_{\mathbb{R}_0} FX(t,z)\widetilde{N}(\delta t, dz)\right] = E\left[\int_0^T \int_{\mathbb{R}_0} FX(t,z)D_{t,z}G\nu(dz)dt\right]$$

$$= E\left[GF \int_0^T \int_{\mathbb{R}_0} X(t,z)\widetilde{N}(\delta t, dz)\right] - E\left[G \int_0^T \int_{\mathbb{R}_0} X(t,z)D_{t,z}F\nu(dz)dt\right]$$

$$- E\left[G \int_0^T \int_{\mathbb{R}_0} X(t,z)D_{t,z}F\widetilde{N}(\delta t, dz)\right]$$

$$= E\left[G\left(F \int_0^T \int_{\mathbb{R}_0} X(t,z)\widetilde{N}(\delta t, dz) - \int_0^T \int_{\mathbb{R}_0} X(t,z)D_{t,z}F\nu(dz)dt\right.\right.$$

$$\left.\left. - \int_0^T \int_{\mathbb{R}_0} X(t,z)D_{t,z}F\widetilde{N}(\delta t, dz)\right)\right].$$

The proof then follows by a density argument applied to F and G.   □

Remark 12.12. Using the Poisson interpretation of Fock space, the formula (12.15) has been shown to be an expression of the multiplication formula for Poisson stochastic integrals. See [125, 220], Proposition 2 and Relation (6) of [197], Definition 7 and Proposition 6 of [201], Proposition 2 of [199], and Proposition 1 of [195]. Moreover, formula (12.15) has been known for some time to quantum probabilists in identical or close formulations. See Proposition 21.6 and Proposition 21.8 in [188], Proposition 18 in [34], and Relation (5.6) in [7], see also [127].

Theorem 12.13. Closability of the Skorohod integral. *Suppose that $X_n(t,z)$, $t \in [0,T], z \in \mathbb{R}$, is a sequence of Skorohod integrable random fields and that the corresponding sequence of integrals*

$$I(X_n) := \int_0^T \int_{\mathbb{R}_0} X_n(t,z)\widetilde{N}(\delta t, dz), \quad n = 1, 2, \dots$$

converges in $L^2(P)$. Moreover, suppose that

$$\lim_{n \to \infty} X_n = 0 \quad in \quad L^2(P \times \lambda \times \nu).$$

Then we have

$$\lim_{n \to \infty} I(X_n) = 0 \quad in \quad L^2(P).$$

Proof By Theorem (12.10) we have that

$$\left(I(X_n), F\right)_{L^2(P)} = \left(X_n, D_{t,z}F\right)_{L^2(P \times \lambda \times \nu)} \longrightarrow 0, \qquad n \to \infty,$$

for all $F \in \mathbb{D}_{1,2}$. Then we conclude that $\lim_{n \to \infty} I(X_n) = 0$ weakly in $L^2(P)$. And since the sequence $I(X_n)$, $n = 1, 2, ...$, is convergent in $L^2(P)$, the result follows. $\square$

Remark 12.14. In view of Theorem 12.13 we can see that if X_n, $n = 1, 2, ...$, is a sequence of Skorohod integrable random fields such that

$$X = \lim_{n \to \infty} X_n \quad in \quad L^2(P \times \lambda \times \nu).$$

Then we can define the *Skorohod integral of* X as

$$I(X) := \int_0^T \int_{\mathbb{R}_0} X(t,z)\widetilde{N}(\delta t, dz) = \lim_{n \to \infty} \int_0^T \int_{\mathbb{R}_0} X_n(t,z)\widetilde{N}(\delta t, dz) =: \lim_{n \to \infty} I(X_n),$$

provided that this limit exists in $L^2(P)$.

12.3.3 Fundamental Theorem of Calculus

The following result is basically Theorem 4.2 in [172], here presented for Poisson random measures, see, for example, [64].

Theorem 12.15. Fundamental theorem of calculus. *Let $X = X(s,y)$, $(s,y) \in [0,T] \times \mathbb{R}_0$, be a stochastic process such that*

$$E\left[\int_0^T \int_{\mathbb{R}_0} X^2(s,y)\nu(dy)ds\right] < \infty.$$

Assume that $X(s,y) \in \mathbb{D}_{1,2}$ for all $(s,y) \in [0,T] \times \mathbb{R}_0$, and that $D_{t,z}X(\cdot, \cdot)$ is Skorohod integrable with

$$E\left[\int_0^T \int_{\mathbb{R}_0} \left(\int_0^T \int_{\mathbb{R}_0} D_{t,z}X(s,y)\widetilde{N}(\delta s, dy)\right)^2 \nu(dz)dt\right] < \infty.$$

Then

$$\int_0^T \int_{\mathbb{R}_0} X(s,y)\widetilde{N}(\delta s, dy) \in \mathbb{D}_{1,2}$$

and

$$D_{t,z} \int_0^T \int_{\mathbb{R}_0} X(s,y)\tilde{N}(\delta s, dy) = \int_0^T \int_{\mathbb{R}_0} D_{t,z}X(s,y)\tilde{N}(\delta s, dy) + X(t,z).$$
$$(12.16)$$

In particular, if $X(s,y) = Y(s)y$, *then*

$$D_{t,z} \int_0^T Y(s)\delta\eta(s) = \int_0^T D_{t,z}Y(s)\delta\eta(s) + zY(t). \qquad (12.17)$$

Proof First suppose that

$$X(s,y) = I_n(f_n(\cdot, s, y)),$$

where $f_n(t_1, z_1, ..., t_n, z_n, s, y)$ is symmetric with respect to $(t_1, z_1), ..., (t_n, z_n)$. By Definition 3.1 we have

$$\int_0^T \int_{\mathbb{R}_0} X(s,y)\tilde{N}(\delta s, dy) = I_{n+1}(\widehat{f_n}), \qquad (12.18)$$

where

$$\widehat{f_n}(t_1, z_1, ..., t_n, z_n, t_{n+1}, z_{n+1})$$

$$= \frac{1}{n+1}[f_n(t_{n+1}, z_{n+1}, \cdot, t_1, z_1) + ... + f_n(t_{n+1}, z_{n+1}, \cdot, t_n, z_n)$$

$$+ f_n(t_1, z_1, \cdot, t_{n+1}, z_{n+1})]$$

is the symmetrization of f_n with respect to the variables $(t_1, z_1), ..., (t_n, z_n)$, $(t_{n+1}, z_{n+1}) = (s, y)$. Therefore, we get

$$D_{t,z}\left(\int_0^T \int_{\mathbb{R}_0} X(s,y)\tilde{N}(\delta s, dy)\right) = I_n(f_n(t, z, \cdot, t_1, z_1) + ... + f_n(t, z, \cdot, t_n, z_n)$$

$$+ f_n(\cdot, t, z)).$$

On the other hand we see that

$$\int_0^T \int_{\mathbb{R}_0} D_{t,z}X(s,y)\tilde{N}(\delta s, dy) \qquad (12.19)$$

$$= \int_0^T \int_{\mathbb{R}_0} nI_{n-1}(f_n(\cdot, t, z, s, y))\tilde{N}(\delta s, dy) = nI_n(\widehat{f_n}(\cdot, t, z, \cdot)),$$

where

$$\widehat{f_n}(t_1, z_1, ..., t_{n-1}, z_{n-1}, t, z, t_n, z_n) = \frac{1}{n}[f_n(t, z, \cdot, t_1, z_1) + ... + f_n(t, z, \cdot, t_n, z_n)]$$

is the symmetrization of $f_n(t_1, z_n, ..., t_{n-1}, z_{n-1}, t, z, t_n, z_n)$ with respect to $(t_1, z_1), ..., (t_{n-1}, z_{n-1})$, $(t_n, z_n) = (s, y)$. A comparison of (12.18) and (12.19) yields formula (12.16).

Next consider the general case

$$X(s, y) = \sum_{n \geq 0} I_n(f_n(\cdot, s, y)).$$

Define

$$X_m(s, y) = \sum_{n=0}^{m} I_n(f_n(\cdot, s, y)), \quad m = 1, 2, \ldots$$

Then (12.16) holds for X_m. Since

$$\left\| \int_0^T \int_{\mathbb{R}_0} D_{t,z} X_m(s, y) \tilde{N}(\delta s, dy) - \int_0^T \int_{\mathbb{R}_0} D_{t,z} X(s, y) \tilde{N}(\delta s, dy) \right\|_{L_2(P \times \lambda \times \nu)}^2$$

$$= \sum_{n \geq m+1} n^2 n! \left\| \widehat{f_n} \right\|_{L_2((\lambda \times \nu)^{n+1})}^2 \longrightarrow 0, \qquad m \longrightarrow \infty,$$

the proof follows by the closedness of $D_{t,z}$. $\square$

12.4 The Clark–Ocone Formula

In this section we state and prove a jump diffusion version of the Clark–Ocone formula (see Theorem 4.1). For this result we refer to, for example, [154].

Theorem 12.16. *Let $F \in \mathbb{D}_{1,2}$. Then*

$$F = E[F] + \int_0^T \int_{\mathbb{R}_0} E[D_{t,z} F | \mathcal{F}_t] \tilde{N}(dt, dz), \qquad (12.20)$$

where we have chosen a predictable version of the conditional expectation process $E[D_{t,z} F | \mathcal{F}_t]$, $t \geq 0$.

Proof The proof is similar to the one for the Brownian motion case (Theorem 4.1). Let us consider the chaos expansion of $F = \sum_{n=0}^{\infty} I_n(f_n)$, where $f_n \in \tilde{L}^2((\lambda \times \nu)^n)$, $n = 1, 2, \ldots$. Then the following equalities hold true:

$$\int_0^T \int_{\mathbb{R}_0} E[D_{t,z} F | \mathcal{F}_t] \tilde{N}(dt, dz) = \int_0^T \int_{\mathbb{R}_0} E\left[\sum_{n=1}^{\infty} n I_{n-1}(f_n(\cdot, t, z)) | \mathcal{F}_t \right] \tilde{N}(dt, dz)$$

$$= \int_0^T \int_{\mathbb{R}_0} \sum_{n=1}^{\infty} n(n-1)! E[J_{n-1}(f_n(\cdot, t, z)) | \mathcal{F}_t] \tilde{N}(dt, dz)$$

$$= \sum_{n=1}^{\infty} n! \int_0^T \int_{\mathbb{R}_0} E\left[\int_0^T \int_{\mathbb{R}_0} \cdots \int_0^{t_2^-} \int_{\mathbb{R}_0} f_n(t_1, z_1, \ldots, t_{n-1}, z_{n-1}, t, z) \tilde{N}(dt_1, dz_1) \right.$$

$$\left. \cdots \tilde{N}(dt_{n-1}, dz_{n-1}) | \mathcal{F}_t \right] \tilde{N}(dt, dz)$$

$$= \sum_{n=1}^{\infty} n! J_n(f_n) = \sum_{n=1}^{\infty} I_n(f_n) = F - E[F]. \quad \square$$

Remark 12.17. Comparing (12.20) with the Itô representation (9.33), we can see that the difference is that (12.20) provides an explicit formula for the process $\Psi(t, z)$, $t \geq 0, z \in \mathbb{R}_0$.

Example 12.18. Suppose $F \in \mathbb{D}_{1,2}$ has the form $F = \varphi(\eta(T))$ for some continuous real function $\varphi(x)$, $x \in \mathbb{R}$. Then by the Clark–Ocone theorem combined with the Markov property of the process η, we get

$$\varphi(\eta(T)) = E\left[\varphi(\eta(T))\right] + \int_0^T \int_{\mathbb{R}_0} E[\varphi(\eta(T) + z) - \varphi(\eta(T))|\mathcal{F}_t]\tilde{N}(dt, dz)$$

$$= E\left[\varphi(\eta(T))\right] + \int_0^T \int_{\mathbb{R}_0} E[\varphi(y + \eta(T - t) + z) \qquad (12.21)$$

$$- \varphi(y + \eta(T - t))]|_{y = \eta(t)}\tilde{N}(dt, dz).$$

12.5 A Combination of Gaussian and Pure Jump Lévy Noises

We now outline how the results of the previous sections can be generalized to the case of combinations of independent Gaussian and pure jump Lévy noise. Let us sketch a framework for treating this combination of noises. Here we follow the ideas in [2], though this work is settled in the white noise framework. The white noise setting will also be treated later in this book (see Chap. 13).

Another approach to deal with the noise generated by general stochastic measures with independent values can be found in [61]. See also [71].

Denote the probability space on which $W = W(t)$, $t \geq 0$, is a Wiener process by $(\Omega_0, \mathcal{F}_T^W, P^W)$ (see Sect. 1.1) and denote the one on which $\tilde{N}(dt, dz) = N(dt, dz) - \nu(dz)dt$ is a compensated Poisson random measure by $(\Omega_0, \mathcal{F}_T^{\tilde{N}}, P^{\tilde{N}})$ (see Sect. 9.1).

Let $(\Omega_1, \mathcal{F}_T^{(1)}, \mu_1), ..., (\Omega_N, \mathcal{F}_T^{(N)}, \mu_N)$ be N independent copies of $(\Omega_0, \mathcal{F}_T^W, P^W)$ and let $(\Omega_{N+1}, \mathcal{F}_T^{(N+1)}, \mu_{N+1}), ..., (\Omega_{N+R}, \mathcal{F}_T^{(N+R)}, \mu_{N+R})$ be R independent copies of $(\Omega_0, \mathcal{F}_T^{\tilde{N}}, P^{\tilde{N}})$, for some $N, R \in \mathbb{N} \cup \{0\}$. We set

$$\Omega = \Omega_1 \times ... \times \Omega_{N+R}, \quad \mathcal{F}_T = \mathcal{F}_T^{(1)} \otimes ... \otimes \mathcal{F}_T^{(N+R)}, \quad P = \mu_1 \times ... \times \mu_{N+R}.$$
$$(12.22)$$

In the sequel, we call the space $(\Omega, \mathcal{F}, P)$ the *Wiener–Poisson space*. We can consider the product of the form

$$\mathbb{H}_\alpha(\omega) := \prod_{k=1}^L I_{\alpha^{(k)}}(f_{k, \alpha^{(k)}})(\omega_k) \qquad (12.23)$$

for any $\alpha \in \mathcal{J}^L$, which is the set of indices of the form $\alpha = (\alpha^{(1)}, ..., \alpha^{(L)})$, with $\alpha^{(k)} = 0, 1, ...$, for $k = 1, ..., L$. Here $I_{\alpha^{(k)}}(f_{k, \alpha^{(k)}})$ is the $\alpha^{(k)}$-fold iterated

Itô integral with respect to the Wiener process, if $k = 1, ..., N$, or to the compensated Poisson random measure, if $k = N + 1, ..., L$.

The elements $\mathbb{H}_\alpha$, $\alpha \in \mathcal{J}^L$, constitute an orthogonal basis in $L^2(P)$. Any real $\mathcal{F}_T$-measurable random variable $F \in L^2(P)$ can be written as

$$F = \sum_{\alpha \in \mathcal{J}^L} \mathbb{H}_\alpha$$

for an appropriate choice of deterministic symmetric integrands in the iterated Itô integrals.

Definition 12.19. *(1) We say that $F \in \mathbb{D}_{1,2}$ if*

$$\|F\|_{\mathbb{D}_{1,2}}^2 := \sum_{k=1}^{N} \sum_{\alpha \in \mathcal{J}^L} \alpha^{(k)} \alpha^{(k)}! \|f_{k,\alpha^{(k)}}\|_{L^2([0,T]^{\alpha^{(k)}})}^2 \tag{12.24}$$

$$+ \sum_{k=N+1}^{L} \sum_{\alpha \in \mathcal{J}^L} \alpha^{(k)} \alpha^{(k)}! \|f_{k,\alpha^{(k)}}\|_{L^2(([0,T]\times\mathbb{R}_0)^{\alpha^{(k)}})}^2 < \infty.$$

(2) If $F \in \mathbb{D}_{1,2}$, we define the Malliavin *derivative DF of F as the gradient*

$$DF = \left(D_{1,t}F, ..., D_{N,t}F, D_{N+1,t,z}F, ..., D_{L,t,z}F\right), \tag{12.25}$$

where

$$D_{k,t}F = \sum_{\alpha \in \mathcal{J}^L} \alpha^{(k)} \mathbb{H}_{\alpha-\epsilon^{(k)}}(t), \quad t \in [0,T] \quad (k = 1, ..., N),$$

and

$$D_{k,t,z}F = \sum_{\alpha \in \mathcal{J}^L} \alpha^{(k)} \mathbb{H}_{\alpha-\epsilon^{(k)}}(t,z), \quad t \in [0,T], z \in \mathbb{R}_0 \quad (k = N+1, ..., L).$$

Here $\epsilon^{(k)} = (0, ...0, 1, 0, ...0)$ with 1 in the kth position, cf. Definition 3.1 and Definition 12.1.

Based on the same concepts and arguments as in the previous sections, one can show the following Clark–Ocone formula:

Theorem 12.20. Clark–Ocone theorem for combined Gaussian-pure jump Lévy noise. *Let $F \in \mathbb{D}_{1,2}$. Then*

$$F = E[F] + \sum_{k=1}^{N} \int_0^T E[D_{k,t}F|\mathcal{F}_t] dW_k(t) + \sum_{k=N+1}^{L} \int_0^T \int_{\mathbb{R}_0} E[D_{k,t,z}F|\mathcal{F}_t] \widetilde{N}_k(dt,dz).$$

$$\tag{12.26}$$

A generalization of this formula to processes with conditionally independent increments can be found in [228].

Similar to the Brownian motion case one can obtain a Clark–Ocone formula under change of measure for Lévy processes. This was first proved by [113] for random variables in $\mathbb{D}_{1,2}$. Subsequently the result was generalized in [185] to all random variables in $L^2(P)$, by means of white noise theory (see Chap. 13). We present the statement of this generalized version later without proof, and refer to the original papers for more information. We first recall the Girsanov theorem for Lévy processes, as presented in [183] (cf. Problem 9.5). See also [149].

Theorem 12.21. Girsanov theorem for Lévy processes. *Let* $\theta(s, x) \leq 1$, $s \in [0, T]$, $x \in \mathbb{R}_0$ *and* $u(s)$, $s \in [0, T]$, *be* $\mathbb{F}$*−predictable processes such that*

$$\int_0^T \int_{\mathbb{R}_0} \{|\log(1 + \theta(s, x))| + \theta^2(s, x)\}\nu(dx)dt < \infty \quad P\text{-}a.e., \qquad (12.27)$$

$$\int_0^T u^2(s)ds < \infty \quad P - a.e. \qquad (12.28)$$

Let

$$Z(t) = \exp\left\{-\int_0^t u(s)dW(s) - \int_0^t u^2(s)ds\right.$$
$$+ \int_0^t \int_{\mathbb{R}_0} \{\log(1 - \theta(s, x)) + \theta(s, x)\}\nu(dx)ds$$
$$\left. + \int_0^t \int_{\mathbb{R}_0} \log(1 - \theta(s, x))\tilde{N}(ds, dx)\right\}, \quad t \in [0, T].$$

Define a measure Q *on* $\mathcal{F}_T$ *by*

$$dQ(\omega) = Z(\omega, T)dP(\omega).$$

Assume that $Z(T)$ *satisfies the Novikov condition, that is,*

$$E\left[\exp\left(\frac{1}{2}\int_0^T u^2(s)ds + \int_0^T \int_{\mathbb{R}_0} \{(1 - \theta(s, x))\log(1 - \theta(s, x))\right.\right.$$
$$\left.\left. + \theta(s, x)\}\nu(dx)ds\right)\right] < \infty.$$

Then $E[Z(T)] = 1$ *and hence* Q *is a probability measure on* $\mathcal{F}_T$. *Define*

$$\tilde{N}_Q(dt, dx) = \theta(t, x)\nu(dx)dt + \tilde{N}(dt, dx)$$

and

$$dW_Q(t) = u(t)dt + dW(t).$$

Then $\tilde{N}_Q(\cdot, \cdot)$ *and* $W_Q(\cdot)$ *are compensated Poisson random measure of* $N(\cdot, \cdot)$ *and Brownian motion under* Q, *respectively.*

In this setting, the Clark–Ocone formula gets the following form. See [113, 185].

Theorem 12.22. Generalized Clark–Ocone theorem under change of measure for Lévy processes. *Let $F \in L^2(P) \cap L^2(Q)$ be $\mathcal{F}_T$-measurable. Assume that u satisfies (4.7), that $\theta \in L^2(P \times \lambda \times \nu)$ and that $(t,x) \to D_{t,x}\theta(s,z)$ is Skorohod integrable for all s,z, with $\delta(D_{t,x}\theta) \in \mathcal{G}^*$. Then the integral representation of F with respect to W_Q and $\tilde{N}_Q$ is as follows:*

$$F = E_Q[F] + \int_0^T E_Q\left[D_t F - F \int_t^T D_t u(s) dW_Q(s) \big| \mathcal{F}_t\right] dW_Q(t)$$

$$+ \int_0^T \int_{\mathbb{R}_0} E_Q\left[F(\tilde{H} - 1) + \tilde{H} D_{t,x} F | \mathcal{F}_t\right] \tilde{N}_Q(dt, dx),$$

where

$$\tilde{H} = \exp\left\{\int_0^t \int_{\mathbb{R}_0} \left[D_{t,x}\theta(s,z) + \log(1 - \frac{D_{t,x}\theta(s,z)}{1 - \theta(s,z)})(1 - \theta(s,z))\right]\nu(dz)\, ds \right.$$

$$\left. + \log(1 - \frac{D_{t,x}\theta(s,z)}{1 - \theta(s,z)})\tilde{N}_Q(ds, dz)\right\}.$$

12.6 Application to Minimal Variance Hedging with Partial Information

Consider a financial market where the unit prices $S_i(t)$, $t \geq 0$, of the assets are as follows:

$$\text{risk free asset} \quad S_0(t) \equiv 1, \quad t \in [0, T],$$

$$\text{risky assets} \quad dS_j(t) = \sigma_j(t) dW(t)$$

$$+ \int_{\mathbb{R}_0^n} \gamma_j(t,z)\tilde{N}(dt, dz), \quad t \in (0, T], \ j = 1, ..., n,$$

where $\sigma(t) = [\sigma_{i,j}(t)] \in \mathbb{R}^{n \times n}$ and $\gamma(t,z) = [\gamma_{i,j}(t,z)] \in \mathbb{R}^{n \times n}$ are predictable processes, which might depend on $S(s) = (S_1(s), ..., S_n(s))$, $s \in [0, t]$.

Let $\mathbb{E} = \{\mathcal{E}_t, t \geq 0\}$ be a given filtration such that

$$\mathcal{E}_t \subseteq \mathcal{F}_t, \quad t \in [0, T].$$

We think of $\mathcal{E}_t$ as the information available to an agent at time t.

Definition 12.23. *A predictable process $\varphi = \varphi(t)$, $t \in [0, T]$, is called admissible if*

(1) $\varphi(t)$ is $\mathbb{E}$-adapted
(2) $E\left[\sum_{j=1}^n \int_0^T \varphi_j^2(t)(\sum_{i=1}^n \sigma_{i,j}^2(t) + \int_{\mathbb{R}_0} \gamma_{i,j}^2(t,z)\nu(dz))dt\right] < \infty.$

The set of all $\mathbb{E}$-admissible portfolios is denoted by $\mathcal{A}_\mathbb{E}$.

We now pose the following question: given a claim $F \in L^2(\mathcal{F}_T)$, how close can we get to F at time T by hedging with portfolios? If we consider closeness in terms of variance, the precise formulation of this question is the following: given $F \in L^2(\mathcal{F}_T)$ find $\varphi^* \in \mathcal{A}_\mathbb{E}$ such that

$$\inf_{\varphi \in \mathcal{A}_\mathbb{E}} E\left[\left(F - E[F] - \int_0^T \varphi(t)dS(t)\right)^2\right] = E\left[\left(F - E[F] - \int_0^T \varphi^*(t)dS(t)\right)^2\right].$$
$$(12.29)$$

Such a portfolio φ^* is called a *partial information minimal variance portfolio*.

We use Malliavin calculus to obtain explicit formulae for such portfolios φ^*. We refer to [25] for the following result. See also [161] and [69] for an extension to the white noise setting and [59, 60] for market models driven by general martingales and random fields.

Theorem 12.24. *Suppose $F \in \mathbb{D}_{1,2}$. Then the partial information minimal variance portfolio $\varphi^* \in \mathcal{A}_\mathbb{E}$ for F is given by*

$$\varphi^*(t) = Q^{-1}(t)R(t), \quad t \in [0, T]. \tag{12.30}$$

Here $Q(t) \in \mathbb{R}^{n \times n}$ has components

$$Q_{ik}(t) = E\big[N_{ik}(t)|\mathcal{E}_t\big],$$

where

$$N_{ik}(t) = \sum_{j=1}^n \sigma_{ij}(t)\sigma_{jk}(t) + \int_{\mathbb{R}_0} \gamma_{ij}(t,z)\gamma_{jk}(t,z)\nu_j(dz) \qquad (i, j = 1, .., n).$$

The matrix $Q^{-1}(t)$ is the inverse of $Q(t)$ (if it exists). The vector $R(t) \in \mathbb{R}^n$ has components

$$R_i(t) = E\big[M_i|\mathcal{E}_t\big],$$

where

$$M_i(t) = \sum_{j=1}^n \sigma_{ij}(t)E\big[D_{j,t}F|\mathcal{F}_t\big] + \int_{\mathbb{R}_0} \gamma_{ij}(t,z)E\big[D_{j,t,z}F|\mathcal{F}_t\big]\nu_j(dz).$$

Moreover, $D_{j,t}F$, $t \in [0, T]$, denotes the Malliavin derivative with respect to W_j and $D_{j,t,z}F$, $t \in [0, T]$, $z \in \mathbb{R}_0$, stands for the Malliavin derivative with respect to $\tilde{N}_j$.

Proof Let φ_j^* be as above and define

$$\hat{F} := E[F] + \sum_{j=1}^n \int_0^T \varphi_j^*(t)dS_j(t)$$

$$= E[F] + \sum_{j=1}^n \int_0^T \varphi_j^*(t)\sigma_j(t)dW_j(t) + \int_0^T \int_{\mathbb{R}_0} \varphi_j^*(t)\gamma_j(t,z)\tilde{N}_j(dt, dz).$$

To prove the statements it is enough to show that

$$E[(F - \hat{F})G] = 0$$

for all $G \in L^2(P)$, $\mathcal{F}_T$-measurable of the form

$$G := E[G] + \sum_{j=1}^{n} \int_0^T \psi_j(t)dS_j(t)$$

$$= E[G] + \sum_{j=1}^{n} \int_0^T \psi_j(t)\sigma_j(t)dW_j(t) + \int_0^T \int_{\mathbb{R}_0} \psi_j(t)\gamma_j(t,z)\tilde{N}_j(t,z),$$

with $\psi \in \mathcal{A}_{\mathbb{E}}$. By the Clark–Ocone theorem (Theorem 12.16) we have

$$F = E[F] + \sum_{j=1}^{n} \int_0^T E[D_{j,t}F|\mathcal{F}_t]dW_j(t) + \int_0^T \int_{\mathbb{R}_0} E[D_{j,t,z}F|\mathcal{F}_t]\tilde{N}_j(t,z).$$

This gives

$$E[(F - \hat{F})G] = E\Big[\Big(\sum_{j=1}^{n} \int_0^T E[D_{j,t}F|\mathcal{F}_t]dW_j(t) - \sum_{j=1}^{n}\sum_{k=1}^{n} \int_0^T \varphi_j^*(t)\sigma_{jk}(t)dW_k(t)$$

$$+ \sum_{j=1}^{n} \int_0^T \int_{\mathbb{R}_0} E[D_{j,t,z}F|\mathcal{F}_t]\tilde{N}_j(dt,dz) - \sum_{j=1}^{n}\sum_{k=1}^{n} \int_0^T \int_{\mathbb{R}_0} \varphi_j^*(t)\gamma_{jk}(t,z)\tilde{N}_k(dt,dz)\Big)$$

$$\cdot \Big(\sum_{j=1}^{n}\sum_{k=1}^{n} \int_0^T \psi_j(t)\sigma_{jk}(t)dW_k(t) + \sum_{j=1}^{n}\sum_{k=1}^{n} \int_0^T \int_{\mathbb{R}_0} \psi_j(t)\gamma_{jk}(t,z)\tilde{N}_k(dt,dz)\Big)\Big]$$

$$= E\Big[\sum_{j=1}^{n} \int_0^T \Big(E[D_{j,t}F|\mathcal{F}_t] - \sum_{k=1}^{n} \varphi_k^*(t)\sigma_{kj}(t)\Big) \cdot \Big(\sum_{k=1}^{n} \psi_k(t)\sigma_{kj}(t)\Big)dt$$

$$+ \sum_{j=1}^{n} \int_0^T \int_{\mathbb{R}_0} \Big(E[D_{j,t,z}F|\mathcal{F}_t] - \sum_{k=1}^{n} \varphi_k^*(t)\gamma_{kj}(t,z)\Big) \cdot \Big(\sum_{k=1}^{n} \psi_k(t)\gamma_{kj}(t,z)\Big)\nu_j(dz)dt$$

$$= E\Big[\sum_{j=1}^{n} \int_0^T \sum_{k=1}^{n} \psi_k(t)\big(\sigma_{kl}(t)E[D_{j,t}F|\mathcal{F}_t] - \sum_{i=1}^{n} \varphi_i^*(t)\sigma_{ij}(t)\sigma_{kj}(t)$$

$$+ \int_{\mathbb{R}_0} (\gamma_{kj}(t,z)E[D_{j,t,z}F|\mathcal{F}_t] - \sum_{i=1}^{n} \varphi_i^*(t)\gamma_{ij}(t,z)\gamma_{kj}(t,z))\nu_j(dz)]dt\Big]$$

$$= E\Big[\sum_{k=1}^{n} \int_0^T \psi_k(t)L_k(t)dt\Big] = 0,$$

where

$$L_k(t) = \sum_{j=1}^{n} \Big(\sigma_{kj}(t)E[D_{j,t}F|\mathcal{F}_t] - \sum_{i=1}^{n} \varphi_i^*(t)\sigma_{ij}(t)\sigma_{kj}(t)$$

$$+ \int_{\mathbb{R}_0} \Big[\gamma_{kj}(t,z)E[D_{j,t,z}F|\mathcal{F}_t] - \sum_{i=1}^{n} \varphi_i^*(t)\gamma_{ij}(t,z)\gamma_{kj}(t,z)\Big]\nu(dz)\Big).$$

This holds for all $\psi \in \mathcal{A}_{\mathbb{E}}$ if and only if

$$E\big[L_k(t)|\mathcal{E}_t\big] = 0, \quad t \in [0,T], \quad k = 1,...,n.$$

We can write

$$L_k(t) = M_k(t) - \sum_{i=1}^{n} \varphi_i^*(t)N_{ik}(t),$$

where

$$M_k(t) = \sum_{k=1}^{n} \Big(\sigma_{kj}(t)E\big[D_{j,t}F|\mathcal{F}_t\big] + \int_{\mathbb{R}_0} \gamma_{kj}(t,z)E\big[D_{j,t,z}F|\mathcal{F}_t\big]\nu_j\Big)$$

and

$$N_{ik}(t) = \sum_{j=1}^{n} \Big(\sigma_{ij}(t)\sigma_{kj}(t) + \int_{\mathbb{R}_0} \gamma_{ij}(t,z)\gamma_{kj}(t,z)\nu_j(dz)\Big).$$

Therefore, we conclude that, for $k = 1,...,n$,

$$E\big[M_k(t)|\mathcal{E}_t\big] - \sum_{i=1}^{n} \varphi_i^*(t)E\big[N_{ik}(t)|\mathcal{E}_t\big] = 0$$

or

$$Q(t)\varphi^*(t) = R(t),$$

where

$$Q \in \mathbb{R}^{n\times n}, \quad Q_{ik}(t) = E\big[N_{ik}(t)|\mathcal{E}_t\big], \quad i,k = 1,...,n,$$

and

$$R(t) \in \mathbb{R}^n, \quad R_i(t) = E\big[M_i(t)|\mathcal{E}_t\big], \quad i = 1,...,n.$$

The solution of this equation is

$$\varphi^*(t) = Q^{-1}(t)R(t), \quad t \geq 0,$$

which completes the proof. $\square$

Corollary 12.25. *(a) Suppose $n = 1$ and $\mathcal{E}_t \subseteq \mathcal{F}_t$, $t \geq 0$. Then*

$$\varphi^*(t) = \frac{E\big[\sigma(t)E\big[D_tF|\mathcal{F}_t\big] + \int_{\mathbb{R}_0} \gamma(t,z)E\big[D_{t,z}F|\mathcal{F}_t\big]\nu(dz)\big|\mathcal{E}_t\big]}{E\big[\sigma^2(t) + \int_{\mathbb{R}_0} \gamma^2(t,z)\nu(dz)|\mathcal{E}_t\big]}. \tag{12.31}$$

(b) Suppose $n = 1$, σ and γ are $\mathbb{E}$-predictable. Then

$$\varphi^*(t) = \frac{\sigma(t)E\big[D_tF|\mathcal{E}_t\big] + \int_{\mathbb{R}_0} \gamma(t,z)E\big[D_{t,z}F|\mathcal{E}_t\big]\nu(dz)}{\sigma^2(t) + \int_{\mathbb{R}_0} \gamma^2(t,z)\nu(dz)}. \tag{12.32}$$

Example 12.26. (1) Suppose $n = 1$ and

$$dS(t) = dW(t) + \int_{\mathbb{R}_0} z\tilde{N}(dt, dz).$$

What is the closest hedge to $F := \int_0^T \int_{\mathbb{R}_0} z\tilde{N}(dt, dz)$ in terms of minimal variance? Since $D_t F \equiv 0$, $t \geq 0$, and $D_{t,z} F = z = E[D_{t,z} F | \mathcal{F}_t]$, $t \geq 0$, $z \in \mathbb{R}_0$, we get

$$\varphi^*(t) = \left(1 + \int_{\mathbb{R}_0} z^2 \nu(dz)\right)^{-1} \int_{\mathbb{R}_0} z^2 \nu(dz), \quad t \in [0, T]. \tag{12.33}$$

We see that this process is actually constant.

(2) Suppose $n = 1$ and $\mathcal{E}_t \subseteq \mathcal{F}_t$, $t \geq 0$, and

$$dS(t) = d\eta(t) = \int_{\mathbb{R}_0} z\tilde{N}(dt, dz).$$

Let us consider $F = S^2(T) = \eta^2(T)$. Then

$$D_{t,z} F = \left(\eta(T) + D_{t,z}\eta(T)\right)^2 - \eta^2(T) = 2\eta(T)z + z^2.$$

Hence the minimal variance portfolio is

$$\begin{aligned}
\varphi^*(t) &= \left(\int_{\mathbb{R}_0} z^2 \nu(dz)\right)^{-1} \int_{\mathbb{R}_0} \left(2z^2 E[S(T)|\mathcal{E}_t] + z^3\right)\nu(dz) \\
&= 2E[S(t)|\mathcal{E}_t] + \frac{\int_{\mathbb{R}_0} z^3 \nu(dz)}{\int_{\mathbb{R}_0} z^2 \nu(dz)}.
\end{aligned} \tag{12.34}$$

(3) Suppose $n = 1$ and $\mathcal{E}_t = \mathcal{F}_t$, $t \geq 0$, and

$$dS(t) = d\eta(t) = \int_{\mathbb{R}_0} z\tilde{N}(dt, dz).$$

Let us consider $F = S^2(T) = \eta^2(T)$. Then, since S is a martingale with respect to $\mathbb{F}$, we get

$$\varphi^*(t) = 2\eta(t^-) + \frac{\int_{\mathbb{R}_0} z^3 \nu(dz)}{\int_{\mathbb{R}_0} z^2 \nu(dz)}. \tag{12.35}$$

The closest hedge $\hat{F}$ in this case is therefore given by

$$\hat{F} - E[F] = \int_0^T 2\eta(t^-)d\eta(t) + \int_0^T \int_{\mathbb{R}_0} z \frac{\int_{\mathbb{R}_0} \zeta^3 \nu(d\zeta)}{\int_{\mathbb{R}_0} \zeta^2 \nu(d\zeta)} \tilde{N}(dt, dz). \tag{12.36}$$

If we compare this to (9.34), that is,

$$F - E[F] = \int_0^T 2\eta(t^-)d\eta(t) + \int_0^T \int_{\mathbb{R}_0} z^2 \tilde{N}(dt, dz),$$

we see that the closest hedge to the non-replicable claim

$$G := \int_0^T \int_{\mathbb{R}_0} z^2 \tilde{N}(dt, dz)$$

is the replicable claim

$$\hat{G} := \int_0^T z \frac{\int_{\mathbb{R}_0} \zeta^3 \nu(d\zeta)}{\int_{\mathbb{R}_0} \zeta^2 \nu(d\zeta)} \tilde{N}(dt, dz) = \int_0^T \frac{\int_{\mathbb{R}_0} \zeta^3 \nu(d\zeta)}{\int_{\mathbb{R}_0} \zeta^2 \nu(d\zeta)} d\eta(t).$$

(4) Suppose $n = 1$ and

$$dS(t) = \int_{\mathbb{R}_0} z \tilde{N}(dt, dz)$$

and

$$F := \int_0^T \int_{\mathbb{R}_0} z^n \tilde{N}(dt, dz).$$

Then the minimal variance portfolio

$$\varphi^*(t) = \left(\int_{\mathbb{R}_0} z^2 \nu(dz) \right)^{-1} \int_{\mathbb{R}_0} z^{n+1} \nu(dz), \quad t \in [0, T],$$

is constant.

(5) Suppose $n = 1$, $\mathcal{E}_t = \mathcal{F}_t$, $t \geq 0$, and

$$dS(t) = S(t^-) \int_{\mathbb{R}_0} z \tilde{N}(dt, dz).$$

Let us consider $F = S^2(T)$. In this case

$$D_{t,z}F = \left(S(T) + D_{t,z}S(T) \right)^2 - S^2(T).$$

Since $S(T) = S(0) \exp U(T)$, where

$$U(T) := \int_0^T \int_{\mathbb{R}_0} \left(\log(1 + z) - z \right) \nu(dz) ds + \int_0^T \int_{\mathbb{R}_0} \log(1 + z) \tilde{N}(ds, dz),$$

then we have

$$
\begin{aligned}
D_{t,z}S(T) &= S(0) \exp\{U + D_{t,z}U\} - S(0) \exp\{U\} \\
&= S(0) \exp U \exp\{\log(1 + z) - 1\} \\
&= S(0) \exp U z = S(T)z.
\end{aligned}
$$

Hence

$$D_{t,z}F = \left(S(T) + S(T)z\right)^2 - S^2(T) = 2S^2(T)z + S^2(T)z^2$$

and the minimal variance hedging portfolio is

$$\varphi^*(t) = \frac{\int_{\mathbb{R}_0} S^2(T)\left(2z^2 + z^3\right)\nu(dz)}{\int_{\mathbb{R}_0} z^2\nu(dz)}.$$

(6) Suppose $n = 1$, $\mathcal{E}_t = \mathcal{F}_t$, $t \geq 0$, and

$$dS(t) = \int_{\mathbb{R}_0} z\tilde{N}(dt, dz).$$

Consider the digital claim

$$F := \chi_{[k,\infty)}(S(T)).$$

The claim F may not belong to $\mathbb{D}_{1,2}$. Then an extended version of Theorem 12.24 can be applied (see Chap. 13). In this case we have

$$D_{t,z}F = \chi_{[k,\infty)}(S(T) + z) - \chi_{[k,\infty)}(S(T)),$$

which yields

$$\varphi^*(t) = \left(\int_{\mathbb{R}_0} z^2\nu(dz)\right)^{-1} \int_{\mathbb{R}_0} zE\left[D_{t,z}F|\mathcal{F}_t\right]\nu(dz)$$

$$= \left(\int_{\mathbb{R}_0} z^2\nu(dz)\right)^{-1} \int_{\mathbb{R}_0} zE\left[\chi_{[k,\infty)}(S(T - t) + z)\right.$$

$$\left. - \chi_{[k,\infty)}(S(T - t))|S(t)\right]\nu(dz).$$

12.7 Computation of "Greeks" in the Case of Jump Diffusions

In Sect. 4.4 it has been demonstrated how Malliavin calculus can be employed to compute option price sensitivities, commonly referred to as the "greeks," for asset price processes modeled by stochastic differential equations driven by a Wiener process. The greeks are in a sense "risk measures", which are used by investors on financial markets to hedge their positions. These greeks measure changes of contract prices with respect to parameters in the underlying model. Roughly speaking, greeks are derivatives with respect to a parameter θ of a risk-neutral price, that is, for example, of the form

$$\frac{\partial}{\partial\theta}E\left[\phi(S(T))\right],$$

where $\phi(S(T))$ is the payoff function and $S(T)$ is the underlying asset, which depends on θ.

The Malliavin approach of [81] for the calculation of greeks has proven to be numerically effective and in many respects sometimes better than other tools such as, the finite difference or the likelihood ratio method [43]. This technique is especially useful if it comes to handling discontinuous payoffs and path dependent options. See for example, [45] for more information and the references therein.

In this section we wish to extend the method of [81] as presented in Sect. 4.4 to the case of Itô jump diffusions. The general idea is to take the Malliavin derivative in the direction of the Wiener process on the Wiener–Poisson space, see Sect. 12.5. This enables us to stay in the framework of a variational calculus for the Wiener process without major changes. However, it should be mentioned that the pure jump case cannot be treated by this approach in the same way, since the Malliavin derivative with respect to the jump component is a difference operator in the sense of Theorem 12.8.

We remark that there are several authors in the literature dealing with jump diffusion models, see, for example, [15, 54, 55].

Rather than striving for the most general setting, we want to illustrate the basic ideas of this method by considering asset prices described by the Barndorff–Nielsen and Shephard model (see [18]). See also [28].

12.7.1 The Barndorff–Nielsen and Shephard Model

Adopting the notation of Sect. 12.5 we assume that the one-dimensional Wiener process $W(t)$ and the compensated Poisson random measure $\widetilde{N}(dt, dz) = N(dt, dz) - \nu(dz)dt$ is constructed on the Wiener–Poisson probability space $(\Omega, \mathcal{F}, P)$ given by

$$(\Omega, \mathcal{F}, P) = (\Omega_0 \times \Omega_0, \mathcal{F}_T^W \otimes \mathcal{F}_T^{\widetilde{N}}, P^W \times P^{\widetilde{N}}),$$

where P^W is the Gaussian and $P^{\widetilde{N}}$ is the white noise Lévy measure on the Schwartz distribution space $\Omega_0 = \mathcal{S}'(\mathbb{R})$. As before, let us denote by D_t and $D_{t,z}$ the Malliavin derivatives in the direction of the Wiener process and the Poisson random measure, respectively. The operator D_t can be defined on the Hilbert space

$$\widetilde{\mathbb{D}}_{1,2}, \tag{12.37}$$

which is the closure of a suitable space of smooth random variables (e.g., the linear span of basis elements $\mathbb{H}_\alpha$ as given in Sect. 12.5) with respect to the semi-norm

$$\|F\|_{1,2} := \left(E\left[F^2\right] + E\left[\int_0^T |D_t F|^2 \, dt\right] \right)^{1/2}.$$

It can be seen from this construction that the results obtained in Chap. 3 still hold for D_t on the Wiener–Poisson space $(\Omega, \mathcal{F}, P)$. So, for example, in this setting the chain rule for D_t reads

$$D_t\phi(F_1, ..., F_m) = \sum_{j=1}^{m} \frac{\partial}{\partial x_j}\phi(F_1, ..., F_m)D_tF_j \ \ \lambda \times P^W \times P^{\widetilde{N}} - \text{a.e.,} \quad (12.38)$$

if $\phi : \mathbb{R}^m \longrightarrow \mathbb{R}$ is a bounded continuously differentiable function and $(F_1, ..., F_m)$ is a random vector with components in $\widetilde{\mathbb{D}}_{1,2}$. See Theorem 3.5.

In this section, we want to briefly discuss the Barndorff–Nielsen and Shephard (BNS) model which was introduced in [18]. This model exhibits nice features and can, for example, be used to fit it to high-frequency stock price data.

In the following, we consider a financial market consisting of a single risk-free asset and a risky asset (stock). Further, let us assume that the stock price $S(t)$ is defined on the Wiener–Poisson space $(\Omega, \mathcal{F}, P)$ and given by

$$S(t) = S^x(t) = x\exp(X(t)), 0 \le t \le T, \quad (12.39)$$

where

$$dX(t) = (\mu + \beta\sigma^2(t))dt + \sigma(t)dW(t) + \rho dZ(\lambda t), X(0) = 0 \quad (12.40)$$

with stochastic volatility $\sigma^2(t)$ given by the Lévy–Ornstein–Uhlenbeck (OU) process

$$d\sigma^2(t) = -\lambda\sigma^2(t)dt + dZ(\lambda t), \sigma^2(0) > 0. \quad (12.41)$$

Here $Z(t)$ is a "background driving" Lévy process given by a *subordinator*, that is, a nondecreasing Lévy process. Such a process has the representation

$$Z(t) = mt + \int_0^t \int_{\mathbb{R}_0} zN(ds, dz) \quad (12.42)$$

for a constant $m \ge 0$ with Lévy measure ν such that $\text{supp}(\nu) \subseteq (0, \infty)$. Further, the law of $Z(t)$, $0 \le t \le T$, is completely determined by its cumulant generating function

$$\kappa(\alpha) := \log\left(E\left[\exp(\alpha Z(1))\right]\right), \quad (12.43)$$

(see, e.g., [32]). The processes $W(t)$ and $Z(t)$ are independent. In addition, $r > 0$ is the constant market interest rate and the constants $\lambda > 0, \rho \le 0$ stand for the mean-reversion rate of the stochastic volatility and the leverage effect of the (log-) price process, respectively. Moreover, μ and β are constant parameters.

Using the Itô formula (Theorem 9.4) one finds that the volatility process in (12.41) has the explicit form

$$\sigma^2(t) = \sigma^2(0)e^{-\lambda t} + \int_0^t e^{\lambda(s-t)}dZ(\lambda s), 0 \le t \le T. \quad (12.44)$$

Throughout the rest of this section we require that the subordinator $Z(t)$ has no drift (i.e., $m = 0$) in (12.42) and that ν has a density w with respect to the Lebsegue measure. The latter implies that

$$\kappa(\alpha) = \int_{\mathbb{R}_0} (e^{\alpha z} - 1)\nu(dz) = \int_{(0,\infty)} (e^{\alpha z} - 1)w(z)dz. \tag{12.45}$$

See [32]. Define

$$\widehat{\alpha} = \sup\{\alpha \in \mathbb{R} : \kappa(\alpha) < \infty\}.$$

In addition, we assume that

$$\widehat{\alpha} > \max\{0, 2\lambda^{-1}(1 + \beta + \rho)\} \text{ and } \lim_{\alpha \longrightarrow \widehat{\alpha}} \kappa(\alpha) = \infty. \tag{12.46}$$

The last condition ensures the square integrability of the asset process $S(t)$ and the existence of an invariant distribution of the volatility process. See [167].

As mentioned the greeks are derivatives of the expected (discounted) payoff under a risk neutral measure Q. However, a measure change from the real world measure to Q might result in a dynamics being different from the BNS model (12.40) and (12.41). Therefore, we are interested in "structure preserving" risk neutral measures Q, which transform BNS models into BNS models with possibly different parameters and Lévy measure for $Z(t)$. It turns out (see [167]) that the risk neutral dynamics of the BNS model under such measures takes the general form

$$dX(t) = (r - \lambda\kappa(\rho) - \frac{1}{2}\sigma^2(t))dt + \sigma(t)dW(t) + \rho dZ(\lambda t) \tag{12.47}$$

and

$$d\sigma^2(t) = -\lambda\sigma^2(t)dt + dZ(\lambda t), \sigma^2(0) > 0. \tag{12.48}$$

From now on we confine ourselves to the risk neutral dynamics of the BNS model given by (12.47) and (12.48), for which conditions (12.45) and (12.46) are satisfied. In what follows we replace Q by our white noise measure P (since we only deal with probability laws under expectations).

12.7.2 Malliavin Weights for "Greeks"

In the sequel we want to compute the Malliavin weights for the delta and gamma of an option. To this end we need the following auxiliary results:

Lemma 12.27. *Suppose that F^θ is a real valued random variable, which depends on a parameter $\theta \in \mathbb{R}$. Further require that the mapping $\theta \longmapsto F^\theta(\omega)$ is continuously differentiable in $[a, b]$ ω-a.e. and that*

$$E\left[\sup_{a \leq \theta \leq b} \left|\frac{\partial}{\partial\theta}F^\theta\right|\right] < \infty.$$

Then $\theta \longmapsto E\left[F^\theta\right]$ is differentiable in (a, b), and for $\theta \in (a, b)$ we have

$$\frac{\partial}{\partial\theta}E\left[F^\theta\right] = E\left[\frac{\partial}{\partial\theta}F^\theta\right].$$

Proof The result follows from the mean value theorem and dominated convergence. The details are left to the reader.   □

Adopting the notation in [28], we denote by $L^2(S)$ the space of locally integrable functions $\phi : \mathbb{R}^m \longrightarrow \mathbb{R}$ such that

$$E\left[\phi^2(S(t_1), ..., S(t_m))\right] < \infty. \tag{12.49}$$

Our asset price process depends on the parameters $\theta = x$, r, ρ, and $\sigma^2(0)$. In the following we write $S(\cdot) = S^\theta(\cdot)$ to indicate this dependency.

Lemma 12.28. *Let $\theta \longmapsto \pi^\theta$ be a process such that $\theta \longmapsto \psi(\theta) := \left\|\pi^\theta\right\|_{L^2(P)}$ is locally bounded. Further assume that*

$$\frac{\partial}{\partial \theta} E\left[\phi(S^\theta(t_1), ..., S^\theta(t_m))\right] = E\left[\phi(S^\theta(t_1), ..., S^\theta(t_m))\pi^\theta\right] \tag{12.50}$$

is valid for all $\phi \in C_c^\infty(\mathbb{R}^m)$ (i.e., ϕ is an infinitely differentiable function with compact support). Then relation (12.50) also holds for all $\phi \in L^2(S)$.

Proof Let ϕ be a bounded function. Then there exists a uniformly bounded sequence of functions $\phi_k, k \geq 1$ satisfying (12.50) such that

$$\phi_k \longrightarrow \phi \text{ a.e.}$$

Using transition probability densities in connection with $X(t)$ in (12.47) one finds that

$$\phi_k(S^\theta(t_1), ..., S^\theta(t_m)) \longrightarrow \phi(S^\theta(t_1), ..., S^\theta(t_m))$$

in $L^2(P)$ uniformly on compact sets. Define

$$u(\theta) = E\left[\phi(S^\theta(t_1), ..., S^\theta(t_m))\right] \text{ and } u_k(\theta) = E\left[\phi_k(S^\theta(t_1), ..., S^\theta(t_m))\right].$$

As above one verifies that $u_k(\theta) \longrightarrow u(\theta)$ for all θ. Further let

$$f(\theta) := E\left[\phi(S^\theta(t_1), ..., S^\theta(t_m))\pi^\theta\right].$$

By the Cauchy–Schwartz inequality we get

$$\left|\frac{\partial}{\partial \theta} u_k(\theta) - f(\theta)\right| \leq \epsilon_k(\theta)\psi(\theta),$$

where

$$\epsilon_k(\theta) = \left(E\left[\left(\phi_k(S^\theta(t_1), ..., S^\theta(t_m)) - \phi(S^\theta(t_1), ..., S^\theta(t_m))\right)^2\right]\right)^{1/2}.$$

Since $\theta \longmapsto \psi(\theta)$ is locally bounded, it follows that

$$\frac{\partial}{\partial \theta} u_k(\theta) \longrightarrow f(\theta) \text{ as } k \longrightarrow \infty$$

uniformly on compact sets. Hence, ϕ also fulfills (12.50). So (12.50) is valid for all bounded measurable functions. The general case finally follows from a truncation argument.   □

Let us remark that $L^2(S)$ contains important options as, for example, the call option.

We are coming to the main result that is due to [28]:

Theorem 12.29. *Let $\phi \in L^2(S)$ and let $a \in L^2([0,T])$ be an adapted process such that*

$$\int_0^{t_i} a(t)dt = 1 \quad P - a.e.$$

for all $i = 1, ..., m$. Then
(1) The delta of the option is given by

$$\frac{\partial}{\partial x} E\left[e^{-rT}\phi(S^x(t_1), ..., S^x(t_m))\right] = E\left[e^{-rT}\phi(S^x(t_1), ..., S^x(t_m))\pi^\Delta\right],$$

where the Malliavin weight π^Δ is given by

$$\pi^\Delta = \int_0^T \frac{a(t)}{x\sigma(t)}dW(t).$$

(2) The gamma of the option is given by

$$\frac{\partial^2}{\partial x^2} E\left[e^{-rT}\phi(S^x(t_1), ..., S^x(t_m))\right] = E\left[e^{-rT}\phi(S^x(t_1), ..., S^x(t_m))\pi^\Gamma\right],$$

where the Malliavin weight π^Γ has the form

$$\pi^\Gamma = \left(\pi^\Delta\right)^2 - \frac{1}{x}\pi^\Delta - \frac{1}{x^2}\int_0^T \left(\frac{a(t)}{\sigma(t)}\right)^2 dt.$$

Proof It is easily seen that $\theta \longmapsto S^\theta$ is pathwise differentiable (with exception of the boundary values $x = 0$ and $\sigma^2(0) = 0$) for the different parameters $\theta = x, r, \rho, \sigma^2(0)$. Further, one checks that the assumptions of Lemma 12.27 and Lemma 12.28 are satisfied. So it remains to verify relation (12.50) for $\phi \in C_0^\infty(\mathbb{R}^m)$.
(1) Using Lemma 12.27, we find

$$\frac{\partial}{\partial x} E\left[e^{-rT}\phi(S^x(t_1), ..., S^x(t_m))\right]$$

$$= E\left[e^{-rT}\frac{\partial}{\partial x}\phi(S^x(t_1), ..., S^x(t_m))\right]$$

$$= E\left[e^{-rT}\sum_{i=1}^{m_i}\phi_{x_i}(S^x(t_1), ..., S^x(t_m))\frac{\partial}{\partial x}S^x(t_i)\right]$$

$$= E\left[e^{-rT}\sum_{i=1}^{m_i}\phi_{x_i}(S^x(t_1), ..., S^x(t_m))\frac{1}{x}S^x(t_i)\right].$$

By applying the chain rule (Theorem 3.5 or (12.38)) and the fundamental theorem of stochastic calculus (Theorem 3.18) in the direction of $W(t)$ we obtain

$$D_t S^x(t_i) = \sigma(t) S^x(t_i) \chi_{[0,t_i]}(t).$$

Since $\int_0^{t_i} a(t)dt = 1$, we get

$$\int_0^T \frac{a(t)}{x\sigma(t)} D_t S^x(t_i)dt = \frac{1}{x} S^x(t_i).$$

Hence,

$$\frac{\partial}{\partial x} E\left[e^{-rT}\phi(S^x(t_1),...,S^x(t_m))\right]$$

$$= E\left[e^{-rT}\int_0^T \sum_{i=1}^{m_i} \phi_{x_i}(S^x(t_1),...,S^x(t_m))\frac{a(t)}{x\sigma(t)}D_t S^x(t_i)dt\right].$$

Then the chain rule (12.38) yields

$$\frac{\partial}{\partial x} E\left[e^{-rT}\phi(S^x(t_1),...,S^x(t_m))\right] = e^{-rT}E\left[\int_0^T \sum_{i=1}^{m_i} D_t\phi(S^x(t_1),...,S^x(t_m))\frac{a(t)}{x\sigma(t)}dt\right].$$

Finally, the duality formula (Theorem 3.14) gives the Malliavin weight $\pi^\Delta = \int_0^T \frac{a(t)}{x\sigma(t)}dW(t).$

(2) Define $F^x = \int_0^T \frac{a(t)}{x\sigma(t)}dW(t)$. Then $\frac{\partial}{\partial x}F^x = -\frac{1}{x}F^x$. Thus

$$\frac{\partial^2}{\partial x^2} E\left[e^{-rT}\phi(S^x(t_1),...,S^x(t_m))\right]$$

$$= \frac{\partial}{\partial x} E\left[e^{-rT}\phi(S^x(t_1),...,S^x(t_m))F^x\right]$$

$$= -\frac{1}{x} E\left[e^{-rT}\phi(S^x(t_1),...,S^x(t_m))F^x\right] \tag{12.51}$$

$$+ E\left[e^{-rT}\sum_{i=1}^{m_i} \phi_{x_i}(S^x(t_1),...,S^x(t_m))\frac{1}{x}S^x(t_i)F^x\right].$$

Repeated use of the arguments of (1) gives

$$E\left[e^{-rT}\sum_{i=1}^{m_i} \phi_{x_i}(S^x(t_1),...,S^x(t_m))\frac{1}{x}S^x(t_i)F^x\right]$$

$$= E\left[e^{-rT}\int_0^T \sum_{i=1}^{m_i} D_t\phi(S^x(t_1),...,S^x(t_m))\frac{a(t)}{x\sigma(t)}F^x dt\right]$$

$$= E\left[e^{-rT}\sum_{i=1}^{m_i} D_t\phi(S^x(t_1),...,S^x(t_m))\delta\left(\frac{a(\cdot)}{x\sigma(\cdot)}F^x\right)\right].$$

Finally, noting that $D_t F^x = \frac{a(t)}{x\sigma(t)}$, it follows from the integration by parts formula (Theorem 3.15) that

$$\delta\left(\frac{a(\cdot)}{x\sigma(\cdot)}F^x\right) = F^x \int_0^T \frac{a(t)}{x\sigma(t)}dW(t) - \int_0^T \left(\frac{a(t)}{x\sigma(t)}\right)^2 dt$$

$$= (F^x)^2 - \int_0^T \left(\frac{a(t)}{x\sigma(t)}\right)^2 dt,$$

which, in connection with (12.52), gives the proof. $\square$

12.8 Exercises

Problem 12.1. (*) Let

$$\eta(t) = \int_0^t \int_{\mathbb{R}_0} z\tilde{N}(dz, dz), \quad t \in [0, T].$$

Use Definition 12.2 and the chaos expansions found in Problem 10.1 to compute the following:

(a) $D_{t,z}\eta^3(T)$, $(t, z) \in [0, T] \times \mathbb{R}_0$,
(b) $D_{t,z}\exp \eta(T)$, $(t, z) \in [0, T] \times \mathbb{R}_0$.

Problem 12.2. (*) Compute the Malliavin derivatives in Problem 12.1 by using the chain rule (see Theorem 12.8) together with (12.5).

Problem 12.3. Let the process $X(t)$, $t \in [0, T]$, be the geometric Lévy process

$$dX(t) = X(t^-)[\alpha(t)dt + \beta(t)dW(t) + \int_{\mathbb{R}_0} \gamma(t, z)\tilde{N}(dt, dz)],$$

where the involved coefficients are deterministic. Find $D_{t,z}X(T)$ for $t \leq T$.

Problem 12.4. Use the integration by parts formula (Theorem 12.11) to compute the Skorohod integrals

$$\int_0^T F\delta\eta(t) = \int_0^T \int_{\mathbb{R}_0} Fz\tilde{N}(\delta t, dz)$$

in the following cases:

(a) $F = \eta(T)$
(b) $F = \eta^2(T)$
(c) $F = \eta^3(T)$
(d) $F = \exp\{\eta(T)\}$
(e) $F = \int_0^T g(s)d\eta(s)$, where $g \in L^2([0, T])$.

Problem 12.5. Solve Problem 9.6 using the Clark–Ocone theorem.

Problem 12.6. Consider the following market

risk free asset: $dS_0(t) = 0, \quad S_0(0) = 1$

risky asset: $dS_1(t) = S_1(t^-) \int_{\mathbb{R}_0} z\tilde{N}(dt, dz), \quad S_1(0) > 0,$

where $z > -1$ ν-a.e. Find the closest hedge in terms of minimal variance for the following claims:

(a) $F = S_1^2(T)$,
(b) $F = \exp\{\lambda S_1(T)\}$, with $\lambda \in \mathbb{R}$ constant.

Problem 12.7. Consider the claim $F = \eta^3(T)$ in the Bachelier–Lévy market

risk free asset: $dS_0(t) = 0, \quad S_0(0) = 1$

risky asset: $dS_1(t) = \int_{\mathbb{R}_0} z\tilde{N}(dt, dz), \quad S_1(0) = 0.$

(a) Is the claim replicable in this market?
(b) If not, what is the closest hedge in terms of minimal variance?

13

Lévy White Noise and Stochastic Distributions

This chapter is dedicated to the white noise analysis for Poisson random measures and its relation with the Malliavin calculus developed in this second part of the book. Just as in the Brownian case (see Chaps. 5 and 6), the Wick product provides a useful multiplication of elements on the distribution spaces. For the presentation we follow [69] and [181]. We can refer to [107] for a presentation of the use of white noise theory in stochastic partial differential equations, see also, for example, [155, 156, 202] and also [145, 166] for the use of the Wick product in the representation of solutions to SDEs. See also, for example, [128].

13.1 The White Noise Probability Space

As in the Brownian motion case (see Chap. 5), the sample space considered is $\Omega = \mathcal{S}'(\mathbb{R})$, the space of tempered distributions on $\mathbb{R}$, which is a topological space (see [79]). We equip this space with the corresponding Borel σ-algebra $\mathcal{F} = \mathcal{B}(\mathbb{R})$. By the Bochner–Minlos–Sazonov theorem, see, for example, [87], there exists a probability measure P such that

$$\int_{\Omega} e^{i\langle\omega,f\rangle} P(d\omega) = \exp\left(\int_{\mathbb{R}} \Psi(f(x))dx\right), \quad f \in \mathcal{S}(\mathbb{R}), \qquad (13.1)$$

where

$$\Psi(w) = \int_{\mathbb{R}} \left(e^{iwz} - 1 - iwz\right)\nu(dz) \quad , \quad i = \sqrt{-1} \qquad (13.2)$$

and $\langle\omega,f\rangle$ denotes the action of $\omega \in \mathcal{S}'(\mathbb{R})$ on $f \in \mathcal{S}(\mathbb{R})$ (see Chap. 5).

The triple $(\Omega, \mathcal{F}, P)$ defined above is called the (pure jump) *Lévy white noise probability space*.

Lemma 13.1. *Let $g \in \mathcal{S}(\mathbb{R})$. Then we have*

$$\mathrm{E}[\langle\cdot,g\rangle] = 0 \qquad (13.3)$$

G.Di Nunno et al., *Malliavin Calculus for Lévy Processes with Applications to Finance*,
© Springer-Verlag Berlin Heidelberg 2009

and

$$\text{Var}[\langle \cdot, g \rangle] := \text{E}[\langle \cdot, g \rangle^2] = M \int_{\mathbb{R}} g^2(y)dy, \tag{13.4}$$

where $M = \int_{\mathbb{R}} z^2 \nu(dz)$, *see* (13.2).

Proof The proof is based on the Taylor expansion of the characteristic function (13.1) applied to the function $f(y) = tg(y)$, for $t \in \mathbb{R}$. Details can be found in [107]. $\square$

Using Lemma 13.1, we can extend the definition of $\langle \omega, f \rangle$ for $f \in \mathcal{S}(\mathbb{R})$ to any $f \in L^2(\mathbb{R})$ as follows. If $f \in L^2(\mathbb{R})$, choose $f_n \in \mathcal{S}(\mathbb{R})$ such that $f_n \to f$ in $L^2(\mathbb{R})$. Then by (13.4) we see that $\{\langle \omega, f_n \rangle\}_{n=1}^{\infty}$ is a Cauchy sequence in $L^2(P)$ and hence convergent in $L^2(P)$. Moreover, the limit depends only on f and not the sequence $\{f_n\}_{n=1}^{\infty}$. We denote this limit by $\langle \omega, f \rangle$.

Now define

$$\tilde{\eta}(t) := \langle \omega, \chi_{[0,t]}(\cdot) \rangle; \quad t \in \mathbb{R}, \tag{13.5}$$

where

$$\chi_{[0,t]}(s) = \begin{cases} 1, & 0 \le s \le t \\ -1, & t \le s \le 0 \text{ except } t = s = 0 \,. \\ 0, & \text{otherwise} \end{cases}$$

Then we have the following result.

Theorem 13.2. *The stochastic process* $\tilde{\eta}(t)$, $t \in \mathbb{R}$, *has a càdlàg version, denoted by* η. *This process* $\eta(t)$, $t \ge 0$, *is a pure jump Lévy process with Lévy measure* ν.

Recall that the process η admits the stochastic integral representation

$$\eta(t) = \int_0^t \int_{\mathbb{R}_0} z\tilde{N}(ds, dz), \quad t \ge 0,$$

with respect to $\tilde{N}$ as integrator, cf. (13.2). Thus it is a martingale. Here $\tilde{N}(ds, dz) = N(ds, dz) - \nu(dz)ds$ is the compensated Poisson random measure associated to η. The process η will be called *pure jump Lévy process*.

13.2 An Alternative Chaos Expansion and the White Noise

From now on we assume that the Lévy measure ν satisfies the following condition:
For all $\varepsilon > 0$ there exists $\lambda > 0$ such that

$$\int_{\mathbb{R}_0 \setminus (-\varepsilon, \varepsilon)} \exp(\lambda |z|)\nu(dz) < \infty. \tag{13.6}$$

This condition implies that ν has finite moments of all orders $n \geq 2$. It is trivially satisfied if ν is supported on $[-R, R]$ for some $R > 0$.

This condition also implies that the polynomials are dense in $L^2(\rho)$, where

$$\rho(dz) = z^2 \nu(dz), \tag{13.7}$$

see [171]. Now let $\{l_m\}_{m \geq 0} = \{1, l_1, l_2, \ldots\}$ be the orthogonalization of $\{1, z, z^2, \ldots\}$ with respect to the inner product of $L^2(\rho)$.

Define

$$p_j(z) := \|l_{j-1}\|_{L^2(\rho)}^{-1} z l_{j-1}(z); \; j = 1, 2, \ldots \tag{13.8}$$

and

$$m_2 := \left(\int_{\mathbb{R}_0} z^2 \nu(dz) \right)^{\frac{1}{2}} = \|l_0\|_{L^2(\rho)} = \|1\|_{L^2(\rho)} . \tag{13.9}$$

In particular,

$$p_1(z) = m_2^{-1} z \text{ or } z = m_2 p_1(z). \tag{13.10}$$

Then $\{p_j(z)\}_{j \geq 1}$ is an *orthonormal basis* for $L^2(\nu)$.

Define the bijection $\kappa : \mathbb{N} \times \mathbb{N} \longrightarrow \mathbb{N}$ by

$$\kappa(i, j) = j + (i + j - 2)(i + j - 1)/2. \tag{13.11}$$

$$
\begin{array}{cccc}
(1) & (2) & (4) & (i) \\
\bullet \longrightarrow \bullet & \bullet \cdots \bullet \longrightarrow \\
(3) \diagup (5) \diagup \\
\bullet & \bullet \\
(6) \diagup \\
\bullet \\
\vdots \\
(j) \\
\bullet \\
\downarrow
\end{array}
$$

As in (5.5), let $\{e_i(t)\}_{i \geq 1}$ be the Hermite functions. Define

$$\delta_{\kappa(i,j)}(t, z) = e_i(t) p_j(z). \tag{13.12}$$

If $\alpha \in \mathcal{J}$ with $Index(\alpha) = j$ and $|\alpha| = m$, we define $\delta^{\otimes \alpha}$ by

$$\delta^{\otimes \alpha}(t_1, z_1, \ldots, t_m, z_m) \tag{13.13}$$

$$= \delta_1^{\otimes \alpha_1} \otimes \cdots \otimes \delta_j^{\otimes \alpha_j}(t_1, z_1, \ldots, t_m, z_m)$$

$$= \underbrace{\delta_1(t_1, z_1) \cdot \ldots \cdot \delta_1(t_{\alpha_1}, z_{\alpha_1})}_{\alpha_1 \text{ factors}} \cdot \ldots \cdot \underbrace{\delta_j(t_{m-\alpha_j+1}, z_{m-\alpha_j+1}) \cdot \ldots \cdot \delta_j(t_m, z_m)}_{\alpha_j \text{ factors}}.$$

We set $\delta_i^{\otimes 0} = 1$. Finally, we let $\delta^{\hat{\otimes}\alpha}$ denote the *symmetrized* tensor product of the δ_k's :

$$\delta^{\hat{\otimes}\alpha}(t_1, z_1, ..., t_m, z_m) = \delta_1^{\hat{\otimes}\alpha_1} \otimes ... \otimes \delta_j^{\hat{\otimes}\alpha_j}(t_1, z_1, ..., t_m, z_m). \qquad (13.14)$$

For $\alpha \in \mathcal{J}$ define

$$K_\alpha := I_{|\alpha|}\left(\delta^{\hat{\otimes}\alpha}\right). \qquad (13.15)$$

Example 13.3. With $\varepsilon^{(k)} = (0, ..., 0, 1, 0, ...)$ with 1 on kth place, we have, writing $\varepsilon^{(\kappa(i,j))} = \varepsilon^{(i,j)}$,

$$K_{\varepsilon^{(i,j)}} = I_1(\delta^{\otimes\varepsilon^{(i,k)}}) = I_1\left(\delta_{\kappa(i,j)}\right) = I_1\left(e_i(t)p_j(z)\right). \qquad (13.16)$$

As in the Brownian motion case, one can prove that $\{K_\alpha\}_{\alpha\in\mathcal{J}}$ are orthogonal in $L^2(P)$ and

$$\|K_\alpha\|_{L^2(P)}^2 = \alpha!$$

Note that, similar to (5.15), we have that, if $|\alpha| = m$,

$$\alpha! = \|K_\alpha\|_{L^2(P)}^2 = m!\|\delta^{\hat{\otimes}\alpha}\|_{L^2((\lambda\times\nu)^m)}^2.$$

By our construction of $\delta^{\hat{\otimes}\alpha}$ we know that any $f \in \widetilde{L}^2((\lambda\times\nu)^n)$ can be written as

$$f(t_1, z_1, ..., t_n, z_n) = \sum_{|\alpha|=n} c_\alpha \delta^{\hat{\otimes}\alpha}(t_1, z_1, ..., t_n, z_n). \qquad (13.17)$$

Hence

$$I_n(f_n) = \sum_{|\alpha|=n} c_\alpha K_\alpha. \qquad (13.18)$$

This gives the following theorem.

Theorem 13.4. Chaos expansion.
Any $F \in L^2(P)$ has a unique expansion of the form

$$F = \sum_{\alpha\in\mathcal{J}} c_\alpha K_\alpha, \qquad (13.19)$$

with $c_\alpha \in \mathbb{R}$. Moreover,

$$\|F\|_{L^2(P)}^2 = \sum_{\alpha\in\mathcal{J}} \alpha! c_\alpha^2. \qquad (13.20)$$

Example 13.5. Let $h \in L^2(\mathbb{R})$ and define

$$F = \int_{\mathbb{R}} h(s)d\eta(s) = I_1(h(s)z).$$

Then F has the expansion

$$F = I_1\left(\sum_{i\geq 1}(h, e_i)e_i(s)z\right) = \sum_{i\geq 1}(h, e_i)I_1(e_i(s)z) \qquad (13.21)$$

$$= \sum_{i\geq 1}\left(\int_{\mathbb{R}} h(s)e_i(s)ds\right)K_{\varepsilon^{(i,1)}}m_2.$$

Recall that $z = m_2 p_1(z)$; see (13.10). In particular,

$$\eta(t) = \sum_{i\geq 1}\left(\int_0^t e_i(s)ds\right)K_{\varepsilon^{(i,1)}}m_2. \qquad (13.22)$$

Definition 13.6. The Lévy–Hida spaces.

(1) **Stochastic test functions $(\mathcal{S})$.** *Let $(\mathcal{S})$ consist of all $\varphi = \sum_{\alpha\in\mathcal{J}} a_\alpha K_\alpha \in L^2(P)$ such that*

$$\|\varphi\|_k^2 := \sum_{\alpha\in\mathcal{J}} a_\alpha^2\alpha!(2\mathbb{N})^{k\alpha} < \infty \text{ for all } k \in \mathbb{N}, \qquad (13.23)$$

equipped with the projective topology, where

$$(2\mathbb{N})^{k\alpha} = \prod_{j\geq 1}(2j)^{k\alpha_j}, \qquad (13.24)$$

if $\alpha = (\alpha_1, \alpha_2, , ...) \in \mathcal{J}$.

(2) **Stochastic distributions $(\mathcal{S})^*$.** *Let $(\mathcal{S})^*$ consist of all expansions $F = \sum_{\alpha\in\mathcal{J}} b_\alpha K_\alpha$ such that*

$$\|F\|_{-q}^2 := \sum_{\alpha\in\mathcal{J}} b_\alpha^2\alpha!(2\mathbb{N})^{-q\alpha} < \infty \text{ for some } q \in \mathbb{N}, \qquad (13.25)$$

endowed with the inductive topology. The space $(\mathcal{S})^*$ is the dual of $(\mathcal{S})$. If $F = \sum_{\alpha\in\mathcal{J}} b_\alpha K_\alpha \in (\mathcal{S})^*$ and $\varphi = \sum_{\alpha\in\mathcal{J}} a_\alpha K_\alpha \in (\mathcal{S})$, then the action of F on φ is

$$\langle F, \varphi\rangle = \sum_{\alpha\in\mathcal{J}} a_\alpha b_\alpha\alpha!. \qquad (13.26)$$

(3) **Generalized expectation.** *If $F = \sum_{\alpha\in\mathcal{J}} a_\alpha K_\alpha \in (\mathcal{S})^*$, we define the generalized expectation $E[F]$ of F by*

$$E[F] = a_0.$$

Note that $E[K_\alpha] = 0$ for all $\alpha \neq 0$. Therefore, the generalized expectation coincides with the usual expectation if $F \in L^2(P)$.

We can now define the white noise $\overset{\bullet}{\eta}(t)$ of the Lévy process

$$\eta(t) = \int_0^t \int_{\mathbb{R}_0} z \widetilde{N}(dt, dz)$$

and the white noise $\overset{\bullet}{\widetilde{N}}(t, z)$ of $\widetilde{N}(dt, dz)$ as follows.

Definition 13.7. The Lévy white noise $\overset{\bullet}{\eta}$ and the white noise of the (compensated) Poisson random measure.

(1) The Lévy white noise process $\overset{\bullet}{\eta}(t)$ *is defined by the expansion*

$$\overset{\bullet}{\eta}(t) = m_2 \sum_{i \geq 1} e_i(t) K_{\varepsilon(i,1)} = m_2 \sum_{i \geq 1} e_i(t) I_1(e_i(t) p_1(z)) \quad (13.27)$$

$$= m_2 \sum_{i \geq 1} e_i(t) I_1(e_i(t) z).$$

(2) The white noise process *(or field)* $\overset{\bullet}{\widetilde{N}}(t, z)$ *of* $\widetilde{N}(dt, dz)$ *is defined by the expansion*

$$\overset{\bullet}{\widetilde{N}}(t, z) = \sum_{i,j \geq 1} e_i(t) p_j(z) K_{\varepsilon(i,j)}(\omega). \quad (13.28)$$

Remark 13.8. Note that for $\overset{\bullet}{\eta}(t)$ we have

$$\sum_{\alpha \in \mathcal{J}} c_\alpha^2 \alpha! (2\mathbb{N})^{-q\alpha} = m_2^2 \sum_{i \geq 1} e_i^2(t) 2^{-q} \kappa(i, 1)^{-q} < \infty$$

for $q \geq 2$, using that $\kappa(i, 1) = 1 + (i-1)i/2 \geq i$, and the following well-known estimate for the Hermite functions:

$$\sup_{x \in \mathbb{R}} |e_n(x)| = O(n^{-\frac{1}{12}}). \quad (13.29)$$

Therefore, $\overset{\bullet}{\eta}(t) \in (\mathcal{S})^*$ for all t. Similarly $\overset{\bullet}{\widetilde{N}}(t, z) \in (\mathcal{S})^*$ for all t, z.

Remark 13.9. (1) Note that by comparing the expansions

$$\eta(t) = \sum_{i \geq 1} \left(\int_0^t e_i(s) ds \right) K_{\varepsilon(i,1)} m_2$$

and

$$\overset{\bullet}{\eta}(t) = m_2 \sum_{i \geq 1} e_i(t) K_{\varepsilon(i,1)},$$

we get formally

$$\overset{\bullet}{\eta}(t) = \frac{d}{dt} \eta(t) \text{ (derivative in } (\mathcal{S})^*). \quad (13.30)$$

This can be proved rigorously. See Problem 13.1.

(2) Choose a Borel set $\mho$ such that $\overline{\mho} \subset \mathbb{R}\backslash\{0\}$. Then

$$\widetilde{N}(t, \mho) = I_1 \left(\chi_{[0,t]}(s)\chi_\mho(z) \right)$$

$$= \sum_{i,j\geq 1} (\chi_{[0,t]}, e_i)_{L^2(\mathbb{R})} (\chi_\mho, p_j)_{L^2(\nu)} I_1 \left(e_i(s), p_j(z) \right)$$

$$= \sum_{i,j\geq 1} \int_0^t e_i(s)ds \cdot \int_\mho p_j(z)\nu(dz)K_{\varepsilon(i,j)}(\omega).$$

This justifies the "relation"

$$\overset{\bullet}{\widetilde{N}}(t, z) = \frac{\widetilde{N}(dt, dz)}{dt \times \nu(dz)} \quad \text{(Radon–Nikodym derivative).} \qquad (13.31)$$

(3) Also note that $\overset{\bullet}{\eta}$ is related to $\overset{\bullet}{\widetilde{N}}$ by

$$\overset{\bullet}{\eta}(t) = \int_{\mathbb{R}_0} \overset{\bullet}{\widetilde{N}}(t, z)z\nu(dz). \qquad (13.32)$$

To see this consider

$$\int_{\mathbb{R}_0} \overset{\bullet}{\widetilde{N}}(t, z)z\nu(dz)$$

$$= \int_{\mathbb{R}_0} \sum_{i,j\geq 1} e_i(t)p_j(z)K_{\varepsilon(i,j)}(\omega)z\nu(dz)$$

$$= \sum_{i\geq 1} e_i(t)I_1 \left(e_i(t)\sum_{j\geq 1} p_j(z) \int_{\mathbb{R}_0} p_j(z)z\nu(dz) \right)$$

$$= \sum_{i\geq 1} e_i(t)I_1 \left(e_i(t)z \right) = \overset{\bullet}{\eta}(t).$$

13.3 The Wick Product

We now proceed as in the Brownian motion case and use the chaos expansion in terms of $\{K_\alpha\}_{\alpha\in\mathcal{J}}$ to define the (Lévy–) Wick product and study some of its properties.

13.3.1 Definition and Properties

Definition 13.10. *Let $F = \sum_{\alpha\in\mathcal{J}} a_\alpha K_\alpha$ and $G = \sum_{\beta\in\mathcal{J}} b_\beta K_\beta$ be two elements of $(\mathcal{S})^*$. Then we define the* Wick product *of F and G by*

$$F \diamond G = \sum_{\alpha,\beta\in\mathcal{J}} a_\alpha b_\beta K_{\alpha+\beta} = \sum_{\gamma\in\mathcal{J}} \left(\sum_{\alpha+\beta=\gamma} a_\alpha b_\beta \right) K_\gamma. \qquad (13.33)$$

We list some properties of the Wick product:

(1) $F, G \in (\mathcal{S})^* \Longrightarrow F \diamond G \in (\mathcal{S})^*$.
(2) $F, G \in (\mathcal{S}) \Longrightarrow F \diamond G \in (\mathcal{S})$.
(3) $F \diamond G = G \diamond F$.
(4) $F \diamond (G \diamond H) = (F \diamond G) \diamond H$.
(5) $F \diamond (G + H) = F \diamond G + F \diamond H$.
(6) $I_n(f_n) \diamond I_m(g_m) = I_{n+m}(f_n \widehat{\otimes} g_m)$

The proofs of these statements are similar to the Brownian motion case and therefore omitted.

In view of the properties (1) and (4) we can define the *Wick powers* $X^{\diamond n}$ $(n = 1, 2, ...)$ of $X \in (\mathcal{S})^*$ as

$$X^{\diamond n} := X \diamond X \diamond \cdots \diamond X \quad (n \text{ times}).$$

We put $X^{\diamond 0} := 1$. Similarly, we define the *Wick exponential* $\exp^\diamond X$ of $X \in (\mathcal{S})^*$ by

$$\exp^\diamond X := \sum_{n=0}^{\infty} \frac{1}{n!} X^{\diamond n} \tag{13.34}$$

if the series converges in $(\mathcal{S})^*$. In view of the aforementioned properties, we have that

$$(X + Y)^{\diamond 2} = X^{\diamond 2} + 2X \diamond Y + Y^{\diamond 2}$$

and also

$$\exp^\diamond(X + Y) = \exp^\diamond X \diamond \exp^\diamond Y, \tag{13.35}$$

for $X, Y \in (\mathcal{S})^*$.

Let $E[\cdot]$ denote the generalized expectation (see Definition 13.6 (3)), then we can see that

$$E[X \diamond Y] = E[X] E[Y], \tag{13.36}$$

for $X, Y \in (\mathcal{S})^*$. Note that independence is not required for this identity to hold.

By induction, it follows that

$$E[\exp^\diamond X] = \exp\{E[X]\}, \tag{13.37}$$

for $X \in (\mathcal{S})^*$.

Example 13.11. (1) Choose $h \in L^2([0, T])$ and define $F = \int_0^T h(t) d\eta(t)$. Then

$$\begin{aligned}
F \diamond F &= I_1(h(t)z) \diamond I_1(h(t)z) \\
&= I_2(h(t_1)h(t_2)z_1 z_2) \\
&= 2 \int_0^T \int_{\mathbb{R}_0} \left(\int_0^T \int_{\mathbb{R}_0} h(t_1)h(t_2)z_1 z_2 \widetilde{N}(dt_1, dz_1) \right) \widetilde{N}(dt_2, dz_2) \\
&= 2 \int_0^T \left(\int_0^{t_2} h(t_1) d\eta(t_1) h(t_2) \right) h(t_2) d\eta(t_2).
\end{aligned}$$

By the Itô formula, if we put $X(t) := \int_0^t h(s)d\eta(s)$,

$$d(X(t))^2 = 2X(t)dX(t) + h^2(t)\int_{\mathbb{R}_0} z^2 N(dt, dz).$$

Hence

$$F \diamond F = 2\int_0^T X(s)dX(s) = X^2(T) - \int_0^T \int_{\mathbb{R}_0} h^2(s)z^2 N(ds, dz). \qquad (13.38)$$

In particular, choosing $h = 1$ we get

$$\eta(T) \diamond \eta(T) = \eta^2(T) - \int_0^T \int_{\mathbb{R}_0} z^2 N(ds, dz). \qquad (13.39)$$

(2) It follows from the above that

$$K_{\varepsilon(i,j)} \diamond K_{\varepsilon(i,j)} = I_1(e_i(t)p_j(z)) \diamond I_1(e_i(t)p_j(z)) \qquad (13.40)$$

$$I_2(e_i(t_1)e_i(t_2)p_j(z_1)p_j(z_2))$$

$$= 2\int_{\mathbb{R}}\int_{\mathbb{R}_0} \left(\int_{-\infty}^{t_2}\int_{\mathbb{R}_0} e_i(t_1)p_j(z_1)\widetilde{N}(dt_1, dz_1)\right) e_i(t_2)p_j(z_2)\widetilde{N}(dt_2, dz_2)$$

$$= K_{\varepsilon(i,j)} \cdot K_{\varepsilon(i,j)} - \int_{\mathbb{R}}\int_{\mathbb{R}_0} e_i^2(s)p_j^2(z)N(ds, dz).$$

Example 13.12. **Wick/Doléans–Dade exponential.** Choose $\gamma \geq -1$ deterministic such that

$$\int_{\mathbb{R}}\int_{\mathbb{R}_0} \{|\log(1 + \gamma(t, z))| + \gamma^2(t, z)\}\nu(dz)dt < \infty,$$

and put

$$F = \exp^\diamond \left(\int_{\mathbb{R}}\int_{\mathbb{R}_0} \gamma(t, z)\widetilde{N}(dt, dz)\right).$$

To find an expression for F not involving the Wick product, we proceed as follows:
Define

$$Y(t) = \exp^\diamond \left(\int_0^t \int_{\mathbb{R}_0} \gamma(s, z)\widetilde{N}(ds, dz)\right)$$

$$= \exp^\diamond \left(\int_0^t \int_{\mathbb{R}_0} \gamma(s, z)\overset{\bullet}{\widetilde{N}}(t, z)\nu(dz)ds\right).$$

Then we have

$$\frac{dY(t)}{dt} = Y(t) \diamond \int_{\mathbb{R}_0} \gamma(t, z)\overset{\bullet}{\widetilde{N}}(t, z)\nu(dz)$$

or

$$dY(t) = Y(t-) \int_{\mathbb{R}_0} \gamma(t, z)\widetilde{N}(dt, dz).$$

Using Itô calculus the solution of this equation is

$$Y(t) = Y(0) \exp\left(\int_0^t \int_{\mathbb{R}_0} \{\log(1 + \gamma(s, z)) - \gamma(s, z)\} \nu(dz)ds \right.$$

$$\left. + \int_0^t \int_{\mathbb{R}_0} \log(1 + \gamma(s, z))\widetilde{N}(ds, dz) \right).$$

Comparing the two expressions for $Y(t)$, we conclude that

$$\exp^\diamond\left(\int_{\mathbb{R}} \int_{\mathbb{R}_0} \gamma(s, z)\widetilde{N}(ds, dz) \right) \tag{13.41}$$

$$= \exp\left(\int_{\mathbb{R}} \int_{\mathbb{R}_0} \{\log(1 + \gamma(s, z)) - \gamma(s, z)\} \nu(dz)ds \right.$$

$$\left. + \int_{\mathbb{R}} \int_{\mathbb{R}_0} \log(1 + \gamma(s, z))\widetilde{N}(ds, dz) \right).$$

In particular, choosing

$$\gamma(s, z) = h(s)z \text{ with } h \in L^2(\mathbb{R})$$

we get

$$\exp^\diamond\left(\int_{\mathbb{R}} h(s)d\eta(s) \right) \tag{13.42}$$

$$= \exp\left(\int_{\mathbb{R}} \int_{\mathbb{R}_0} \{\log(1 + h(s)z) - h(s)z\} \nu(dz)ds \right.$$

$$\left. + \int_{\mathbb{R}} \int_{\mathbb{R}_0} \log(1 + h(s)z)\widetilde{N}(ds, dz) \right).$$

13.3.2 Wick Product and Skorohod Integral

One of the reasons for the importance of the Wick product is the following result.

Theorem 13.13. *(1) Let $Y(t)$ be Skorohod integrable with respect to η. Then $Y(t) \diamond \overset{\bullet}{\eta}(t)$ is dt-integrable in the space $(\mathcal{S})^*$ and*

$$\int_{\mathbb{R}} Y(t)\delta\eta(t) = \int_{\mathbb{R}} Y(t) \diamond \overset{\bullet}{\eta}(t)dt. \tag{13.43}$$

(2) Let $X(t, z)$ be Skorohod-integrable with respect to $\widetilde{N}(\cdot, \cdot)$. Then $X(t, z) \diamond \overset{\bullet}{\widetilde{N}}(t, z)$ is $\nu(dz)dt$-integrable in $(\mathcal{S})^*$ and

$$\int_{\mathbb{R}} \int_{\mathbb{R}_0} X(t,z)\widetilde{N}(\delta t, dz) = \int_{\mathbb{R}} \int_{\mathbb{R}_0} X(t,z) \diamond \overset{\bullet}{\widetilde{N}}(t,z)\nu(dz)dt. \tag{13.44}$$

The proof is similar to the Brownian motion case and therefore omitted.

Example 13.14. In Example 11.4 we proved that

$$\int_0^T \eta(T)\delta\eta(t) = 2\int_0^T \eta(t)d\eta(t)$$

$$= \eta^2(T) - \int_0^T \int_{\mathbb{R}_0} z^2 N(dt,dz).$$

Using Theorem 13.13 and (13.39) we can also get the following:

$$\int_0^T \eta(T)\delta\eta(t) = \int_0^T \eta(t) \diamond \overset{\bullet}{\eta}(t)dt = \eta(T) \diamond \eta(T)$$

$$= \eta^2(T) - \int_0^T \int_{\mathbb{R}_0} z^2 N(dt,dz).$$

The following result corresponds to Theorem 4.1 in [172], presented here in the setting of Poisson random measures; see, for example, [64].

Theorem 13.15. The Lévy–Skorohod isometry. *Let $X \in L^2(P \times \lambda \times \nu)$ and $DX \in L^2(P \times (\lambda \times \nu)^2)$. Then the following isometry holds*

$$E\left[\left(\int_0^\infty \int_{\mathbb{R}_0} X(t,z)\widetilde{N}(\delta t, dz)\right)^2\right] \tag{13.45}$$

$$= E\left[\int_0^\infty \int_{\mathbb{R}_0} X^2(t,z)\nu(dz)dt\right]$$

$$+ E\left[\int_0^\infty \int_{\mathbb{R}_0} \int_0^\infty \int_{\mathbb{R}_0} D_{t,z}X(s,y)D_{s,y}X(t,z)\nu(dy)ds\nu(dz)dt\right].$$

Proof Consider
$$X(t,z) = \sum_{\alpha \in \mathcal{J}} c_\alpha(t,z)K_\alpha.$$

Define

$$S_1 = \sum_{\alpha \in \mathcal{J}} \alpha! \|c_\alpha\|^2_{L^2(\lambda \times \nu)}, \qquad S_2 = \sum_{\alpha \in \mathcal{J}, i,j \in \mathbb{N}} \alpha_{\gamma(i,j)}\alpha!(c_\alpha, \xi_j\psi_i)^2,$$

and

$$S_3 = \sum_{\substack{\alpha,\beta \in \mathcal{J}, i,j,k,l \in \mathbb{N} \\ (i,j) \neq (k,l)}} (\alpha_{\gamma(i,j)} + 1)\alpha!(c_\alpha, \xi_j\psi_i)(c_\beta, \xi_k\psi_l)\chi_{\{\alpha + \epsilon^{\gamma(i,j)} = \epsilon^{\gamma(k,l)}\}},$$

where $(\cdot,\cdot) = (\cdot,\cdot)_{L^2(\lambda \times \nu)}$. Note that by the assumption and Lemma 3.12 in [181], the aforementioned sums are convergent. First it follows that

$$
E\left[\left(\int_0^\infty \int_{\mathbb{R}_0} X(t,z)\tilde{N}(\delta t, dz)\right)^2\right] = E\left[\left(\int_0^\infty \int_{\mathbb{R}_0} X(t,z) \diamond \overset{\bullet}{\tilde{N}}(t,z)\nu(dz)dt\right)^2\right]
$$

$$
= E\left[\left(\int_0^\infty \int_{\mathbb{R}_0} \left(\sum_{\alpha \in \mathcal{J}} c_\alpha(t,z)K_\alpha\right) \diamond \left(\sum_{i,j} \xi_j(t)\psi_i(z)K_{\epsilon^{\gamma(i,j)}}\right)\nu(dz)dt\right)^2\right]
$$

$$
= E\left[\left(\sum_{\alpha \in \mathcal{J},i,j} (c_\alpha, \xi_j\psi_i)K_{\alpha+\epsilon^{\gamma(i,j)}}\right)^2\right]
$$

$$
= \sum_{\substack{\alpha,\beta \in \mathcal{J},i,j,k,l \in \mathbb{N} \\ (i,j)\neq(k,l)}} (\alpha+\epsilon^{\gamma(i,j)})!(c_\alpha,\xi_j\psi_i)(c_\beta,\xi_k\psi_l)\chi_{\{\alpha+\epsilon^{\gamma(i,j)}=\epsilon^{\gamma(k,l)}\}} = S_1 + S_2 + S_3,
$$

since $(\alpha + \epsilon^{\gamma(i,j)})! = (\alpha_{\gamma(i,j)}+1)\alpha!$. Next, we have

$$
E\left[\int_0^\infty \int_{\mathbb{R}_0} X^2(t,z)\nu(dz)dt\right] = E\left[\int_0^\infty \int_{\mathbb{R}_0} \left(\sum_{\alpha \in \mathcal{J}} c_\alpha(t,z)K_\alpha\right)^2 \nu(dz)dt\right]
$$

$$
= \sum_{\alpha \in \mathcal{I}} \int_0^\infty \int_{\mathbb{R}_0} c_\alpha^2(t,z)\alpha!\nu(dz)dt = S_1.
$$

Finally, we get

$$
E\left[\int_0^\infty \int_{\mathbb{R}_0} \int_0^\infty \int_{\mathbb{R}_0} D_{t,z}X(s,y)D_{s,y}X(t,z)\nu(dy)ds\nu(dz)dt\right]
$$

$$
= E\left[\int_0^\infty \int_{\mathbb{R}_0} \int_0^\infty \int_{\mathbb{R}_0} \left(\sum_{\alpha,k,l} c_\alpha(s,y)\xi_k(t)\psi_l(z)\alpha_{\epsilon^{\gamma(k,l)}} K_{\alpha-\epsilon^{\gamma(k,l)}}\right)\right.
$$

$$
\left. \cdot\left(\sum_{\beta,i,j} c_\alpha(t,z)\xi_k(s)\psi_l(y)\alpha_{\epsilon^{\gamma(i,j)}} K_{\beta-\epsilon^{\gamma(i,j)}}\right)\nu(dy)ds\nu(dz)dt\right]
$$

$$
= \sum_{\alpha,\beta \in \mathcal{J},i,j,k,l \in \mathbb{N}} (c_\alpha,\xi_j\psi_i)(c_\beta,\xi_k\psi_l)\beta_{\gamma(i,j)}\alpha!\chi_{\{\alpha+\epsilon^{\gamma(i,j)}=\epsilon^{\gamma(k,l)}\}} = S_2 + S_3.
$$

Combining the three steps of the proof the desired result follows. □

Remark 13.16. Formula (13.45) can also be obtained as a consequence of the Poisson interpretation of Fock space. See Proposition 17 in [34] and Proposition 1 in [201]. For an isometry of this type, which is not based on Fock space, see Proposition 3.3 in [198].

13.3.3 Wick Product vs. Ordinary Product

We conclude this section with a result on the relationship between Wick products and ordinary products in the line of Theorem 6.8. The result reads as follows.

Theorem 13.17. Relation between Wick products and ordinary products.
Suppose $\gamma(t, z)$ is deterministic and

$$\int_{\mathbb{R}} \int_{\mathbb{R}_0} \gamma^2(t, z)\nu(dz)ds < \infty.$$

Further require that $F \in \mathbb{D}_{1,2}$. Then

$$F \diamond \int_{\mathbb{R}} \int_{\mathbb{R}_0} \gamma(t, z)\widetilde{N}(dt, dz) + \int_{\mathbb{R}} \int_{\mathbb{R}_0} \gamma(t, z)D_{t,z}F\widetilde{N}(\delta t, dz) \quad (13.46)$$

$$= F \cdot \int_{\mathbb{R}} \int_{\mathbb{R}_0} \gamma(t, z)\widetilde{N}(dt, dz) - \int_{\mathbb{R}} \int_{\mathbb{R}_0} \gamma(t, z)D_{t,z}F\nu(dz)dt.$$

Proof Let $\varphi(t, z) \geq -1$ be deterministic and $\int_{\mathbb{R}} \int_{\mathbb{R}_0} \{|\log(1 + \varphi(t, z))| + \varphi^2(t, z)\}\nu(dz)ds < \infty$. It suffices to prove the result when

$$F = \exp^\diamond \left(\int_{\mathbb{R}} \int_{\mathbb{R}_0} \varphi(s, z)\widetilde{N}(ds, dz) \right)$$

$$= \exp \left(\int_{\mathbb{R}} \int_{\mathbb{R}_0} \{\log(1 + \varphi(s, z)) - \varphi(s, z)\} \nu(dz)ds \right.$$

$$\left. + \int_{\mathbb{R}} \int_{\mathbb{R}_0} \log(1 + \varphi(s, z))\widetilde{N}(ds, dz) \right)$$

(see Example 13.12). For $y \in \mathbb{R}$ define

$$G_y = \exp^\diamond \left(\int_{\mathbb{R}} \int_{\mathbb{R}_0} y(1 + \varphi(s, z))\gamma(s, z)\widetilde{N}(ds, dz) \right)$$

$$= \exp \left(\int_{\mathbb{R}} \int_{\mathbb{R}_0} \{\log(1 + y(1 + \varphi(s, z))\gamma(s, z)) - y(1 + \varphi(s, z))\gamma(s, z)\} \nu(dz)ds \right.$$

$$\left. + \int_{\mathbb{R}} \int_{\mathbb{R}_0} \log(1 + y(1 + \varphi(s, z))\gamma(s, z))\widetilde{N}(ds, dz) \right).$$

Then

$$F \diamond G_y = \exp^\diamond \left(\int_{\mathbb{R}} \int_{\mathbb{R}_0} \varphi(s, z) + y(1 + \varphi(s, z))\gamma(s, z)\widetilde{N}(ds, dz) \right)$$

$$= \exp \left(\int_{\mathbb{R}} \int_{\mathbb{R}_0} \{\log(1 + \varphi(s, z) + y(1 + \varphi(s, z))\gamma(s, z)) \right.$$

$$- \varphi(s, z) - y(1 + \varphi(s, z))\gamma(s, z)\} \nu(dz)ds$$

$$\left. + \int_{\mathbb{R}} \int_{\mathbb{R}_0} \log(1 + \varphi(s, z) + y(1 + \varphi(s, z))\gamma(s, z))\widetilde{N}(ds, dz) \right).$$

On differentiating with respect to y, we get

$$\frac{d}{dy}(F \diamond G_y) = F \diamond \frac{d}{dy} G_y = F \diamond G_y \diamond \int_{\mathbb{R}} \int_{\mathbb{R}_0} (1 + \varphi(s,z)) \gamma(s,z) \tilde{N}(ds, dz),$$

and using the other expression,

$$\frac{d}{dy}(F \diamond G_y)$$

$$= F \diamond G_y \cdot \left(\int_{\mathbb{R}} \int_{\mathbb{R}_0} \left\{ \frac{(1 + \varphi(s,z))\gamma(s,z)}{1 + \varphi(s,z) + y(1 + \varphi(s,z))\gamma(s,z)} \right. \right.$$

$$- (1 + \varphi(s,z))\gamma(s,z) \right\} \nu(dz) ds$$

$$+ \int_{\mathbb{R}} \int_{\mathbb{R}_0} \left\{ \frac{(1 + \varphi(s,z))\gamma(s,z)}{1 + \varphi(s,z) + y(1 + \varphi(s,z))\gamma(s,z)} \right\} \tilde{N}(ds, dz).$$

Putting $y = 0$ gives the identity

$$F \diamond \int_{\mathbb{R}} \int_{\mathbb{R}_0} (1 + \varphi(s,z)) \gamma(s,z) \tilde{N}(ds, dz)$$

$$= F \cdot \int_{\mathbb{R}} \int_{\mathbb{R}_0} (-\varphi(s,z)) \gamma(s,z) \nu(dz) ds + F \cdot \int_{\mathbb{R}} \int_{\mathbb{R}_0} \gamma(s,z) \tilde{N}(ds, dz).$$

Since

$$D_{s,z} F = \varphi(s,z) F,$$

this can be written as

$$F \diamond \int_{\mathbb{R}} \int_{\mathbb{R}_0} \gamma(t,z) \tilde{N}(dt, dz) + \int_{\mathbb{R}} \int_{\mathbb{R}_0} \gamma(t,z) D_{t,z} F \tilde{N}(\delta t, dz)$$

$$= F \cdot \int_{\mathbb{R}} \int_{\mathbb{R}_0} \gamma(t,z) \tilde{N}(dt, dz) - \int_{\mathbb{R}} \int_{\mathbb{R}_0} \gamma(t,z) D_{t,z} F \nu(dz) dt. \quad \square$$

13.3.4 Lévy–Hermite Transform

It is useful to remark that, just as in the Brownian motion case (see Sect. 5.3), the Hermite transform can be introduced. More precisely, the *Lévy–Hermite transform* of an element $X = \sum_\alpha c_\alpha K_\alpha \in (\mathcal{S})_{-1}$ is defined as

$$\mathcal{H} X(z) = \tilde{X}(z) = \sum_\alpha c_\alpha z^\alpha \in \mathbb{C}, \tag{13.47}$$

where $z = (z_1, z_2 \dots) \in \mathbb{C}^n$ and $z^\alpha = z_1^{\alpha_1} z_2^{\alpha_2} \dots$. Here $(\mathcal{S})_{-1}$ is the Lévy version of the Kondratiev distribution space treated in Section 5.3.

An important property of the Lévy–Hermite transform is that it transforms the Lévy–Wick product into an ordinary product, that is,

$$\mathcal{H}(X \diamond Y)(z) = \mathcal{H}(X) \cdot \mathcal{H}(Y), \tag{13.48}$$

cf. (5.49).

We mention that all the results related to the Hermite transform in Sect. 5.3 carry over to the case of the Lévy–Hermite transform. For example, the Lévy–Hermite transform can be used to characterize the Lévy–Hida distribution space $(\mathcal{S})_{-1}$, cf. Theorem 5.46.

13.4 Spaces of Smooth and Generalized Random Variables: $\mathcal{G}$ and $\mathcal{G}^*$

As we saw in Sect. 12.6, the Clark–Ocone formula serves as a useful tool to compute the closest hedge of contingent claims in the sense of minimal variance. However, this formula exhibits the deficiency to fail if the claims are not Malliavin differentiable. To overcome this problem we establish in Sect. 13.6 a generalized Clark–Ocone formula on a space of generalized random variables. An appropriate candidate for such a space that meets the requirement to comprise a rich class of claims is given by a Lévy version of the space $\mathcal{G}^*$ introduced in Sect. 5.3.3.

The spaces of smooth random variables, $\mathcal{G}$, and $\mathcal{G}^*$ have been studied in the Gaussian case by Potthoff and Timpel [194]; see also [96]. Later on the Poissonian case has been treated in [26] and [106]. The Lévy versions of these spaces are defined, analogously.

Definition 13.18. *(1) Let $k \in \mathbb{N}_0$. We say that $f = \sum_{m \geq 0} I_m(f_m) \in L^2(P)$ belongs to the space $\mathcal{G}_k$ if*

$$\|f\|_{\mathcal{G}_k}^2 := \sum_{m \geq 0} m! \, \|f_m\|_{L^2((\lambda \otimes \nu)^m)}^2 \, e^{2km} < \infty. \qquad (13.49)$$

We define the space of smooth random variables $\mathcal{G}$ *as*

$$\mathcal{G} = \bigcap_{k \in \mathbb{N}_0} \mathcal{G}_k.$$

The space $\mathcal{G}$ is endowed with the projective topology.
(2) We say that a formal expansion

$$G = \sum_{m \geq 0} I_m(g_m)$$

belongs to the space $\mathcal{G}_{-q}$ $(q \in \mathbb{N}_0)$ if

$$\|G\|_{\mathcal{G}_{-q}}^2 := \sum_{m \geq 0} m! \, \|g_m\|_{L^2((\lambda \otimes \nu)^m)}^2 \, e^{-2qm} < \infty. \qquad (13.50)$$

The space of generalized random variables $\mathcal{G}^*$ *is defined as*

$$\mathcal{G}^* = \bigcup_{q \in \mathbb{N}_0} \mathcal{G}_{-q}.$$

We equip $\mathcal{G}^*$ with the inductive topology. Note that $\mathcal{G}^*$ is the dual of $\mathcal{G}$, with action

$$\langle G, f \rangle = \sum_{m \geq 0} m! (f_m, g_m)_{L^2((\lambda \otimes \nu)^m)}$$

if $G \in \mathcal{G}^*$ and $f \in \mathcal{G}$.

Also note that using (10.4) and (13.15) the connection between the expansions

$$F = \sum_{m \geq 0} I_m(f_m)$$

and

$$F = \sum_{\alpha \in \mathcal{J}} c_\alpha K_\alpha$$

is given by

$$f_m = \sum_{|\alpha| = m} c_\alpha \delta^{\widehat{\otimes} \alpha}, \quad m \geq 0, \tag{13.51}$$

with the functions $\delta^{\widehat{\otimes} \alpha}$ as in (13.17). Since this gives

$$\|I_m(f_m)\|^2_{L^2(P)} = m! \|f_m\|^2_{L^2((\lambda \otimes \nu)^m)} = \sum_{|\alpha| = m} c_\alpha^2 \|K_\alpha\|^2_{L^2(P)}, \tag{13.52}$$

it follows that we can express the $\mathcal{G}_r$-norm of F as follows:

$$\|F\|^2_{\mathcal{G}_r} = \sum_{m \geq 0} \left(\sum_{|\alpha| = m} c_\alpha^2 \|K_\alpha\|^2_{L^2(P)} \right) e^{2rm}, \quad r \in \mathbb{Z}. \tag{13.53}$$

By inspecting Theorem 13.6, we find the following chain of continuous inclusions

$$(\mathcal{S}) \subset \mathcal{G} \subset L^2(P) \subset \mathcal{G}^* \subset (\mathcal{S})^*.$$

Just as in the Gaussian case, it can be seen that $\mathcal{G}$ and $\mathcal{G}^*$ constitute topological algebras with respect to the Wick product.

In [186] the authors show that a larger class of jump SDEs belongs to the space of smooth random variables $\mathcal{G}$.

13.5 The Malliavin Derivative on $\mathcal{G}^*$

Definition 12.2 together with (13.15) and (13.51) motivates the following generalization of the stochastic derivative $D_{t,z}$ to $\mathcal{G}^*$. See Remark 13.21.

Definition 13.19. *Let $F = \sum_\alpha c_\alpha K_\alpha \in \mathcal{G}^*$. Then define the stochastic derivative of F at (t, z) by*

$$D_{t,z}F := \sum_\alpha c_\alpha \sum_i \alpha_i K_{\alpha-\epsilon^i} \cdot \delta^{\widehat{\otimes}\epsilon^i}(t,z) \tag{13.54}$$

$$= \sum_\alpha c_\alpha \sum_{k,m} \alpha_{\kappa(k,m)} K_{\alpha-\epsilon^{\kappa(k,m)}} \cdot e_k(t)p_m(z)$$

$$= \sum_\beta (\sum_{k,m} c_{\beta+\epsilon^{\kappa(k,m)}}(\beta_{\kappa(k,m)}+1)e_k(t)p_m(z))K_\beta,$$

with the map $\kappa(i,j)$ in (13.11) and $\epsilon^l = \epsilon^{(l)}$ as in (5.9).

We need the following result.

Lemma 13.20. *(1) Let $F \in \mathcal{G}^*$. Then $D_{t,z}F \in \mathcal{G}^*$ $\lambda \times \nu$ – a.e.*
(2) Suppose $F, F_n \in \mathcal{G}^*$ for all $n \in \mathbb{N}$ and

$$F_n \longrightarrow F \text{ in } \mathcal{G}^*.$$

Then there exists a subsequence $\{F_{n_k}\}_{k\geq 1}$ such that

$$D_{t,z}F_{n_k} \longrightarrow D_{t,z}F$$

in $\mathcal{G}^*$ $\lambda \times \nu$ – a.e.

Proof It can be shown in same way as in Lemma 6.34 that for all $F \in \mathcal{G}^*$ there exists a $p < \infty$ such that

$$\int_{\mathbb{R}_+} \int_{\mathbb{R}_0} \|D_{t,z}F\|^2_{\mathcal{G}_{-p}} \nu(dz)dt \leq const. \cdot \|F\|^2_{\mathcal{G}_{-p+1}}. \tag{13.55}$$

The result then follows immediately.   □

Remark 13.21. Definition 13.19 coincides with Definition 12.2, if $F \in \mathbb{D}_{1,2} \subset \mathcal{G}^*$. This follows with the help of the closability of the operators and a duality formula for the operators. A duality or integration by parts formula for $D_{t,z}$ in Definition 13.19 can be stated as

$$\int_{\mathbb{R}_+} \int_{\mathbb{R}_0} <Y(t,z), D_{t,z}F> \nu(dz)dt =< \int_{\mathbb{R}_+} \int_{\mathbb{R}_0} Y(t,z) \diamond \dot{\tilde{N}}(t,z)\nu(dz)dt, F >,$$

where $F \in (\mathcal{S})$ and $\int_{\mathbb{R}_+} \int_{\mathbb{R}_0} \|Y(t,z)\|^2_{\mathcal{G}_{-q}} \nu(dz)dt < \infty$ for some $q \geq 0$. The latter can be easily seen from the definitions.

If

$$P(x) = \sum_\alpha c_\alpha x^\alpha; \ x \in \mathbb{R}^\mathbb{N}, c_\alpha \in \mathbb{R}$$

is a polynomial, where $x^\alpha := x_1^{\alpha_1} x_2^{\alpha_2}...$ and $x_j^0 := 1$, we define its *Wick version* at $X = (X_1, ..., X_m)$ by

$$P^\diamond(X) = \sum_\alpha c_\alpha X^{\diamond\alpha}.$$

In the sequel the differentiability of a process $X : \mathbb{R}_+ \longrightarrow (\mathcal{S})^*$ is understood in the sense of the topology of $(\mathcal{S})^*$. We denote the derivative of $X(t)$ by $\frac{d}{dt}X(t)$. Define

$$X_{k,m}^{(t)} = \int_0^t \int_{\mathbb{R}_0} e_k(s)p_m(z)\widetilde{N}(ds, dz), k, m \geq 1. \tag{13.56}$$

It follows from Theorem 13.13 that $X_{k,m}^{(t)} = \int_0^t \int_{\mathbb{R}_0} e_k(s)p_m(z)\overset{\bullet}{\widetilde{N}}(t, z)\nu(dz)ds$. We deduce from the proof of Lemma 2.8.4 in [107] that $X_{k,m}^{(t)}$ is differentiable and

$$\frac{d}{dt}X_{k,m}^{(t)} = e_k(t)L_m(t) \in (\mathcal{S})^*, \tag{13.57}$$

with

$$L_m(t) := \int_{\mathbb{R}_0} p_m(z)\overset{\bullet}{\widetilde{N}}(t, z)\nu(dz),$$

where we use a Bochner integral with respect to ν.

We obtain by induction the following *Wick chain rule* for polynomials.

Lemma 13.22. *Let*

$$P(x) = \sum_\alpha c_\alpha x^\alpha$$

be a polynomial in $\mathbb{R}^n$. Suppose $k_i, m_i \geq 1$ for all $i = 1, ..., n$ and let

$$X^{(t)} = (X_{k_1,m_1}^{(t)}, ..., X_{k_n,m_n}^{(t)}),$$

with $X_{k,m}^{(t)}$ as in (13.56). Then

$$\frac{d}{dt}P^\diamond(X^{(t)}) = \sum_{i=1}^n \left(\frac{\partial P}{\partial x_i}\right)^\diamond (X^{(t)}) \diamond L_{m_i}(t) \cdot e_{k_i}(t).$$

13.6 A Generalization of the Clark–Ocone Theorem

In this section we aim at deriving a Clark–Ocone formula on the space generalized random variables $\mathcal{G}^*$ by employing the Lévy white noise framework developed in the previous section. The classical *Clark–Ocone theorem* (see Theorem 12.16) states that if $F \in L^2(P)$ is $\mathcal{F}_T$-measurable and $F \in \mathbb{D}_{1,2}$, then

$$F = E[F] + \int_0^T E[D_t F | \mathcal{F}_t]dW(t), \tag{13.58}$$

where D_t denotes the Malliavin derivative with domain $\mathbb{D}_{1,2}$. This result and its generalizations have important applications in finance. As seen in Sect. 4.3, the conditional expectation $E[D_t F | \mathcal{F}_t]$ can be regarded as the replicating portfolio of a given T-claim F (see, e.g., [174]).

The requirement that $F \in \mathbb{D}_{1,2}$ excludes interesting applications. For example, the digital option of the form

$$F = \chi_{[K,\infty)}(B_T)$$

is not in $\mathbb{D}_{1,2}$ and (13.58) cannot be applied to compute the hedging portfolio of the T-claim F. Relation (13.58) was generalized to a Clark–Ocone formula in the setting of white noise analysis in Sect. 6.5. This Clark–Ocone theorem is valid for all $\mathcal{F}_T$-measurable F in the space of generalized random variables as defined in Sect. 6.5. The generalization has the form

$$F = E[F] + \int_0^T E[D_t F \,|\, \mathcal{F}_t] \diamond \overset{\bullet}{W}_t dt, \qquad (13.59)$$

where $\diamond$ denotes a Wick product and $\overset{\bullet}{W}$ is the white noise of W. Further, $E[F]$ is the generalized expectation of F and $E[D_t F \,|\, \mathcal{F}_t]$ is the generalized conditional expectation. The integral on the right hand side is a Bochner integral; see Chap. 5. In the Wiener space setting another generalization of (13.58) for *Meyer–Watanabe distributions* $F \in \mathbb{D}_{-\infty} \subset \mathcal{G}^*$ has been obtained by Üstünel [223, 224]. Similar results for the Poisson process can be found in [2].

In the sequel, we prove a corresponding version of (13.59) for Poisson random measures (see Theorem 13.26 and Theorem 13.27).

Financial markets (with no trading constraints or transition costs) modeled by a Brownian motion or a Poisson process are complete. In this case, the Clark–Ocone theorem gives an almost direct formula for the replicating portfolio of a contingent claim in terms of the conditional expectation of its Malliavin derivative. However, as mentioned, financial markets modeled by Lévy processes are in general not complete and hence the Clark–Ocone formula does not permit a similar interpretation. On the other hand, it can be used to obtain an explicit formula for the closest hedge, in the sense of minimal variance. See [25, 69, 181]. As in the Brownian motion case, the extension from $\mathbb{D}_{1,2}$ to $L^2(P)$ is important for the applications to finance. For example, the digital option

$$F = \chi_{[K,\infty)}(\eta_T), \qquad (13.60)$$

where η_t is a Lévy process and is not contained in $\mathbb{D}_{1,2}$, in general. To see this, take, for example, a driftless pure jump Lévy process η with Lévy measure $\nu(dz) = \chi_{(0,1)}(z)\frac{1}{z^2}dz$. So η is of unbounded variation. We get then for the real part of its characteristic exponent Ψ that $\operatorname{Re}\Psi(\lambda) = |\lambda| \int_0^{|\lambda|} (1 - \cos z)\frac{1}{z^2}dz$. This yields $\lim_{|\lambda| \to \infty} |\lambda|^{-\frac{1}{2}} \operatorname{Re}\Psi(\lambda) = \infty$, so that the semigroup of η has the strong Feller property. The latter is equivalent to the law of $\eta(T)$ being absolutely continuous with respect to the Lebesgue measure (see [32]), that is, we have $P(\eta(T) \leq K) = \int_{(-\infty,K]} f dt$ for some integrable f. So there exists a $K_0 > 0$ such that $f(K_0) > 0$ and $P(\eta(T) \leq K)$ is differentiable in K_0. Then it is not difficult to see that $\rho := \inf_{z \in (0, K_0/2]} \frac{P(K_0 - z \leq \eta(T) < K_0)}{|z|} > 0$. This and the unbounded variation of η imply

$$\int_{\mathbb{R}_0} P(K_0 - z \leq \eta(T) < K_0)\nu(dz) \geq \int_{(-\infty, \frac{K_0}{2}]} \rho |z| \nu(dz) = \infty.$$

Suppose $F \in \mathbb{D}_{1,2}$, then Theorem 12.8 gives

$$\int_0^T \int_{\mathbb{R}_0} E[(D_{t,z}F)^2]\nu(dz)dz$$

$$= \int_0^T \int_{\mathbb{R}_0} E[(\chi_{[K,\infty)}(\eta_T(\omega) + z) - \chi_{[K,\infty)}(\eta_T(\omega)))^2]\nu(dz)dt$$

$$= T \int_{\mathbb{R}_0} \mu(K_0 - z \leq \eta(T) < K_0)\nu(dz) < \infty,$$

but this leads to a contradiction. Thus $F \notin \mathbb{D}_{1,2}$.

We need the following definition of conditional expectation in the space $\mathcal{G}^*$ (see [69] or [181]).

Definition 13.23. *Let* $F = \sum_{m \geq 0} I_m(f_m) \in \mathcal{G}^*$. *Then the conditional expectation of* F *with respect to* $\mathcal{F}_t$ *is defined by*

$$E[F | \mathcal{F}_t] = \sum_{m \geq 0} I_m(f_m \cdot \chi_{[0,t]^m}). \tag{13.61}$$

Note that this notion coincides with the usual conditional expectation if $F \in L^2(P)$. Obviously, we get

$$\|E[F | \mathcal{F}_t]\|_{\mathcal{G}_r} \leq \|F\|_{\mathcal{G}_r} \tag{13.62}$$

for all $r \in \mathbb{Z}$. Thus

$$E[F | \mathcal{F}_t] \in \mathcal{G}^* \tag{13.63}$$

for all t.

Lemma 13.24. *For* $F, G \in \mathcal{G}^*$ *we have*

$$E[F \diamond G | \mathcal{F}_t] = E[F | \mathcal{F}_t] \diamond E[G | \mathcal{F}_t]. \tag{13.64}$$

Proof The proof follows the same line of the one for the Brownian motion case, see Lemma 6.20. See also [69].   □

Lemma 13.25. The Clark–Ocone formula for polynomials. *Let* $F = P^\diamond(X^{(T)})$ *for some polynomial* $P(x) = \sum_\alpha c_\alpha x^\alpha$ *and the variable* $X^{(T)}$ *as in Lemma 13.22. Then*

$$F = E[F] + \int_0^T \int_{\mathbb{R}_0} E[D_{t,z}F | \mathcal{F}_t]\tilde{N}(dt, dz).$$

Proof Without loss of generality, let us assume that $F = K_\alpha^{(T)}$ with $K_\alpha^{(t)} := E[K_\alpha | \mathcal{F}_t]$ and

$$\alpha = \alpha_{\kappa(k_1, m_1)} \epsilon^{\kappa(k_1, m_1)} + \ldots + \alpha_{\kappa(k_n, m_n)} \epsilon^{\kappa(k_n, m_n)}$$

with $(k_i, m_i) \neq (k_j, m_j)$ for $i \neq j$. Note that $K_\alpha^{(T)} = (X_{k_1, m_1}^{(T)})^{\diamond \alpha_{\kappa(k_1, m_1)}} \diamond \ldots \diamond (X_{k_n, m_n}^{(T)})^{\diamond \alpha_{\kappa(k_n, m_n)}}$, where $X_{k_i, m_i}^{(T)}$ is as in (13.56). We conclude with the help of Definition 13.19, Theorem 13.13, and Lemma 13.22 that

$$\int_0^T \int_{\mathbb{R}_0} E[D_{t,z} F | \mathcal{F}_t] \tilde{N}(dt, dz)$$

$$= \int_0^T \int_{\mathbb{R}_0} (E[\sum_{k,m \geq 1} \alpha_{\kappa(k,m)} K_{\alpha - \epsilon^{\kappa(k,m)}} \cdot e_k(t) p_m(z) | \mathcal{F}_t]) \diamond \overset{\bullet}{\tilde{N}}(t, z) \nu(dz) dt$$

$$= \int_0^T \int_{\mathbb{R}_0} (\sum_{k,m \geq 1} \alpha_{\kappa(k,m)} K_{\alpha - \epsilon^{\kappa(k,m)}}^{(t)} \cdot e_k(t) l_m(z)) \diamond \overset{\bullet}{\tilde{N}}(t, z) \nu(dz) dt$$

$$= \int_0^T \sum_{k,m \geq 1} \alpha_{\kappa(k,m)} K_{\alpha - \epsilon^{\kappa(k,m)}}^{(t)} \diamond L_m(t) \cdot e_k(t) dt$$

$$= \int_0^T \frac{d}{dt} K_\alpha^{(t)} dt = K_\alpha^{(T)} - K_\alpha^{(0)} = F - E[F]. \quad \square$$

We can now prove a Clark–Ocone theorem for $L^2(P)$. This result was first proved in [69, 181]. It is an extension to $L^2(P)$ of Theorem 12.16 and a Poisson random measure version of Theorem 3.11 in [2].

Theorem 13.26. Clark–Ocone theorem for $L^2(P)$. *Let $F \in L^2(P)$ be $\mathcal{F}_T$-measurable. Then*

$$E[D_{t,z} F | \mathcal{F}_t], \quad t \in [0, T], z \in \mathbb{R}_0,$$

is an element in $L^2(\lambda \times \nu \times P)$ and

$$F = E[F] + \int_0^T \int_{\mathbb{R}_0} E[D_{t,z} F | \mathcal{F}_t] \tilde{N}(dt, dz).$$

Proof There exists a sequence of $\mathcal{F}_T$-measurable random variables F_n as in Lemma 13.25 such that $F_n \longrightarrow F$ in $L^2(P)$. Then by Lemma 13.25, we get

$$F_n = E[F_n] + \int_0^T \int_{\mathbb{R}_0} E[D_{t,z} F_n | \mathcal{F}_t] \tilde{N}(dt, dz)$$

for all n. By the representation theorem of Itô there exists a unique predictable process $u(t, z)$, $t \in [0, T]$, $z \in \mathbb{R}_0$, such that

$$E[\int_0^T \int_{\mathbb{R}_0} u^2(t,z)\nu(dz)dt] < \infty$$

and

$$F = E[F] + \int_0^T \int_{\mathbb{R}_0} u(t,z)\tilde{N}(dt,dz).$$

Further, we have

$$E[\int_0^T \int_{\mathbb{R}_0} (E[D_{t,z}F_n \,|\, \mathcal{F}_t] - u(t,z)^2\nu(dz)dt]$$
$$= E[(F_n - F - E[F_n] + E[F])^2] \to 0, \quad n \to \infty.$$

So

$$E[D_{t,z}F_n \,|\, \mathcal{F}_t] \longrightarrow u, \quad n \to \infty, \quad \text{in } L^2(\lambda \times \nu \times P).$$

On the other hand, by taking a subsequence $\{F_{n_k}\}_{k \geq 1}$, we know from Lemma 13.20 that

$$E[D_{t,z}F_{n_k} \,|\, \mathcal{F}_t] \longrightarrow E[D_{t,z}F \,|\, \mathcal{F}_t], \quad k \to \infty, \quad \text{in } \mathcal{G}^*, \ \lambda \times \nu - \text{a.e.}$$

Taking again a subsequence, we have that

$$E[D_{t,z}F_{n_k} \,|\, \mathcal{F}_t] \longrightarrow u(t,z), \quad k \to \infty, \quad \text{in } L^2(P), \lambda \times \nu - \text{a.e.}$$

It follows that

$$F = E[F] + \int_0^T \int_{\mathbb{R}_0} E[D_{t,z}F \,|\, \mathcal{F}_t]\tilde{N}(dt,dz). \quad \square$$

Recall the definition of generalized expectation as given in Definition 13.6. We say that $F \in \mathcal{G}^*$ is $\mathcal{F}_T$-*measurable* if

$$E[F \,|\, \mathcal{F}_T] = F. \tag{13.65}$$

Hereafter we prove a Clark–Ocone formula for $\mathcal{G}^*$. This result was first proved in [69] and [181]. It is a Poisson random measure version of Theorem 3.15 in [2].

Theorem 13.27. Clark–Ocone theorem for $\mathcal{G}^*$. *Let* $F \in \mathcal{G}^*$ *be* $\mathcal{F}_T$-*measurable. Then*

$$F = E[F] + \int_0^T \int_{\mathbb{R}_0} E[D_{t,z}F \,|\, \mathcal{F}_t] \diamond \overset{\bullet}{\tilde{N}}(t,z)\nu(dz)dt,$$

where $E[F]$ *denotes the generalized expectation of* F.

Proof Let $F_n \in L^2(P)$, $n = 1, 2, ...$, be $\mathcal{F}_T$-measurable random variables such that $F_n \longrightarrow F$, $n \to \infty$, in $\mathcal{G}^*$. Then there exists a p such that $\|F_n - F\|^2_{\mathcal{G}_{-p+1}} \longrightarrow 0$ as $n \to \infty$. By Theorem 13.26 we obtain

$$F_n = E[F_n] + \int_0^T \int_{\mathbb{R}_0} E[D_{t,z}F_n \,|\mathcal{F}_t] \diamond \overset{\bullet}{\tilde{N}}(t,z)\nu(dz)dt \quad \text{for all } n.$$

For going to the limit in the last relation let us proceed as follows. First we get that

$$\left\| E[D_{t,z}\tilde{F}\,|\mathcal{F}_t] \diamond \overset{\bullet}{\tilde{N}}(t,z) \right\|_{0,-(3p+5)}^2 \leq const. \cdot \left\| E[D_{t,z}\tilde{F}\,|\mathcal{F}_t] \right\|_{0,-3p}^2 \| \overset{\bullet}{\tilde{N}}(t,z) \|_{0,-3}^2$$

for $\tilde{F} \in \mathcal{G}^*$, with $\left\| \tilde{F} \right\|_{\mathcal{G}_{-p+1}} < \infty$. So it follows that

$$|< E[D_{t,z}\tilde{F}\,|\mathcal{F}_t] \diamond \overset{\bullet}{\tilde{N}}(t,z), f >| \leq const. \cdot \left\| D_{t,z}\tilde{F} \right\|_{\mathcal{G}_{-p}} \| \overset{\bullet}{\tilde{N}}(t,z) \|_{0,-3} \|f\|_{0,3p+5} \,,$$

for $f \in (\mathcal{S})$. Thus this gives

$$|< \int_0^T \int_{\mathbb{R}_0} E[D_{t,z}\tilde{F}\,|\mathcal{F}_t] \diamond \overset{\bullet}{\tilde{N}}(t,z)\nu(dz)dt, f >|$$

$$\leq \int_0^T \int_{\mathbb{R}_0} |< E[D_{t,z}\tilde{F}\,|\mathcal{F}_t] \diamond \overset{\bullet}{\tilde{N}}(t,z), f >| \,\nu(dz)dt$$

$$\leq const. \cdot \|f\|_{0,3p+5} \Big(\int_0^T \int_{\mathbb{R}_0} \left\| D_{t,z}\tilde{F} \right\|_{\mathcal{G}_{-p}}^2 \nu(dz)dt \Big)^{1/2}$$

$$\times \Big(\int_0^T \int_{\mathbb{R}_0} \| \overset{\bullet}{\tilde{N}}(t,z) \|_{0,-3}^2 \,\nu(dz)dt \Big)^{1/2}.$$

Together with (13.55) we get the following inequality

$$|< \int_0^T \int_{\mathbb{R}_0} E[D_{t,z}\tilde{F}\,|\mathcal{F}_t] \diamond \overset{\bullet}{\tilde{N}}(t,z)\nu(dz)dt, f >| \leq const. \cdot \|f\|_{0,3p+5} \left\| \tilde{F} \right\|_{\mathcal{G}_{-p+1}}.$$

The last estimate implies that the expression

$$\int_0^T \int_{\mathbb{R}_0} E[D_{t,z}(F_n - F)\,|\mathcal{F}_t] \diamond \overset{\bullet}{\tilde{N}}(t,z)\nu(dz)dt$$

converges to 0 in $(\mathcal{S})^*$ as $n \to \infty$. The result follows.   □

For a version of this result under change of measure, see Theorem 12.22.

13.7 A Combination of Gaussian and Pure Jump Lévy Noises in the White Noise Setting

Let us sketch the framework for the combination of Gaussian noise and pure jump Lévy noise in the white noise setting (cf. also [2]). For the classical setting this was done in Sect. 13.5. Denote by P_W the Gaussian white noise measure on $(\Omega_0, \mathcal{F}_T^W)$. Recall that $\Omega_0 = \mathcal{S}'(\mathbb{R})$; see Chap. 5. Fix N, $R \in \mathbb{N}_0$. Then let $(\Omega_1, \mathcal{F}_T^{(1)}, \mu_1), ..., (\Omega_N, \mathcal{F}_T^{(N)}, \mu_N)$ be N independent copies of $(\Omega, \mathcal{F}_T^W, P_W)$ and let

$$(\Omega_{N+1}, \mathcal{F}_T^{(N+1)}, \mu_{N+1}), ..., (\Omega_{N+R}, \mathcal{F}_T^{(N+R)}, \mu_{N+R})$$

be R independent copies of $(\Omega_0, \mathcal{F}_T^{\widetilde{N}}, P^{\widetilde{N}})$, where $P^{\widetilde{N}}$ is the pure jump Lévy white noise measure (see Sect. 13.1).

We set

$$\Omega = \Omega_1 \times ... \times \Omega_{N+R}, \quad \mathcal{F}_T = \mathcal{F}_T^{(1)} \otimes ... \otimes \mathcal{F}_T^{(L)}, \quad P = \mu_1 \times ... \times \mu_{N+R}. \quad (13.66)$$

Set $L = N + R$. We denote by $\mathcal{J}^L$ the set of all $\alpha = (\alpha^{(1)}, ..., \alpha^{(L)})$ with $\alpha^{(j)} \in \mathcal{J}$, $j = 1, ..., L$. Further, we put

$$\mathbb{H}_\alpha(\omega) = H_{\alpha^{(1)}}(\omega_1) \cdot ... \cdot H_{\alpha^{(N)}}(\omega_N) \cdot K_{\alpha^{(N+1)}}(\omega_{N+1}) \cdot ... \cdot K_{\alpha^{(L)}}(\omega_L). \quad (13.67)$$

Then the family $\mathbb{H}_\alpha$, $\alpha \in \mathcal{J}^L$, constitutes an orthogonal basis for $L^2(P)$ with

$$E[(\mathbb{H}_\alpha)^2] = \alpha! := \alpha^{(1)}! \cdot ... \cdot \alpha^{(L)}!.$$

We can define the corresponding test function spaces $\mathcal{G}$, $(\mathcal{S})$ and the stochastic distribution spaces $\mathcal{G}^*$, $(\mathcal{S})^*$ as in the previous sections. The Wick product is defined by

$$\mathbb{H}_\alpha(\omega) \diamond \mathbb{H}_\beta(\omega) = \mathbb{H}_{\alpha+\beta}(\omega) \quad (13.68)$$

and extended linearly. For each n, we introduce the stochastic derivatives $D_{t,n}$, $D_{t,z,n}$ as mappings $\mathcal{G}^* \longrightarrow \mathcal{G}^*$ by

$$D_{t,n}F = \sum_{\alpha \in \mathcal{J}^L} \sum_{i \geq 1} c_\alpha \alpha_i^{(n)} \mathbb{H}_{\alpha-\epsilon^{(n,i)}} e_i(t), \quad n = 1, ..., N \quad (13.69)$$

and

$$D_{t,z,n}F = \sum_{\alpha \in \mathcal{J}^L} \sum_{i \geq 1} c_\alpha \alpha_i^{(n)} \mathbb{H}_{\alpha-\epsilon^{(n,i)}} \delta^{\widehat{\otimes}\epsilon^i}(t,z), \quad n = N+1, ..., L, \quad (13.70)$$

where e_i, $i = 1, 2, ...$, denote the Hermite functions and $\delta^{\widehat{\otimes}\epsilon^i}$, $i = 1, 2, ...$, are as in (13.14) and where

$$\epsilon^{(n,i)} = (0, ..., 0, \epsilon^i, 0, ..., 0)$$

with ϵ^i on the nth place. Next, let W be the Wiener process and

$$\widetilde{N}(dt, dz) = N(dt, dz) - \nu(dz)dt$$

the compensated Poisson random measure on Ω. Define the independent Wiener processes and respectively the compensated Poisson random measures as follows

$$W_j(t, \omega) = W(t, \omega_j), \quad j = 1, ..., N, \quad \text{and}$$
$$\tilde{N}_j(dt, dz)(\omega) = \tilde{N}(dt, dz)(\omega_j), \quad j = N+1, ..., L.$$

The Clark–Ocone theorem for the combined Gaussian and pure jump Lévy stochastic measures (cf. Theorem 12.20) can be extended to hold for all $F \in L^2(P)$ in the same way as in Theorem 13.26. The result is the following:

Theorem 13.28. Clark–Ocone theorem for $L^2(P)$. *Let $F \in L^2(P)$, $\mathcal{F}_T$-measurable random variable. Then*

$$F = E[F] + \sum_{j=1}^{N} \int_0^T E[D_{t,j}F \,|\, \mathcal{F}_t]\,dW_j(t) + \sum_{j=N+1}^{L} \int_0^T E[D_{t,z,j}F \,|\, \mathcal{F}_t]\tilde{N}_j(dt, dz),$$

(13.71)

where $E[D_{t,j}F \,|\, \mathcal{F}_T] \in L^2(\lambda \times P)$ and $E[D_{t,z,j}F \,|\, \mathcal{F}_T] \in L^2(\lambda \times \nu \times P)$.

Remark 13.29. Using the result and framework above, one can take applications to minimal variance hedging into consideration (see Sect. 12.6) and extend the results to all random variables in $L^2(P)$.

13.8 Generalized Chain Rules for the Malliavin Derivative

In this section, which appears at first separated from the main stream of the chapter, we present some alternative chain rules that will be applied later on in the book. These results are settled in the frame of the calculus with respect to Brownian motion and aim at weakening the conditions assumed in the original result, Theorem 3.5. These are introduced in the sequel and were first proved in [66]. Recall that the probability space is the same as in Sect. 13.7.

Definition 13.30. *Let $F(\cdot, \omega_2) \in L^2(P^W)$ be $\mathcal{F}_T$-measurable and let $\gamma \in L^2(\mathbb{R})$ (deterministic).*

(a) For any fixed $\omega_2 \in \Omega_2$, we say that F has a directional derivative in the direction γ if

$$D_\gamma F(\omega_1, \omega_2) := \lim_{\varepsilon \to 0} \frac{F(\omega_1 + \varepsilon\gamma, \omega_2) - F(\omega_1, \omega_2)}{\varepsilon}$$

exists with convergence in probability. Then $D_\gamma F$ is called the directional derivative of F in the direction γ.

(b) For any fixed $\omega_2 \in \Omega_2$, we say that F is Malliavin differentiable in probability if there exists a process $\Psi(t)$, $t \geq 0$, such that

$$\int_{\mathbb{R}} \Psi^2(t, \omega_1, \omega_2)dt < \infty \qquad P - a.s.$$

and

$$D_\gamma F(\omega_1, \omega_2) = \int_{\mathbb{R}} \Psi(t, \omega_1, \omega_2) \gamma(t) dt \qquad \text{for all } \gamma \in L^2(\mathbb{R}).$$

If this is the case, we put $\Psi(t) = \mathcal{D}_t F$ *and call this* the Malliavin derivative in probability *of* F *with respect to* W.

Lemma 13.31. Chain rule. *Suppose* $D_\gamma F$ *exists for some* $\gamma \in L^2(\mathbb{R})$ *and that* $g \in C^1(\mathbb{R})$. *Then* $g(F)$ *has a directional derivative in the direction* γ *and*

$$D_\gamma g(F) = g'(F) D_\gamma F.$$

Proof Consider the equations

$$\lim_{\varepsilon \to 0} \frac{g(F(\omega_1 + \varepsilon\gamma, \omega_2)) - g(F(\omega_1, \omega_2))}{\varepsilon} = \lim_{\varepsilon \to 0} \frac{g(F(\omega_1, \omega_2) + \varepsilon D_\gamma F(\omega_1, \omega_2)) - g(F(\omega_1, \omega_2))}{\varepsilon}$$

$$= \lim_{\varepsilon \to 0} \frac{g'(F(\omega_1, \omega_2)) \varepsilon D_\gamma F(\omega_1, \omega_2)}{\varepsilon}$$

$$= g'(F(\omega_1, \omega_2)) D_\gamma F(\omega_1, \omega_2),$$

with convergence in probability. $\square$

Theorem 13.32. Chain rule. *Suppose* F *is Malliavin differentiable in probability. Let* $g \in C^1(\mathbb{R})$. *Then* $g(F)$ *is Malliavin differentiable in probability and*

$$\mathcal{D}_t g(F) = g'(F) \mathcal{D}_t F \qquad (t, \omega) - a.e.$$

Proof By Lemma 13.31 we have

$$D_\gamma g(F) = g'(F) D_\gamma F$$

$$= g'(F) \int_{\mathbb{R}} \gamma(t) \mathcal{D}_t F dt$$

$$= \int_{\mathbb{R}} \gamma(t) g'(F) \mathcal{D}_t F dt \qquad \text{for all } \gamma \in L^2(\mathbb{R}),$$

and the result follows. $\square$

We now proceed to prove that $\mathcal{D}F$ is an extension of DF.

Theorem 13.33. *Let* $\omega_2 \in \Omega_2$ *be fixed. Suppose* $F(\cdot, \omega_2)$ *is Malliavin differentiable (with respect to the Brownian motion), that is,* $F \in \mathbb{D}_{1,2}^W$, *then* F *is Malliavin differentiable in probability and*

$$\mathcal{D}_t F = D_t F.$$

Proof Let $e \in L^2(\mathbb{R})$ with $\int_{\mathbb{R}} e^2(t) dt = 1$. Consider the iterated Itô integral

$$I_n(e^{\otimes n}) := n! \int_{\mathbb{R}} \int_{-\infty}^{t_n} \cdots \int_{-\infty}^{t_2} e(t_1) \cdots e(t_n) dW(t_1) \cdots dW(t_n).$$

By (1.15), we have that

$$I_n(e^{\otimes n}) = h_n\Big(\int_{\mathbb{R}} e(t)dW(t)\Big),$$

where h_n is the Hermite polynomial of order n. Hence, by the chain rule and a basic property of Hermite polynomials, we have

$$\mathcal{D}_t(I_n(e^{\otimes n})) = h'_n\Big(\int_{\mathbb{R}} e(t)dW(t)\Big)e(t)$$

$$= nh_{n-1}\Big(\int_{\mathbb{R}} e(t)dW(t)\Big)e(t)$$

$$= nI_{n-1}(e^{\otimes(n-1)}))e(t)$$

$$= D_t(I_n(e^{\otimes n})).$$

It follows that $\mathcal{D}_t$ coincides with D_t on the finite sums of iterated integrals and, hence, also that $\mathcal{D}_t F$ exists for all $F \in \mathbb{D}_{1,2}^W$ and

$$\mathcal{D}_t F = D_t F \qquad \text{for all } F \in \mathbb{D}_{1,2}^W.$$

And the result follows.   □

Corollary 13.34. *If F is Malliavin differentiable in probability and*

$$E_{P^W}\left[\int_{\mathbb{R}} \big(\mathcal{D}_t F(\cdot, \omega_2)\big)^2 dt\right] < \infty, \tag{13.72}$$

then $F \in \mathbb{D}_{1,2}^W$.

Proof Let $F = \sum_{n=0}^{\infty} I_n(f_n)$ be the chaos expansion of $F(\cdot, \omega_2) \in L^2(P^W)$. By Theorem 13.33 we have

$$\mathcal{D}_t(I_n(f_n)) = D_t(I_n(f_n)) \qquad \text{for all } n.$$

Hence we have

$$\|F\|_{\mathbb{D}_{1,2}^W} = \lim_{N\to\infty} \sum_{n=1}^{N} E_{P^W}\left[\int_0^T \big(D_t(I_n(f_n))\big)^2 dt\right]$$

$$= E_{P^W}\left[\int_{\mathbb{R}} \big(\mathcal{D}_t F\big)^2 dt\right] < \infty.$$

Thus $F \in \mathbb{D}_{1,2}^W$.   □

In view of this last theorem, in the sequel we denote the Malliavin derivative and the Malliavin derivative in probability by the same notation D_t.

13.9 Exercises

Problem 13.1. (*) Prove that

$$\frac{d}{dt}\eta(t) = \overset{\bullet}{\eta}(t) \quad \text{in } (\mathcal{S})^*.$$

Problem 13.2. Let $\gamma(t, z)$, $t \in [0, T]$, $z \in \mathbb{R}_0$, be a deterministic function such that

$$\int_{\mathbb{R}} \int_{\mathbb{R}_0} \gamma^2(s, z)\nu(dz)ds < \infty.$$

Use (13.41) and (13.37) to prove that

$$E\left[\exp\left\{\int_{\mathbb{R}} \int_{\mathbb{R}_0} \log\left(1 + \gamma(s, z)\right)\widetilde{N}(ds, dz)\right\}\right]$$

$$= \exp\left\{-\int_{\mathbb{R}} \int_{\mathbb{R}_0} \left[\log\left(1 + \gamma(s, z)\right) - \gamma(s, z)\right]\nu(dz)ds\right\}.$$

14

The Donsker Delta Function of a Lévy Process and Applications

Similar to the Brownian motion case (see Chap. 7), we can introduce the Donsker delta function of a Lévy process $\eta = \eta(t)$, $t \geq 0$, that is, a stochastic concept of delta function that is intimately linked to a remarkable increasing process $L_t(x)$, $t \geq 0$, $(x \in \mathbb{R})$ called local time. The local time of a Lévy process can be considered a nontrivial measure for the amount of time spent by η in the infinitesimal neighborhood of a point. It exhibits the amazing property that it exactly increases on the closed zero sets of the Lévy process. Just as in the Brownian case, one can view the Donsker delta function of η as a time derivative of local time.

Let $\eta(t) = \int_0^t \int_{\mathbb{R}_0} z\widetilde{N}(ds, dz)$, $t \geq 0$, be a pure jump Lévy process without drift on $(\Omega, \mathcal{F}, P)$, where P is the probability law of η. The main purpose of this chapter is to present the *Donsker delta function* $\delta_x(\eta(t))$ of $\eta(t)$ and some related results. For example, in Theorem 14.4, we show that, for a certain class of pure jump Lévy processes, $\delta_x(\eta(t))$ exists as an element of the Lévy–Hida stochastic distribution space $(\mathcal{S})^*$ and admits the representation

$$\delta_x(\eta(t)) = \frac{1}{2\pi} \int_{\mathbb{R}} \exp^{\diamond} \Big(\int_0^t \int_{\mathbb{R}_0} (e^{i\lambda z} - 1)\widetilde{N}(ds, dz)$$

$$+ t \int_{\mathbb{R}_0} (e^{i\lambda z} - 1 - i\lambda z)\nu(dz) - i\lambda x \Big) d\lambda.$$

Note that the Donsker delta function for Lévy processes was also defined in [146], though in a different setting.

As mentioned, we also study the relationship between the Donsker delta function and the local time of a Lévy process. It is known (see, e.g., [32]) that, under certain conditions on the characteristic exponent Ψ (see (9.10)) of η, the local time $L_t(x) = L_t(x, \omega)$ of η at the point $x \in \mathbb{R}$ up to time t exists and the mapping

$$(x, \omega) \longmapsto L_t(x, \omega)$$

belongs to $L^2(P \times \lambda)$ for all t. We show that $L_t(x)$ is related to $\delta_x(\eta(t))$ by the formula

$$L_T(x) = \int_0^T \delta_x(\eta(t))dt,$$

just as in the Brownian motion case (see Chap. 7). Moreover, we also present an explicit chaos expansion for $L_t(\cdot)$ in terms of iterated integrals with respect to the associated compensated Poisson random measure $\widetilde{N}(dt, dz)$. See Theorem 14.8. Let us mention that the study is inspired by the methods in [110], where the chaos expansion of local time of fractional Brownian motion is obtained. We would also like to mention that the first use of a version of Malliavin calculus for jump processes in the study of local time was initiated in [19].

The Donsker delta function has applications in finance within the hedging of portfolios and sensitivity analysis, as presented in the last two sections.

Our presentation is based on [67, 68, 162].

14.1 The Donsker Delta Function of a Pure Jump Lévy Process

In the following, we require that the Lévy measure ν satisfies the following integrability condition. For every $\varepsilon > 0$ there exists a $\lambda > 0$ such that

$$\int_{\mathbb{R} \setminus (-\varepsilon, \varepsilon)} \exp(\lambda |z|)\nu(dz) < \infty. \tag{14.1}$$

This condition implies that the Lévy measure ν has moments $\int_{\mathbb{R}_0} z^n \nu(dz) < \infty$ for all $n \geq 2$. Just as in the Gaussian case, we introduce the Donsker delta function of a Lévy process as follows. For notation and framework we refer to Chap. 13.

Definition 14.1. *Suppose that $X : \Omega \to \mathbb{R}$ is a random variable belonging to the Lévy–Hida distribution space $(\mathcal{S})^*$. The Donsker delta function of X is a continuous function $\delta.(X) : \mathbb{R} \to (\mathcal{S})^*$ such that*

$$\int_{\mathbb{R}} h(y)\delta_y(X)dy = h(X) \tag{14.2}$$

for all measurable functions $h : \mathbb{R} \longrightarrow \mathbb{R}$ such that the integral converges in $(\mathcal{S})^*$.

14.2 An Explicit Formula for the Donsker Delta Function

The main result of this section is an explicit formula for the Donsker delta function in the case of a certain class of pure jump Lévy processes. In fact, from now on we limit ourselves to consider Lévy measures whose characteristic exponent Ψ satisfies the following condition:

- There exists $\varepsilon \in (0,1)$ such that

$$\lim_{|\lambda| \longrightarrow \infty} |\lambda|^{-(1+\varepsilon)} \operatorname{Re} \Psi(\lambda) = \infty, \tag{14.3}$$

where $\operatorname{Re} \Psi$ is the real part of Ψ.

Remark 14.2. Condition (14.3) entails the strong Feller property of the semigroup of our Lévy process $\eta(t)$, implying that the probability law of $\eta(t)$ is absolutely continuous with respect to the Lebesgue measure. The assumption covers, e.g., the following Lévy process $\eta(t)$ of unbounded variation with Lévy measure ν, given by

$$\nu(dz) = \chi_{(0,1)}(z) z^{-(2+\alpha)} dz,$$

where $0 < \alpha < 1$. We also emphasize that various other conditions of the type (14.3) are conceivable as the proof of Theorem 14.5 unveils.

Our main result is based on the following lemma.

Lemma 14.3. *Let $\lambda \in \mathbb{C}$, $t \geq 0$, then*

$$\exp(\lambda \eta(t)) = \exp^\circ \left(\int_0^t \int_{\mathbb{R}_0} (e^{\lambda z} - 1) \widetilde{N}(ds, dz) + t \int_{\mathbb{R}_0} (e^{\lambda z} - 1 - \lambda z) \nu(dz) \right). \tag{14.4}$$

Proof Define

$$Y(t) = \exp\left(\lambda \eta(t) - t \int_{\mathbb{R}_0} (e^{\lambda z} - 1 - \lambda z) \nu(dz) \right). \tag{14.5}$$

Then Itô formula shows that Y satisfies the stochastic differential equation

$$dY(t) = Y(t_-) \int_{\mathbb{R}_0} (e^{\lambda z} - 1) \widetilde{N}(dt, dz), \qquad Y(0) = 1.$$

By relation (13.43), the last equation can be rewritten as

$$\frac{d}{dt} Y(t) = Y(t_-) \diamond \int_{\mathbb{R}_0} (e^{\lambda z} - 1) \overset{\bullet}{\widetilde{N}}(t, z) \nu(dz); \ Y(0) = 1.$$

By means of a version of the Wick chain rule (see Proposition 5.14), we can see that the solution of the last equation is given by

$$Y(t) = \exp^\circ \left(\int_0^t \int_{\mathbb{R}_0} (e^{\lambda z} - 1) \overset{\bullet}{\widetilde{N}}(t, z) \nu(dz) dt \right) \tag{14.6}$$

$$= \exp^\circ \left(\int_0^t \int_{\mathbb{R}_0} (e^{\lambda z} - 1) \widetilde{N}(ds, dz) \right).$$

This solution is unique. Moreover, if we compare (14.5) and (14.6) we receive the desired formula.   □

The following result is due to [162].

Theorem 14.4. *The Donsker delta function $\delta_y(\eta(t))$ of $\eta(t)$ exists uniquely. Moreover, $\delta_y(\eta(t))$ takes the explicit form*

$$\delta_y(\eta(t)) \hspace{7cm} (14.7)$$
$$= \frac{1}{2\pi} \int_{\mathbb{R}} \exp^{\diamond} \left(\int_0^t \int_{\mathbb{R}_0} (e^{i\lambda z} - 1)\widetilde{N}(ds, dz) + t \int_{\mathbb{R}_0} (e^{i\lambda z} - 1 - i\lambda z)\nu(dz) - i\lambda y \right) d\lambda.$$

Proof The proof is essentially based on the application of the Lévy–Hermite transform $\mathcal{H}$ (see Sect. 13.3) and the use of the Fourier inversion formula. To ease the notation we define

$$X_t = X_t(\lambda) = \int_0^t \int_{\mathbb{R}_0} (e^{i\lambda z} - 1)\widetilde{N}(ds, dz)$$

and

$$f(\lambda) = \exp(X_t + t \int_{\mathbb{R}_0} (e^{i\lambda z} - 1 - i\lambda z)\nu(dz)), \quad \lambda \in \mathbb{R}.$$

Further we set

$$g(\lambda, \zeta) = (\mathcal{H}f^{\diamond}(\lambda))(\zeta), \quad \zeta \in \mathbb{C},$$

where $f^{\diamond}$ is the Lévy–Wick version of f. Because of the properties of the Lévy–Hermite transform we can write

$$g(\lambda, \zeta) = \exp((\mathcal{H}X_t)(\zeta) + t \int_{\mathbb{R}_0} (e^{i\lambda z} - 1 - i\lambda z)\nu(dz))$$
$$= \exp((\mathcal{H}X_t)(\zeta) - t\Psi(\lambda)).$$

We subdivide the proof into the following steps.

Step (1). We want to show that $g(\cdot, \zeta)$ is an element of the Schwartz space $\mathcal{S}(\mathbb{R})$. To this end we provide some reasonable upper bound for $g(\cdot, \zeta)$. Let $t \in [0, T]$. By means of (13.44) and (13.28), we have

$$X_t = \int_0^t \int_{\mathbb{R}_0} (e^{i\lambda z} - 1)\widetilde{N}(ds, dz)$$
$$= \int_0^t \int_{\mathbb{R}_0} (e^{i\lambda z} - 1) \diamond \overset{\bullet}{\widetilde{N}}(s, z)\nu(dz)ds$$
$$= \int_0^t \int_{\mathbb{R}_0} (e^{i\lambda z} - 1) \diamond \sum_{k,j} e_k(s)p_j(z)K_{e^{\kappa(k,j)}}\nu(dz)ds$$
$$= \sum_{k,j} \int_0^t \int_{\mathbb{R}_0} (e^{i\lambda z} - 1)e_k(s)p_j(z)\nu(dz)ds \cdot K_{e^{\kappa(k,j)}},$$

with the basis elements e_k and p_j as in Sect. 13.2. Then we can find the estimate

$$|(\mathcal{H}X_t)(z)| \leq |\lambda| \left(\sum_{k,j}\left(\int_0^t\int_{\mathbb{R}_0}|z|\,|e_k(s)p_j(z)|\,\nu(dz)ds\right)^2(2\mathbb{N})^{-2\kappa(k,j)}\right)^{\frac{1}{2}}$$

$$\cdot \left(\sum_\alpha |z^\alpha|^2 (2\mathbb{N})^{2\alpha}\right)^{\frac{1}{2}} \tag{14.8}$$

$$\leq |\lambda| \left(\sum_{k,j}\left(\int_0^t\int_{\mathbb{R}_0}|z|^2\,\nu(dz)ds\right)(2\mathbb{N})^{-2\kappa(k,j)}\right)^{\frac{1}{2}} \cdot \left(\sum_\alpha |z^\alpha|^2 (2\mathbb{N})^{2\alpha}\right)^{\frac{1}{2}}$$

$$\leq const. \cdot |\lambda|$$

for all $z \in \mathbb{K}_2(R)$ for some $R < \infty$ (see (5.47)), where we used that $\sum_{k,j}(2\mathbb{N})^{-2\kappa(k,j)} < \infty$ (see [107, Proposition 2.3.3]). Therefore, using the definition of $g(\lambda, \zeta)$ we get

$$|g(\lambda,\zeta)| \leq e^{const.|\lambda|}e^{-t\,\mathrm{Re}\,\Psi(\lambda)}$$

for all $\zeta \in \mathbb{K}_2(R)$. By condition (14.3) we require $|\lambda|^{-(1+\varepsilon)}\,\mathrm{Re}\,\Psi(\lambda) \geq 1$ for $|\lambda| \geq L \geq 1$. This implies

$$|g(\lambda,\zeta)| \leq e^{const.|\lambda|}e^{-t|\lambda|^{(1+\varepsilon)}|\lambda|^{-(1+\varepsilon)}\,\mathrm{Re}\,\Psi(\lambda)} \leq e^{const.|\lambda|-t|\lambda|^{(1+\varepsilon)}}$$

$$\leq e^{-tC|\lambda|^{(1+\varepsilon)}}$$

for $\zeta \in \mathbb{K}_2(R)$ and $|\lambda| \geq M$ with positive constants M, C. Next we cast a glance at the derivatives $\frac{\partial^n}{\partial\lambda^n}g(\lambda,\zeta)$. Since a similar estimate to (14.8) yields

$$\left|\int_0^t\int_{\mathbb{R}_0}(iz)^n e^{i\lambda z}\, e_k(s)p_j(z)\nu(dz)ds \cdot \zeta_{\kappa(k,j)}\right|$$

$$\leq \int_0^t\int_{\mathbb{R}_0}|z|^n\,|e_k(s)p_j(z)|\,\nu(dz)ds \cdot \left|\zeta_{\kappa(k,j)}\right| \in L^1(\mathbb{N}\times\mathbb{N})$$

for all fixed $\zeta \in \mathbb{K}_2(R)$ and $n \in \mathbb{N}$, we obtain that for all $n \in \mathbb{N}$ there exists a constant C_n such that

$$\left|\frac{\partial^n}{\partial\lambda^n}(\mathcal{H}X_t)(\lambda,\zeta)\right| \leq C_n$$

for all $\lambda \in \mathbb{R}$, $\zeta \in \mathbb{K}_2(R)$. Further, we observe that

$$\left|\frac{d}{d\lambda}\left(t\int_{\mathbb{R}_0}(e^{i\lambda z}-1-i\lambda z)\nu(dz)\right)\right| = \left|t\int_{\mathbb{R}_0}iz(e^{i\lambda z}-1)\nu(dz)\right|$$

$$\leq |\lambda|\,t\int_{\mathbb{R}_0}|z|^2\,\nu(dz)$$

or more generally that

$$\left|\frac{d^n}{d\lambda^n}\left(t\int_{\mathbb{R}_0}(e^{i\lambda z}-1-i\lambda z)\nu(dz)\right)\right| \leq C_n\,|\lambda|.$$

Altogether, we conclude that for all $k, n \in \mathbb{N}_0$ there exists a polynomial $p_{k,n}(\lambda)$ with positive coefficients such that

$$\sup_{\zeta \in \mathbb{K}_2(R)} \left| (1 + |\lambda|^k) \frac{\partial^n}{\partial \lambda^n} g(\lambda, \zeta) \right| \le p_{k,n}(|\lambda|) e^{-tC|\lambda|^{(1+\varepsilon)}} \tag{14.9}$$

for all $|\lambda| \ge M$. This implies that $g(\cdot, \zeta) \in \mathcal{S}(\mathbb{R})$ for all $\zeta \in \mathbb{K}_2(R)$. Since the (inverse) Fourier transform maps $\mathcal{S}(\mathbb{R})$ onto itself, we get that

$$\widehat{g}(y, \zeta) = \frac{1}{2\pi} \int_{\mathbb{R}} e^{-iy\lambda} g(\lambda, \zeta) d\lambda$$

is L^1-integrable and that

$$\int_{\mathbb{R}} e^{iy\lambda} \widehat{g}(y, \zeta) dy = g(\lambda, \zeta) \tag{14.10}$$

for all $\zeta \in \mathbb{K}_2(R)$. Using the properties of the Lévy–Hermite transform, Lemma 14.3, and condition (14.1), we can see that relation (14.10) gives rise to defining the Donsker delta function of $\eta(t)$ as

$$\delta_y(\eta(t)) = \frac{1}{2\pi} \int_{\mathbb{R}} e^{-iy\lambda} f^\diamond(\lambda) d\lambda, \tag{14.11}$$

i.e.,

$$\mathcal{H}(\delta_y(\eta(t)))(\zeta) = \frac{1}{2\pi} \int_{\mathbb{R}} e^{-iy\lambda} g(\lambda, \zeta) d\lambda = \widehat{g}(y, \zeta).$$

To complete the proof, we proceed as follows. We check that the Lévy–Hermite transform $\mathcal{H}(\delta_y(\eta(t)))(\zeta) = \widehat{g}(y, \zeta)$ in the integrand of (14.10) can be extracted outside the integral and that all occurring expressions are well-defined. Then we can apply the inverse Lévy–Hermite transform and Lemma 14.3 to show that $\delta_y(\eta(t))$ in (14.11) fulfils the property (14.1) for $h(x) = e^{i\lambda x}$. Finally, the proof follows from a well-known density argument, using trigonometric polynomials.

Step (2). Let us verify that $\delta_y(\eta(t))$ exists in $(\mathcal{S})^*$ for all y. By a Lévy version of [107, Lemma 2.8.5] it follows that

$$Y_n(y) := \frac{1}{2\pi} \int_{-n}^{n} e^{-iy\lambda} f^\diamond(\lambda) d\lambda$$

exists in $(\mathcal{S})^*$ and that

$$\mathcal{H}Y_n(y, \zeta) = \frac{1}{2\pi} \int_{-n}^{n} e^{-iy\lambda} g(\lambda, \zeta) d\lambda$$

for all $n \in \mathbb{N}$. Further, the bound (14.9) gives

$$\sup_{n \in \mathbb{N}, \zeta \in \mathbb{K}_2(R)} |\mathcal{H}Y_n(y, \zeta)| \le \frac{1}{2\pi} \int_{\mathbb{R}} \sup_{\zeta \in \mathbb{K}_2(R)} |g(\lambda, \zeta)| \, d\lambda < \infty. \tag{14.12}$$

Thus, with the help of a Lévy analogue of Theorem 5.12, it follows that $\delta_y(\eta(t)) \in (\mathcal{S})^*$ for all y.

Step (3). We check that the integral

$$\int_{\mathbb{R}} e^{iy\lambda} \delta_y(\eta(t)) dy \tag{14.13}$$

converges in $(\mathcal{S})^*$. Because of the estimate (14.12) we also get that

$$X_n(\lambda) := \int_{-n}^{n} e^{iy\lambda} \delta_y(\eta(t)) dy$$

exists in $(\mathcal{S})^*$. By (14.9) and integration by parts we deduce

$$\sup_{n \in \mathbb{N}, \zeta \in \mathbb{K}_2(R)} |\mathcal{H} X_n(y, \zeta)| \leq \sup_{\zeta \in \mathbb{K}_2(R)} \int_{\mathbb{R}} \frac{1}{1+y^2} \cdot (1+y^2) |\mathcal{H}(\delta_y(\eta(t)))(\zeta)| \, dy$$

$$\leq const. \cdot \sup_{\lambda \in \mathbb{R}, \zeta \in \mathbb{K}_2(R)} \left(\left| (1+\lambda^2) \frac{\partial^2}{\partial\lambda^2} g(\lambda, \zeta) \right| \right.$$

$$\left. + \left| (1+\lambda^2) g(\lambda, \zeta) \right| \right)$$

$$\leq M < \infty,$$

where we have used that

$$y^2 \mathcal{H}(\delta_y(\eta(t)))(\zeta) = y^2 \widehat{g}(y, \zeta) = \frac{1}{2\pi} \int_{\mathbb{R}} e^{-iy\lambda} \frac{\partial^2}{\partial\lambda^2} g(\lambda, \zeta) d\lambda.$$

Once again by a Lévy version of Theorem 5.12, we see that the integral (14.13) is well-defined in $(\mathcal{S})^*$. Finally, by using the inverse Lévy–Hermite transform and Lemma 14.3 we obtain

$$\int_{\mathbb{R}} e^{iy\lambda} \delta_y(\eta(t)) dy = f^\diamond(\lambda) = e^{i\lambda\eta(t)} \tag{14.14}$$

Thus, we have proved relation (14.1) for $h(y) = e^{i\lambda y}$. Since (14.1) still holds for linear combinations of such functions, the general case is attained by a well-known density argument. The continuity of $y \longmapsto \delta_y(\eta(t))$ is also a direct consequence of a Lévy version of Theorem 5.12. $\square$

14.3 Chaos Expansion of Local Time for Lévy Processes

In this section we investigate the local time $L_T(x)$ ($x \in \mathbb{R}$) of a certain class of pure jump Lévy processes $\eta(t) = \int_0^t \int_{\mathbb{R}_0} z \widetilde{N}(ds, dz)$, $t \geq 0$. This local time can be heuristically described by

$$L_T(x) = \int_0^T \delta(\eta(t) - x) dt, \tag{14.15}$$

where $\delta(u)$ is the *Dirac delta function*, which can be approximated by

$$P_\varepsilon(u) = \frac{1}{\sqrt{2\pi\varepsilon}} e^{-\frac{|u|^2}{2\varepsilon}} = \frac{1}{\sqrt{2\pi}} \int_{\mathbb{R}} e^{iyu - \frac{1}{2}\varepsilon y^2} dy, \quad u \in \mathbb{R}, \tag{14.16}$$

with $i = \sqrt{-1}$. Formally, this implies

$$\delta(u) = \lim_{\varepsilon \to 0} P_\varepsilon(u) = \frac{1}{\sqrt{2\pi}} \int_{\mathbb{R}} e^{iyu} dy. \tag{14.17}$$

A justification for this heuristical reasoning above comes from the Gaussian case (see [4] and [102]). In the sequel we make the above considerations rigorous by showing that the *Donsker delta function* $\delta(\eta(t) - x)$ of $\eta(t)$ may be realized as a generalized Lévy functional. Furthermore, we provide an explicit formula for $\delta(\eta(t) - x)$. We also prove that the identity (14.15) makes sense in terms of a Bochner integral. To demonstrate an application, we use the Donsker delta function to derive a chaos expansion of $L_T(x)$ with explicit kernels.

First of all, let us recall the definition of a particular version of the density of an occupation measure, which is referred to as the *local time*.

Definition 14.5. *Let* $t \geq 0$, $x \in \mathbb{R}$. *The* local time *of* $\eta(t)$ *at level* x *and time* t, *denoted by* $L_t(x)$, *is defined by*

$$L_t(x) := \lim_{\varepsilon \to 0+} \frac{1}{2\varepsilon} \int_0^t \chi_{\{|\eta(s) - x| < \varepsilon\}} ds$$

$$= \lim_{\varepsilon \to 0+} \frac{1}{2\varepsilon} \lambda(\{s \in [0, t] : \eta(s) \in (x - \varepsilon, x + \varepsilon)\}),$$

with convergence in $L^2(P \times \lambda)$. *Recall that* λ *is the Lebesgue measure.*

Here, we recall that the local time of $\eta(t)$ exists if the integrability condition

$$\int_{\mathbb{R}} \operatorname{Re}\left(\frac{1}{1 + \Psi(\lambda)}\right) d\lambda < \infty \tag{14.18}$$

holds, where Ψ denotes the characteristic exponent of $\eta(t)$ (see e.g., [32] for details). Since we have the inequality

$$\operatorname{Re}\left(\frac{1}{1 + \Psi(\lambda)}\right) \leq \frac{1}{1 + \operatorname{Re}\Psi(\lambda)},$$

condition (14.3) entails (14.18), giving the existence of $L_t(x)$. We point out that $(x, \omega) \longmapsto L_t(x)(\omega)$ belongs to $L^2(P \times \lambda)$ for all $t \geq 0$ and that

$$\int_0^t f(\eta(s)) ds = \int_{\mathbb{R}} f(x) L_t(x) dx \tag{14.19}$$

for all measurable bounded functions $f \geq 0$ a.s. The relation (14.19) is called the *occupation density formula*. Note that $t \longmapsto L_t(x)$ is an increasing process and that $L_t(\cdot)$ has compact support for all $t > 0$ a.e. Furthermore, (14.3) implies that $x \longmapsto L_t(x)$ is Hölder-continuous a.e. for every $t > 0$ (see [32, p. 151]).

Next, we give a rigorous proof of relation (14.15).

Proposition 14.6. *Fix $T > 0$. Then*

$$L_T(x) = \int_0^T \delta_x(\eta(s))ds$$

for all $x \in \mathbb{R}$ a.e.

Proof Let f be a continuous function with compact support in $[-r, r] \subset \mathbb{R}$. Define the function $Z : [0, T] \times \mathbb{R} \times [-r, r] \longrightarrow (\mathcal{S})^*$ by

$$Z(t, \lambda, y) = e^{i\lambda(\eta(t) - y)}.$$

First, we want to show that Z is Bochner-integrable in $(\mathcal{S})^*$ with respect to the measure $\lambda^3 = \lambda \times \lambda \times \lambda$. One observes that $(t, \lambda, y) \longmapsto \langle Z(t, \lambda, y), l \rangle$ is $\mathcal{B}([0, T]) \otimes \mathcal{B}(\mathbb{R}) \otimes \mathcal{B}([-r, r]), \mathcal{B}(\mathbb{R})$−measurable for all $l \in (\mathcal{S})$. Further we consider the exponential process

$$Y(t) := \exp\left(\int_0^t \varphi(s)d\eta(s) - \int_0^t \int_{\mathbb{R}_0} (e^{\varphi(s)z} - 1 - \varphi(s)z)\nu(dz)ds\right)$$

for deterministic functions φ such that $e^{\varphi(s)z} - 1 \in L^2([0, T] \times \mathbb{R}, \lambda \times \nu)$. Then by the Lévy–Itô formula we have

$$dY(t) = Y(t^-) \int_{\mathbb{R}_0} (e^{\varphi(t)z} - 1)\widetilde{N}(dt, dz),$$

so that

$$Y(t) = 1 + \int_0^t Y(s_1^-) \int_{\mathbb{R}_0} (e^{\varphi(s_1)z_1} - 1)\widetilde{N}(ds_1, dz_1).$$

If we iterate this process, we get the following chaos expansion for $Y(t)$:

$$Y(t) = \sum_{n \geq 0} I_n(g_n) \text{ in } L^2(P)$$

with $g_n(s_1, z_1, ..., s_n, z_n) = \frac{1}{n!} \prod_{j=1}^{n}(e^{\varphi(s_j)z_j} - 1)\chi_{t > \max(s_i)}$, where I_n denotes the iterated integral in (10.1). Thus, we obtain for $\varphi(s) \equiv i\lambda$

$$Z(t, \lambda, y) = \sum_{n \geq 0} I_n(g_n \cdot h) \text{ in } L^2(P) \tag{14.20}$$

with the function $h(t, \lambda, y) = \exp(\int_0^t \int_{\mathbb{R}_0} (e^{i\lambda z} - 1 - i\lambda z)\nu(dz)ds - i\lambda y)$. We also get the isometry

$$E\left[(Z(t,\lambda,y))^2\right] = \sum_{n\geq 0} n!\, \|g_n \cdot h\|^2_{L^2((\lambda\times\nu)^n)}\,.$$

Now, let us have a look at the weighted sum

$$\|Z(t,\lambda,y)\|^2_{-q} := \sum_{n\geq 0} n!\, \|g_n \cdot h\|^2_{L^2((\lambda\times\nu)^n)}\, e^{-qn} < \infty$$

for $q \geq 0$. Then

$$\|Z(t,\lambda,y)\|^2_{-q} \leq \sum_{n\geq 0} \frac{1}{n!} \int_{\mathbb{R}_0^n} \int_{[0,T]^n}$$

$$\cdot \left(\prod_{j=1}^{n} \left|e^{i\lambda z_j}-1\right| \left|e^{-t\,\mathrm{Re}\,\Psi(\lambda)}\right)^2 \chi_{t>\max(s_i)}\, ds_1...ds_n \nu(dz_1)...\nu(dz_n)\cdot e^{-qn}$$

$$\leq \sum_{n\geq 0} \frac{1}{n!} t^n 2^n (\mathrm{Re}\,\Psi(\lambda))^n e^{-2t\,\mathrm{Re}\,\Psi(\lambda)}\cdot e^{-qn},$$

where we used the fact that

$$\int_{\mathbb{R}_0} \left|e^{i\lambda z}-1\right|^2 \nu(dz) = 2\,\mathrm{Re}\,\Psi(\lambda).$$

So it follows that

$$\|Z(t,\lambda,y)\|_{-q} \leq \sum_{n\geq 0} \frac{1}{(n!)^{\frac{1}{2}}} t^{\frac{n}{2}} 2^{\frac{n}{2}} (\mathrm{Re}\,\Psi(\lambda))^{\frac{n}{2}} e^{-t\,\mathrm{Re}\,\Psi(\lambda)}\cdot e^{-\frac{q}{2}n}. \qquad (14.21)$$

Next, suppose that $|\lambda|^{-(1+\varepsilon)}\,\mathrm{Re}\,\Psi(\lambda) \geq 1$ for $|\lambda| \geq K \geq 0$. Then

$$\int_{-r}^{r} \int_{|\lambda|\geq K} \int_0^T \|Z(t,\lambda,y)\|_{-q}\, dt d\lambda dy$$

$$\leq 2r \sum_{n\geq 0} \frac{1}{(n!)^{\frac{1}{2}}} 2^{\frac{n}{2}} e^{-\frac{q}{2}n} \int_{|\lambda|\geq K} (\mathrm{Re}\,\Psi(\lambda))^{\frac{n}{2}} \int_0^T t^{\frac{n}{2}} e^{-t\,\mathrm{Re}\,\Psi(\lambda)} dt d\lambda$$

$$\leq 2r \sum_{n\geq 0} \frac{1}{(n!)^{\frac{1}{2}}} 2^{\frac{n}{2}} e^{-\frac{q}{2}n} \int_{|\lambda|\geq K} \frac{1}{\mathrm{Re}\,\Psi(\lambda)} d\lambda \cdot \Gamma(\frac{n}{2}+1),$$

where Γ denotes the Gamma function. Together with (14.3) we obtain

$$\int_{-r}^{r} \int_{|\lambda|\geq K} \int_0^T \|Z(t,\lambda,y)\|_{-q}\, dt d\lambda dy$$

$$\leq const. \sum_{n\geq 0} \frac{(\frac{n}{2})!}{(n!)^{\frac{1}{2}}} 2^{\frac{n}{2}} e^{-\frac{q}{2}n} \int_{|\lambda|\geq K} \frac{1}{\mathrm{Re}\,\Psi(\lambda)} d\lambda$$

$$\leq const. \sum_{n\geq 0} \frac{(\frac{n}{2})!}{(n!)^{\frac{1}{2}}} 2^{\frac{n}{2}} e^{-\frac{q}{2}n}.$$

Since

$$\frac{((\frac{n}{2})!)^2}{n!} \leq \frac{2^{[\frac{n}{2}]}}{2^{2[\frac{n+1}{2}]}} \leq 1$$

for all $n \in \mathbb{N}$, we conclude that

$$\int_{-r}^{r} \int_{|\lambda| \geq K} \int_0^T \|Z(t, \lambda, y)\|_{-q} \, dt d\lambda dy < \infty$$

for all $q \geq 2$. It also follows directly from (14.21) that

$$\int_{-r}^{r} \int_{|\lambda| \leq K} \int_0^T \|Z(t, \lambda, y)\|_{-q} \, dt d\lambda dy < \infty$$

for all $q \geq 2$. Therefore, we have

$$\int_{-r}^{r} \int_{\mathbb{R}} \int_0^T \|Z(t, \lambda, y)\|_{-q} \, dt d\lambda dy < \infty \tag{14.22}$$

for all $q \geq 2$. Let $l \in (\mathcal{S})$. Then it is not difficult to verify that

$$|\langle Z(t, \lambda, y), l \rangle| \leq const. \|Z(t, \lambda, y)\|_{-2} \quad (t, \lambda, y) - a.e.$$

The latter gives

$$\int_{-r}^{r} \int_{\mathbb{R}} \int_0^T |\langle Z(t, \lambda, y), l \rangle| \, dt d\lambda dy < \infty$$

for all $l \in (\mathcal{S})$. Note that $\|Z(t, \lambda, y)\|_{-q}$ is measurable. Thus we have proved the Bochner-integrability of Z in $(\mathcal{S})^*$. By using Fubini theorem, Lemma 14.3, and Theorem 14.4 we get

$$\int_{\mathbb{R}} f(y) \int_0^T \delta_y(\eta(t)) dt dy = \int_0^T f(\eta(t)) dt.$$

Then we can deduce from relation (14.19) that

$$\int_{\mathbb{R}} f(y) \int_0^T \delta_y(\eta(t)) dt dy = \int_{\mathbb{R}} f(y) L_T(y) dy \tag{14.23}$$

for all continuous $f : \mathbb{R} \longrightarrow \mathbb{R}$ with compact support. Using a density argument we find

$$L_T(x) = \int_0^T \delta_x(\eta(t)) dt \text{ for a.a. } x \in \mathbb{R} \quad P - a.e. \tag{14.24}$$

Let $l \in (\mathcal{S})$. The map $x \longmapsto \delta_x(\eta(t))$ is continuous by Theorem 14.4. Thus $\langle \delta_x(\eta(t)), l \rangle = \frac{1}{2\pi} \int_0^T \int_{\mathbb{R}} \langle Z(t, \lambda, x), l \rangle \, d\lambda dt$ is continuous too. Then, based on (14.21), one shows that $x \longmapsto \int_0^T \delta_x(\eta(t)) dt$ is continuous in $(\mathcal{S})^*$. Since $x \longmapsto L_T(x)$ is Hölder-continuous, relation (14.24) is valid for *all* $x \in \mathbb{R}$ P-a.e. $\square$

Remark 14.7. Relation (14.22) in the proof of Proposition 14.6 shows that the Donsker delta function $\delta_x(\eta(t))$ even takes values in the space of generalized random variables $\mathcal{G}^* \subseteq (\mathcal{S})^*$. See Sect. 13.4.

We are now arriving at one of the main results of this chapter. See [162].

Theorem 14.8. *The chaos expansion of the local time $L_T(x)$ of $\eta(t)$ is given by*

$$L_T(x) = \sum_{n \geq 0} I_n(f_n)$$

$$= \sum_{n \geq 0} n! \int_0^\infty \int_{\mathbb{R}_0} \cdots \int_0^{t_2} \int_{\mathbb{R}_0} f_n(s_1, z_1, ..., s_n, z_n) \widetilde{N}(ds_1, dz_1)...\widetilde{N}(ds_n, dz_n)$$

in $L^2(P)$ with the symmetric functions

$$f_n(s_1, z_1, ..., s_n, z_n) = \frac{1}{2\pi} \frac{1}{n!} \int_0^T \int_{\mathbb{R}} (\prod_{j=1}^n (e^{i\lambda z_j} - 1))h(t, \lambda, x)\chi_{t > \max(s_i)} d\lambda dt$$

for

$$h(t, \lambda, x) = \exp(t \int_{\mathbb{R}_0} (e^{i\lambda z} - 1 - i\lambda z)\nu(dz) - i\lambda x).$$

Proof By Proposition 14.6 we can write $L_T(x) = \int_0^T \delta_x(\eta(t))dt$. Using the definition of the function $Z(t, \lambda, y) = e^{i\lambda(\eta(t) - y)}$ in the proof of Proposition 14.6 and its chaos representation (14.20), we get

$$L_T(x) = \frac{1}{2\pi} \int_0^T \int_{\mathbb{R}} \sum_{n \geq 0} I_n(g_n \cdot h)d\lambda dt, \tag{14.25}$$

with $g_n(s_1, z_1, ..., s_n, z_n) = \frac{1}{n!} \prod_{j=1}^n (e^{i\lambda z_j} - 1)\chi_{t > \max(s_i)}$ and h as in the statement of this theorem. Because of the inequality (14.21) and similar estimates directly after this relation in the proof of Proposition 14.6, we can take the sum sign outside the double integral in (14.25). Thus, we obtain

$$\frac{1}{2\pi} \int_0^T \int_{\mathbb{R}} \sum_{n \geq 0} I_n(g_n \cdot h)d\lambda dt = \frac{1}{2\pi} \sum_{n \geq 0} \int_0^T \int_{\mathbb{R}} I_n(g_n \cdot h)d\lambda dt \text{ in } (\mathcal{S})^*. \tag{14.26}$$

Further, we can interchange the integrals in (14.26), so that we obtain

$$L_T(x) = \frac{1}{2\pi} \sum_{n \geq 0} I_n(\int_0^T \int_{\mathbb{R}} g_n \cdot h d\lambda dt) \text{ in } (\mathcal{S})^*.$$

Note that this is a consequence of the integrability condition (14.22). Since $L_T(x)$ is in $L^2(P)$, the result follows. $\square$

14.4 Application to Hedging in Incomplete Markets

Let us consider once again a pure jump Lévy process

$$\eta(t) = \int_0^t \int_{\mathbb{R}_0} z \tilde{N}(ds, dz), \quad t \geq 0.$$

For any $\mathcal{F}_T$-measurable $\xi \in L^2(P)$ (where $T > 0$ is fixed) there exists a unique $\mathcal{F}_t$-adapted process $\varphi \in L^2(P \times \lambda \times \nu)$, i.e., $E\big[\int_0^T \int_{\mathbb{R}_0} \varphi^2(t, z)\nu(dz)dt\big] < \infty$, such that

$$\xi = E[\xi] + \int_0^T \int_{\mathbb{R}_0} \varphi(t, z)\tilde{N}(dt, dz). \tag{14.27}$$

In view of financial applications, it is of interest to find φ explicitly. This can be done, for instance, via Malliavin calculus and a Lévy version of the Clark–Ocone theorem (Sect. 4.1, and Sect. 12.4). Another approach based on nonanticipating differentiation is, e.g., proposed in [59, 60].

The purpose of this section is to show how the Donsker delta function for Lévy processes can be used to find an explicit formula for φ. See the forthcoming Theorem 14.9. This method does not rely on Malliavin calculus for Lévy processes nor does it involve conditional expectations, as the previous methods do.

Theorem 14.9. *Let $g : \mathbb{R} \longrightarrow \mathbb{R}$ be a given function in $L^1(\lambda)$ such that its Fourier transform*

$$\hat{g}(u) := \frac{1}{2\pi} \int_{\mathbb{R}} g(x)e^{-iux}\,dx$$

belongs to $L^1(\lambda)$. Then, for $T \geq 0$, we have

$$g(\eta(T)) = \int_{\mathbb{R}} \hat{g}(u) \exp\left\{ T \int_{\mathbb{R}_0} \left(e^{iuz} - 1 - iuz \right) \nu(dz) \right\} du + \int_0^T \int_{\mathbb{R}_0} \varphi(t, z)\tilde{N}(dt, dz),$$
$$\tag{14.28}$$

where

$$\varphi(t, z) = \int_{\mathbb{R}} \hat{g}(u) \exp\left\{ (T - t) \int_{\mathbb{R}_0} \left(e^{iuz} - 1 - iuz \right) \nu(dz) \right\} e^{iu\eta(t)} \left(e^{iuz} - 1 \right) du.$$
$$\tag{14.29}$$

Proof Define

$$Y_u(t) := \exp^{\diamond} \left\{ \int_0^t \int_{\mathbb{R}_0} \left(e^{iuz} - 1 \right) \tilde{N}(ds, dz) \right\}, \quad u \in \mathbb{R}, \, t \in [0, T].$$

First let us assume that g is continuous with compact support. Then, by (14.1), (14.4), (13.44), and Lévy–Wick calculus we have

$$g(\eta(T)) = \int_{\mathbb{R}} g(y)\delta_y(\eta(T))dy$$

$$= \int_{\mathbb{R}} \frac{1}{2\pi} g(y) \left(\int_{\mathbb{R}} Y_u(T) \exp\left\{ T \int_{\mathbb{R}_0} \left(e^{iuz} - 1 - iuz\right) \nu(dz) - iuy \right\} du \right) dy$$

$$= \int_{\mathbb{R}} Y_u(T) \left(\int_{\mathbb{R}} \frac{1}{2\pi} g(y) e^{-iuy} dy \right) \exp\left\{ T \int_{\mathbb{R}_0} \left(e^{iuz} - 1 - iuz\right) \nu(dz) \right\} du$$

$$= \int_{\mathbb{R}} \hat{g}(u) \exp\left\{ T \int_{\mathbb{R}_0} \left(e^{iuz} - 1 - iuz\right) \nu(dz) \right\} \left(Y_u(0) + \int_0^T \frac{\partial}{\partial t} Y_u(t)dt \right) du$$

$$= \int_{\mathbb{R}} \hat{g}(u) \exp\left\{ T \int_{\mathbb{R}_0} \left(e^{iuz} - 1 - iuz\right) \nu(dz) \right\} du$$

$$+ \int_{\mathbb{R}} \hat{g}(u) \exp\left\{ T \int_{\mathbb{R}_0} \left(e^{iuz} - 1 - iuz\right) \nu(dz) \right\} \int_0^T Y_u(t)$$

$$\diamond \int_{\mathbb{R}_0} \left(e^{iuz} - 1\right) \overset{\bullet}{\tilde{N}}(t,z)\nu(dz)dt\, du$$

$$= \int_{\mathbb{R}} \hat{g}(u) \exp\left\{ T \int_{\mathbb{R}_0} \left(e^{iuz} - 1 - iuz\right) \nu(dz) \right\} du$$

$$+ \int_{\mathbb{R}} \hat{g}(u) \exp\left\{ T \int_{\mathbb{R}_0} \left(e^{iuz} - 1 - iuz\right) \nu(dz) \right\} \int_0^T \int_{\mathbb{R}_0} Y_u(t) \left(e^{iuz} - 1\right) \tilde{N}(dt,dz) du$$

$$= \int_{\mathbb{R}} \hat{g}(u) \exp\left\{ T \int_{\mathbb{R}_0} \left(e^{iuz} - 1 - iuz\right) \nu(dz) \right\} du + \int_0^T \int_{\mathbb{R}_0} \varphi(t,z) \tilde{N}(dt,dz),$$

where

$$\varphi(t,z) = \int_{\mathbb{R}} \hat{g}(u) \exp\left\{ T \int_{\mathbb{R}_0} \left(e^{iuz} - 1 - iuz\right) \nu(dz) \right\} Y_u(t) \left(e^{iuz} - 1\right) du.$$

By (14.4) we have

$$Y_u(t) \exp\left\{ t \int_{\mathbb{R}_0} \left(e^{iuz} - 1 - iuz\right) \nu(dz) \right\} = \exp\left(iu\eta(t)\right).$$

Hence, we get

$$\varphi(t,z) = \int_{\mathbb{R}} \hat{g}(u) \exp\left\{ (T-t) \int_{\mathbb{R}_0} \left(e^{iuz} - 1 - iuz\right) \nu(dz) \right\} \exp\left(iu\eta(t)\right) \left(e^{iuz} - 1\right) du.$$

For the general case, let g be as in Theorem 14.9. Then there exists a sequence $g_n \leq g$, $n = 1, 2, ...$, of continuous functions with compact support such that $g_n \longrightarrow g$ in $L^1(\lambda)$. Then $\hat{g}_n(u) \longrightarrow \hat{g}(u)$ pointwise dominatedly. Since (14.28) holds for all g_n, it follows by dominated convergence that it holds for g, as required. □

As an application of Theorem 14.9, we consider a financial market where there are two investment possibilities:

- A risk free asset (e.g., bond), where the price $S_0(t)$, $t \geq 0$, is constantly equal to 1.

- A risky asset (e.g., stock), where the price $S(t)$ at time t is given by

$$dS(t) = S(t^-)d\eta(t), \qquad S(0) > 0,$$

where $\eta(t) = \int_0^t\int_{\mathbb{R}_0} z\widetilde{N}(ds, dz)$ as before. Assume that $z > -1 + \varepsilon$ for a.a. z with respect to ν, for some $\varepsilon > 0$. This guarantees that $S(t) > 0$ for all $t \geq 0$.

As in Sect. 9.4 we say that an $\mathcal{F}_T$-measurable random variable $\xi \in L^2(P)$ (called a *claim*) is *hedgeable* or *replicable* in this market if there exists an $\mathcal{F}_t$-adapted process $\psi(t)$, $t \in [0, T]$, in $L^2(P \times \lambda)$ such that

$$\xi = E[\xi] + \int_0^T \psi(t)dS(t) = E[\xi] + \int_0^T \psi(t)S(t^-)d\eta(t). \qquad (14.30)$$

The market is called *complete* if all claims are replicable and *incomplete* otherwise. We can write (14.30) as

$$\xi = E[\xi] + \int_0^T\int_{\mathbb{R}_0} \psi(t)S(t^-)z\widetilde{N}(dt, dz). \qquad (14.31)$$

Thus, by comparing (14.31) with (14.27), we see that ξ is hedgeable if and only if the process φ in (14.27) has the product form

$$\varphi(t, z) = \psi(t)S(t^-)z \quad \text{for almost all } t, z \text{ with respect to } \lambda \times \nu. \qquad (14.32)$$

It is well known that, unless η is the centered Poisson process, the market is *incomplete*. Hence, there exist claims ξ that are not hedgeable. It is of interest to find out which claims are replicable and which claims are not. Our next result gives information in this direction. See [67]

Theorem 14.10. *Let g be as in Theorem 14.9 and define*

$$f(t, u) - \hat{g}(u)\exp\left\{(T-t)\int_{\mathbb{R}_0} \left(e^{iuz} - 1 - iuz\right)\nu(dz)\right\}\exp\left(iu\eta(t)\right). \qquad (14.33)$$

Suppose that $\xi := g(\eta(T))$ is replicable and let $\psi(t)$ be as in (14.30). Then for almost all $t \in [0, T]$, we have

$$\hat{f}(t, \cdot)(z) := \frac{1}{2\pi}\int_{\mathbb{R}} f(t, u)e^{-iuz}du = \frac{1}{2\pi}\int_{\mathbb{R}} f(t, u)du - \frac{1}{2\pi}\psi(t)S(t^-)z \qquad (14.34)$$

for all $z \in -\mathrm{supp}\,\nu$, where $\mathrm{supp}\,\nu$ is the smallest closed set in $\mathbb{R}$ in which ν has all its mass.

Proof By Theorem 14.9 and (14.32), we have

$$\varphi(t, z) = \int_{\mathbb{R}} f(t, u)\left(e^{iuz} - 1\right)du = \psi(t)S(t^-)z,$$

for all $z \in supp\,\nu$ and almost all $t \in [0, T]$. Hence,

$$\frac{1}{2\pi} \int_{\mathbb{R}} f(t, u)e^{-iuz}\,du = \frac{1}{2\pi} \int_{\mathbb{R}} f(t, u)du - \frac{1}{2\pi}\psi(t)S(t^-)z \quad \text{for all } z \in -supp\,\nu.$$

By this we end the proof. $\square$

See [67] for the next result.

Theorem 14.11. *Let g be as in Theorem 14.9 and suppose $supp\,\nu$ is unbounded. Then $g(\eta(T))$ is not replicable unless $g = 0$.*

Proof Suppose $\xi = g(\eta(T))$ is replicable. Since the Fourier transform vanishes at $\pm\infty$, we get from (14.34)

$$0 = \lim_{\substack{|z|\to\infty \\ z\in -supp\,\nu}} \hat{f}(t, \cdot)(z) = \lim_{\substack{|z|\to\infty \\ z\in -supp\,\nu}} \left(\frac{1}{2\pi} \int_{\mathbb{R}} f(t, u)du - \frac{1}{2\pi}\psi(t)S(t^-)z \right),$$

for almost all $t \in [0, T]$. This is only possible if $\psi(t) = 0$ almost everywhere. Hence, $\xi = E[\xi]$, so g is constant and hence $g = 0$ (since $g \in L^1(\lambda)$). $\square$

14.5 A Sensitivity Result for Jump Diffusions

In this section we derive a representation theorem for functions of a class of jump diffusions, which contains the Brownian motion and the pure jump Lévy process of Sect. 14.2 as special cases. As an alternative approach to [81, 200] (see Sects. 4.4 and 12.7), we illustrate how this formula can be applied to the computation of "greeks" in finance. Our presentation is based on [68]. Another application of this theorem is an explicit formula for the Donsker delta function of a special class of jump diffusions, which can be used as in Sects. 7.1 and 14.4 to compute hedging strategies of portfolios.

14.5.1 A Representation Theorem for Functions of a Class of Jump Diffusions

We aim at presenting a representation formula for functions of a special class of jump diffusions (see Theorem 14.13), which if applied to the sensitivity analysis context, gives a computationally efficient formula for the greek "delta." Further, we employ this result to establish an explicit representation of the Donsker delta function. Our approach rests on the arguments of Sect. 14.2.

On the given probability space $(\Omega, \mathcal{F}, P)$ we consider the Itô–Lévy process $\xi(t)$, $t \geq 0$, of the form

$$\xi(t) = \xi(0) + \int_0^t \alpha(s)ds + \int_0^t \beta(s)dW(s) + \int_0^t \int_{\mathbb{R}_0} \gamma(s, z)\tilde{N}(ds, dz), \quad (14.35)$$

where $\alpha(t), \beta(t)$, and $\gamma(t,z)$, $t \geq 0$, $z \in \mathbb{R}_0$, are *deterministic* functions such that

$$\int\limits_0^\infty \left[|\alpha(t)| + \beta^2(t) + \int\limits_{\mathbb{R}_0} \gamma^2(t,z)\nu(dz) \right] dt < \infty \tag{14.36}$$

holds.

On the other hand, we could consider a representation of the type (14.35) on the product space introduced in Sect. 12.5. We recall the construction here. Let $(\Omega, \mathcal{F}, P)$ be a product of two complete probability spaces, i.e.,

$$\Omega = \Omega_1 \times \Omega_2, \quad \mathcal{F} = \mathcal{F}_1 \otimes \mathcal{F}_2 \quad P = P_1 \otimes P_2. \tag{14.37}$$

In such a framework, we could consider stochastic processes $\xi(t)$, $t \geq 0$, on $(\Omega, \mathcal{F}, P)$ such that

$$\xi(t, \omega_1, \omega_2) = y + \int\limits_0^t \alpha(s)ds + \int\limits_0^t \beta(s)dW(s, \omega_1)$$

$$+ \int\limits_0^t \int\limits_{\mathbb{R}_0} \gamma(s,z)\tilde{N}(ds, dz, \omega_2) \tag{14.38}$$

for $y \in \mathbb{R}$ constant and $\alpha(t), \beta(t)$ and $\gamma(t,z)$, $t \geq 0$, $z \in \mathbb{R}_0$, deterministic functions fulfilling (14.36).

We equip the probability space $(\Omega_1, \mathcal{F}_1, P_1)$ with the filtration $\mathcal{F}_t^1$, $t \geq 0$, $(\mathcal{F}_\infty^1 = \mathcal{F}_1)$ generated by $W(t)$, $t \geq 0$, augmented by all P_1-null sets and the space $(\Omega_2, \mathcal{F}_2, P_2)$ with the filtration $\mathcal{F}_t^2$, $t \geq 0$, $(\mathcal{F}_\infty^2 = \mathcal{F}_2)$ generated by the values of $\tilde{N}(dt, dz)$, $t \geq 0$, $z \in \mathbb{R}_0$, augmented of all P_2-null sets. Then on the product $(\Omega, \mathcal{F}, P)$ we fix the filtration

$$\mathcal{F}_t := \mathcal{F}_t^1 \otimes \mathcal{F}_t^2, \qquad t \geq 0.$$

In the sequel we apply white noise analysis for a combination of Gaussian and pure jump Lévy noise as introduced in Sect. 13.7. Thus, we choose $(\Omega_1, \mathcal{F}_1, P_1)$ to be a Gaussian white noise and $(\Omega_2, \mathcal{F}_2, P_2)$ a Lévy white noise probability space. Recall that we have assumed (14.1), see Sect. 14.1.

Let $\xi(t) = \xi^y(t)$, $t \in [0,T]$, be a stochastic process on $(\Omega, \mathcal{F}, P)$ of the form

$$\begin{cases} d\xi(t) = \alpha(t)dt + \beta(t)dW(t) + \int\limits_{\mathbb{R}_0} \gamma(t,z)\tilde{N}(dt, dz), & t \in [0,T], \\ \xi(0) = y \in \mathbb{R}, \end{cases} \tag{14.39}$$

where $\alpha(t), \beta(t)$, and $\gamma(t,z)$; $t \in [0,T]$, $z \in \mathbb{R}$, are deterministic functions satisfying (14.36). For $\lambda \in \mathbb{R}$ define

$$Y(t) = \exp(\lambda \xi(t)), \qquad t \in [0,T]. \tag{14.40}$$

Then by the Itô formula (see Theorem 9.4) we have

$$dY(t) = Y(t^-)\Big[(\lambda\alpha(t) + \tfrac{1}{2}\lambda^2\beta^2(t))dt + \lambda\beta(t)dW(t)$$

$$+ \int_{\mathbb{R}_0} \{\exp(\lambda\gamma(t,z)) - 1 - \lambda\gamma(t,z)\}\nu(dz)dt$$

$$+ \int_{\mathbb{R}_0} \{\exp(\lambda\gamma(t,z)) - 1\}\tilde{N}(dt,dz)\Big]. \tag{14.41}$$

Using the white noise notation and Wick calculus for a combination of Gaussian and pure jump Lévy noise (see Sect. 12.5), the last equation can be written as

$$\frac{dY(t)}{dt} = Y(t^-) \diamond \Big\{ \lambda\alpha(t) + \tfrac{1}{2}\lambda^2\beta^2(t) + \lambda\beta(t)\dot{W}(t)$$

$$+ \int_{\mathbb{R}_0} [\exp\{\lambda\gamma(t,z)\} - 1 - \lambda\gamma(t,z)]\nu(dz)$$

$$+ \int_{\mathbb{R}_0} \{\exp\{\lambda\gamma(t,z)\} - 1\}\dot{\tilde{N}}(t,z)\nu(dz)\Big\}, \qquad Y(0) = e^{\lambda y}. \tag{14.42}$$

By applying Wick calculus in $(\mathcal{S})^*$ as developed in Sect. 14.7 to (14.42), we obtain

$$Y(t) = Y(0)\exp^\diamond\Big\{ \int_0^t \Big[\lambda\alpha(s) + \tfrac{1}{2}\lambda^2\beta^2(s)$$

$$+ \int_{\mathbb{R}_0} [\{e^{\lambda\gamma(s,z)} - 1 - \lambda\gamma(s,z)]\nu(dz)\Big]ds$$

$$+ \int_0^t \lambda\beta(s)dW(s) + \int_0^t\int_{\mathbb{R}_0} \{e^{\lambda\gamma(s,z)} - 1\}\tilde{N}(ds,dz)\Big\}. \tag{14.43}$$

Comparing (14.43) with (14.40) we get the following formula for the Wick exponential.

Lemma 14.12. *Let $\xi(t)$ be as in (14.39) and let $\lambda \in \mathbb{R}$. Then we have*

$$e^{\lambda\xi(t)} = e^{\lambda y}\exp^\diamond\Big\{ \int_0^t \Big[\lambda\alpha(s) + \tfrac{1}{2}\lambda^2\beta^2(s)$$

$$+ \int_{\mathbb{R}_0} \Big[e^{\lambda\gamma(s,z)} - 1 - \lambda\gamma(s,z)\Big]\nu(dz)\Big]ds + \int_0^t \lambda\beta(s)dW(s)$$

$$+ \int_0^t\int_{\mathbb{R}_0} [e^{\lambda\gamma(s,z)} - 1]\tilde{N}(ds,dz)\Big\}. \tag{14.44}$$

Using this we get the following result. See [68].

Theorem 14.13. Representation theorem for functions of a jump diffusion.
Let $g : \mathbb{R} \to \mathbb{R}$ be a function with Fourier transform

$$\hat{g}(\lambda) = \tfrac{1}{2\pi} \int_{\mathbb{R}} e^{-i\lambda x} g(x) dx, \qquad \lambda \in \mathbb{R},$$

satisfying the Fourier inversion property

$$g(u) = \int_{\mathbb{R}} e^{i\lambda u} \hat{g}(\lambda) d\lambda, \qquad u \in \mathbb{R}.$$

Then

$$g(\xi^y(t)) = \int_{\mathbb{R}} \hat{g}(\lambda) \exp^{\circ}\{X_\lambda^y(t)\} d\lambda, \qquad t \in [0, T], \tag{14.45}$$

where

$$X_\lambda^y(t) = i\lambda y + \int_0^t i\lambda \beta(s) dW(s)$$

$$+ \int_0^t \int_{\mathbb{R}_0} [e^{i\lambda\gamma(s,z)} - 1] \tilde{N}(ds, dz) + \int_0^t \Big[i\lambda\alpha(s) - \tfrac{1}{2}\lambda^2\beta^2(s)$$

$$+ \int_{\mathbb{R}_0} [e^{i\lambda\gamma(s,z)} - 1 - i\lambda\gamma(s,z)]\nu(dz) \Big] ds, \qquad t \in [0, T]. \tag{14.46}$$

Proof Applying (14.44) with $i\lambda$ instead of λ we obtain

$$g(\xi^y(t)) = \tfrac{1}{2\pi} \int_{\mathbb{R}} e^{i\lambda\xi(t)} \hat{g}(\lambda) d\lambda = \tfrac{1}{2\pi} \int_{\mathbb{R}} \hat{g}(\lambda) \exp^{\circ}(X_\lambda^y(t)) d\lambda. \quad \square$$

See [68] for the following result.

Corollary 14.14. *Let g be a real function as in Theorem 14.13. Then we have*

$$E[g(\xi^y(t))] = \int_{\mathbb{R}} \hat{g}(\lambda) \cdot \exp(i\lambda y + G_\lambda(t)) d\lambda, \tag{14.47}$$

where

$$G_\lambda(t) = \int_0^t \Big[i\lambda\alpha(s) - \tfrac{1}{2}\lambda^2\beta^2(s) + \int_{\mathbb{R}_0} (e^{i\lambda\gamma(s,z)} - 1 - i\lambda\gamma(s,z))\nu(dz) \Big] ds. \tag{14.48}$$

Proof This follows from Theorem 14.13 in connection with the fact that

$$E[\exp^\diamond X^y_\lambda(t)] = \exp(E[X^y_\lambda(t)]).$$

Compare with Sect. 5.3.   □

We conclude this section with an explicit formula for the Donsker delta function of $\xi(t) = \xi^y(t)$, $t \geq 0$. This is derived as an application of Theorem 14.13. Adopting the white noise framework of Sect. 12.5, we consider the Donsker delta function of a random variable $X \in (\mathcal{S})^*$, see Definition 14.1.

In the following, we retain condition (14.3). Using Theorem 14.13 we can obtain an explicit formula for the Donsker delta function of $\xi(t)$. See [68].

Theorem 14.15. Representation of the Donsker delta function. *Assume that (14.3) holds. Then the Donsker delta function $\delta_u(\xi^y(t))$, $u \in \mathbb{R}$, of $\xi^y(t)$, $t \in [0, T]$, exists in $(\mathcal{S})^*$ and is given by*

$$\delta_u(\xi^y(t)) = \frac{1}{2\pi} \int_{\mathbb{R}} \exp^\diamond \Big\{ \int_0^t \int_{\mathbb{R}_0} (e^{i\lambda\gamma(s,z)} - 1)\tilde{N}(ds, dz) + \int_0^t i\lambda\beta(s)dW(s)$$

$$+ \int_0^t \Big[\int_{\mathbb{R}_0} (e^{i\lambda\gamma(s,z)} - 1 - i\lambda\gamma(s,z))\nu(dz)$$

$$+ i\lambda\alpha(s) - \tfrac{1}{2}\lambda^2\beta^2(s) \Big] ds + i\lambda y - i\lambda u \Big\} d\lambda. \tag{14.49}$$

Proof Here, we sketch only the basic ideas. Formally this follows from (14.45) by using the Fubini theorem in $(\mathcal{S})^*$, as follows. By (14.45) we have

$$g(\xi^y(t)) = \int_{\mathbb{R}} \hat{g}(\lambda) \exp^\diamond(X^y_\lambda(t))d\lambda$$

$$= \frac{1}{2\pi} \int_{\mathbb{R}} \Big(\int_{\mathbb{R}} e^{-i\lambda u} g(u)du \Big) \exp^\diamond(X^y_\lambda(t))d\lambda$$

$$= \frac{1}{2\pi} \int_{\mathbb{R}} g(u) \int_{\mathbb{R}} \exp^\diamond(-i\lambda u + X^y_\lambda(t))d\lambda\, du.$$

For justification and more details, we refer to the proof of Theorem 14.4.   □

Remark 14.16. Using white noise techniques, an explicit representation of the Donsker delta function of solutions to a general SDE driven by a Brownian motion is given in [145] . See also [166] in the case of Lévy processes.

14.5.2 Application: Computation of the "Greeks"

Let $X(t) = X^x(t)$, $t \in [0, T]$, be a jump diffusion of the form

$$
\begin{cases}
dX(t) = X(t^-)\Big[\mu(t)dt + \sigma(t)dW(t) + \int_{\mathbb{R}_0} \theta(t, z)\tilde{N}(dt, dz)\Big], \\
X(0) = x > 0,
\end{cases}
\tag{14.50}
$$

where $\mu(t), \sigma(t)$, and $\theta(t, z), t \in [0, T], z \in \mathbb{R}$, are deterministic, $\theta(t, z) > -1 + \varepsilon$ for a.a. t, z, for some $\varepsilon > 0$, and

$$
\int_0^T \Big[|\mu(t)| + \sigma^2(t) + \int_{\mathbb{R}_0} \theta^2(t, z)\nu(dz) \Big] dt < \infty.
$$

By the Itô formula for Lévy processes (see Theorem 9.4), the solution of this equation is given by

$$
X^x(t) = x \exp \Big\{ \int_0^t \Big[\mu(s) - \tfrac{1}{2}\sigma^2(s) + \int_{\mathbb{R}_0} (\log(1 + \theta(s, z)) - \theta(s, z))\nu(dz) \Big] ds
$$

$$
+ \int_0^t \sigma(s)dW(s) + \int_0^t \int_{\mathbb{R}_0} \log(1 + \theta(s, z))\tilde{N}(ds, dz) \Big\}
\tag{14.51}
$$

$$
= \exp(\xi^y(t)),
$$

where

$$
d\xi^y(t) = \alpha(t)dt + \beta(t)dW(t) + \int_{\mathbb{R}_0} \gamma(t, z)\tilde{N}(dt, dz)
$$

with

$$
\alpha(t) = \mu(t) - \tfrac{1}{2}\sigma^2(t) + \int_{\mathbb{R}_0} (\log(1 + \theta(t, z)) - \theta(t, z))\nu(dz),
$$

$$
\beta(t) = \sigma(t),
\tag{14.52}
$$

$$
\gamma(t, z) = \log(1 + \theta(t, z)).
$$

See (14.39). Therefore, if $h : \mathbb{R} \to \mathbb{R}$ then

$$
E[h(X^x(T))] = E[h(\exp(\xi^y(T)))] = E[g(\xi^y(T))],
$$

where

$$
g(u) := h(\exp(u)), \qquad u \in \mathbb{R}.
$$

If this g satisfies the conditions of Theorem 14.13 then

$$\tfrac{d}{dx}E[h(X^x(T))] = \tfrac{d}{dx}[Eg(\xi^{\log x}(T))] = \tfrac{d}{dx}\int_{\mathbb{R}} \hat{g}(\lambda)\exp(i\lambda\log x + G_\lambda(T))d\lambda$$

$$= \int_{\mathbb{R}} \hat{g}(\lambda)\tfrac{i\lambda}{x}\exp(i\lambda\log x + G_\lambda(T))d\lambda,$$

where $G_\lambda(T)$ is given by (14.48). We have proved the following result [68].

Theorem 14.17. Sensitivity formula. *Suppose* $h : \mathbb{R} \to \mathbb{R}$ *is such that* $g(u) := h(\exp(u))$, $u \in \mathbb{R}$, *and satisfies the conditions of Theorem 14.13 and that*

$$\int_{\mathbb{R}} |\hat{g}(\lambda)\lambda\exp\{\operatorname{Re} G_\lambda(T)\}|\, d\lambda < \infty.$$

Then

$$\Delta = \tfrac{d}{dx}E[h(X^x(T))] = \int_{\mathbb{R}} \hat{g}(\lambda)\tfrac{i\lambda}{x}\exp(i\lambda\log x + G_\lambda(T))d\lambda. \qquad (14.53)$$

Example 14.18. Choose $h(u) = \chi_{[H,K]}(u)$, $u \in \mathbb{R}$ $(H, K > 0)$. Then $h(X^x(T))$ may be regarded as the payoff of a digital option on a stock with price $X^x(T)$. In this case

$$g(u) = \chi_{[H,K]}(e^u), \qquad u \in \mathbb{R},$$

and

$$2\pi\hat{g}(\lambda) = \int_{\mathbb{R}} e^{-i\lambda u}g(u)du = \int_{\mathbb{R}} e^{-i\lambda u}\chi_{[H,K]}(e^u)du = \int_{\log H}^{\log K} e^{-i\lambda u}du = \frac{H^{-i\lambda} - K^{-i\lambda}}{i\lambda}.$$

Therefore,

$$\tfrac{d}{dx}E\left[\chi_{[H,K]}(X^x(T))\right] = \int_{\mathbb{R}} \frac{H^{-i\lambda} - K^{-i\lambda}}{x}\exp(i\lambda\log x + G_\lambda(T))d\lambda, \quad (14.54)$$

provided that the integral converges. A sufficient condition for this is that, for some $\delta > 0$,

$$\lambda^2 \int_0^T \left[\beta^2(s) + \int_{\mathbb{R}_0}(1 - \cos(\lambda\gamma(s,z)))\nu(dz)\right]ds \geq \delta\lambda^2 \qquad (14.55)$$

(where β and γ are given by (14.52)), which is a weak form of nondegeneracy of the equation (14.50). Thus, in spite of the fact that h is not even continuous, (14.53) is a computationally efficient formula for $\tfrac{d}{dx}E^x[h(X^x(T))]$. Note that we may have $\beta = \sigma = 0$, as long as (14.55) holds.

14.6 Exercises

Problem 14.1. Consider a path-dependent option given by an *Asian option* that has the form

$$A(T_1, T_2) := \frac{1}{T_1 - T_2} \int_{T_1}^{T_2} h(X^x(t)) dt,$$

where $X^x(t)$ is the diffusion of (14.50). Hence, the Asian option is an average asset value over the period of the life time of the option. Such an option has proved to be useful for asset classes (e.g., commodities) that are sparsely traded on a market. Assume that

$$\int_{T_1}^{T_2} \int_{\mathbb{R}} |\hat{g}(\lambda) \lambda \exp\{\operatorname{Re} G_\lambda(t)\}| \, d\lambda dt < \infty,$$

where $g(\lambda) = h(\exp(\lambda))$ and $G_\lambda(t)$ is as in (14.48). Calculate the delta of this option.

15

The Forward Integral

The original forward integral (see [209]) was introduced with respect to the Brownian motion only. See Chap. 8. In this chapter we present a form of forward integration with respect to the Poisson random measure $\widetilde{N}$. We prove an Itô formula for the corresponding forward processes and apply this to obtain an Itô formula for Skorohod integrals. This presentation is mainly based on [64].

15.1 Definition of Forward Integral and its Relation with the Skorohod Integral

Definition 15.1. *The* forward integral

$$J(\theta) := \int_0^T \int_{\mathbb{R}_0} \theta(t, z) \widetilde{N}(d^- t, dz)$$

with respect to the Poisson random measure $\widetilde{N}$ of a stochastic function (or random field) $\theta(t, z)$, $t \in [0, T]$, $z \in \mathbb{R}_0$ $(T > 0)$, with

$$\theta(t, z) := \theta(\omega, t, z), \qquad \omega \in \Omega,$$

and càglàd with respect to t, is defined as

$$J(\theta) = \lim_{m \to \infty} \int_0^T \int_{\mathbb{R}_0} \theta(t, z) 1_{U_m} \widetilde{N}(dt, dz)$$

if the limit exists in $L^2(P)$. Here, U_m, $m = 1, 2, ...$, is an increasing sequence of compact sets $U_m \subseteq \mathbb{R}_0 := \mathbb{R} \backslash \{0\}$ with $\nu(U_m) < \infty$ such that $\lim_{m \to \infty} U_m = \mathbb{R}_0$.

Remark 15.2. Note that if $\mathbb{G} := \{\mathcal{G}_t, t \in [0, T]\}$ is a filtration such that

(1) $\mathcal{F}_t \subseteq \mathcal{G}_t$ for all t

G.Di Nunno et al., *Malliavin Calculus for Lévy Processes with Applications to Finance*,
© Springer-Verlag Berlin Heidelberg 2009

(2) The process $\eta(t) = \int_0^t \int_{\mathbb{R}_0} z\widetilde{N}(ds, dz)$, $t \in [0, T]$, is a semimartingale with respect to $\mathbb{G}$

(3) The stochastic process $\theta = \theta(t, z)$, $t \in [0, T]$, $z \in \mathbb{R}_0$ is $\mathbb{G}$-predictable and

(4) The integral $\int_0^T \int_{\mathbb{R}_0} \theta(t, z)\widetilde{N}(dt, dz)$ exists as a classical Itô integral

then the forward integral of θ with respect to $\widetilde{N}$ also exists and we have

$$\int_0^T \int_{\mathbb{R}_0} \theta(t, z)\widetilde{N}(d^- t, dz) = \int_0^T \int_{\mathbb{R}_0} \theta(t, z)\widetilde{N}(dt, dz).$$

This follows from the basic construction of the semimartingale integral. See, e.g., [204]. Thus, the forward integral can be regarded as an extension of the Itô integral to possibly nonsemimartingale contexts.

Remark 15.3. Directly from the definition we can see that if G is a random variable then

$$G\int_0^T \int_{\mathbb{R}_0} \theta(t, z)\widetilde{N}(d^- t, dz) = \int_0^T \int_{\mathbb{R}_0} G\theta(t, z)\widetilde{N}(d^- t, dz), \qquad (15.1)$$

a property that does not hold for the Skorohod integrals.

Definition 15.4. *In the sequel we let $\mathcal{M}$ denote the set of the stochastic functions $\theta(t, z)$, $t \in [0, T]$, $z \in \mathbb{R}_0$, such that*

(1) $\theta(\omega, t, z) = \theta_1(\omega, t)\theta_2(\omega, t, z)$, where $\theta_1(\cdot, t) \in \mathbb{D}_{1,2}$ for all t, $\theta_1(\omega, \cdot)$ is càglàd P-a.s. and $\theta_2(\omega, t, z)$ is adapted and such that

$$E\left[\int_0^T \int_{\mathbb{R}_0} \theta_2^2(t, z)\nu(dz)dt\right] < \infty.$$

(2) $D_{t+, z}\theta(t, z) := \lim_{s \to t+} D_{s, z}\theta(t, z)$ exists in $L^2(P \times \lambda \times \nu)$.

(3) $\theta(t, z) + D_{t+, z}\theta(t, z)$ is Skorohod integrable.

Let $\mathbb{M}_{1,2}$ be the closure of the linear span of $\mathcal{M}$ with respect to the norm given by

$$\|\theta\|_{\mathbb{M}_{1,2}}^2 := \|\theta\|_{L^2(P \times \lambda \times \nu)}^2 + \|D_{t+, z}\theta(t, z)\|_{L^2(P \times \lambda \times \nu)}^2.$$

We can now show the relation between the forward integral and the Skorohod integral.

Lemma 15.5. *If $\theta \in \mathbb{M}_{1,2}$, then its forward integral exists and*

$$\int_0^T \int_{\mathbb{R}_0} \theta(t, z)\widetilde{N}(d^- t, dz) = \int_0^T \int_{\mathbb{R}_0} D_{t+, z}\theta(t, z)\nu(dz)dt$$

$$+ \int_0^T \int_{\mathbb{R}_0} \left(\theta(t, z) + D_{t+, z}\theta(t, z)\right)\widetilde{N}(\delta t, dz).$$

Proof First consider the case when $\theta(\omega, t, z) = \theta_1(\omega, t)\theta_2(\omega, t, z)$. Let us take a sequence of partitions of $[0, T]$ of the form $0 = t_0^n < t_1^n < ... < t_{J_n}^n = T$ with $|\Delta t| := \max(t_j^n - t_{j-1}^n) \longrightarrow 0$, for $n \to \infty$, into account. By the integration by parts formula (see Theorem 12.11) we have

$$F \int_{t_{i-1}^n}^{t_i^n} \int_{\mathbb{R}_0} \theta(t, z)\widetilde{N}(\delta t, dz) = \int_{t_{i-1}^n}^{t_i^n} \int_{\mathbb{R}_0} F\theta(t, z)\widetilde{N}(\delta t, dz)$$

$$+ \int_{t_{i-1}^n}^{t_i^n} \int_{\mathbb{R}_0} \theta(t, z)D_{t,z}F\nu(dz)dt + \int_{t_{i-1}^n}^{t_i^n} \int_{\mathbb{R}_0} \theta(t, z)D_{t,z}F\widetilde{N}(\delta t, dz).$$

Hence

$$\int_0^T \int_{\mathbb{R}_0} \theta(t, z)\widetilde{N}(d^-t, dz) = \lim_{|\Delta t| \longrightarrow 0} \sum_{i=1}^{J_n} \theta_1(t_{i-1}^n) \int_{t_{i-1}^n}^{t_i^n} \int_{\mathbb{R}_0} \theta_2(t, z)\widetilde{N}(dt, dz)$$

$$= \lim_{|\Delta t| \longrightarrow 0} \sum_{i=1}^{J_n} \theta_1(t_{i-1}^n) \int_{t_{i-1}^n}^{t_i^n} \int_{\mathbb{R}_0} \theta_2(t, z)\widetilde{N}(\delta t, dz)$$

$$= \lim_{|\Delta t| \longrightarrow 0} \sum_{i=1}^{J_n} \int_{t_{i-1}^n}^{t_i^n} \int_{\mathbb{R}_0} [\theta_1(t_{i-1}^n) + D_{t,z}\theta_1(t_{i-1}^n)]\theta_2(t, z)\widetilde{N}(\delta t, dz)$$

$$+ \lim_{|\Delta t| \longrightarrow 0} \sum_{i=1}^{J_n} \int_{t_{i-1}^n}^{t_i^n} \int_{\mathbb{R}_0} D_{t,z}\theta_1(t_{i-1}^n) \cdot \theta_2(t, z)\nu(dz)dt$$

$$= \int_0^T \int_{\mathbb{R}_0} \theta(t, z)\widetilde{N}(\delta t, dz) + \int_0^T \int_{\mathbb{R}_0} D_{t+,z}\theta(t, z)\nu(dz)dt$$

$$+ \int_0^T \int_{\mathbb{R}_0} D_{t+,z}\theta(t, z)\widetilde{N}(\delta t, dz).$$

The proof is then completed by a limit argument in view of Remark 12.14 and Definition 15.4. □

Corollary 15.6. *If $\theta \in \mathbb{M}_{1,2}$, then*

$$E\left[\int_0^T \int_{\mathbb{R}_0} \theta(t, z)\widetilde{N}(d^-t, dz)\right] = E\left[\int_0^T \int_{\mathbb{R}_0} D_{t+,z}\theta(t, z)\nu(dz)dt\right]. \quad (15.2)$$

Proof This follows from (11.4) and Lemma 15.5. □

15.2 Itô Formula for Forward and Skorohod Integrals

Definition 15.7. *A* forward process *is a measurable stochastic function* $X(t) = X(t, \omega)$, $t \in [0, T]$, $\omega \in \Omega$, *that admits the representation*

$$X(t) = x + \int_0^t \int_{\mathbb{R}_0} \theta(s, z) \widetilde{N}(d^-s, dz) + \int_0^t \alpha(s)ds, \qquad (15.3)$$

where $x = X(0)$ *is a constant. A shorthand notation for* (15.3) *is*

$$d^- X(t) = \int_{\mathbb{R}_0} \theta(t, z) \widetilde{N}(d^-t, dz) + \alpha(t)dt, \qquad X(0) = x. \qquad (15.4)$$

We call $d^- X(t)$ *the* forward differential *of* $X(t)$, $t \in [0, T]$.

The next result was first proved in [64].

Theorem 15.8. Itô formula for forward integrals. *Let* $X(t)$, $t \in [0, T]$, *be a forward process of the form* (15.3), *where* $\theta(t, z)$, $t \in [0, T], z \in \mathbb{R}_0$, *is locally bounded in* z *near* $z = 0$ $P \times \lambda$- *a.e., such that*

$$\int_0^T \int_{\mathbb{R}_0} |\theta(t, z)|^2 \nu(dz)dt < \infty \qquad P - a.s.$$

Also suppose that $|\theta(t, z)|$, $t \in [0, T], z \in \mathbb{R}_0$, *is forward integrable. For any function* $f \in C^2(\mathbb{R})$, *the forward differential of* $Y(t) = f(X(t))$, $t \in [0, T]$, *is given by the following formula:*

$$d^- Y(t) = f'(X(t))\alpha(t)dt \qquad (15.5)$$

$$+ \int_{\mathbb{R}_0} \left(f(X(t^-) + \theta(t, z)) - f(X(t^-)) - f'(X(t^-))\theta(t, z) \right) \nu(dz)dt$$

$$+ \int_{\mathbb{R}_0} \left(f(X(t^-) + \theta(t, z)) - f(X(t^-)) \right) \widetilde{N}(d^-t, dz).$$

Proof The proof follows the same line of the one in the classical Itô formula (see [115] Chap. 2, Sect. 5). For simplicity, we assume $x = 0$ and $\alpha \equiv 0$. Then we can write

$$X_m(t) := \int_0^t \int_{\mathbb{R}_0} \theta(s, z) 1_{U_m}(z) N(ds, dz) - \int_0^t \int_{\mathbb{R}_0} \theta(s, z) 1_{U_m}(z) \nu(dz)ds.$$

Let σ_i, $i = 1, 2, ...$, be the stopping times for which the jumps of the Lévy process occur. Thus, we obtain

$$f(X_m(t)) - f(X_m(0)) = \sum_i \left[f(X_m(\sigma_i \wedge t)) - f(X_m(\sigma_i \wedge t^-)) \right]$$

$$+ \sum_i \left[f(X_m(\sigma_i \wedge t^-)) - f(X_m(\sigma_{i-1} \wedge t)) \right] =: \mathcal{J}_1(t) + \mathcal{J}_2(t),$$

with

$$f(X_m(\sigma_i \wedge t^-)) = \begin{cases} f(X_m(\sigma_i^-)), & \sigma_i \leq t, \\ f(X_m(t)), & \sigma_i > t. \end{cases}$$

By the change of variable formula for finite variation processes, it follows that

$$\mathcal{J}_2(t) = - \int_0^t \int_{\mathbb{R}_0} f'(X_m(s)) \theta(s,z) 1_{U_m}(z) \nu(dz) ds.$$

Moreover, it is

$$\mathcal{J}_1(t) = \sum_i \left[f(X_m(\sigma_i)) - f(X_m(\sigma_i^-)) \right] 1_{\{\sigma_i \leq t\}}$$

$$= \int_0^t \int_{\mathbb{R}_0} \left[f(X_m(s^-) + \theta(s,z) 1_{U_m}(z)) - f(X_m(s^-)) \right] N(ds, dz)$$

$$= \int_0^t \int_{\mathbb{R}_0} \left[f(X_m(s^-) + \theta(s,z) 1_{U_m}(z)) - f(X_m(s^-)) \right] \tilde{N}(d^- s, dz)$$

$$+ \int_0^t \int_{\mathbb{R}_0} \left[f(X_{m,n}(s^-) + \theta(s,z) 1_{U_m}(z)) - f(X_m(s^-)) \right] \nu(dz) ds.$$

By letting $m \to \infty$, formula (15.5) follows. $\square$

To state an Itô formula for Skorohod integrals we need to combine Lemma 15.5 and Theorem 15.8. To this end we go into the technical step of solving equations of the following type: given a random variable G, find the stochastic function $F(t,z)$, $t \in [0,T]$, $z \in \mathbb{R}_0$, such that

$$F(t,z) + D_{t^+,z} F(t,z) = G, \tag{15.6}$$

for almost all $(t,z) \in \mathbb{R}_+ \times \mathbb{R}$. For example, if $G = g(\eta(T))$, for some measurable function $g : \mathbb{R} \longrightarrow \mathbb{R}$ and

$$\eta(t) = \int_0^t \int_{\mathbb{R}_0} z \tilde{N}(dt, dz), \qquad t \in [0,T],$$

then

$$F(t,z) := g(\eta(T) - z\chi_{[0,T)}(t))$$

does the job. Indeed, with this choice of $F(t,z)$, $t \in [0,T]$, $z \in \mathbb{R}_0$, we have

$$F(t,z) + D_{t^+,z} F(t,z) = g(\eta(T) - z\chi_{[0,T)}) + g(\eta(T)) - g(\eta(T) - z\chi_{[0,T)}) = G.$$

The above observation motivates the following definition.

Definition 15.9. *The linear operator S is defined on the space of all $\mathcal{F}_T$-measurable random variables G as follows. If $G = \prod_{i=1}^{k} g_i(\eta(t_i))$, for some $t_i \in [0,T]$, $i = 1, ..., k$, we define*

$$S_{t,z}\left(\prod_{i=1}^{k} g_i(\eta(t_i)) \right) = \prod_{i=1}^{k} g_i(\eta(t_i) - z\chi_{[0,t_i)}(t)). \tag{15.7}$$

Note that via this definition the solution of (15.6) can be written as $F(t,z) = S_{t,z}G$, i.e.,

$$S_{t,z}G + D_{t^+,z}(S_{t,z}G) = G. \tag{15.8}$$

Combining the above facts with Lemma 15.5 and Theorem 15.8, we obtain the following result originally proved in this form in [64].

Theorem 15.10. Itô formula for Skorohod integrals. *Let*

$$X(t) = \int_0^t \int_{\mathbb{R}_0} \gamma(s,z)\widetilde{N}(\delta s, dz) + \int_0^t \alpha(s)ds, \qquad t \in [0,T],$$

or, in shorthand notation,

$$\delta X(t) = \int_{\mathbb{R}_0} \gamma(t,z)\widetilde{N}(\delta t, dz) + \alpha(t)dt, \qquad t \in [0,T].$$

Let $f \in C^2(\mathbb{R})$ and let $Y(t) = f(X(t))$. Set

$$\theta(t,z) := S_{t,z}\gamma(t,z) \tag{15.9}$$

for all $t \in [0,T]$, $z \in \mathbb{R}_0$, and assume $\theta \in \mathbb{M}_{1,2}$. Then

$$\delta Y(t) = f'(X(t))\alpha(t)dt + \int_{\mathbb{R}_0} \left\{ f(X(t^-) + \theta(t,z)) - f(X(t^-)) \right. \tag{15.10}$$

$$+ D_{t^+,z}\left[f(X(t^-) + \theta(t,z)) - f(X(t^-)) \right] \right\}\widetilde{N}(\delta t, dz)$$

$$+ \int_{\mathbb{R}_0} \left\{ f(X(t^-) + \theta(t,z)) - f(X(t^-)) - f'(X(t^-))\theta(t,z) \right.$$

$$+ D_{t^+,z}\left[f(X(t^-) + \theta(t,z)) - f(X(t^-)) \right] - f'(X(t^-))D_{t^+,z}\theta(t,z) \right\}\nu(dz)dt.$$

Remark 15.11. Note that if γ and α are adapted, then $\theta(t,z) = \gamma(t,z)$, $t \in [0,T], z \in \mathbb{R}_0$, and

$$D_{t^+,z}\theta(t,z) = D_{t^+,z}\left[f(X(t^-) + \theta(t,z)) - f(X(t^-)) \right].$$

Therefore, Theorem 15.10 reduces to the classical adapted Itô formula.

Proof For simplicity, we assume $\alpha \equiv 0$. By (15.8) we have

$$\theta(t, z) + D_{t+,z}\theta(t, z) = \gamma(t, z).$$

Hence by Lemma 15.5 we have

$$X(t) = \int_0^t \int_{\mathbb{R}_0} \theta(s, z)\widetilde{N}(d^- s, dz) - \int_0^t \int_{\mathbb{R}_0} D_{s+,z}\theta(s, z)\nu(dz)ds.$$

We can, therefore, apply Theorem 15.8 and get

$$Y(t) - Y(0) = \int_0^t f'(X(s))\left(-\int_{\mathbb{R}_0} D_{s+,z}\theta(s, z)\nu(dz)\right)ds$$

$$+ \int_0^t \int_{\mathbb{R}_0} \left\{f(X(s^-) + \theta(s, z)) - f(X(s^-)) - f'(X(s^-))\theta(s, z)\right\}\nu(dz)ds$$

$$+ \int_0^t \int_{\mathbb{R}_0} \left\{f(X(s^-) + \theta(s, z)) - f(X(s^-))\right\}\widetilde{N}(d^- s, dz)$$

$$- \int_0^t \int_{\mathbb{R}_0} f'(X(s^-))D_{s+,z}\theta(s, z)\nu(dz)ds$$

$$+ \int_0^t \int_{\mathbb{R}_0} \left\{f(X(s^-) + \theta(s, z)) - f(X(s^-)) - f'(X(s^-))\theta(s, z)\right\}\nu(dz)ds$$

$$+ \int_0^t \int_{\mathbb{R}_0} D_{s+,z}\{f(X(s^-) + \theta(s, z)) - f(X(s^-))\}\nu(dz)dt$$

$$+ \int_0^t \int_{\mathbb{R}_0} \left\{f(X(s^-) + \theta(s, z)) - f(X(s^-)) + D_{s+,z}\{f(X(s^-) + \theta(s, z))\right.$$

$$\left.- f(X(s^-))\}\right\}\widetilde{N}(\delta s, dz)$$

$$= \int_0^t \int_{\mathbb{R}_0} \left\{f(X(s^-) + \theta(s, z)) - f(X(s^-)) - f'(X(s^-))\theta(s, z)\right.$$

$$+ D_{s+,z}\left[f(X(s^-) + \theta(s, z)) - f(X(s^-))\right] - f'(X(s^-))D_{s+,z}\theta(s, z)\Big\}\nu(dz)ds$$

$$+ \int_0^t \int_{\mathbb{R}_0} \left\{f(X(s^-) + \theta(s, z)) - f(X(s^-)) + D_{s+,z}\left[f(X(s^-) + \theta(s, z))\right.\right.$$

$$\left.\left.- f(X(s^-))\right]\right\}\widetilde{N}(\delta s, dz).$$

This completes the proof.   □

We would like to mention that an early version of the Itô formula for Skorohod integrals was proved in [124, 125]. Another anticipative Itô formula valid for polynomials is obtained in [195].

Remark 15.12. Following the same scheme as in Sect. 12.5, the Itô formula can be extended to cover the mixed case, involving a combination of Gaussian and compensated Poisson random measures.

15.3 Exercises

Problem 15.1. (*) Let

$$\eta(t) = \int_0^t \int_{\mathbb{R}_0} z\widetilde{N}(ds, dz), \quad t \in [0, T],$$

and define

$$X(t) = \int_0^t \eta(T)\delta\eta(s) = \int_0^t \int_{\mathbb{R}_0} \eta(T)z\widetilde{N}(\delta s, dz), \quad t \in [0, T].$$

(a) Use Lemma 15.5 to show that

$$X(t) = \eta(t)\eta(T) - \int_0^t \int_{\mathbb{R}_0} z^2 N(ds, dz), \quad t \in [0, T].$$

(b) Find $D_{t+,z}X(t)$.
(c) With $\gamma(t, z) := \eta(T)z$, $t \in [0, T]$, $z \in \mathbb{R}_0$, find $\theta(t, z) := S_{t,z}\gamma(t, z)$, see
  Definition 15.9.
(d) Use Itô formula for Skorohod integrals to find $\delta Y(t)$, where $Y(t) = X^2(t)$,
  $t \in [0, T]$.

Applications to Stochastic Control: Partial and Inside Information

16.1 The Importance of Information in Portfolio Optimization

The purpose of this chapter is to present an overview of recent results on stochastic control, in particular, portfolio optimization, achieved via techniques of Malliavin calculus and forward integration. Quite surprisingly, a characterization of the existence of an optimal portfolio can be given in very general settings. In particular, we are interested in studying portfolio optimization problems from the point of view of a trader who may have:

- *Less* information than the one produced by the market noises, here called *partial information*
- *More* information than the one produced by the market noises. Such a trader has some form of anticipation of future event, here called *inside information*

We will see that the two situations are substantially different, in fact the formulation of the problems intrinsically requires different settings. However, in both cases, some techniques of anticipating calculus are going to be applied.

The role of information in portfolio optimization has gained more and more importance according to the common sense knowledge that *the more the information available, the better the performance in the market*. While some occasional cases of insider trading have been detected and been confirmed in the recent years proving that the above statement is a fact, mathematics has started searching for consistent models that are able to describe these situations. In the same line of research, but opposite direction, is the situation of traders having less information than the one usually assumed to be in the market. This is quite a common situation that includes, for example, the case of delayed information. Moreover, it is also interesting to see how a trader with partial information can optimize his portfolio in a financial world where insiders of large capacity are present.

This chapter is an extension of topics treated in Sect. 8.6. A different way to treat portfolio optimization from the minimal variance point of view has already been presented in Sect. 12.6. This chapter is mainly based on [63, 65, 66, 182].

16.2 Optimal Portfolio Problem under Partial Information

We consider the probability space $(\Omega, \mathcal{F}, P)$ as described in Sect. 12.5. We say that a trader has *full information* if he can make his decisions relying on the flow of information generated by the market noises, i.e., the filtration

$$\mathbb{F} := \{\mathcal{F}_t, t \in [0, T]\}.$$

Here, we put $\mathcal{F} = \mathcal{F}_T$. In this section we consider an agent having strictly less information available during his decision making process. We assume that his information is represented by a filtration

$$\mathbb{E} := \{\mathcal{E}_t, t \in [0, T]\}, \qquad \mathcal{E}_t \subseteq \mathcal{F}_t.$$

In this setting of *partial information*, the trader is faced with the problem of finding a portfolio (within an admissible class of portfolios) that maximizes a personal given utility of the corresponding terminal wealth. This is a *stochastic control* problem.

Traditionally, there have been two approaches to this type of problem:

(a) A *dynamic programming* or *stochastic maximum principle* approach. These approaches usually require that the system is *Markovian*. See, e.g., [182, 228], and references therein, for more information about these methods.

(b) The *martingale approach* or *duality approach*. This is an efficient method if the financial market equations have a special "multiplicative" form (see (16.1)–(16.2)) and the market is complete. An advantage of the method is that the system need not be Markovian. However, in the incomplete market case the duality method transforms the original problem into a dual problem usually of equal difficulty. We refer to, e.g., [130], and references therein, for a presentation of the martingale method.

The problem under study, however, cannot be approached easily, if at all, by any of the two methods (a) and (b) above. In fact the system is, in general, neither Markovian nor complete. For example, consider the case of *delayed information*: in case $\delta > 0$ is a fixed *time delay*, at any time $t \geq \delta$, the actual information available would be $\mathcal{E}_t = \mathcal{F}_{t-\delta}$, while it would be $\mathcal{E}_t = \mathcal{E}_0$ for $0 \leq t < \delta$.

We emphasize the distinction between the *partial information* control problem studied in this section and the *partial observation* control problem, where the control is based on noisy observations of the state. The latter problem has been studied by many authors, but the methods and results of the partial observation problem do not apply to our situation.

16.2.1 Formalization of the Optimization Problem: General Utility Function

We consider a market model with finite time horizon $T > 0$ where two investments possibilities are available:

- A risk free asset with price dynamics

$$\begin{cases} dS_0(t) = \rho(t)S_0(t)dt, & t \in (0, T], \\ S_0(0) = 1. \end{cases} \tag{16.1}$$

- A risky asset with price dynamics

$$\begin{cases} dS_1(t) = S_1(t^-)\big[\mu(t)dt + \sigma(t)dW(t) + \int_{\mathbb{R}_0}\theta(t, z)\widetilde{N}(dt, dz)\big], & t \in (0, T], \\ S_1(0) > 0. \end{cases}$$
$$\tag{16.2}$$

The driving noises in the model are the standard Brownian motion $W(t)$, $t \in [0, T]$, and the compensated Poisson random measure $\widetilde{N}(dt, dz)$, $(t, z) \in [0, T] \times \mathbb{R}_0$. These are assumed to be independent. Recall that $E\big[\widetilde{N}(dt, dz)^2\big] = \nu(dz)dt$, where $\nu(dz)$, $z \in \mathbb{R}_0$, is a σ-finite Borel measure, which we assume to satisfy

$$\int_{\mathbb{R}_0} z^2\nu(dz) < \infty.$$

The parameters $\rho(t)$, $\mu(t)$, $\sigma(t)$, and $\theta(t, z)$, $t \in [0, T]$, $z \in \mathbb{R}_0$, are measurable càglàd stochastic processes adapted to the filtration $\mathbb{F} := \{\mathcal{F}_t \subseteq \mathcal{F}, t \in [0, T]\}$ that is generated by the Brownian motion and the Poisson random measure, see Sect. 12.5. We also assume that $|\sigma(t)| \leq K_\sigma$ and $-1 + \epsilon_\theta \leq \theta(t, z) \leq K_\theta$, $\nu(dz)dt$-a.e., for some $\epsilon_\theta \in (0, 1)$ and $K_\sigma, K_\theta < \infty$, and

$$E\Big[\int_0^T \Big\{|\rho(t)| + |\mu(t)| + \sigma^2(t) + \int_{\mathbb{R}_0}\theta^2(t, z)\nu(dz)\Big\}dt\Big] < \infty.$$

As mentioned earlier, at any time t, the σ-algebra $\mathcal{E}_t \subseteq \mathcal{F}_t$ represents the information at his disposal. We consider the collection of σ-algebras $\mathbb{E} := \{\mathcal{E}_t, t \in [0, T]\}$ to be a filtration.

In view of this lack of information, the portfolio $\pi(t)$ that resulted from the trader's decisions taken at time t is an $\mathcal{E}_t$-measurable random variable. Here, the process π represents the fraction of wealth invested in the risky asset.

Taking the point of view of such a trader, we are interested in studying the optimization problem

$$u(x) := \sup_{\pi \in \mathcal{A}_{\mathbb{E}}} E\left[U(X_\pi^x(T))\right] = E\left[U(X_{\pi^*}^x(T))\right], \qquad (16.3)$$

for a given *utility function*

$$U : [0, \infty) \longrightarrow [-\infty, \infty)$$

that is, an increasing, concave, and lower semi-continuous function, which we assume to be continuously differentiable on $(0, \infty)$. The process $X_\pi(t) = X_\pi^x(t)$, $t \in [0, T]$, given by

$$dX_\pi(t) = X_\pi(t^-)\Big\{\big[\rho(t) + \big(\mu(t) - \rho(t)\big)\pi(t)\big]dt$$
$$+ \pi(t)\sigma(t)dW(t) + \int_{\mathbb{R}_0} \pi(t)\theta(t, z)\widetilde{N}(dt, dz)\Big\}, \qquad (16.4)$$

represents the *value process* of the admissible portfolio π. The initial capital $X_\pi(0) = X_\pi^x(0) = x > 0$ is fixed. By the Itô formula, the unique solution of the equation earlier is

$$X_\pi(t) = x \exp\Big\{\int_0^t \Big[\rho(s) + \big(\mu(s) - \rho(s)\big)\pi(s) - \frac{1}{2}\sigma^2(s)\pi^2(s)\Big]ds$$
$$+ \int_0^t \int_{\mathbb{R}_0} \Big[\log\big(1 + \pi(s)\theta(s, z)\big) - \pi(s)\theta(s, z)\Big]\nu(dz)ds \qquad (16.5)$$
$$+ \int_0^t \pi(s)\sigma(s)dW(s) + \int_0^t \int_{\mathbb{R}_0} \log\big(1 + \pi(s)\theta(s, z)\big)\widetilde{N}(ds, dz)\Big\},$$

with $\pi \in \mathcal{A}_{\mathbb{E}}$. The set of admissible portfolios $\mathcal{A}_{\mathbb{E}}$ is defined as follows:

Definition 16.1. *The set $\mathcal{A}_{\mathbb{E}}$ of admissible portfolios consists of all processes* $\pi = \pi(t)$, $t \in [0, T]$, *such that*

▷ π *is càglàd and adapted to the filtration $\mathbb{E}$.* $\qquad (16.6)$

▷ $\pi(t)\theta(t, z) > -1 + \epsilon_\pi$ *for a.a. (t, z) with respect to $dt \times \nu(dz)$,*
  for some $\epsilon_\pi \in (0, 1)$ depending on π. $\qquad (16.7)$

▷ $E\int_0^T \Big\{|\mu(s) - \rho(s)||\pi(s)| + \sigma^2(s)\pi^2(s) + \int_{\mathbb{R}_0} \pi^2(s)\theta^2(s, z)\nu(dz)\Big\}ds < \infty.$
$$\qquad (16.8)$$

16.2.2 Characterization of an Optimal Portfolio Under Partial Information

Our forthcoming discussion aims at giving results on the existence of locally optimal portfolios for problem (16.3), i.e. $\pi \in \mathcal{A}_{\mathbb{E}}$ is *locally optimal* if

$$E\Big[U(X_{\pi+\beta y}(T))\Big] \le E\Big[U(X_\pi(T))\Big]$$

for all bounded $\beta \in \mathcal{A}_{\mathbb{E}}$ and all $y \in (-\delta, \delta)$ for some $\delta > 0$. In the sequel of the chapter whenever we refer to optimal portfolio, we intend locally optimal in the sense given here above.

For given $\pi \in \mathcal{A}_{\mathbb{E}}$, we define the stochastic process $Y_\pi(t)$, $t \in [0, T]$, as

$$
\begin{aligned}
Y_\pi(t) := & \int_0^t \left\{ \mu(s) - \rho(s) - \sigma^2(s)\pi(s) - \int_{\mathbb{R}_0} \frac{\pi(s)\theta^2(s, z)}{1 + \pi(s)\theta(s, z)} \nu(dz) \right\} ds \\
& + \int_0^t \sigma(s)dW(s) + \int_0^t \int_{\mathbb{R}_0} \frac{\theta(s, z)}{1 + \pi(s)\theta(s, z)} \widetilde{N}(ds, dz), \quad t \in [0, T].
\end{aligned}
\tag{16.9}
$$

Then our result can be stated as follows, see [66].

Theorem 16.2. *Let* $\pi \in \mathcal{A}_{\mathbb{E}}$ *be such that*

▷ *The random variable* $F_\pi(T) := U'(X_\pi(T))X_\pi(T)$ *belongs to* $\mathbb{D}_{1,2} :=$

$$
\mathbb{D}_{1,2}^W \cap \mathbb{D}_{1,2}^{\widetilde{N}},
\tag{16.10}
$$

i.e., Malliavin differentiable both with respect to the Brownian motion

and the compensated Poisson random measure.

▷ *For all* $\beta \in \mathcal{A}_{\mathbb{E}}$, *with* β *bounded, there exists a* $\delta > 0$ *such that the family*
$$
\tag{16.11}
$$

$$
\left\{ U'(X_{\pi + y\beta}(T))X_{\pi + y\beta}(T)Y_{\pi + y\beta}(T) \right\}_{y \in (-\delta, \delta)} \quad \textit{is uniformly integrable.}
$$

(Here, $U'(x) = \dfrac{d}{dx}U(x)$*).*

Suppose π *is a local maximum point for the problem* (16.3)*. Then* π *satisfies the equation*

$$
E\left[\left(\mu(s) - \rho(s) - \sigma^2(s)\pi(s) \right) F_\pi(T) + \sigma(s)D_s\big(F_\pi(T)\big) \Big| \mathcal{E}_s \right]
$$

$$
+ E\left[\int_{\mathbb{R}_0} \frac{\theta(s, z)D_{s,z}\big(F_\pi(T)\big) - \pi(s)\theta^2(s, z)F_\pi(T)}{1 + \pi(s)\theta(s, z)} \nu(dz) \Big| \mathcal{E}_s \right] = 0, \quad s \in [0, T].
\tag{16.12}
$$

Conversely, suppose that (16.12) *holds and that*

$$
xU''(x) + U'(x) \le 0, \qquad x > 0,
\tag{16.13}
$$

then π *is a local maximum for the problem* (16.3)*.*

Remark 16.3. (1) Note that the *buy–hold–sell* portfolios β, i.e.,

$$
\beta(\omega, s) := \alpha(\omega) 1_{(t, t+h]}(s), \qquad s \in [0, T], \omega \in \Omega,
$$

belong to $\mathcal{A}_\mathbb{E}$ provided that α is $\mathcal{E}_t$-measurable and $|\alpha| \leq K_\alpha$ for some sufficiently small $K_\alpha > 0$.

(2) If $\pi, \beta \in \mathcal{A}_\mathbb{E}$ with β bounded, then there exists $\delta > 0$, such that $\pi + y\beta \in \mathcal{A}_\mathbb{E}$ for all $y \in (-\delta, \delta)$.

(3) Condition (16.13) holds if, for example, $U(x) = \log x$ or $U(x) = \frac{1}{\gamma}x^\gamma$ $(\gamma < 0)$.

Proof Let us suppose that π gives a *local maximum for the problem* (16.3), in the sense that

$$E[U(X_{\pi+y\beta}(T))] \leq E[U(X_\pi(T))]$$

for all bounded $\beta \in \mathcal{A}_\mathbb{E}$ and all $y \in (-\delta, \delta)$, where $\delta > 0$ is such that $\pi + y\beta \in \mathcal{A}_\mathbb{E}$ (see Remark 16.3 (2)) and (16.11) holds. For convenience, let us define

$$g(y) := E[U(X_{\pi+y\beta}(T))], \quad y \in (-\delta, \delta).$$

Since the function g is locally maximal at $y = 0$, we have

$$0 = \frac{d}{dy}g(y)|_{y=0}$$

$$= E\Big[U'(X_\pi(T))X_\pi(T)\Big\{ \int_0^T \beta(s)\big[\mu(s) - \rho(s) - \sigma^2(s)\pi(s)$$

$$- \int_{\mathbb{R}_0} \Big\{\theta(s,z) - \frac{\theta(s,z)}{1+\pi(s)\theta(s,z)}\Big\}\nu(dz)\big]ds$$

$$+ \int_0^T \beta(s)\sigma(s)dW(s) + \int_0^T \int_{\mathbb{R}_0} \frac{\beta(s)\theta(s,z)}{1+\pi(s)\theta(s,z)}\tilde{N}(ds,dz)\Big\}\Big].$$

Let us choose the portfolio $\beta \in \mathcal{A}_\mathbb{E}$ to be of the form buy–hold–sell, i.e.,

$$\beta(s) = \alpha 1_{(t,t+h]}(s), \quad 0 \leq s \leq T,$$

for $t \in [0, T)$ and $h > 0$ such that $t + h \leq T$ – see Remark 16.3 (1). Then from the above mentioned equations we have

$$0 = E\Big[U'(X_\pi(T))X_\pi(T) \cdot \alpha \cdot \Big\{ \int_t^{t+h} [\mu(s) - \rho(s) - \sigma^2(s)\pi(s)$$

$$- \int_{\mathbb{R}_0} \frac{\pi(s)\theta^2(s,z)}{1+\pi(s)\theta(s,z)}\nu(dz)]ds + \int_t^{t+h} \sigma(s)dW(s)$$

$$+ \int_t^{t+h} \int_{\mathbb{R}_0} \frac{\theta(s,z)}{1+\pi(s)\theta(s,z)}\tilde{N}(ds,dz)\Big\},\Big],$$

which holds for all choices of α in the buy–hold–sell portfolios. Define

$$F_\pi(T) = U'(X_\pi(T))X_\pi(T), \quad \text{for } \pi \in \mathcal{A}_\mathbb{E}.$$

Then we have

$$E\big[F_\pi(T)\big(Y_\pi(t+h) - Y_\pi(t)\big) \cdot \alpha\big] = 0 \qquad (16.14)$$

for the process $Y_\pi(t)$, $t \in [0,T]$, given in (16.9). By our assumption (16.10) and the duality formulae, see Corollay 4.4 and Theorem 12.10, we have

$$E\Big[F_\pi(T) \int_t^{t+h} \alpha\sigma(s)dW(s)\Big] = E\Big[\int_t^{t+h} \alpha\sigma(s)D_s\big(F_\pi(T)\big)ds\Big] \qquad (16.15)$$

and

$$E\Big[F_\pi(T) \int_t^{t+h} \int_{\mathbb{R}_0} \frac{\alpha\theta(s,z)}{1+\pi(s)\theta(s,z)} \widetilde{N}(ds,dz)\Big]$$

$$= E\Big[\int_t^{t+h} \int_{\mathbb{R}_0} \frac{\alpha\theta(s,z)D_{s,z}\big(F_\pi(T)\big)}{1+\pi(s)\theta(s,z)}\nu(dz)ds\Big]. \qquad (16.16)$$

Thus, substituting these two equations into (16.14) and noting that the σ-algebra generated by all the α's in the buy–hold–sell portfolios coincides with $\mathcal{E}_t$, we obtain

$$E\Big[\int_t^{t+h} \Big\{\big(\mu(s) - \rho(s) - \sigma^2(s)\pi(s)\big)F_\pi(T) - \int_{\mathbb{R}_0} \frac{\pi(s)\theta^2(s,z)F_\pi(T)}{1+\pi(s)\theta(s,z)}\nu(dz) +$$

$$\sigma(s)D_s\big(F_\pi(T)\big) + \int_{\mathbb{R}_0} \frac{\theta(s,z)D_{s,z}\big(F_\pi(T)\big)}{1+\pi(s)\theta(s,z)}\nu(dz)\Big\}ds\Big|\mathcal{E}_t\Big] = 0.$$

Since this holds for all t, h such that $0 \leq t < t+h \leq T$, we conclude that

$$E\Big[\big(\mu(s) - \rho(s) - \sigma^2(s)\pi(s)\big)F_\pi(T) + \sigma(s)D_s\big(F_\pi(T)\big)\Big|\mathcal{E}_s\Big]$$

$$+ E\Big[\int_{\mathbb{R}_0} \frac{\theta(s,z)D_{s,z}\big(F_\pi(T)\big) - \pi(s)\theta^2(s,z)F_\pi(T)}{1+\pi(s)\theta(s,z)}\nu(dz)\Big|\mathcal{E}_s\Big] = 0, \quad s \in [0,T]$$

$$(16.17)$$

– see (16.12)

On the other side, the argument can be reversed as follows. If we assume that (16.12) holds, then

$$E\big[F_\pi(T)\big(Y_\pi(t+h) - Y_\pi(t)\big)|\mathcal{E}_t\big] = 0,$$

which is equivalent to

$$E\Big[F_\pi(T) \cdot \int_0^T \beta(s)\Big[\mu(s) - \rho(s) - \sigma^2(s)\pi(s) - \int_{\mathbb{R}_0} \Big\{\frac{\pi(s)\theta^2(s,z)}{1+\pi(s)\theta(s,z)}\Big\}\nu(dz)\Big]ds$$

$$+ \int_0^T \beta(s)\sigma(s)dW(s) + \int_0^T \int_{\mathbb{R}_0} \frac{\beta(s)\theta(s,z)}{1+\pi(s)\theta(s,z)}\widetilde{N}(ds,dz)\Big] = 0$$

for all buy–hold–sell portfolios $\beta(s) = \alpha 1_{(t,t+h]}(s)$, $0 \leq s \leq T$. By (16.11) the function $g(y) = E[U(X_{\pi+y\beta}(T))]$, $y \in (-\delta, \delta)$, satisfies

$$g'(0) = \frac{d}{dy}g(y)_{|y=0} = E\Big[U'(X_\pi(T))\frac{d}{dy}X_{\pi+y\beta}(T)_{|y=0}\Big] = 0$$

for all bounded $\beta \in \mathcal{A}_{\mathbb{E}}$. We want to conclude that $y = 0$ gives a local maximum for g and hence $\pi \in \mathcal{A}_{\mathbb{E}}$ is locally optimal for the problem (16.3). To this end, fix

$$h(y):=\int_0^T \{\rho(s) + (\mu(s) - \rho(s))(\pi(s) + y\beta(s)) - \frac{1}{2}\sigma^2(s)(\pi(s) + y\beta(s))^2\}ds$$

$$-\int_0^T\int_{\mathbb{R}_0}\{(\pi(s) + y\beta(s))\theta(s,z) - \log(1 + (\pi(s) + y\beta)\theta(s,z))\}\nu(dz)ds$$

$$+\int_0^T (\pi(s) + y\beta(s))\sigma(s)dW(s) + \int_0^T\int_{\mathbb{R}_0} \log(1+(\pi(s) + y\beta)\theta(s,z))\tilde{N}(ds,dz).$$

Then

$$g'(y) = E\Big[U'(X_{\pi+y\beta}(T))\frac{d}{dy}X_{\pi+y\beta}(T)\Big]$$

and

$$g''(y) = E\Big[U''(X_{\pi+y\beta}(T))\Big(\frac{d}{dy}X_{\pi+y\beta}(T)\Big)^2 + U'(X_{\pi+y\beta}(T))\frac{d^2}{dy^2}X_{\pi+y\beta}(T)\Big],$$

where

$$\frac{d^2}{dy^2}X_{\pi+y\beta}(T) = \frac{d}{dy}\Big(X_{\pi+y\beta}(T)h'(y)\Big)$$

$$= X_{\pi+y\beta}(T)h''(y) + X_{\pi+y\beta}(T)(h'(y))^2.$$

Note that

$$h''(0) = -\int_0^T \sigma^2(s)\beta^2(s)ds - \int_0^T\int_{\mathbb{R}_0}\frac{\beta^2(s)\theta^2(s,z)}{(1 + \pi(s)\theta(s,z))^2}N(ds,dz) \le 0.$$

Therefore,

$$g''(0) = E\Big[U''(X_\pi(T))X_\pi^2(T)(h'(0))^2 + U'(X_\pi(T))X_\pi(T)\big(h''(0) + (h'(0))^2\big)\Big]$$

and we see that $g''(0) \le 0$ for all bounded $\beta \in \mathcal{A}$ if

$$xU''(x) + U'(x) \le 0, \qquad x > 0. \quad \square$$

Remark 16.4. Condition (16.10) depends on the choice of utility function. Here, we give some examples:

(1) By the chain rule (see Theorem 3.5 and Theorem 12.8), condition (16.10) holds if $X_\pi(T) \in \mathbb{D}_{1,2} := \mathbb{D}_{1,2}^W \cap \mathbb{D}_{1,2}^{\tilde{N}}$ (for the optimal π) and, if $\sigma \ne 0$,

$$\frac{d}{dx}U'(x)x, \qquad x \in (0,\infty),$$

is bounded. In particular, this is the case for

$$U(x) = \log x, \quad x > 0$$

and

$$U(x) = -\exp\{-\lambda x\}, \quad x > 0 \quad (\lambda > 0).$$

(2) Suppose

$$U(x) = \frac{1}{\gamma}x^\gamma, \quad x > 0 \quad (\gamma \in (-\infty, 1) \setminus \{0\}).$$

Then the classical chain rule for Brownian motion cannot be applied and we need a similar result under weaker conditions, see Sect. 16.2.1. Then by Theorem 13.32, we have

$$D_t F_\pi(T) = D_t(X_\pi^\gamma(T)) = \gamma X_\pi^\gamma(T) D_t(\log X_\pi(T)),$$

and, by Theorem 12.8, we have

$$D_{t,z} F_\pi(T) = \left(X_\pi(T) + D_{t,z} X_\pi(T)\right)^\gamma - X_\pi^\gamma(T)$$
$$= X_\pi^\gamma(T)\left[\exp\left\{\gamma D_{t,z} \log X_\pi(T)\right\} - 1\right],$$

provided that $\log X_\pi(T)$ is Malliavin differentiable in probability with respect to W and belongs to $D_{1,2}^{\widetilde{N}}$. For this it suffices that $\rho(s), \mu(s), \sigma(s), \theta(s,z)$, and $\pi(s)$ are Malliavin differentiable for each s and (s, z). Note that by Corollary 13.34 we then get that $F \in \mathbb{D}_{1,2}^W$ if

$$E\left[X_\pi^{2\gamma}(T) \int_0^T \left(D_t(\log X_\pi(T))\right)^2 dt\right] < \infty.$$

Remark 16.5. Using the white noise framework, condition (16.10) on $F_\pi(T)$ could be relaxed and replaced by

$$E[(F_\pi(T))^2] < \infty.$$

The result still holds in the same form.

Remark 16.6. Condition (16.11) depends on the choice of utility function and may be difficult to verify. Here, we give some examples under which it holds:

(1) First consider the case in which $U'(x)x$ is uniformly bounded for $x \in (0, \infty)$. In particular, it holds for the logarithmic utility $U(x) - \log x$ and the exponential utility $U(x) = -\exp\{-\lambda x\}$ ($\lambda > 0$). The condition (16.11) holds if $Y(y) := Y_{\pi + y\beta}(T)$, $y \in (-\delta, \delta)$, is uniformly integrable. The uniform integrability of $\{Y(y)\}_{y \in (-\delta, \delta)}$ is assured by

$$\sup_{y \in (-\delta, \delta)} E\left[|Y(y)|^p\right] < \infty \quad \text{for some } p > 1.$$

Since $\pi, \beta \in \mathcal{A}_\mathbb{E}$ (see (16.7)), we have that $1 + (\pi(s) + y\beta(s))\theta(s, z) \geq \epsilon_\pi - \delta$ $dt \times \nu(dz)$-a.e. for some δ small enough. Thus

$$E\left[\left(\int_0^T \int_{\mathbb{R}_0} \frac{\theta(s, z)}{1 + (\pi(s) + y\beta(s))\theta(s, z)} \widetilde{N}(ds, dz)\right)^2\right]$$

$$\leq \frac{1}{(\epsilon_\pi - \delta)^2} E\left[\int_0^T \int_{\mathbb{R}_0} \theta^2(s, z)\nu(dz)ds\right] < \infty.$$

Therefore, we have that $E[Y^2(y)]$ is uniformly bounded in $y \in (-\delta, \delta)$.
(2) In the case of power utility function

$$U(x) = \frac{1}{\gamma}x^\gamma, \quad x > 0 \qquad \text{for some } \gamma \in (0, 1),$$

we can see that $U'(X_{\pi+y\beta}(T))X_{\pi+y\beta}(T)|Y(y)| = X_{\pi+y\beta}^\gamma(T)|Y(y)|$ and condition (16.11) would be satisfied if

$$\sup_{y \in (-\delta, \delta)} E[(X_{\pi+y\beta}^\gamma(T)|Y(y)|)^p] < \infty \qquad \text{for some } p > 1.$$

We can write

$$X_{\pi+y\beta}(T) = X_\pi(T)Z(y),$$

where

$$Z(y) := \exp\left\{\int_0^T \left[(\mu(s) - \rho(s))y\beta(s) - \sigma^2(s)y\beta(s)\pi(s) - \frac{1}{2}\sigma^2(s)y^2\beta^2(s)\right]ds\right.$$

$$+ \int_0^T y\sigma(s)\beta(s)dW(s)$$

$$+ \int_0^T \int_{\mathbb{R}_0} \left[\log(1 + (\pi(s) + y\beta(s))\theta(s, z)) - \log(1 + \pi(s)\theta(s, z))\right.$$

$$\left. - y\beta(s)\theta(s, z)\right]\nu(dz)ds$$

$$+ \int_0^T \int_{\mathbb{R}_0} \left[\log(1 + (\pi(s) + y\beta(s))\theta(s, z))\right.$$

$$\left.\left. - \log(1 + \pi(s)\theta(s, z))\right]\widetilde{N}(ds, dz)\right\}.$$

From the repeated application of the Hölder inequality we have

$$E[(X_{\pi+y\beta}^\gamma(T)|Y(y)|)^p]$$

$$\leq \left(E[(X_\pi(T))^{\gamma p a_1 b_1}]\right)^{\frac{1}{a_1 b_1}} \left(E[(Z(y))^{\gamma p a_1 b_2}]\right)^{\frac{1}{a_1 b_2}} \left(E[(|Y(y)|)^{p a_2}]\right)^{\frac{1}{a_2}},$$

where a_1, a_2: $\frac{1}{a_1} + \frac{1}{a_2} = 1$ and b_1, b_2: $\frac{1}{b_1} + \frac{1}{b_2} = 1$. Then we can choose $a_1 = \frac{2}{2-p}$, $a_2 = \frac{2}{p}$ and also $b_1 = \frac{2-p}{\gamma p}$, $b_2 = \frac{2-p}{2-p-\gamma p}$ for some $p \in (1, \frac{2}{\gamma+1})$.
Hence

$$E\left[(X_{\pi+y\beta}^{\gamma}(T)|Y(y)|)^{p}\right]$$

$$\leq \left(E\left[(X_{\pi}(T))^{2}\right]\right)^{\frac{\gamma p}{2}} \left(E\left[(Z(y))^{\frac{2\gamma p}{2-p-\gamma p}}\right]\right)^{\frac{2-p-\gamma p}{2}} \left(E\left[(|Y(y)|)^{2}\right]\right)^{\frac{p}{2}}.$$

If the value $X_{\pi}(T)$ in (16.5) satisfies

$$E\left[(X_{\pi}(T))^{2}\right] < \infty, \qquad (16.18)$$

then the condition (16.11) holds if

$$\sup_{y\in(-\delta,\delta)} E\left[(Z(y))^{\frac{2\gamma p}{2-p-\gamma p}}\right] < \infty. \qquad (16.19)$$

Condition (16.19) holds if, e.g.,

$$E\left[\exp\left\{K\left(\int_{0}^{T} [|\mu(s) - \rho(s)| + |\pi(s)|ds\right)\right\}\right] < \infty \text{ for all } K > 0.$$

Note that condition (16.18) is verified, for example, if for all $K > 0$

$$E\left[\exp\left\{K\left(\int_{0}^{T} [|\mu(s) - \rho(s)| + |\pi(s)|]ds + |\int_{0}^{T} \pi(s)\sigma(s)dW(s)|\right.\right.$$
$$\left.\left. + |\int_{0}^{T}\int_{\mathbb{R}} \log(1 + \pi(s)\theta(s,z))\tilde{N}(ds,dz)|\right)\right\}\right] < \infty.$$

By similar arguments we can also treat the case of a utility function with $U'(x)$ uniformly bounded for $x \in (0,\infty)$. We omit the details.

16.2.3 Examples

Example 16.7. Let us consider problem (16.3) with *logarithmic utility*, i.e.,

$$U(x) = \log x, \quad x > 0.$$

In this case we have

$$F_{\pi}(T) = U'(X_{\pi}(T))X_{\pi}(T) = 1$$

and (16.12) simplifies to

$$\hat{\mu}(s) - \hat{\rho}(s) - \hat{\sigma}^2(s)\pi(s) - \pi(s)E\left[\int_{\mathbb{R}_0} \frac{\theta^2(s,z)}{1 + \pi(s)\theta(s,z)}\nu(dz)\Big|\mathcal{E}_s\right] = 0, \quad s \in [0,T],$$
$$(16.20)$$

where, for convenience in notation, we have put $\hat{\mu}(s) := E[\mu(s)|\mathcal{E}_s]$, $\hat{\rho}(s) := E[\rho(s)|\mathcal{E}_s]$, $\hat{\sigma}^2(s) := E[\sigma^2(s)|\mathcal{E}_s]$, $s \in [0,T]$. Compare this result with Corollary 16.16 and Theorem 16.20 in the forthcoming section dedicated to the study of a portfolio optimization problem under partial observation in the logarithmic utility case when in the market there is presence of insiders. In particular,

(1) If $\theta \equiv 0$ and $\sigma > 0$ in the price dynamics, then the process

$$\pi^*(s) = \frac{\hat{\mu}(s) - \hat{\rho}(s)}{\hat{\sigma}^2(s)}, \qquad s \in [0, T],$$

belongs to $\mathcal{A}_\mathbb{E}$ and is an optimal portfolio.

(2) If the price dynamics (16.2) are driven by a Brownian motion and a centered Poisson process, i.e., $\sigma > 0$, $\nu(dz) = \delta_1(dz)$, and $\theta(t, z) = z$, then (16.20) can be written as

$$\hat{\mu} - \hat{\rho} - \hat{\sigma}^2 \pi - \frac{\pi}{1 + \pi} \equiv 0.$$

Thus the processes

$$\pi^* \equiv \frac{1}{2\hat{\sigma}^2}\left\{ (\hat{\mu} - \hat{\rho}) - \hat{\sigma}^2 - 1 \pm \sqrt{\left[(\hat{\mu} - \hat{\rho}) - \hat{\sigma}^2 - 1\right]^2 + 4\hat{\sigma}^2(\hat{\mu} - \hat{\rho})} \right\}$$

are optimal for the problem (16.3) if $\pi \geq -1 + \epsilon_\pi$ for some $\epsilon_\pi > 0$ (this depends on the choices of the coefficients in the price dynamics). If, in this setting, the price dynamics of the risky asset is driven by only the centered Poisson process, i.e., $\sigma \equiv 0$, then (16.20) leads to

$$\hat{\mu} - \hat{\rho} - \frac{\pi}{1 + \pi} \equiv 0.$$

Hence

$$\pi^* \equiv \frac{\hat{\mu} - \hat{\rho}}{1 - (\hat{\mu} - \hat{\rho})} \qquad (\mu - \rho < 1 - \epsilon, \text{ for some } \epsilon \in (0, 1))$$

belongs to $\mathcal{A}_\mathbb{E}$ and is optimal for (16.3).

Example 16.8. Next let us consider the case with *power utility*, i.e.,

$$U(x) = \frac{1}{\gamma} x^\gamma, \quad x > 0,$$

where $\gamma \in (0, 1)$ is a constant. In this case we get $F_\pi(T) = X_\pi^\gamma(T)$, thus, by the chain rules (see Theorem 12.8 and Theorem 13.32),

$$D_t F_\pi(T) = \gamma X_\pi^\gamma(T) D_t \log X_\pi(T)$$

and

$$D_{t,z} F_\pi(T) = \left(X_\pi(T) + D_{t,z} X_\pi(T)\right)^\gamma - X_\pi^\gamma(T)$$
$$= X_\pi^\gamma(T)\left[\exp\left\{\gamma D_{t,z} \log X_\pi(T)\right\} - 1\right],$$

if $\log X_\pi(T) \in \mathbb{D}_{1,2}^W \cap \mathbb{D}_{1,2}^{\tilde{N}}$ – see Remark 16.4. Then (16.12) becomes

$$E\Big[X_\pi^\gamma(T)\Big(\mu(t) - \rho(t) - \sigma^2(t)\pi(t)$$

$$+ \int_{\mathbb{R}_0} \frac{\theta(t,z)}{1 + \pi(t)\theta(t,z)} \big(\exp\{\gamma D_{t,z}\log X_\pi(T)\} - \pi(t)\theta(t,z) - 1\big)\nu(dz)$$

$$+ \gamma\sigma(t)D_t\log X_\pi(T)\Big)\Big|\mathcal{E}_t\Big] = 0.$$

$$(16.21)$$

In particular, if the coefficients $\mu(t), \rho(t), \sigma(t)$, and $\theta(t,z)$, $t \in [0,T]$, $z \in \mathbb{R}_0$, are all *deterministic* and we would like to have a $\pi(t)$, $t \in [0,T]$, deterministic also, then π must satisfy the equation

$$\mu(t) - \rho(t) + (\gamma - 1)\sigma^2(t)\pi(t) + \int_{\mathbb{R}_0} \theta(t,z)\big[(1 + \pi(t)\theta(t,z))^{\gamma-1} - 1\big]\nu(dz) = 0.$$

$$(16.22)$$

Conversely, any solution π of the equation above is an optimal deterministic portfolio.

Remark 16.9. The main feature of (16.12) is that it gives an explicit relation between the optimal portfolio π and the corresponding optimal *terminal* wealth $\hat{X}_\pi(T) = X_\pi(T)$. The following examples illustrate this:

Example 16.10. In the price dynamics, let us assume that $\theta = 0$, $\sigma(t) \neq 0$, $t \in [0,T]$ and

$$E\Big[\exp\Big\{\frac{1}{2}\int_0^T \big(\frac{\mu(s) - \rho(s)}{\sigma(s)}\big)^2 ds\Big\}\Big] < \infty.$$

Moreover, we set $\mathcal{E}_t = \mathcal{F}_t$ for all t. In this context the market model is complete. It is known that the optimal terminal wealth $\hat{X}_\pi(T)$ is given by

$$\hat{X}_\pi(T) = I(\mathcal{Y}(x)H_0(T)),$$

$$(16.23)$$

where $I := (U')^{-1}$ is the inverse of $U'(u) = \frac{d}{du}U(u)$, and $\mathcal{Y}(x) = \mathcal{X}^{-1}(x)$ is the inverse of the function $\mathcal{X}$ that is defined by

$$\mathcal{X}(y) = E\big\lfloor H_0(T)I(yH_0(T))\big\rfloor,$$

$$(16.24)$$

with

$$H_0(T) = \exp\Big\{-\int_0^T \frac{\mu(s) - \rho(s)}{\sigma(s)}dW(s) - \int_0^T \big[\rho(s) + \frac{1}{2}\big(\frac{\mu(s) - \rho(s)}{\sigma(s)}\big)^2\big]ds\Big\}$$

$$(16.25)$$

– see, e.g., [130], Chap. 3. Hence

$$F_\pi(T) = U'(X_\pi(T))X_\pi(T) = \mathcal{Y}(x)H_0(T)I(\mathcal{Y}(x)H_0(T))$$

(cf. (16.10)) is known in this case and, since $H_0(T) \in \mathbb{D}_{1,2}^W$ (see assumptions on the coefficients in the price dynamics), we can solve (16.12) for π as follows:

$$\pi(s) = \frac{\mu(s) - \rho(s)}{\sigma^2(s)} + \frac{E[D_s F_\pi(T)|\mathcal{F}_s]}{\sigma(s)E[F_\pi(T)|\mathcal{F}_s]}, \qquad s \in [0, T]. \tag{16.26}$$

Thus, any solution π of (16.26) is optimal for the problem (16.3). Note that if the utility function is logarithmic, then $F_\pi(T) = 1$ and hence we find directly the classical solution to the optimization problem under full information, cf. also Example 16.7.

Example 16.11. Here we consider an extension of Example 16.10 to the general case of (16.1) and (16.2), where $\sigma(t) \neq 0$, $\theta(t, z) \neq 0$, $t \in [0, T]$, $z \in \mathbb{R}_0$, and hence the market is possibly incomplete. As in the previous case we assume $\mathcal{E}_t = \mathcal{F}_t$, for all t.

Let $u(x)$ be as in (16.3) and consider the associated *dual problem*

$$v(y) := \inf_{Q \in \mathcal{M}_a} E\left[V\left(y\frac{dQ}{dP}\right)\right], \qquad y > 0, \tag{16.27}$$

where

$$V(\lambda) := \sup_{\xi \in \mathbb{R}} \{U(\xi) - \lambda\xi\}, \qquad \lambda > 0, \tag{16.28}$$

is the *Legendre transform* of U and $\mathcal{M}_a$ is the set of measures Q absolutely continuous with respect to P.

Then – under certain conditions – the optimal terminal wealth $\hat{X}_\pi(T)$ is given by

$$\hat{X}_\pi(T) = I\left(y(x)\frac{d\hat{Q}(y(x))}{dP}\right), \tag{16.29}$$

where $I := (U')^{-1}$ (as in Example 16.10). Here $x > 0$ is related to $y = y(x) > 0$ via $u'(x) = y$ or, equivalently, $x = -v'(y)$, and the measure $\hat{Q} = \hat{Q}(y) \in \mathcal{M}_a$ is the optimal measure for the dual problem (16.27). We refer to, e.g., [136] and the survey [214] and the references therein for more details. Therefore, in terms of $\hat{Q}$ we get

$$F_\pi(T) = U'(\hat{X}_\pi(T))\hat{X}_\pi(T) = y(x)\frac{d\hat{Q}(y(x))}{dP}I\left(y(x)\frac{d\hat{Q}(y(x))}{dP}\right)$$

– cf. (16.10). With this expression for $F_\pi(T)$ in hands, we can see that if $F_\pi(T) \in \mathbb{D}_{1,2}^W \cap \mathbb{D}_{1,2}^{\widetilde{N}}$, then a portfolio π is optimal if and only if it satisfies the following equation:

$$(\mu(s) - \rho(s) - \sigma^2(s)\pi(s))E[F_\pi(T)|\mathcal{F}_s] + \sigma(s)E[D_s(F_\pi(T))|\mathcal{F}_s]$$

$$+ \int_{\mathbb{R}_0} \frac{\theta(s,z)E[D_{s,z}(F_\pi(T))|\mathcal{F}_s] - \pi(s)\theta^2(s,z)E[F_\pi(T)|\mathcal{F}_s]}{1 + \pi(s)\theta(s,z)}\nu(dz) = 0, \quad s \in [0, T]. \tag{16.30}$$

16.3 Optimal Portfolio under Partial Information in an Anticipating Environment

On the probability space $(\Omega, \mathcal{F}, P)$, (see Sect. 12.5) we consider filtrations $\mathbb{E} = \{\mathcal{E}_t, t \in [0, T]\}$, $\mathbb{F} = \{\mathcal{F}_t, t \in [0, T]\}$, and $\mathbb{G} = \{\mathcal{G}_t, t \in [0, T]\}$ such that

$$\mathcal{E}_t \subseteq \mathcal{F}_t \subseteq \mathcal{G}_t \subseteq \mathcal{F} \qquad \text{for all } t \in [0, T]. \tag{16.31}$$

This represents the structure of the flow of information considered in this section.

The market model considered has two investment possibilities:

- Price of a risk free asset with price dynamics

$$\begin{cases} dS_0(t) = \rho(t)S_0(t)dt, & t \in (0, T], \\ S_0(0) = 1 \end{cases} \tag{16.32}$$

- A risky asset with price dynamics

$$\begin{cases} dS_1(t) = S_1(t^-)\left[\mu(t)dt + \sigma(t)d^-W(t) + \int_{\mathbb{R}_0}\theta(t, z)\widetilde{N}(d^-t, dz)\right], & t \in (0, T], \\ S_1(0) > 0. \end{cases}$$
$$\tag{16.33}$$

The driving noises in the model are a standard Brownian motion $W(t)$, $t \in [0, T]$, and an independent compensated Poisson random measure $\widetilde{N}(dt, dz)$, $(t, z) \in [0, T] \times \mathbb{R}_0$. Recall once again that we assumed

$$\int_{\mathbb{R}_0} z^2 \nu(dz) < \infty.$$

In this section, we assume that the coefficients ρ, μ, σ, and θ may be influenced by other noises than the two driving the price dynamics. Thus we assume that they are in general $\mathbb{G}$-measurable. This leads to a different situation from the one considered in (16.1) and (16.2). In fact, Itô nonanticipating calculus cannot be applied any more and we have to enter the domain of anticipating integration. We choose to interpret these integrals as *forward* integrals, because this is what the integrals would be identical to if we happen to be in a semimartingale context, i.e., if the integrators would be semimartingales with respect to the filtration $\mathbb{G}$ – see Lemma 8.9 and Remark 15.2. Moreover, in the Brownian motion case the forward integrals can be regarded as limits of Riemann sums, which makes the forward integrals natural for modeling gain processes of insiders in finance. See Lemma 8.4.

Remark 16.12. Note that in the wide literature of the so-called *enlargement of filtrations*, some conditions (often difficult to verify) have to be given to ensure that the integrator is a semimartingale with respect to the enlarged filtration. This is an unavoidable point if one does not want to leave the domain of semimartingale integration. The approach presented here does not need these assumptions of semimartingality, and hence provides a consistent framework for the study of these problems with respect to a *general* larger filtration. For example, in Sect. 16.4 we prove a general insider optimal consumption result, valid without any semimartingale assumptions about the filtration/information available to the insider. This illustrates that in some situations the assumption of semimartingality may be irrelevant. On the other side, we prove that if an optimal portfolio exists, then the integrators are actually semimartingales with respect to the general enlarged filtration. See Theorem 8.26 and the forthcoming Theorem 16.34.

The literature with respect to enlargement of filtrations is wide indeed. Without being able to or aiming at being complete, we can here refer to, e.g., [204] and references therein. Related works with respect to the application to insider trading are, e.g., [6, 49, 76, 77, 91, 92, 93, 94, 117, 131, 132, 147, 192].

Remark 16.13. Market models of type (16.32) and (16.33) may appear also in a different context, such as stochastic volatility models. In fact, in general, the coefficients σ and θ need not be $\mathbb{F}$-adapted, but can possibly be influenced by other noises as well. The same can be said about μ and also ρ. See [182].

An example of a situation in which the price dynamics are of type (16.32) and (16.33) is the case when there are "large" investors in the market (i.e., investors with influential capacity) and these investors have *inside* information; this means that they have access to a larger filtration $\mathcal{G}_t \supset \mathcal{F}_t$ when making their decisions.

In this context a trader with only a partial information $\mathbb{E}$ at disposal will be able to optimize his portfolio relying only on his own knowledge, thus producing an $\mathbb{E}$-measurable portfolio π. At any time t, $\pi(t)$ represents the fraction of the total wealth $X_\pi(t) = X_\pi^x(t)$ of the agent invested in the risky asset. The value of such a portfolio is given by

$$dX_\pi(t) = X_\pi(t^-)\Big[(\rho(t) + (\mu(t) - \rho(t))\pi(t))dt + \pi(t)\sigma(t)d^-W(t)$$

$$+ \pi(t)\int_{\mathbb{R}_0} \theta(t,z)\tilde{N}(d^-t,dz)\Big]; \qquad X(0) = x > 0.$$

(16.34)

Note that this equation is substantially different from (16.4), in fact the forward integrals are coming in.

In this section, we are dealing with the optimal portfolio problem

$$u(x) := \sup_{\pi \in \mathcal{A}_\mathbb{E}} E\left[U(X_\pi^x(T))\right] = E\left[U(X_{\pi^*}^x(T))\right], \qquad (16.35)$$

for a given *utility function* $U : [0, \infty) \longrightarrow [-\infty, \infty)$. Here $\mathcal{A}_{\mathbb{E}}$ is the set of all admissible portfolios for an agent with partial information $\mathbb{E}$ at disposal.

In particular, we focus on the case of logarithmic utility function, i.e.,

$$U(x) = \log x, \quad x > 0.$$

For simplicity, we split the discussion into two cases:
(1) The continuous case: $\sigma \neq 0, \theta = 0$.
(2) The pure jump case: $\sigma = 0, \theta \neq 0$.

16.3.1 The Continuous Case: Logarithmic Utility

Here we study the market model given by

(Bond price) $\quad dS_0(t) = \rho(t)S_0(t)dt; \quad S_0(0) = 1$ $\qquad$ (16.36)

(Stock price) $\quad dS_1(t) = S_1(t)[\mu(t)dt + \sigma(t)d^- W(t)], \quad S_1(0) > 0,$
$\qquad\qquad$ (16.37)

where we assume that $\rho(t), \mu(t),$ and $\sigma(t)$ satisfy the following conditions:

▷ $\rho(t), \mu(t), \sigma(t)$ are $\mathcal{G}_t$-adapted. $\qquad\qquad\qquad$ (16.38)

▷ $E\left[\int_0^T \{|\rho(t)| + |\mu(t)| + \sigma^2(t)\}dt \right] < \infty.$ $\qquad\qquad$ (16.39)

▷ $\sigma(t)$ is Malliavin differentiable and $\quad D_{t^+}\sigma(t) = \lim_{s \to t^+} D_s\sigma(t)$ exists
$\qquad\qquad\qquad\qquad\qquad\qquad\qquad\qquad\qquad\qquad$ (16.40)

$\quad$ for a.a. $t \in [0, T]$.

▷ Equation (16.37) has a unique $\mathcal{G}_t$-adapted solution $S_1(t), t \in [0, T]$.
$\qquad\qquad\qquad\qquad\qquad\qquad\qquad\qquad\qquad\qquad$ (16.41)

As before, $\mathbb{E} = \{\mathcal{E}_t, t \in [0, T]\}$ and $\mathbb{G} = \{\mathcal{G}_t, t \in [0, T]\}$ are given filtrations such that

$$\mathcal{E}_t \subseteq \mathcal{F}_t \subseteq \mathcal{G}_t \subseteq \mathcal{F} \qquad \text{for all } t \in [0, T]. \qquad (16.42)$$

Definition 16.14. *The set $\mathcal{A}_{\mathbb{E}}$ of admissible portfolios consists of all processes $\pi = \pi(t), t \in [0, T]$, satisfying the following conditions:*

▷ π *is $\mathcal{E}_t$-adapted.* $\qquad\qquad\qquad\qquad\qquad\qquad\qquad$ (16.43)

▷ $\pi(t)\sigma(t), t \in [0, T],$ *is Skorohod integrable and* càglàd. $\qquad$ (16.44)

▷ $E\left[\int_0^T |\pi(t)D_{t^+}\sigma(t)|dt \right] < \infty.$ $\qquad\qquad\qquad\qquad$ (16.45)

▷ $E\left[\int_0^T |\mu(t) - \rho(t)| \cdot |\pi(t)|dt \right] < \infty.$ $\qquad\qquad\qquad$ (16.46)

In this framework, problem (16.35) can be formulated as follows. Find $u(x)$, $x > 0$, and $\pi^* \in \mathcal{A}_\mathbb{E}$ such that

$$u(x) = \sup_{\pi \in \mathcal{A}_\mathbb{E}} E[\log(X^x_\pi(T))] = E[\log(X^x_{\pi^*}(T))], \qquad (16.47)$$

where $X_\pi(t) = X^x_\pi(t)$, $t \in [0, T]$, is given by

$$dX_\pi(t) = X_\pi(t)[(\rho(t) + (\mu(t) - \rho(t))\pi(t))dt + \pi(t)\sigma(t)d^-W(t)] \qquad (16.48)$$

and $X_\pi(0) = X^x_\pi(0) = x > 0$. We now proceed to solve our problem (16.47). Applying Theorem 8.12 to the forward differential equation (16.48), we get the (unique) solution

$$X_\pi(T) = x \exp\left\{ \int_0^T (\rho(t) + (\mu(t) - \rho(t))\pi(t) \right.$$

$$\left. - \tfrac{1}{2}\pi^2(t)\sigma^2(t))dt + \int_0^T \pi(t)\sigma(t)d^-W(t) \right\}. \qquad (16.49)$$

Hence, using Corollary 8.19,

$$E[\log X_\pi(T)] - \log x =$$

$$= E\left[\int_0^T (\rho(t) + (\mu(t) - \rho(t))\pi(t) - \tfrac{1}{2}\pi^2(t)\sigma^2(t))dt + \int_0^T \pi(t)\sigma(t)d^-W(t) \right]$$

$$= E\left[\int_0^T \{\rho(t) + (\mu(t) - \rho(t))\pi(t) - \tfrac{1}{2}\pi^2(t)\sigma^2(t) + D_{t^+}(\pi(t)\sigma(t))\}dt \right].$$

$$(16.50)$$

Since $\pi(t)$ is $\mathcal{E}_t$-measurable and $\mathcal{E}_t \subseteq \mathcal{F}_t$, we have

$$D_s\pi(t) = 0 \qquad \text{for all } s > t.$$

Therefore, by the chain rule for the Malliavin derivative

$$D_{t^+}(\pi(t)\sigma(t)) = \sigma(t)D_{t^+}\pi(t) + \pi(t)D_{t^+}\sigma(t) = \pi(t)D_{t^+}\sigma(t),$$

when substituted into (16.50) gives

$$E[\log X_\pi(T)] - \log x = E\left[\int_0^T \{\rho(s) + \beta(s)\pi(s) - \tfrac{1}{2}\sigma^2(s)\pi^2(s)\}ds \right], \quad (16.51)$$

where

$$\beta(s) := \mu(s) - \rho(s) + D_{s+}\sigma(s). \tag{16.52}$$

Equation (16.51) can also be written as

$$E\big[\log X_\pi(T)\big] - \log x = E\Big[\int_0^T \{\hat\rho(s) + \hat\beta(s)\pi(s) - \tfrac{1}{2}\hat\sigma^2(s)\pi^2(s)\}ds\Big], \tag{16.53}$$

where

$$\hat\rho(s) = E[\rho(s)|\mathcal{E}_s],$$

and similarly we obtain for $\hat\sigma$, $\hat\beta$, $\hat{\sigma^2}$. We can now maximize pointwise for each s with respect to π under the integral sign. We obtain

$$\pi^*(s)\hat{\sigma^2}(s) = \hat\beta(s).$$

Summarizing the above we get the following result. See [182].

Theorem 16.15. *(a) Suppose that $\sigma(t) \neq 0$ for a.a. (ω, t) and*

$$E\Big[\int_0^T \frac{\hat\beta^2(s)}{\hat\sigma^2(s)}ds\Big] < \infty, \tag{16.54}$$

where $\beta(s)$ is defined in (16.52). Then the value function u of problem (16.47) is

$$u(x) = \log x + E\Big[\int_0^T \Big\{\rho(s) + \frac{\beta(s)\hat\beta(s)}{\hat\sigma^2(s)} - \frac{\sigma^2(s)}{2}\Big(\frac{\hat\beta(s)}{\hat\sigma^2(s)}\Big)^2\Big\}ds\Big].$$

It is also equal to

$$u(x) = \log x + E\Big[\int_0^T \Big\{\rho(s) + \frac{\hat\beta(s)^2}{2\hat\sigma^2(s)}\Big\}ds\Big] < \infty. \tag{16.55}$$

(b) Suppose that $\sigma(t) \neq 0$ for a.a. (ω, t) and that

$$\hat\pi(s) := \frac{\hat\beta(s)}{\hat\sigma^2(s)} \in \mathcal{A}_\mathbb{E}. \tag{16.56}$$

Then $\pi^*(s) := \hat\pi(s)$ is an optimal control for problem (16.47).
(c) Suppose there exists an optimal portfolio $\pi^* \in \mathcal{A}_\mathbb{E}$ for problem (16.47). Then

$$\pi^*(s)\hat{\sigma^2}(s) = \hat\beta(s). \tag{16.57}$$

Corollary 16.16. *(a) Suppose*

$$\sigma(s) \quad \text{is } \mathcal{F}_s\text{-measurable for all } s \in [0, T].$$

Then

$$D_{s+}\sigma(s) = 0 \quad \text{for all } s \in [0, T]$$

and hence

$$\beta(s) = \mu(s) - \rho(s). \tag{16.58}$$

This gives, under the conditions of Theorem 16.15,

$$\pi^*(s) = \frac{\hat{\mu}(s) - \hat{\rho}(s)}{\hat{\sigma}^2(s)} \tag{16.59}$$

(b) In particular, if we assume that

$$\mathcal{E}_t = \mathcal{F}_t = \mathcal{G}_t \quad \text{for all } t \in [0, T]$$

then we get the well-known result

$$\pi^*(s) = \frac{\mu(s) - \rho(s)}{\sigma^2(s)} \tag{16.60}$$

and

$$u(x) = \log x + E\left[\int_0^T \left\{ \rho(s) + \tfrac{1}{2}\left(\frac{\mu(s) - \rho(s)}{\sigma(s)} \right)^2 \right\} ds \right], \tag{16.61}$$

provided that

$$E\left[\int_0^T \left(\frac{\mu(s) - \rho(s)}{\sigma(s)} \right)^2 ds \right] < \infty.$$

Remark 16.17. This result earlier should be compared with the ones in Example 16.7.

Example 16.18. **Delayed noise effect.** Suppose we have a market where the stock price dynamics (16.37) is given by

$$d^- S_1(t) = S_i(t)\left[\mu(t)dt + \sigma(t)d^- W(t - \delta) \right]. \tag{16.62}$$

We assume that $\mu(t)$ and $\sigma(t)$ are $\mathbb{F}$-adapted. However, in this model we allow for a *delay* $\delta \geq 0$ in the effect on $S_1(\cdot)$ of the noise coming from $W(\cdot)$.
    Integrating (16.62) we get

$$S_1(t) = S_1(0) + \int\limits_0^t S_1(s)\mu(s)ds + \int\limits_0^t S_1(s)\sigma(s)d^-W(s-\delta)$$

$$= S_1(0) + \int\limits_{-\delta}^{t-\delta} S_1(r+\delta)\mu(r+\delta)dr + \int\limits_{-\delta}^{t-\delta} S_1(r+\delta)\sigma(r+\delta)d^-W(r).$$

$$(16.63)$$

Define
$$\tilde{S}_1(t) = S_1(t+\delta); \qquad -\delta \le t. \tag{16.64}$$

Then (16.63) can be written as

$$\tilde{S}_1(t) = S_1(0) + \int\limits_{-\delta}^t \tilde{S}_1(r)\mu(r+\delta)dr + \int\limits_{-\delta}^t \tilde{S}_1(r)\sigma(r+\delta)d^-W(r)$$

$$= \tilde{S}_1(0) + \int\limits_0^t \tilde{S}_1(r)\mu(r+\delta)dr + \int\limits_0^t \tilde{S}_1(r)\sigma(r+\delta)d^-W(r).$$

Or, equivalently,

$$d\tilde{S}_1(t) = \tilde{S}_1(t)\Big[\tilde{\mu}(t)dt + \tilde{\sigma}(t)d^-W(t)\Big]; \qquad \tilde{S}_1(0) = S_1(\delta), \tag{16.65}$$

where $\tilde{\mu}(t) = \mu(t+\delta)$, $\tilde{\sigma}(t) = \sigma(t+\delta)$.

Note that this is a price equation of the same type as in (16.33), where the coefficients $\tilde{\mu}(t), \tilde{\sigma}(t)$ are adapted to the filtration

$$\mathcal{G}_t := \mathcal{F}_{t+\delta}.$$

Suppose $\mathcal{E}_t = \mathcal{F}_t$. Let $\mu(s)$ and $\rho(s)$ be bounded $\mathcal{F}_{s+\delta}$-measurable and choose

$$\sigma(s) = \exp(W(s+\delta)); \qquad s \in [0,T].$$

Then $D_{s+}\sigma(s) = \sigma(s)$ and hence the corresponding optimal portfolio is, by Theorem 16.15,

$$\pi_\delta^*(s) = \frac{E[\mu(s) - \rho(s) + \sigma(s)|\mathcal{F}_s]}{E[\sigma^2(s)|\mathcal{F}_s]} \qquad \text{for } \delta > 0. \tag{16.66}$$

On the other hand, if $\mathcal{E}_t = \mathcal{F}_t = \mathcal{G}_t$ (corresponding to $\delta = 0$) then $D_{s+}\sigma(s) = 0$ and we know by Corollary 16.16 that the optimal portfolio is

$$\pi_0^*(s) = \frac{\mu(s) - \rho(s)}{\sigma^2(s)}. \tag{16.67}$$

Comparing (16.66) and (16.67) we see that, perhaps surprisingly,

$$\lim_{\delta \to 0^+} \pi_\delta^*(s) \neq \pi_0^*(s). \tag{16.68}$$

Similarly, if the corresponding value functions are denoted by $u_\delta(s)$ and $u_0(x)$, respectively, we get

$$\lim_{\delta \to 0^+} u_\delta(x) = \log x + E\Big[\int_0^T \Big\{\rho(s) + \tfrac{1}{2}\Big(\frac{\mu(s) - \rho(s)}{\sigma(s)} + 1\Big)^2\Big\} ds\Big] \neq u_0(x). \tag{16.69}$$

We conclude that any positive delay δ in the information, no matter how small, has a substantial effect on the optimal control and the value function.

16.3.2 The Pure Jump Case: Logarithmic Utility

We now consider the market given by

(Bond price) $dS_0(t) = \rho(t)S_0(t)dt, \qquad S_0(0) = 1. \tag{16.70}$

(Stock price) $dS_1(t) = S_1(t^-)[\mu(t)dt + \int_{\mathbb{R}_0} \theta(t, z)\tilde{N}(d^-t, dz)], \qquad S_1(0) > 0,$
$$\tag{16.71}$$

where we assume that $\rho(t), \mu(t)$, and $\theta(t, z)$ satisfy the following conditions:

▷ $\rho(t), \mu(t)$, and $\theta(t, z)$ are $\mathbb{G}$-measurable, for all $t \in [0, T], z \in \mathbb{R}$. $\tag{16.72}$
▷ $\theta(t, z)$ is bounded and Malliavin differentiable and $D_{t^+,z}\theta(t, z) :=$
$$\lim_{s \to t^+} D_{s,z}\sigma(t, z) \tag{16.73}$$

exists for a.a. t, z and is bounded.

▷ $E[\int_0^T \{|\rho(s)| + |\mu(s)| + \int_{\mathbb{R}_0} (|\theta(s, z)| + |D_{s^+,z}\theta(s, z)|)\nu(dz)\} ds] < \infty. \tag{16.74}$

▷ The equation (16.71) has a unique $\mathbb{G}$-adapted solution $S_1(t), t \in [0, T]. \tag{16.75}$

As before, $\{\mathbb{E} = \mathcal{E}_t, t \in [0, T]\}$ and $\{\mathbb{G} = \mathcal{G}_t, t \in [0, T]\}$ are given filtrations such that
$$\mathcal{E}_t \subseteq \mathcal{F}_t \subseteq \mathcal{G}_t \subseteq \mathcal{F} \qquad \text{for all } t \in [0, T].$$

Definition 16.19. *The set $\mathcal{A}_\mathbb{E}$ of admissible portfolios consists of all processes $\pi = \pi(t), t \in [0, T]$, satisfying the following conditions:*

▷ π *is $\mathbb{E}$-adapted.* $\tag{16.76}$
▷ $\pi(t)\theta(t, z), t \in [0, T], z \in \mathbb{R}_0$, *is Skorohod integrable with respect to $\tilde{N}$.*
$$\tag{16.77}$$

▷ $\pi(t)\theta(t, z) > -1 + \epsilon$ for a.a. t, z (where $\epsilon > 0$ may depend on π), and

(16.78)

$$E[\int_0^T \int_{\mathbb{R}_0} |\log(1 + \pi(s)\theta(s, z))|\nu(dz)dt] < \infty$$

▷ $\pi(t)(\theta(t, z) + D_{t+,z}\theta(t, z)) > -1 + \epsilon$ for a.a. t, z (where $\epsilon > 0$ may depend

(16.79)

on π), and

$$E\left[\int_0^T \int_{\mathbb{R}_0} |\log(1 + \pi(t)(\theta(t, z) + D_{t+,z}\theta(t, z)))|\nu(dz)dt\right] < \infty .$$

In this framework, problem (16.35) can be formulated as follows. Find $u(x)$, $x > 0$, and $\pi^* \in \mathcal{A}_{\mathbb{E}}$ such that

$$u(x) = \sup_{\pi \in \mathcal{A}_{\mathbb{E}}} E[\log X_\pi^x(T)] = E[\log(X_{\pi^*}^x(T))], \qquad (16.80)$$

where $X_\pi(t) = X_\pi^x(t)$, $t \in [0, T]$, is given by

$$dX_\pi(t) = X_\pi((t^-))\left[(\rho(t) + (\mu(t) - \rho(t))\pi(t))dt + \pi(t)\int_{\mathbb{R}_0} \theta(t, z)\tilde{N}(d^-t, dz)\right]$$

(16.81)

and $X_\pi(0) = X_\pi^x(0) = x > 0$.

We now proceed with the study of the solution of the above problem. First note that if we apply the Itô formula for forward integrals (Theorem 15.8), we get that the solution of (16.81) is given by

$$X(t) = x \exp\Big[\int_0^t \{\rho(s) + (\mu(s) - \rho(s))\pi(s)$$

$$+ \int_{\mathbb{R}_0} [\log(1 + \pi(s)\theta(s, z)) - \pi(s)\theta(s, z)]\nu(dz)\}ds$$

$$+ \int_0^t \int_{\mathbb{R}_0} \log(1 + \pi(s)\theta(s, z))\tilde{N}(d^-s, dz)\Big]. \qquad (16.82)$$

Hence, using Corollary 15.6 we get

$$E\left[\log\frac{X(T)}{x}\right] = E\left[\int_0^T \{\rho(s) + (\mu(s) - \rho(s))\pi(s)\right. \qquad (16.83)$$

$$+ \int_{\mathbb{R}_0} [\log(1 + \pi(s)\theta(s, z)) - \pi(s)\theta(s, z)]\nu(dz)\Big\}ds$$

$$+ \int_0^T \int_{\mathbb{R}_0} \log(1 + \pi(s)\theta(s, z))\tilde{N}(d^- s, dz)\Big]$$

$$= E\Big[\int_0^T \Big\{\rho(s) + (\mu(s) - \rho(s))\pi(s)$$

$$+ \int_{\mathbb{R}_0} [\log(1 + \pi(s)\theta(s, z)) - \pi(s)\theta(s, z)$$

$$+ D_{s+,z}\log(1 + \pi(s)\theta(s, z))]\nu(dz)\Big\}ds\Big] =: F_\pi. \qquad (16.84)$$

By the chain rule (Theorem 12.8) we get

$$D_{s+,z}\log(1 + \pi(s)\theta(s, z))$$

$$= \log(1 + \pi(s)\theta(s, z) + D_{s+,z}(\pi(s)\theta(s, z))) - \log(1 + \pi(s)\theta(s, z))$$

$$= \log(1 + \pi(s)(\theta(s, z) + D_{s+,z}\theta(s, z)) - \log(1 + \pi(s)\theta(s, z))$$

$$= \log\Big(1 + \frac{\pi(s)D_{s+,z}\theta(s, z)}{1 + \pi(s)\theta(s, z)}\Big).$$

When substituted into (16.84) this gives

$$F_\pi := E\Big[\int_0^T \Big\{\rho(s) + (\mu(s) - \rho(s))\pi(s)$$

$$+ \int_{\mathbb{R}_0} [\log(1 + \pi(s)(\theta(s, z) + D_{s+,z}\theta(s, z))) - \pi(s)\theta(s, z)]\nu(ds)\Big\}ds\Big].$$

$$(16.85)$$

We want to maximize the function

$$\pi \to F_\pi, \qquad \pi \in \mathcal{A}_\mathbb{E}.$$

Suppose that an optimal $\pi^* \in \mathcal{A}_\mathbb{E}$ exists. Then for all bounded $\eta \in \mathcal{A}_\mathbb{E}$, there exists $\delta > 0$ such that $\pi^* + r\eta \in \mathcal{A}_\mathbb{E}$ for $r \in (-\delta, \delta)$ and the function

$$f(r) := F_{\pi^* + r\eta}, \qquad r \in (-\delta, \delta),$$

is maximal for $r = 0$. Therefore,

$$0 = f'(0) = E\Bigg[\int\limits_0^T \Big\{(\mu(s) - \rho(s))\eta(s)$$

$$+ \int\limits_{\mathbb{R}_0}[(1 + \pi^*(s)\tilde{\theta}(s,z))^{-1}\tilde{\theta}(s,z)\eta(s) - \theta(s,z)\eta(s)]\nu(dz)\Big\}ds\Bigg],$$

where we have put

$$\tilde{\theta}(s,z) = \theta(s,z) + D_{s^+,z}\theta(s,z).$$

Hence

$$\int\limits_0^T E\Bigg[\Big\{\mu(s) - \rho(s) + \int\limits_{\mathbb{R}_0}[(1 + \pi^*(s)\tilde{\theta}(s,z))^{-1}\tilde{\theta}(s,z) - \theta(s,z)]\nu(dz)\Big\}\eta(s)\Bigg]ds = 0.$$

Since for each s the random variables $\eta(s)$, $\eta \in \mathcal{A}_\mathbb{E}$, generate the whole σ-algebra $\mathcal{E}_s$, we conclude that, for all $s \in [0,T]$,

$$E\Bigg[\Big\{\mu(s) - \rho(s) + \int\limits_{\mathbb{R}_0}[(1 + \pi^*(s)\tilde{\theta}(s,z))^{-1}\tilde{\theta}(s,z) - \theta(s,z)]\nu(dz)\Big\}\Big|\mathcal{E}_s\Bigg] = 0.$$

This proves part (a) of the following result – see [182].

Theorem 16.20. *(a) Suppose there exists an optimal portfolio $\pi^* \in \mathcal{A}_\mathbb{E}$ for problem (16.80). Then $y = \pi^*(s)$ satisfies the equation*

$$E\Bigg[\int\limits_{\mathbb{R}_0}\frac{\theta(s,z) + D_{s^+,z}\theta(s,z)}{1 + y(\theta(s,z) + D_{s^+,z}\theta(s,z))}\nu(dz)\Big|\mathcal{E}_s\Bigg]$$

$$= E\Bigg[\Big\{\rho(s) - \mu(s) + \int\limits_{\mathbb{R}_0}\theta(s,z)\nu(dz)\Big\}\Big|\mathcal{E}_s\Bigg], \qquad s \in [0,T]. \qquad (16.86)$$

(b) Suppose

$$\theta(s,z) + D_{s^+,z}\theta(s,z) \geq 0 \qquad \text{for a.a. } s,z \qquad (16.87)$$

and that for all s there exists a solution

$$y =: \hat{\pi}(s)$$

of (16.86). Suppose

$$\hat{\pi}(s) \in \mathcal{A}_\mathbb{E}.$$

Then $\hat{\pi}$ is an optimal portfolio for problem (16.80).

Proof Part (a) of the theorem is already discussed. As for Part (b) it is enough to observe that if (16.87) holds, then the function F_π given by (16.85) is concave. □

Remark 16.21. This earlier result should be compared with those in Example 16.7.

Example 16.22. **The Poisson process.** Suppose $\eta(t) = \int_0^t \int_{\mathbb{R}_0} z \tilde{N}(ds, dz)$, $t \in [0, T]$, is a compensated Poisson process. Then the Lévy measure $\nu(dz)$ is the point mass at $z = 1$ and (16.86) gets the form

$$E\left[\frac{\tilde{\theta}(s)}{1 + y\tilde{\theta}(s)}\Big|\mathcal{E}_s\right] = E[\rho(s) - \mu(s) + \theta(s, 1)|\mathcal{E}_s], \qquad (16.88)$$

where

$$\tilde{\theta}(s) = \theta(s, 1) + D_{s^+, 1}\theta(s, 1).$$

Assume in addition that

$$\tilde{\theta}(s) \quad \text{is } \mathcal{E}_s\text{-measurable}.$$

Then (16.88) has the solution

$$y = \hat{\pi}(s) = \pi^*(s) = E[\rho(s) - \mu(s) + \theta(s, 1)|\mathcal{E}_s]^{-1} - (\tilde{\theta}(s))^{-1}, \qquad (16.89)$$

provided that

$$E[\rho(s) - \mu(s) + \theta(s, 1)|\mathcal{E}_s] \neq 0 \quad \text{and} \quad \tilde{\theta}(s) \neq 0; \quad s \in [0, T].$$

Corollary 16.23. Full information case. *Suppose*

$$\mathcal{E}_t = \mathcal{F}_t = \mathcal{G}_t \qquad \text{for all } t \in [0, T]$$

and that there exists an optimal portfolio $\pi^* \in \mathcal{A}_{\mathbb{E}}$ for problem (16.80). Then $y = \pi^*(s)$ solves the equation

$$\int_{\mathbb{R}_0} \frac{\theta(s, z)}{1 + y\theta(s, z)}\nu(dz) = \rho(s) - \mu(s) + \int_{\mathbb{R}_0} \theta(s, z)\nu(dz). \qquad (16.90)$$

In the special case of Markovian coefficients, this result could have been obtained by dynamic programming.

16.4 A Universal Optimal Consumption Rate for an Insider

Suppose the cash flow at time t is modeled by a geometric Brownian motion given by

$$dX(t) = X(t)[\mu dt + \sigma\, dW(t)]; \qquad X(0) = x > 0, \quad t \geq 0. \qquad (16.91)$$

Here μ, σ, and x are known constants. Suppose that at any time t we are free to take out consumption (or dividends) at a rate $c(t) = c(\omega, t) \geq 0$. The corresponding cash flow equation is then

$$dX^{(c)}(t) = X^{(c)}[\mu dt + \sigma dW(t)] - c(t)dt; \qquad X^{(c)}(0) = x > 0. \qquad (16.92)$$

In the classical setup it is assumed that $c(t)$ is adapted with respect to the filtration $\mathcal{F}_t$ generated by $W(s)$; $s \leq t$. This ensures that (16.92) still makes sense as an Itô stochastic differential equation. Moreover, this is a natural assumption from a modeling point of view. The decision about the consumption $c(t)$ rate at time t should only depend upon the information obtained from observing the market up to time t and not upon any future event.

Now assume that the consumer has a logarithmic utility of his consumption rate. The expected total discounted utility of a chosen consumption rate $c(\cdot)$ is then

$$J(c) = E\left[\int_0^\infty e^{-\delta t} \log c(t) dt \right], \qquad (16.93)$$

where $\delta > 0$ (constant) is a given discounting exponent. Let the set $\mathcal{A}_{\mathbb{F}}$ of admissible controls (consumption rates) be the set of non-negative $\mathbb{F}$-adapted processes $c = c(t)$, $t \geq 0$, such that $X^{(c)}(t) \geq 0$ for all t. Consider the following problem:

- Find $c^* \in \mathcal{A}_{\mathbb{F}}$ such that

$$J(c^*) = \sup_{c \in \mathcal{A}_{\mathbb{F}}} J(c). \qquad (16.94)$$

Such a control c^* (if it exists) is called an *optimal* control for problem (16.94).

It is well-known, and easy to prove by using stochastic control theory, that the optimal control c^* for (16.94) is given by (in feedback form)

$$c^*(t) = \delta X^{(c^*)}(t). \qquad (16.95)$$

In other words, it is optimal to consume at a rate proportional to current cash amount, with a constant of proportionality equal to the discounting exponent.

This is a remarkably simple result. Note in particular that c^* does not depend on the parameters μ and σ. It is natural to ask if this result remains valid in a more general setting. More precisely, we ask what happens if we allow the following generalizations:

(i) We add a jump term (represented by a Poissonian random measure) in (16.92).

(ii) We replace the constant coefficients by measurable stochastic processes: $\delta = \delta(\omega, t)$, $\mu = \mu(\omega, t)$ etc.

(iii) We do *not* assume that these coefficient processes are $\mathbb{F}$-adapted, but we allow them to be arbitrary measurable processes.

(iv) We introduce a stochastic terminal time (or *default* time) τ, with values in $[0, \infty]$. This random time is not necessarily a stopping time with respect to $\mathbb{F}$, but just assumed to be measurable.

(v) We assume that the consumption rate $c(t)$ is adapted to some filtration $\mathbb{H} = \{\mathcal{H}_t, t \geq 0\}$, without any prior assumption about the relation between $\mathcal{F}_t$ and $\mathcal{H}_t$.

Two special cases in (v) are the following:

(v1) $$\mathcal{H}_t \subset \mathcal{F}_t \quad \text{for all } t \geq 0.$$

In this case the consumer has *less* information than what is represented by the filtration of the underlying driving process(es). This is often reduced to a *partial observation* control problem. We refer to [22] for more information about this topic.

(v2) $$\mathcal{F}_t \subset \mathcal{H}_t \quad \text{for all } t \geq 0.$$

In this case the consumer has more information than what can be obtained by observing the driving processes. This is the case of *inside information* and in our terminology the consumer is an *insider*. For example, in the original model (16.92) the extra information available could be the future value $W(T)$ of the underlying Brownian motion at some time $T > 0$.

In the cases (i), (iii), (iv) and (v2) it is clear that the corresponding equation for the cash flow $X^{(c)}(t)$ is no longer an Itô stochastic differential equation, because of the anticipating coefficients, the default time, and the control. We choose to model the integral

$$\int_0^t \varphi(s) dW(s) \tag{16.96}$$

when $\varphi(s) = X^{(c)}(s)\sigma(s)$, $s \in [0, T]$, is not $\mathbb{F}$-adapted, by using *forward integrals*, which are defined and motivated in Sect. 8.2.

Similarly, in order to deal with (i) we use *forward integrals with respect to Poisson random measures* when modeling strategies in such markets. These integrals and their properties were discussed in Chap. 15. We now apply this forward stochastic calculus machinery to study the extension of problem (16.94) given in (1)–(5) above. This presentation is based on [180].

16.4.1 Formalization of a General Optimal Consumption Problem

We now consider a cash flow that after being subject to a consumption/dividend rate $c(t) \geq 0$, is modeled by the equation

$$dX^{(c)}(t) = X^{(c)}(t^-)\Big[\mu(t)dt + \sigma(t)d^-W(t) + \int_{\mathbb{R}_0} \theta(t,z)\tilde{N}(d^-t, dz)\Big]$$

$$- c(t)dt, \quad t \geq 0 \tag{16.97}$$

$$X^{(c)} = x > 0.$$

Let $\mathbb{H} = \{\mathcal{H}_t, t \geq 0\}$ be another filtration, with no a priori relation to $\mathbb{F} = \{\mathcal{F}_t, t \geq 0\}$. The σ-algebra $\mathcal{H}_t$ represents the information available to the agent at time t. We make no a priori adaptedness conditions on the given processes $\mu(t), \sigma(t)$, and $\theta(t, z)$, $t \in [0, T]$, $z \in \mathbb{R}_0$, except that they are measurable. This allows us to model the situation where the cash flow may be influenced by the actions of other traders who are insiders.

The corresponding stochastic differential equation (16.97) is interpreted as the *forward* stochastic integral equation

$$X^{(c)}(t) = x + \int_0^t X^{(c)}(s)\mu(s)ds + \int_0^t X^{(c)}(s)\sigma(s)d^-W(s)$$

$$+ \int_0^t\int_{\mathbb{R}_0} X^{(c)}(s^-)\theta(s, z)\tilde{N}(d^-s, dz) - \int_0^t c(s)ds; \quad t \geq 0, \tag{16.98}$$

where $d^-W(s)$ and $\tilde{N}(d^-s, dz)$ indicate that *forward integral* interpretation is being used. See Sect. 12.5 for the probability space description and Chaps. 8 and 15 for definition and properties of the forward integrals.

Let $\delta(t) \geq 0$ be a given measurable process, modeling a discounting exponent, and let $\tau : \Omega \to [0, \infty]$ be a given measurable random variable, representing a *terminal* or *default time* for the cash flow. We consider the problem to maximize the sum of the expected total discounted logarithmic utility of the consumption rate $c(t)$ up to the default time τ and the logarithmic utility of the terminal cash amount $X^{(c)}(\tau)$, given by

$$J(c) = E\Big[\int_0^\tau e^{-\delta(t)} \log c(t)dt + \gamma e^{-\delta(\tau)} \log X^{(c)}(\tau)\Big], \tag{16.99}$$

subject to the condition that $X^{(c)}(t) > 0$ for all $t < \tau$. Here $\gamma \geq 0$ is a constant. We assume that the choice of $c(\cdot)$ has no influence on $\delta, \mu, \sigma, \theta$, or τ.

In Sect. 16.4.2 we show that, under some conditions, the optimal consumption/dividend rate $c^*(t)$ for (16.99) has the simple feedback form

$$c^*(t) = \lambda^*(t)X^{(c^*)}(t), \tag{16.100}$$

where

$$\lambda^*(t) = \frac{E[\mathcal{X}_{[0,\tau]}(t)e^{-\delta(t)}|\mathcal{H}_t]}{E[\mathcal{X}_{[0,\tau]}(t)(\int_t^T e^{-\delta(s)}ds + \gamma e^{-\delta(\tau)})|\mathcal{H}_t]}. \tag{16.101}$$

See Theorem 16.25. Note that $\lambda^*(t)$ does not depend on any of the coefficients $\mu(\cdot), \sigma(\cdot)$, and $\theta(\cdot)$.

The problem discussed in this section is related to the optimal consumption and portfolio problems associated with a random time horizon studied in [40], [41]. However, our approach is different.

16.4.2 Characterization of an Optimal Consumption Rate

We assume the following about the market:

$\triangleright$ $\mu(t), \sigma(t), \theta(t, z), \delta(t)$ and $\tau : \Omega \to [0, \infty]$ (16.102)

   are measurable for all t, z.

$\triangleright$ $\theta(t, z) > -1$ for a.a. t, z with respect to $dt \times \nu(dz)$.   (16.103)

$\triangleright$ $\sigma(s)$ and $\theta(s, z)$ are càglàd (with respect to s)   (16.104)

   and the forward integrals

$$\int_0^\tau \sigma(s) d^- W(s) \quad \text{and} \quad \int_0^\tau \int_{\mathbb{R}_0} \log(1 + \theta(s, z)) \tilde{N}(d^- s, dz)$$

   exist and belong to $L^1(P)$.

$\triangleright$ $E\left[\int_0^\tau e^{-\delta(t)} k(t) dt + e^{-\delta(\tau)} k(\tau) \right] < \infty$,   where   (16.105)

$$k(t) = \int_0^t \left\{ |\mu(s)| + \sigma^2(s) + \int_{\mathbb{R}_0} |\log(1 + \theta(s, z)) - \theta(s, z)| \nu(dz) \right\} ds$$

$$+ \left| \int_0^t \sigma(s) d^- W(s) \right| + \left| \int_0^t \int_{\mathbb{R}_0} \log(1 + \theta(s, z)) \tilde{N}(d^- s, dz) \right|.$$

We now represent the consumption rate c by its *fraction* λ of the total wealth, i.e., we put

$$\lambda(t) = \frac{c(t)}{X^{(c)}(t)}, \quad t \geq 0. \qquad (16.106)$$

We call $\lambda = \lambda(t), t \geq 0$, the *relative* consumption rate. If $X^{(c)}(\tau) = 0$, we put $\lambda(\tau) = 0$.

If $\int_0^t \lambda(s) ds < \infty$ a.s., for all $t < \tau$, then the solution $X(t) = X^{(\lambda)}(t)$ of the corresponding wealth equation is, by the Itô formula for forward integrals, given by

$$X^{(\lambda)}(t) = x \exp \left[\int_0^t \{\mu(s) - \lambda(s) - \tfrac{1}{2}\sigma^2(s) \right.$$

$$+ \int_{\mathbb{R}_0} (\log(1 + \theta(s,z)) - \theta(s,z))\nu(dz)\}ds$$

$$\left. + \int_0^t \sigma(s)d^-W(s) + \int_0^t \int_{\mathbb{R}_0} \log(1 + \theta(s,z))\tilde{N}(d^-s,dz) \right]. \quad (16.107)$$

Definition 16.24. *The set* $\mathcal{A}_{\mathbb{H}}$ *of admissible controls or admissible relative consumption rate is the set of non-negative* $\mathbb{H}$*-adapted processes* $\lambda = \lambda(t)$, $t \geq 0$, *such that*

$$\int_0^\tau \lambda(s)ds < \infty \quad P - a.s. \quad (16.108)$$

and

$$E\left[\int_0^\tau e^{-\delta(t)} |\log \lambda(t)|dt + e^{-\delta(\tau)} |\log X^{(\lambda)}(\tau)| \right] < \infty. \quad (16.109)$$

To each $\lambda \in \mathcal{A}_{\mathbb{H}}$, we associate the consumption/dividend rate

$$c_\lambda(t) := \lambda(t)X^{(\lambda)}(t), t \geq 0.$$

The problem we study can now be formulated as follows:
Find $\lambda^* \in \mathcal{A}_{\mathbb{H}}$ such that

$$J(c_{\lambda^*}) = \sup_{\lambda \in \mathcal{A}_{\mathbb{H}}} J(c_\lambda), \quad (16.110)$$

where

$$J(c_\lambda) = E\left[\int_0^\tau e^{-\delta(t)} \log c_\lambda(t)dt + \gamma e^{-\delta(\tau)} \log X^{(\lambda)}(\tau) \right]$$

Such λ^* – if it exists – is called an *optimal control* for problem (16.110).
We can now state and prove our main result.

Theorem 16.25. *Define*

$$\hat{\lambda}(t) = \frac{E[\mathcal{X}_{[0,\tau]}(t)e^{-\delta(t)}|\mathcal{H}_t]}{E[\mathcal{X}_{[0,\tau]}(t)(\int_t^\tau e^{-\delta(s)}ds + \gamma e^{-\delta(\tau)})|\mathcal{H}_t]}, \quad t \geq 0. \quad (16.111)$$

If $\hat{\lambda} \in \mathcal{A}_{\mathbb{H}}$ *then* $\hat{\lambda} = \lambda^*$ *is the optimal control for problem (16.110). If* $\hat{\lambda} \notin \mathcal{A}_{\mathbb{H}}$ *then an optimal control does not exist.*

Proof Choose $\lambda \in \mathcal{A}_{\mathbb{H}}$ and put $c(t) = \lambda(t)X^{(\lambda)}(t)$. Then

$$J(c) = E\left[\int_0^\tau e^{-\delta(t)} \log c(t)dt + \gamma e^{-\delta(\tau)} \log X^{(\lambda)}(\tau)\right]$$

$$= E\left[\int_0^\tau e^{-\delta(t)}\left(\log \lambda(t) - \int_0^t \lambda(s)ds\right)dt - \gamma e^{-\delta(\tau)}\int_0^\tau \lambda(t)dt\right] + K,$$

$$\tag{16.112}$$

where

$$K = E\left[\int_0^\tau e^{-\delta(t)}h(t)dt + \gamma e^{-\delta(\tau)}(\log x + h(\tau))\right],$$

with

$$h(t) = \int_0^t \left\{\mu(s) - \tfrac{1}{2}\sigma^2(s) + \int_{\mathbb{R}_0}(\log(1 + \theta(s,z)) - \theta(s,z))\nu(dz)\right\}ds$$

$$+ \int_0^t \sigma(s)d^-W(s) + \int_0^t\int_{\mathbb{R}_0} \log(1 + \theta(s,z))\tilde{N}(d^-s,dz). \tag{16.113}$$

Note that K does not depend on λ. Now, by the Fubini theorem,

$$\int_0^\tau e^{-\delta(t)} \int_0^t \lambda(s)ds\,dt = \int_0^\tau \left(\int_t^\tau e^{-\delta(s)}ds\right)\lambda(t)dt.$$

Substituting this into (16.112) we get

$$J(c) - K = E\left[\int_0^\tau \left\{e^{-\delta(t)}\log\lambda(t) - \lambda(t)\int_t^\tau e^{-\delta(s)}ds - \gamma e^{-\delta(\tau)}\lambda(t)\right\}dt\right]$$

$$= E\left[\int_0^\infty \mathcal{X}_{[0,\tau]}(t)\left\{e^{-\delta(t)}\log\lambda(t) - \lambda(t)\int_t^\tau e^{-\delta(s)}ds - \gamma e^{-\delta(\tau)}\lambda(t)\right\}dt\right]$$

$$= E\left[\int_0^\infty E\left[\mathcal{X}_{[0,\tau]}(t)\left\{e^{-\delta(t)}\log\lambda(t) - \lambda(t)\int_t^\tau e^{-\delta(s)}ds - \gamma e^{-\delta(\tau)}\lambda(t)\right\}\Big|\mathcal{H}_t\right]dt\right]$$

$$= E\left[\int_0^\infty \left\{\log\lambda(t)E[\mathcal{X}_{[0,\tau]}(t)e^{-\delta(t)}|\mathcal{H}_t]\right.\right.$$

$$\left.\left. - \lambda(t)E\left[\mathcal{X}_{[0,\tau]}(t)\left(\int_t^\tau e^{-\delta(s)}ds + \gamma e^{-\delta(\tau)}\right)\Big|\mathcal{H}_t\right]\right\}dt\right].$$

We can maximize this pointwise, for each t, ω: The concave function

$$f(\lambda) := \log \lambda \cdot E\left[\mathcal{X}_{[0,\tau]}(t)e^{-\delta(t)}|\mathcal{H}_t\right]$$
$$- \lambda E\left[\mathcal{X}_{[0,\tau]}(t)\left(\int_t^\tau e^{-\delta(s)}ds + \gamma e^{-\delta(\tau)}\right)|\mathcal{H}_t\right]$$

is maximal when

$$0 = f'(\lambda) = \lambda^{-1}E\left[\mathcal{X}_{[0,\tau]}(t)e^{-\delta(t)}|\mathcal{H}_t\right]$$
$$- E\left[\mathcal{X}_{[0,\tau]}(t)\left(\int_t^\tau e^{-\delta(s)}ds + \gamma e^{-\delta(\tau)}\right)|\mathcal{H}_t\right].$$

This gives the only possible optimal dividend candidate

$$\hat{\lambda}(t) = \frac{E[\mathcal{X}_{[0,\tau]}(t)e^{-\delta(t)}|\mathcal{H}_t]}{E[\mathcal{X}_{[0,\tau]}(t)(\int_t^\tau e^{-\delta(s)}ds + \gamma e^{-\delta(\tau)})|\mathcal{H}_t]}. \qquad \square$$

In particular, we get the following extension of the result (16.95).

Corollary 16.26. *Suppose $\tau = \infty$, $\gamma = 0$, and $\delta(t) = \delta_0 t$ for some constant $\delta_0 > 0$. Then the optimal relative consumption rate is*

$$\lambda^*(t) = \delta_0.$$

16.4.3 Optimal Consumption and Portfolio

In this section we apply the result from Sect. 16.4.2 to study the problem of joint optimal consumption and portfolio for a trader (possibly with inside information) in a market possibly influenced by other traders with inside information.

Suppose we have a financial market with the following two investment possibilities:

(1) A *bond*, with price $S_0(t)$ at time t given by

$$dS_0(t) = \rho(t)S_0(t)dt, \qquad S_0(0) = 1. \tag{16.114}$$

(2) A *stock*, with price $S_1(t)$ at time t given by

$$dS_1(t) = S_1(t^-)\left[\alpha(t)dt + \beta(t)d^-W(t) + \int_{\mathbb{R}_0} \xi(t,z)\tilde{N}(d^-t,dz)\right], \tag{16.115}$$
$$S_1(0) > 0.$$

In addition, we assume as before that we are given a *discounting exponent process* $\delta(t) \geq 0$ and a *default* (or bankruptcy) *time* $\tau : \Omega \rightarrow [0, \infty]$.

We make the similar assumptions as we did for the cash flow in Sect. 16.4.2, i.e., we assume the following:

> $\rho(t), \alpha(t), \beta(t), \xi(t, z), \delta(t)$ and τ are measurable for all t, z.   (16.116)

> $\xi(t, z) > -1$ for a.a. t, z with respect to $dt \times \nu(dz)$.   (16.117)

> $\beta(s)$ and $\xi(s, z)$ are caglad (with respect to s) for all z (16.118)

and the forward integrals

$$\int_0^\tau \sigma(s)d^-W(s) \quad \text{and} \quad \int_0^\tau \int_{\mathbb{R}_0} \log(1 + \xi(s, z))\tilde{N}(d^-s, dz)$$

exist and belong to $L^1(P)$.

Now suppose that a trader is free to choose at any time t both the *relative consumption rate* $\lambda(t) = c(t)/X(t) \geq 0$ and the *fraction* $\pi(t)$ of the current total wealth $X(t)$ to be invested in the stocks. The wealth process $X(t) = X^{(\lambda, \pi)}(t)$ corresponding to the consumption–portfolio pair (λ, π) is given by

$$dX(t) = \rho(t)(1 - \pi(t))X(t)dt - \lambda(t)X(t)dt$$
$$+ \pi(t)X(t^-)\Big[\alpha(t)dt + \beta(t)d^-W(t) + \int_{\mathbb{R}_0} \xi(t, z)\tilde{N}(d^-t, dz)\Big]$$
$$= X(t^-)\Big[\{\rho(t) + (\alpha(t) - \rho(t))\pi(t) - \lambda(t)\}dt$$
$$+ \pi(t)\beta(t)d^-W(t) + \int_{\mathbb{R}_0} \pi(t)\xi(t, z)\tilde{N}(d^-t, dz)\Big], \quad (16.119)$$

$$X(0) = x > 0.$$

As in Sect. 16.4.2 we assume that the information available to the trader at time t is represented by a filtration $\mathbb{H} = \{\mathcal{H}_t, t \geq 0\}$, with no a priori relation to $\mathbb{F} = \{\mathcal{F}_t, t \geq 0\}$.

Note that for each given portfolio choice $\pi(t)$, (16.119) has the same form as the cash flow equation (16.98), with

$$\mu(s) = \rho(s) + (\alpha(s) - \rho(s))\pi(s), \quad (16.120)$$
$$\sigma(s) = \pi(s)\beta(s), \quad (16.121)$$
$$\theta(s, z) = \pi(s)\xi(s, z). \quad (16.122)$$

In view of this, the following definition is natural:

Definition 16.27. *We say that a consumption–portfolio pair (λ, π) is admissible if*

$\triangleright$ $\lambda(s)$ and $\pi(s)$ are $\mathcal{H}_s$-measurable, $s \geq 0$. (16.123)

$\triangleright$ The processes $\mu(\cdot), \sigma(\cdot)$, and $\theta(\cdot, \cdot)$ defined by (4.7)–(4.9) (16.124)
   satisfy conditions (3.2), (3.3), and (3.4).

$\triangleright$ $\lambda(s) \geq 0$ and $\displaystyle\int_0^\tau \lambda(s)ds < \infty$ a.s. (16.125)

$\triangleright$ $\displaystyle E\left[\int_0^\tau e^{-\delta(t)}|\log(\lambda(t)X^{(\lambda,\pi)}(t)|dt + \gamma e^{-\delta(\tau)}|\log X^{(\lambda,\pi)}(\tau)|\right] < \infty.$ (16.126)

The set of all admissible pairs (λ, π) is denoted by $\mathcal{A}_{\mathbb{H}}$.

We now consider the following problem:
Find $(\lambda^*, \pi^*) \in \mathcal{A}_{\mathbb{H}}$ such that

$$J(\lambda^*, \pi^*) = \sup_{(\lambda,\pi)\in\mathcal{A}_{\mathbb{H}}} J(\lambda, \pi), \qquad (16.127)$$

where

$$J(\lambda, \pi) = E\left[\int_0^\tau e^{-\delta(t)} \log(\lambda(t)X^{(\lambda,\pi)}(t))dt + \gamma e^{-\delta(\tau)} \log X^{(\lambda,\pi)}(\tau)\right], \quad (16.128)$$

with $\gamma \geq 0$ a constant.

By applying Theorem 16.25 to the case when the coefficients μ, σ, and θ are given by (16.120), (16.121), and (16.122) for each given π, we obtain the following result:

Theorem 16.28. *Define $\hat{\lambda}(t)$ as in (16.111). Then $\hat{\lambda}(t)$ is an optimal relative consumption rate independent of the portfolio chosen, in the sense that*

$$J(\hat{\lambda}, \pi) \geq J(\lambda, \pi)$$

for all λ, π such that $(\hat{\lambda}, \pi) \in \mathcal{A}_{\mathbb{H}}$ and $(\lambda.\pi) \in \mathcal{A}_{\mathbb{H}}$.

Corollary 16.29. *Suppose there exists an optimal pair $(\lambda^*, \pi^*) \in \mathcal{A}_{\mathbb{H}}$ for problem (16.127). Then*

$$\lambda^*(t) = \hat{\lambda}(t)   \text{ for all } t \geq 0,$$

where $\hat{\lambda}(t)$ is given by (16.111).

Thus we see that the optimal consumption–portfolio problem splits into an optimal consumption problem (with solution $\lambda^* = \hat{\lambda}$) and then – by substituting $\lambda = \lambda^*$ into (16.119) – an optimal portfolio problem. For the solution of optimal portfolio problems for an insider see the following sections.

16.5 Optimal Portfolio Problem under Inside Information

In this section, we take the point of view of a trader who has some larger information at his disposal during his portfolio selection process and would like to take advantage of it.

16.5.1 Formalization of the Optimization Problem: General Utility Function

Let us consider the following market model with two investment possibilities:

- A bond with price dynamics

$$\begin{cases} dS_0(t) = \rho(t)S_0(t)dt, & t \in (0,T], \\ S_0(0) = 1. \end{cases} \tag{16.129}$$

- A stock with price dynamics

$$\begin{cases} dS_1(t) = S_1(t^-)\big[\mu(t,\pi(t))dt + \sigma(t)d^-W(t) + \int_{\mathbb{R}_0} \theta(t,z)\widetilde{N}(d^-t,dz)\big], t \in (0,T], \\ S_1(0) > 0 \end{cases} \tag{16.130}$$

on the complete probability space $(\Omega, \mathcal{F}, P)$ (see Sect. 12.5). The stochastic coefficients $\rho(t)$, $\mu(t,\pi)$, $\sigma(t)$, and $\theta(t,z)$, $t \in [0,T]$, $z \in \mathbb{R}_0$, are measurable, càglàd processes with respect to the parameter t, adapted to some given filtration $\mathbb{G}$, for each constant value of π. Here $\mathbb{G} := \{\mathcal{G}_t \subset \mathcal{F}, t \in [0,T]\}$ is a filtration with

$$\mathcal{G}_t \supset \mathcal{F}_t, \quad t \in [0,T].$$

We recall that $\mathbb{F} := \{\mathcal{F}_t \subset \mathcal{F}, t \in [0,T]\}$ is the filtration generated by the development of the noise events, i.e., the driving noises $dW(t)$ and $\widetilde{N}(dt,dz)$, $t \in [0,T], z \in \mathbb{R}_0$. We also assume that $\theta(t,z) > -1$, $dt \times \nu(dz)$-a.e., and that

$$E \int_0^T \Big\{ |\rho(t)| + |\mu(t)| + \sigma^2(t) + \int_{\mathbb{R}_0} \theta^2(t,z)\nu(dz) \Big\}dt < \infty.$$

In this model the coefficient $\mu(t)$, $t \in [0,T]$, depends on the portfolio choice $\pi(t)$, $t \in [0,T]$, of an insider who has access to the information represented by the filtration $\mathbb{H} := \{\mathcal{H}_t \subset \mathcal{F}, t \in [0,T]\}$ with

$$\mathcal{H}_t \supset \mathcal{G}_t \supset \mathcal{F}_t, \quad t \in [0,T].$$

Accordingly, the insider's portfolio $\pi = \pi(t)$, $t \in [0,T]$, is a stochastic process adapted to $\mathbb{H}$. With the above conditions on μ, we intend to model a possible situation in which an insider is so influential in the market to affect the prices with his choices. In this sense we talk about a "large" insider.

This exogenous model for the price dynamics (16.129) and (16.130) is in line with [51]. In [51] a dependence of the coefficient r on the portfolio π is also considered. In our paper, this can also be mathematically carried through without substantial change; however, the assumption that the return of the bond depends on the agent's portfolio could be considered unrealistic.

We consider the *insider's wealth process* $X_\pi(t) = X_\pi^x(t)$, $t \in [0, T]$, to be given by

$$
dX_\pi(t) = X_\pi(t^-)\Big\{ \big[\rho(t) + (\mu(t, \pi(t)) - \rho(t))\pi(t)\big]dt
$$
$$
+ \pi(t)\sigma(t)d^-W(t) + \pi(t)\int_{\mathbb{R}_0}\theta(t, z)\widetilde{N}(d^-t, dz)\Big\},
$$
(16.131)

with initial capital $X_\pi(0) = X_\pi^x(0) = x > 0$. By the Itô formula for forward integrals, see Theorem 8.12 and Theorem 15.8, the final wealth of the admissible portfolio π is the unique solution of (16.131):

$$
X_\pi(t) = x\exp\Big\{ \int_0^t \Big[\rho(s) + (\mu(s, \pi(s)) - \rho(s))\pi(s)
$$
$$
- \frac{1}{2}\sigma^2(s)\pi^2(s)\Big]ds - \int_0^t\int_{\mathbb{R}_0}\Big[\pi(s)\theta(s, z) - \log\big(1 + \pi(s)\theta(s, z)\big)\Big]\nu(dz)ds
$$
$$
+ \int_0^t \pi(s)\sigma(s)d^-W(s) + \int_0^t\int_{\mathbb{R}_0}\log\big(1 + \pi(s)\theta(s, z)\big)\widetilde{N}(d^-s, dz)\Big\}.
$$
(16.132)

Taking the point of view of an insider, with the only purpose of understanding his opportunities in the market, we are interested in solving the optimization problem

$$
u(x) := \sup_{\pi \in \mathcal{A}_\mathbb{H}} E\left[U(X_\pi^x(T))\right] = E\left[U(X_{\pi^*}^x(T))\right],
$$
(16.133)

for the given *utility function*

$$
U : [0, \infty) \longrightarrow [-\infty, \infty)
$$

that is a nondecreasing, concave, and lower semi-continuous function that we assume to be continuously differentiable on $(0, \infty)$. Here the controls belonging to the set $\mathcal{A}_\mathbb{H}$ of admissible portfolios are characterized as follows:

Definition 16.30. *The set $\mathcal{A}_\mathbb{H}$ of* admissible portfolios *consists of all processes $\pi = \pi(t)$, $t \in [0, T]$, such that*

▷ π *is càglàd and adapted to the filtration $\mathbb{H}$.* (16.134)

▷ $\pi(t)\sigma(t)$, $t \in [0,T]$, is forward integrable with respect to $d^-W(t)$.

$$(16.135)$$

▷ $\pi(t)\theta(t,z)$, $t \in [0,T], z \in \mathbb{R}_0$, is forward integrable with respect to $\widetilde{N}(d^-t,dz)$.

$$(16.136)$$

▷ $\pi(t)\theta(t,z) > -1 + \epsilon_\pi$ for a.a. (t,z) with respect to $dt \times \nu(dz)$, for some $\epsilon_\pi \in (0,1)$ depending on π.

$$(16.137)$$

▷ $E\int_0^T \left\{ |\mu(s,\pi(s)) - \rho(s)||\pi(s)| + (1+\sigma^2(s))\pi^2(s) + \int_{\mathbb{R}_0} \pi^2(s)\theta^2(s,z)\nu(dz) \right\} ds < \infty$

and $E\left[\exp\left\{ K \int_0^T |\pi(s)|ds \right\} \right] < \infty$ for all $K > 0$.

$$(16.138)$$

▷ $\log\left(1 + \pi(t)\theta(t,z)\right)$ is forward integrable with respect to $\widetilde{N}(d^-t,dz)$.

$$(16.139)$$

▷ $E[U(X_\pi(T))] < \infty$ and $0 < E[U'(X_\pi(T))X_\pi(T)] < \infty$,

where $U'(w) = \dfrac{d}{dw}U(w), w \geq 0.$

$$(16.140)$$

▷ For all $\pi, \beta \in \mathcal{A}$, with β bounded, there exists a $\zeta > 0$ such that the family

$$(16.141)$$

$$\left\{ U'(X_{\pi+\delta\beta}(T))X_{\pi+\delta\beta}(T) \middle| M_{\pi+\delta\beta}(T) \middle| \right\}_{\delta\in(-\zeta,\zeta)}$$

is uniformly integrable.

Note that, for $\pi \in \mathcal{A}_{\mathbb{H}}$ and $\beta \in \mathcal{A}_{\mathbb{H}}$ bounded, $\pi + \delta\beta \in \mathcal{A}_{\mathbb{H}}$ for any $\delta \in (-\zeta,\zeta)$ with ζ small enough. Here the stochastic process $M_\pi(t)$, $t \in [0,T]$, is defined as

$$M_\pi(t) := \int_0^t \left\{ \mu(s,\pi(s)) - \rho(s) + \mu'(s,\pi(s))\pi(s) \right.$$

$$\left. - \sigma^2(s)\pi(s) - \int_{\mathbb{R}_0} \frac{\pi(s)\theta^2(s,z)}{1+\pi(s)\theta(s,z)}\nu(dz) \right\} ds$$

$$+ \int_0^t \sigma(s)d^-W(s) + \int_0^t\int_{\mathbb{R}_0} \frac{\theta(s,z)}{1+\pi(s)\theta(s,z)}\widetilde{N}(d^-s,dz),$$

$$(16.142)$$

where $\mu'(s,\pi) = \frac{\partial}{\partial\pi}\mu(s,\pi)$.

Remark 16.31. Condition (16.141) may be difficult to verify. Here we give some examples of conditions under which it holds.

First, consider $M(\delta) := M_{\pi+\delta\beta}(T)$. The uniformly integrability of $\{M(\delta)\}_{\delta\in(-\zeta,\zeta)}$ is assured by

$$\sup_{\delta\in(-\zeta,\zeta)} E[|M|^p(\delta)] < \infty \quad \text{for some } p > 1.$$

Observe that, since $\pi, \beta \in \mathcal{A}_{\mathbb{H}}$ (see (16.137)), we have $1 + \big(\pi(s) + \delta\beta(s)\big)\theta(s, z) \geq \epsilon_\pi - \zeta$ $dt \times \nu(dz)$-a.e. for some $\zeta \in (0, \epsilon_\pi)$. Moreover, for $\epsilon > 0$,

$$\int_0^T \int_{|z| \geq \epsilon} \frac{\theta(s, z)}{1 + (\pi(s) + \delta\beta(s))\theta(s, z)} \widetilde{N}(d^- s, dz)$$

$$= \int_0^T \int_{|z| \geq \epsilon} \frac{\theta(s, z)}{1 + (\pi(s) + \delta\beta(s))\theta(s, z)} \widetilde{N}(ds, dz).$$

Thus we have

$$E\bigg[\bigg(\int_0^T \int_{|z| \geq \epsilon} \frac{\theta(s, z)}{1 + (\pi(s) + \delta\beta(s))\theta(s, z)} \widetilde{N}(d^- s, dz)\bigg)^2\bigg]$$

$$\leq \frac{1}{(\epsilon_\pi - \zeta)^2} E\bigg[\int_0^T \int_{|z| \geq \epsilon} \theta^2(s, z)\nu(dz)ds\bigg] < \infty.$$

So, if

$$E\bigg[\bigg(\int_0^T \sigma(s)d^- W(s)\bigg)^2\bigg] < \infty \text{ and } E\bigg[\bigg(\int_0^T \int_{|z| < \epsilon} |\theta(s, z)|\widetilde{N}(d^- s, dz)\bigg)^2\bigg] < \infty,$$

we have $E[M^2(\delta)] < \infty$ uniformly in $\delta \in (-\zeta, \zeta)$ if, for example, the coefficients μ, μ', r, σ are bounded. This shows that (16.141) holds if $U'(x)x$ is uniformly bounded for $x \in (0, \infty)$. This is the case, for example, logarithmic utility of $U(x) = \log x$ and exponential utility $U(x) = -\exp\{-\lambda x\}$ ($\lambda > 0$).

Similarly, in the case of power utility function

$$U(x) = \frac{1}{\gamma}x^\gamma, \quad x > 0 \qquad \text{for some } \gamma \in (0, 1),$$

we see that $U'(X_{\pi+\delta\beta}(T))X_{\pi+\delta\beta}(T)|M(\delta)| = X_{\pi+\delta\beta}^\gamma(T)|M(\delta)|$ and condition (16.141) would be satisfied if

$$\sup_{\delta \in (-\zeta, \zeta)} E\big[(X_{\pi+\delta\beta}^\gamma(T)|M(\delta)|)^p\big] < \infty \qquad \text{for some } p > 1.$$

Note that we can write

$$X_{\pi+\delta\beta}(T) = X_\pi(T)N(\delta),$$

where

$$N(\delta) := \exp\bigg\{\int_0^T \big[(\mu(s, \pi(s) + \delta\beta(s)) - \rho(s))\delta\beta(s) + (\mu(s, \pi(s) + \delta\beta(s))$$

$$- \mu(s, \pi(s))\pi(s) - \sigma^2(s)\delta\beta(s)\pi(s) - \frac{1}{2}\sigma^2(s)\delta^2\beta^2(s)\big]ds$$

$$+ \int_0^T \delta\sigma(s)\beta(s)d^- W(s) + \int_0^T \int_{\mathbb{R}_0} \big[\log(1 + (\pi(s) + \delta\beta(s))\theta(s,z))$$

$$- \log(1 + \pi(s)\theta(s,z)) - \delta\beta(s)\theta(s,z)\big]\nu(dz)ds$$

$$+ \int_0^T \int_{\mathbb{R}_0} \big[\log(1 + (\pi(s) + \delta\beta(s))\theta(s,z))$$

$$- \log(1 + \pi(s)\theta(s,z))\big]\widetilde{N}(d^- s, dz)\Big\}.$$

From the iterated application of the Hölder inequality, we have

$$E\big[(X_{\pi+\delta\beta}^\gamma(T)|M(\delta)|)^p\big]$$

$$\leq \big(E\big[(X_\pi(T))^{\gamma p a_1 b_1}\big]\big)^{\frac{1}{a_1 b_1}} \big(E\big[(N(\delta))^{\gamma p a_1 b_2}\big]\big)^{\frac{1}{a_1 b_2}} \big(E\big[(|M(\delta)|)^{p a_2}\big]\big)^{\frac{1}{a_2}},$$

where a_1, a_2: $\frac{1}{a_1} + \frac{1}{a_2} = 1$ and b_1, b_2: $\frac{1}{b_1} + \frac{1}{b_2} = 1$. Then we can choose $a_1 = \frac{2}{2-p}$, $a_2 = \frac{2}{p}$ and also $b_1 = \frac{2-p}{\gamma p}$, $b_2 = \frac{2-p}{2-p-\gamma p}$ for some $p \in (1, \frac{2}{\gamma+1})$. Hence

$$E\big[(X_{\pi+\delta\beta}^\gamma(T)|M(\delta)|)^p\big]$$

$$\leq \big(E\big[(X_\pi(T))^2\big]\big)^{\frac{\gamma p}{2}} \big(E\big[(N(\delta))^{\frac{2\gamma p}{2-p-\gamma p}}\big]\big)^{\frac{2-p-\gamma p}{2}} \big(E\big[(|M(\delta)|)^2\big]\big)^{\frac{p}{2}}.$$

If the value $X_\pi(T)$ in (16.132) satisfies

$$E\big[(X_\pi(T))^2\big] < \infty, \tag{16.143}$$

then the condition (16.141) holds if

$$\sup_{\delta\in(-\zeta,\zeta)} E\big[(N(\delta))^{\frac{2\gamma p}{2-p-\gamma p}}\}\big] < \infty.$$

Since (16.138) holds, it is enough, e.g., that μ, μ', r, σ are bounded to have $E\big[(N(\delta))^{\frac{2\gamma p}{2-p-\gamma p}}\}\big] < \infty$ uniformly in $\delta \in (-\zeta, \zeta)$. Note that condition (16.143) is verified, for example, if for all $K > 0$

$$E\Big[\exp\Big\{K\Big(\int_0^T |\pi(s)|ds + |\int_0^T \pi(s)\sigma(s)d^- W(s)|$$

$$+ |\int_0^T \int_{\mathbb{R}_0} \log(1 + \pi(s)\theta(s,z))\widetilde{N}(d^- s, dz)|\Big)\Big\}\Big] < \infty.$$

By similar arguments, we can also treat the case of a utility function such with $U'(x)$ is uniformly bounded for $x \in (0, \infty)$. We omit the details. □

16.5.2 Characterization of an Optimal Portfolio under Inside Information

The forward stochastic calculus gives an adequate mathematical framework in which we can proceed to solve the optimization problem (16.133). This approach was first used in [33], for the Brownian motion case only. See Sect. 9.6. The following is an extension to Lévy processes. Define

$$J(\pi) := E\big[U\big(X_\pi(T)\big)\big], \quad \pi \in \mathcal{A}_{\mathbb{H}}.$$

First, let us suppose that π is locally optimal for the insider, in the sense that $J(\pi) \geq J(\pi + \delta\beta)$ for all $\beta \in \mathcal{A}$ bounded, and $\pi + \delta\beta \in \mathcal{A}_{\mathbb{H}}$ for all δ small enough. Recall that this is the concept of optimality considered in this chapter. Since the function $J(\pi + \delta\beta)$ is maximal at π, by (16.141) and (15.1), we have

$$
\begin{aligned}
0 &= \frac{d}{d\delta} J(\pi + \delta\beta)_{|\delta=0} \\
&= E\Big[U'(X_\pi(T))X_\pi(T)\Big\{ \int_0^T \beta(s)\big[\mu(s, \pi(s)) - \rho(s) \\
&\quad + \mu'(s, \pi(s))\pi(s) - \sigma^2(s)\pi(s) \\
&\quad - \int_{\mathbb{R}_0} \{\theta(s, z) - \frac{\theta(s, z)}{1 + \pi(s)\theta(s, z)}\}\nu(dz)\big]ds \\
&\quad + \int_0^T \beta(s)\sigma(s)d^-W(s) + \int_0^T \int_{\mathbb{R}_0} \frac{\beta(s)\theta(s, z)}{1 + \pi(s)\theta(s, z)}\widetilde{N}(d^-s, dz)\Big\}\Big].
\end{aligned}
$$
(16.144)

Now let us fix $t \in [0, T)$ and $h > 0$ such that $t + h \leq T$. We can choose $\beta \in \mathcal{A}$ of the form

$$\beta(s) = \alpha\chi_{(t,t+h]}(s), \quad 0 \leq s \leq T,$$

where α is an arbitrary bounded $\mathcal{H}_t$-measurable random variable. Then (16.144) gives

$$
\begin{aligned}
0 &= E\Big[U'(X_\pi(T))X_\pi(T)\Big\{ \int_t^{t+h} [\mu(s, \pi(s)) - \rho(s) \\
&\quad + \mu'(s, \pi(s))\pi(s) - \sigma^2(s)\pi(s) \\
&\quad - \int_{\mathbb{R}_0} \frac{\pi(s)\theta^2(s, z)}{1 + \pi(s)\theta(s, z)}\nu(dz)]ds \\
&\quad + \int_t^{t+h} \sigma(s)d^-W(s) + \int_t^{t+h} \int_{\mathbb{R}_0} \frac{\theta(s, z)}{1 + \pi(s)\theta(s, z)}\widetilde{N}(d^-s, dz)\Big\} \cdot \alpha\Big].
\end{aligned}
$$
(16.145)

Since this holds for all such α, we can conclude that

$$E\big[F_\pi(T)\big(M_\pi(t + h) - M_\pi(t)\big)|\mathcal{H}_t\big] = 0, \tag{16.146}$$

where

$$F_\pi(T) := \frac{U'(X_\pi(T))X_\pi(T)}{E[U'(X_\pi(T))X_\pi(T)]} \tag{16.147}$$

and

$$M_\pi(t) := \int_0^t \Big\{ \mu(s, \pi(s)) - \rho(s) + \mu'(s, \pi(s))\pi(s)$$

$$- \sigma^2(s)\pi(s) - \int_{\mathbb{R}_0} \frac{\pi(s)\theta^2(s, z)}{1 + \pi(s)\theta(s, z)} \nu(dz) \Big\} ds \qquad (16.148)$$

$$+ \int_0^t \sigma(s) d^- W(s) + \int_0^t \int_{\mathbb{R}_0} \frac{\theta(s, z)}{1 + \pi(s)\theta(s, z)} \tilde{N}(d^- s, dz), \quad t \in [0, T],$$

cf. (16.142). Define the probability measure Q_π on $(\Omega, \mathcal{H}_T)$ by

$$Q_\pi(d\omega) := F_\pi(T) P(d\omega) \qquad (16.149)$$

and denote the expectation with respect to the measure Q_π by E_{Q_π}. Then, by (16.147), we have

$$E_{Q_\pi}\big[M_\pi(t + h) - M_\pi(t)|\mathcal{H}_t\big] = \frac{E\big[F_\pi(T)\big(M_\pi(t + h) - M_\pi(t)\big)|\mathcal{H}_t\big]}{E\big[F_\pi(T)|\mathcal{H}_t\big]} = 0.$$

Hence the process $M_\pi(t)$, $t \in [0, T]$, is a $(\mathbb{H}, Q_\pi)$-*martingale* (i.e., a martingale with respect to the filtration $\mathbb{H}$ and under the probability measure Q_π).
On the other hand, the argument can be reversed as follows. If $M_\pi(t)$, $t \in [0, T]$, is a $(\mathbb{H}, Q_\pi)$-martingale, then

$$E\big[F_\pi(T)\big(M_\pi(t + h) - M_\pi(t)\big)|\mathcal{H}_t\big] = 0,$$

for all $h > 0$ such that $0 \leq t < t + h \leq T$, which is (16.146). Or equivalently,

$$E\big[\alpha F_\pi(T)\big(M_\pi(t + h) - M_\pi(t)\big)\big] = 0$$

for all bounded $\mathcal{H}_t$-measurable $\alpha \in \mathcal{A}_{\mathbb{H}}$. Hence (16.145) holds for all such α. Taking linear combinations we see that (16.144) holds for all càglàd step processes $\beta \in \mathcal{A}_{\mathbb{H}}$. By our assumptions (16.135) and (16.136) on $\mathcal{A}_{\mathbb{H}}$ we get, by an approximation argument, that (16.144) holds for all $\beta \in \mathcal{A}_{\mathbb{H}}$. If the function $g(\delta) := E\big[U(X_{\pi+\delta\beta}(T))\big]$, $\delta \in (-\zeta, \zeta)$, is concave for each $\beta \in \mathcal{A}_{\mathbb{H}}$, we conclude that its maximum is achieved at $\delta = 0$. Hence we have proved the following result – see [63].

Theorem 16.32. *(1) If the stochastic process $\pi \in \mathcal{A}_{\mathbb{H}}$ is optimal for the problem (16.133), then the stochastic process $M_\pi(t)$, $t \in [0, T]$, is a $(\mathbb{H}, Q_\pi)$-martingale.*
(2) Conversely, if the function $g(\delta) := E\big[U(X_{\pi+\delta\beta}(T))\big]$, $\delta \in (-\zeta, \zeta)$, is concave for each $\beta \in \mathcal{A}_{\mathbb{H}}$ and $M_\pi(t)$, $t \in [0, T]$, is a $(\mathbb{H}, Q_\pi)$-martingale, then $\pi \in \mathcal{A}_{\mathbb{H}}$ is optimal for the problem (16.133).

Remark 16.33. Since the composition of a concave increasing function with a concave function is concave, we can see that a sufficient condition for the function $g(\delta)$, $\delta \in (-\zeta, \zeta)$, to be concave is that the function

$$\Lambda(s) : \pi \longrightarrow \rho(s) + (\mu(s, \pi) - \rho(s))\pi - \frac{1}{2}\sigma^2(s)\pi^2 \qquad (16.150)$$

is concave for all $s \in [0, T]$. For this it is sufficient that $\mu(s, \cdot)$ are C^2 for all s and that

$$\mu''(s, \pi)\pi + 2\mu'(s, \pi) - \sigma^2 \leq 0 \qquad (16.151)$$

for all s, π. Here we have set $\mu' = \frac{\partial \mu}{\partial \pi}$ and $\mu'' = \frac{\partial^2 \mu}{\partial \pi^2}$.

Moreover, we also obtain the following result – see [63].

Theorem 16.34. *(1) A stochastic process $\pi \in \mathcal{A}_{\mathbb{H}}$ is optimal for the problem (16.133) only if the process*

$$\hat{M}_\pi(t) := M_\pi(t) - \int_0^t \frac{d[M_\pi, Z_\pi](s)}{Z_\pi(s)}, \quad t \in [0, T], \qquad (16.152)$$

is a $(\mathbb{H}, P)$-martingale (i.e., a martingale with respect to the filtration $\mathbb{H}$ and under the probability measure P). Here

$$Z_\pi(t) := E_{Q_\pi}\left[\frac{dP}{dQ_\pi}|\mathcal{H}_t\right] = (E[F_\pi(T)|\mathcal{H}_t])^{-1}, \quad t \in [0, T]. \qquad (16.153)$$

(2) Conversely, if $g(\delta) := E[U(X_{\pi+\delta\beta}(T))]$, $\delta \in (-\zeta, \zeta)$, is concave and (16.152) is a $(\mathbb{H}, P)$-martingale, then $\pi \in \mathcal{A}_{\mathbb{H}}$ is optimal for the problem (16.133).

Proof If $\pi \in \mathcal{A}_{\mathbb{H}}$ is an optimal portfolio for an insider, then by Theorem 16.32 we know that $M_\pi(t)$, $t \in [0, T]$, is a $(\mathbb{H}, Q_\pi)$-martingale. Applying the Girsanov theorem (see e.g., [204] Theorem III.39) we obtain that

$$\hat{M}_\pi(t) := M_\pi(t) - \int_0^t \frac{d[M_\pi, Z_\pi](s)}{Z_\pi(s)}, \quad t \in [0, T],$$

is a $(\mathbb{H}, P)$-martingale with

$$Z_\pi(t) = E_{Q_\pi}\left[\frac{dP}{dQ_\pi}|\mathcal{H}_t\right] = E\left[(F_\pi(T))^{-1}\frac{F_\pi(T)}{E[F_\pi(T)|\mathcal{H}_t]}|\mathcal{H}_t\right] = (E[F_\pi(T)|\mathcal{H}_t])^{-1}.$$

Conversely, if $\hat{M}_\pi(t)$, $t \in [0, T]$, is $(\mathbb{H}, P)$-martingale, then $M_\pi(t)$, $t \in [0, T]$, is a $(\mathbb{H}, Q_\pi)$-martingale. Hence π is optimal by Theorem 16.32.

16.5.3 Examples: General Utility and Enlargement of Filtration

We now give some examples to illustrate the contents of the main results:

Example 16.35. Suppose that

$$\sigma(t) \neq 0, \quad \theta = 0 \quad \text{and} \quad \mathcal{H}_t = \mathcal{F}_t \vee \sigma(W(T_0)), \quad \text{for all } t \in [0, T] \text{ (for some } T_0 > T),$$
$$(16.154)$$

i.e., we consider a market driven by the Brownian motion only and where the insider's filtration is a classical example of enlargement of the filtration $\mathbb{F}$ by the knowledge derived from the value of the Brownian motion at some future time $T_0 > T$.

Then we obtain the following result – see [63].

Theorem 16.36. *Suppose that the function Λ in (16.150) is concave for all $s \in [0, T]$. A portfolio $\pi \in \mathcal{A}_{\mathbb{H}}$ is optimal for the problem (16.133) if and only if $d[M_\pi, Z_\pi](t)$ is absolutely continuous with respect to the Lebesgue measure dt and*

$$\mu'(t, \pi(t))\pi(t) + \mu(t, \pi(t)) - \rho(t)$$
$$- \sigma^2(t)\pi(t) + \sigma(t)\left[\frac{W(T_0) - W(t)}{T_0 - t} - \frac{1}{Z_\pi(t)}\frac{d}{dt}[W, Z_\pi](t)\right] = 0. \quad (16.155)$$

Proof By Theorem 16.34, the portfolio $\pi \in \mathcal{A}_{\mathbb{H}}$ is optimal for the problem (16.133) if and only if the process

$$\hat{M}_\pi(t) = \int_0^t \{\mu'(s, \pi(s))\pi(s)$$
$$+ \mu(s, \pi(s)) - \rho(s) - \sigma^2(s)\pi(s)\}ds + \int_0^t \sigma(s)d^-W(s) - \int_0^t \frac{d[M_\pi, Z_\pi](s)}{Z_\pi(s)}$$
$$(16.156)$$

is a $(\mathbb{H}, P)$-martingale. Since $\hat{M}_\pi(t)$ is continuous and has quadratic variation

$$[\hat{M}_\pi, \hat{M}_\pi](t) = \int_0^t \sigma^2(s)ds,$$

we conclude that $\hat{M}_\pi(t)$ can be written

$$\hat{M}_\pi(t) = \int_0^t \sigma(s)d\hat{W}(s) \quad (16.157)$$

for some $(\mathbb{H}, P)$-Brownian motion $\hat{W}$.

On the other hand, by a result of Itô [121] we know that $W(t)$, $t \in [0, T]$, is a semimartingale with respect to $(\mathbb{H}, P)$ with decomposition

$$W(t) = \tilde{W}(t) + \int_0^t \frac{W(T_0) - W(s)}{T_0 - s}ds, \quad 0 \leq t \leq T, \quad (16.158)$$

for some $(\mathbb{H}, P)$-Brownian motion $\tilde{W}(t)$. Combining (16.156), (16.157), and (16.158), we get

$$
\sigma(t)d\hat{W}(t) = d\hat{M}_\pi(t) = \{\mu'(t, \pi(t))\pi(t)
$$

$$
+ \mu(t, \pi(t)) - \rho(t) - \sigma^2(t)\pi(t)\}dt + \sigma(t)d\tilde{W}(t) \tag{16.159}
$$

$$
+ \sigma(t)\frac{W(T_0) - W(t)}{T_0 - t}dt - \frac{d[M_\pi, Z_\pi](t)}{Z_\pi(t)}.
$$

By uniqueness of the semimartingale decomposition of $\hat{M}_\pi(t)$ with respect to $(\mathbb{H}, P)$, we conclude that $\hat{W}(t) = \tilde{W}(t)$ and

$$
\{\mu'(t, \pi(t))\pi(t) + \mu(t, \pi(t)) - \rho(t) - \sigma^2(t)\pi(t)
$$

$$
\sigma(t)\frac{W(T_0) - W(t)}{T_0 - t}\}dt - \frac{d[M_\pi, Z_\pi](t)}{Z_\pi(t)} = 0. \tag{16.160}
$$

From this we deduce that $d[M_\pi, Z_\pi](t) = \sigma(t)d[W, Z_\pi](t)$ is absolutely continuous with respect to dt and (16.155) follows. $\qquad\square$

Corollary 16.37. *Assume that (16.154) holds and, in addition, that*

$$
\mu(t, \pi) = \mu_0(t) + a(t)\pi \tag{16.161}
$$

for some $\mathbb{F}$-adapted processes μ_0 and a with $0 \le a(t) \le \frac{1}{2}\sigma^2(t)$, $t \in [0, T]$, which do not depend on π. Then $\pi \in \mathcal{A}_{\mathbb{H}}$ is optimal if and only if $d[M_\pi, Z_\pi](t)$ is absolutely continuous with respect to dt and

$$
(\sigma^2(t) - 2a(t))\pi(t) = \mu_0(t) - \rho(t) + \sigma(t)\left[\frac{W(T_0) - W(t)}{T_0 - t} - \frac{1}{Z_\pi(t)}\frac{d[W, Z_\pi](t)}{dt}\right]. \tag{16.162}
$$

Proof In this case we have that $\mu'(t, \pi(t)) = a(t)$. Therefore, the function Λ defined in (16.150) is concave (by (16.151)) and the result follows from Theorem 16.36. $\qquad\square$

Next we give an example for a pure jump financial market.

Example 16.38. Suppose that

$$
\sigma(t) = 0 \quad \text{and} \quad \theta(t, z) = \beta z, \tag{16.163}
$$

where $\beta z > -1$ $\nu(dz)$-a.e. $(\beta > 0)$ and that

$$
\mathcal{H}_t = \mathcal{F}_t \vee \sigma(\eta(T_0)) \quad \text{for some } T_0 > T, \tag{16.164}
$$

where

$$
\eta(t) = \int_0^t \int_{\mathbb{R}_0} z\tilde{N}(ds, dz)
$$

(i.e., the insider's filtration is the enlargement of $\mathbb{F}$ by the knowledge derived from some future value $\eta(T_0)$ of the market driving process). Then by a result of Itô, as extended by Kurtz (see [204] p. 256), the process

$$\hat{\eta}(t) := \eta(t) - \int_0^t \frac{\eta(T_0) - \eta(s)}{T_0 - s} ds \qquad (16.165)$$

is a $(\mathbb{H}, P)$-martingale. By Proposition 5.2 in [65] the $\mathbb{H}$-compensating measure $\nu_{\mathbb{H}}$ of the jump measure N is given by

$$\nu_{\mathbb{H}}(ds, dz) = \nu_{\mathbb{F}}(dz)ds + E\left[\frac{1}{T_0 - s} \int_s^{T_0} \tilde{N}(dr, dz) \Big| \mathcal{H}_s\right] ds \qquad (16.166)$$

$$= E\left[\frac{1}{T_0 - s} \int_s^{T_0} N(dr, dz) \Big| \mathcal{H}_s\right] ds,$$

where $\nu_{\mathbb{F}} = \nu$. This implies that the $\mathbb{H}$-compensated random measure $\tilde{N}_{\mathbb{H}}$ is related to $\tilde{N}_{\mathbb{F}} = \tilde{N}$ by

$$\tilde{N}_{\mathbb{H}}(ds, dz) = N(ds, dz) - \nu_{\mathbb{H}}(ds, dz) = \tilde{N}(ds, dz) - E\left[\frac{1}{T_0 - s} \int_s^{T_0} \tilde{N}(dr, dz) \Big| \mathcal{H}_s\right] ds. \qquad (16.167)$$

(Note that, in general, the random measure $\tilde{N}_{\mathbb{H}}(ds, dz)$ is not a compensated Poisson random measure and its compensator $\nu_{\mathbb{H}}(ds, dz)$ is stochastic. We can refer to [123] for a short introduction to random measures with integer values and their compensators.)

Directly from the definition of the forward integral, we have

$$\int_0^t \int_{\mathbb{R}_0} \frac{\beta z}{1 + \pi(s)\beta z} \tilde{N}(d^- s, dz) = \int_0^t \int_{\mathbb{R}_0} \frac{\beta z}{1 + \pi(s)\beta z} \tilde{N}_{\mathbb{H}}(ds, dz)$$

$$+ \int_0^t \int_{\mathbb{R}_0} \frac{\beta z}{1 + \pi(s)\beta z} E\left[\frac{1}{T_0 - s} \int_s^{T_0} \tilde{N}(dr, dz) \Big| \mathcal{H}_s\right] ds. \qquad (16.168)$$

By Theorem 16.34, a portfolio $\pi \in \mathcal{A}_{\mathbb{H}}$ is optimal if and only if the process

$$\hat{M}_\pi(t) = \int_0^t \Big\{ \mu(s, \pi(s)) - \rho(s) + \mu'(s, \pi(s))\pi(s)$$

$$- \int_{\mathbb{R}_0} \frac{\beta^2 z^2 \pi(s)}{1 + \pi(s)\beta z} \nu(dz) \Big\} ds \qquad (16.169)$$

$$+ \int_0^t \int_{\mathbb{R}_0} \frac{\beta z}{1 + \pi(s)\beta z} \tilde{N}(d^- s, dz) - \int_0^t \frac{d[M_\pi, Z_\pi](s)}{Z_\pi(s)}$$

is a $(\mathbb{H}, P)$-martingale. Therefore, if we put

$$G_\pi(s) := \mu(s, \pi(s)) - \rho(s) + \mu'(s, \pi(s))\pi(s)$$
$$- \int_{\mathbb{R}_0} \frac{\beta^2 z^2 \pi(s)}{1 + \pi(s)\beta z} \nu(dz) \tag{16.170}$$
$$+ \int_{\mathbb{R}_0} \frac{\beta z}{1 + \pi(s)\beta z} E\Big[\frac{1}{T_0 - s} \int_s^{T_0} \tilde{N}(dr, dz)\big|\mathcal{H}_s\Big],$$

and combining (16.168) and (16.169), we obtain that the process

$$\hat{M}_\pi(t) = \int_0^t G_\pi(s)ds - \int_0^t \frac{d[M_\pi, Z_\pi](s)}{Z_\pi(s)} + \int_0^t \int_{\mathbb{R}_0} \frac{\beta z}{1 + \pi(s)\beta z} \tilde{N}_{\mathbb{H}}(ds, dz)$$

is a $(\mathbb{H}, P)$-martingale. This is possible if and only if

$$\int_0^t G_\pi(s)ds - \int_0^t \frac{d[M_\pi, Z_\pi](s)}{Z_\pi(s)} = 0, \quad \text{for all } t \in [0, T].$$

This implies that $d[M_\pi, Z_\pi](t)$ is absolutely continuous with respect to the Lebesgue measure dt. We have thus proved the following statement – see [63].

Theorem 16.39. *Assume that (16.163) and (16.164) hold. Then $\pi \in \mathcal{A}_{\mathbb{H}}$ is optimal if and only if $d[M_\pi, Z_\pi](t)$ is absolutely continuous with respect to the Lebesgue measure dt and*

$$G_\pi(t) = \frac{1}{Z_\pi(t)} \frac{d}{dt}[M_\pi, Z_\pi](t) \quad \text{for almost all } t \in [0, T] \tag{16.171}$$

where G_π is given by (16.170).

In analogy with Corollary 16.37, we get the following result in the special case when the influence of the trader on the market is given by (16.161).

Corollary 16.40. *Assume that (16.163) and (16.164) hold and, in addition, that also (16.161) holds. Then $\pi \in \mathcal{A}_{\mathbb{H}}$ is optimal if and only if $d[M_\pi, Z_\pi](t)$ is absolutely continuous with respect to the Lebesgue measure dt and*

$$\pi(s) \int_{\mathbb{R}_0} \frac{\beta^2 z^2}{1 + \pi(s)\beta z} \nu(dz) - \int_{\mathbb{R}_0} \frac{\beta z}{1 + \pi(s)\beta z} E\Big[\frac{1}{T_0 - s} \int_s^{T_0} \tilde{N}(dr, dz)\big|\mathcal{H}_s\Big]$$
$$- 2a(s)\pi(s) = \mu_0(s) - \rho(s) - \frac{1}{Z_\pi(s)} \frac{d}{ds}[M_\pi, Z_\pi](s). \tag{16.172}$$

Corollary 16.41. *Suppose that (16.161), (16.163), and (16.164) hold and that*

$$U(x) = \log x, \quad x \geq 0. \tag{16.173}$$

Then $\pi \in \mathcal{A}_{\mathbb{H}}$ is optimal if and only if

$$\pi(s)\int_{\mathbb{R}_0}\frac{\beta^2 z^2}{1+\pi(s)\beta z}\nu(dz)-\int_{\mathbb{R}_0}\frac{\beta z}{1+\pi(s)\beta z}E\Big[\frac{1}{T_0-s}\int_s^{T_0}\tilde{N}(dr,dz)\big|\mathcal{H}_s\Big]-2a(s)\pi(s)$$

$$=\mu_0(s)-\rho(s).$$

Proof If $U(x)=\log x$ then $F_\pi(T)=1=Z_\pi(t)$, $t\in[0,T]$. Hence $[M_\pi,Z_\pi]=0$.
$\qquad\qquad\qquad\qquad\qquad\qquad\qquad\qquad\qquad\qquad\qquad\qquad\qquad\qquad\qquad\quad\square$

16.6 Optimal Portfolio Problem under Inside Information: Logarithmic Utility

In this sequel, we illustrate the study of the previous section to the case of logarithmic utility. We proceed discussing the pure jump case alone first and then add the Brownian component later.

16.6.1 The Pure Jump Case

Suppose now that our financial market is of the form (16.129) and (16.130) with the coefficient $\sigma\equiv 0$. In addition, we assume $\theta(\omega,t,z)$, $\omega\in\Omega$, $t\in[0,T]$, $z\in\mathbb{R}_0$, to be $\mathbb{F}$-adapted and càglàd.

The optimization problem (16.133) we study is

$$u(x):=\sup_{\pi\in\mathcal{A}_{\mathbb{H}}}E\left[\log X_\pi^x(T)\right]=E\left[\log X^x(\pi^*)(T)\right],\qquad(16.174)$$

where, as usual, the wealth $X_\pi(t)=X_\pi^x(t)$, $t\in[0,T]$, of the insider is described by the equation

$$dX_\pi(t)=X_\pi(t^-)\left[\{\rho(t)(1-\pi(t))+\pi(t)\mu(t))\}\,dt+\int_{\mathbb{R}_0}\pi(t)\theta(t,z)\tilde{N}(d^-t,dz)\right]$$

and $X_\pi(0)=X_\pi^x(0)=x$.

In the logarithmic case, we can describe the set of admissible portfolios as follows:

Definition 16.42. *The set $\mathcal{A}_{\mathbb{H}}$ of admissible portfolios consists of all processes $\pi(t)$, $t\in[0,T]$, such that*

▷ π *is $\mathbb{H}$-adapted.* $\qquad\qquad\qquad\qquad\qquad\qquad\qquad\qquad\qquad\qquad\qquad\quad$ (16.175)

▷ $\pi(t)\theta(t,z)$, $t\in[0,T]$, $z\in\mathbb{R}_0$, *is càglàd and forward integrable with*

$\quad$ *respect to $\tilde{N}$.* $\qquad\qquad\qquad\qquad\qquad\qquad\qquad\qquad\qquad\qquad\qquad\qquad$ (16.176)

▷ $\pi(t)\theta(t,z) > -1 + \epsilon_\pi$ for $\nu(dz)dt$-a.a. (t,z) for some $\epsilon_\pi > 0$.   (16.177)

▷ $E\left[\int_0^T \int_{\mathbb{R}_0} (\pi(t)\theta(t,z))^2 \, \nu_{\mathcal{F}}(dz)dt\right] < \infty$.   (16.178)

▷ π is Malliavin differentiable and $D_{t^+,z}\pi(t) = \lim_{s\to t^+} D_{s,z}\pi(t)$ exists

  for a.a. (t,z).   (16.179)

▷ $\theta(t,z)(\pi(t) + D_{t^+,z}\pi(t)) > -1 + \varepsilon_\pi$ for a.a. (t,z) for some $\varepsilon_\pi > 0$.
     (16.180)

▷ $E\left[\int_0^T \int_{\mathbb{R}_0} |\theta(t,z)D_{t^+,z}\pi(t)|\nu_{\mathcal{F}}(dz)dt\right] < \infty$.   (16.181)

Recall that $D_{t^+,z}F = 0$ whenever F is $\mathcal{F}_t$-measurable. Then we have

$$D_{t^+,z}\log\left(1 + \pi(t)\theta(t,z)\right) = \log\left(1 + \theta(t,z)(\pi(t) + D_{t^+,z}\pi(t))\right) - \log\left(1 + \pi(t)\theta(t,z)\right).$$

The expression above together with conditions (16.176) and (16.180) yield

▷ $\log\left(1 + \pi(t)\theta(t,z)\right)$ is càglàd and forward integrable.
▷ $\log\left(1 + \pi(t)\theta(t,z)\right) \in \mathbb{D}_{1,2}$ and $D_{t^+,z}\log\left(1 + \pi(t)\theta(t,z)\right)$ exists
  for a.a. (t,z).
▷ $E\left[\int_0^T \int_{\mathbb{R}_0} |D_{t^+,z}\log\left(1 + \pi(t)\theta(t,z)\right)|\nu_{\mathcal{F}}(dz)dt\right] < \infty$.

Compare with Definition 16.30.
Following the same lines of the proof of Theorem 16.32, we obtain the following result.

Theorem 16.43. *Suppose* $\pi(s) = \pi^*(t)$, $t \in [0,T]$, *is optimal for problem* (16.174). *Then* $M_\pi(t)$, $t \in [0,T]$, *given by*

$$M_\pi(t) = \int_0^t \left\{\mu(s) - \rho(s) - \int_{\mathbb{R}_0} \frac{\pi(s)\theta^2(s,z)}{1 + \pi(s)\theta(s,z)}\nu_{\mathcal{F}}(dz)\right\} ds$$

$$+ \int_0^t \int_{\mathbb{R}_0} \frac{\theta(s,z)}{1 + \pi(s)\theta(s,z)}\tilde{N}(d^-s,dz),$$ (16.182)

is a martingale with respect to the filtration $\mathbb{H}$.

Remark 16.44. Note that the process

$$R_t := \int_0^t \int_{\mathbb{R}_0} \frac{\theta(s,z)}{1 + \pi(s)\theta(s,z)}\tilde{N}(d^-s,dz), \quad t \in [0,T],$$

is a special $\mathbb{H}$-semimartingale with decomposition given by (16.182) (for the definition of a special semimartingale see, e.g., [204] p. 129).

We recall that the integer valued random measure $N(dt, dz)$ has a unique predictable compensator $\nu_{\mathbb{H}}(dt, dz)$ with respect to $\mathbb{H}$ (see [123], p.66). Note, however, that this alone would not imply that R_t, $t \in [0, T]$, is a $\mathbb{H}$-semimartingale, because the integrals with respect to $\nu_{\mathbb{H}}(dt, dz)$ need not be processes of finite variation. In any case, we may write

$$
\begin{aligned}
M_\pi(t) = & \int_0^t \int_{\mathbb{R}_0} \frac{\theta(s, z)}{1 + \pi(s)\theta(s, z)} (N - \nu_{\mathbb{H}})(ds, dz) \\
& + \int_0^t \int_{\mathbb{R}_0} \frac{\theta(s, z)}{1 + \pi(s)\theta(s, z)} (\nu_{\mathbb{H}} - \nu_{\mathbb{F}})(ds, dz) \\
& + \int_0^t \left\{ \mu(s) - \rho(s) - \int_{\mathbb{R}_0} \frac{\pi(s)\theta^2(s, z)}{1 + \pi(s)\theta(s, z)} \nu_{\mathcal{F}}(dz) \right\} ds,
\end{aligned}
$$

where $\nu_{\mathbb{F}}(ds, dz) = \nu(dz)dt$. Hence by uniqueness of the semimartingale decomposition of the $\mathbb{H}$-semimartingale $M_\pi(t)$, $t \in [0, T]$ (see, e.g., [204], Theorem 30, Chap. 7) we conclude that the finite variation part above must be 0. Therefore, we get the following result – see [65].

Theorem 16.45. *Suppose $\pi \in \mathcal{A}_{\mathbb{H}}$ is optimal for problem (16.174). Then π solves the equation*

$$
\begin{aligned}
\int_0^t & \left[\mu(s) - \rho(s) - \int_{\mathbb{R}_0} \frac{\pi(s)\theta^2(s, z)}{1 + \pi(s)\theta(s, z)} \nu(dz) \right] ds \\
& = \int_0^t \int_{\mathbb{R}_0} \frac{\theta(s, z)}{1 + \pi(s)\theta(s, z)} (\nu_{\mathbb{F}} - \nu_{\mathbb{H}})(ds, dz).
\end{aligned}
\tag{16.183}
$$

In particular, we get

Corollary 16.46. *Suppose $\mathbb{H} = \mathbb{F}$. Then a necessary condition for π to be optimal is that for a.a. s*

$$
\mu(s) - \rho(s) - \int_{\mathbb{R}_0} \frac{\pi(s)\theta^2(s, z)}{1 + \pi(s)\theta(s, z)} \nu_{\mathcal{F}}(dz) = 0.
\tag{16.184}
$$

Note that this could also be seen by direct computation.

The following result may be regarded as a variant (in the Malliavin calculus setting) of the result of [56], stating that if $S_1(t)$, $t \in [0, T]$, is a given locally bounded, adapted càdlàg price process with respect to the filtration $\mathbb{H}$ and there is no arbitrage by simple strategies on $S_1(t)$, $t \in [0, T]$, then $S_1(t)$, $t \in [0, T]$, is a $\mathbb{H}$-semimartingale. The result reads as follows – see [65].

Theorem 16.47. *Suppose there exists an optimal portfolio for problem (16.174). Then the process*

$$
\int_0^t \int_{\mathbb{R}_0} \theta(s, z)\widetilde{N}(ds, dz), \qquad t \in [0, T],
$$

is a $\mathbb{H}$-semimartingale.

Proof We only need to show that $\int_0^t \int_{\mathbb{R}_0} \theta(s,z)(\nu_{\mathbb{F}} - \nu_{\mathbb{H}})(ds,dz)$, $t \in [0,T]$, exists and is of finite variation. In this case, in fact, the $\mathbb{H}$-martingale

$$\int_0^t \int_{\mathbb{R}_0} \theta(s,z)(N - \nu_{\mathbb{H}})(ds,dz) = \int_0^t \int_{\mathbb{R}_0} \theta(s,z)(N - \nu_{\mathbb{F}})(ds,dz)$$

$$+ \int_0^t \int_{\mathbb{R}_0} \theta(s,z)(\nu_{\mathbb{F}} - \nu_{\mathbb{H}})(ds,dz)$$

exists and $\int_{\mathbb{R}_0} \theta(t,z)\widetilde{N}(dt,dz)$, $t \in [0,T]$, is a $\mathbb{H}$-semimartingale. By Theorem 16.32 we know that

$$\int_0^t \int_{\mathbb{R}_0} \frac{\theta(s,z)}{1 + \pi(s)\theta(s,z)}(\nu_{\mathbb{F}} - \nu_{\mathbb{H}})(ds,dz), t \in [0,T],$$

is of finite variation. So by our assumption (16.177) it follows that

$$\int_0^t \int_{\mathbb{R}_0} \theta(s,z)(\nu_{\mathbb{F}} - \nu_{\mathbb{H}})(ds,dz), t \in [0,T],$$

is of finite variation also. $\square$

16.6.2 A Mixed Market Case

We consider once again our market model of form (16.129) and (16.130) where, in addition, we assume $\theta(\omega, t, z)$, $\omega \in \Omega$, $t \in [0,T]$, $z \in \mathbb{R}_0$, to be $\mathbb{F}$-adapted and continuous in z around zero for a.a. (ω, t). The flow of information available is the same as before with $\mathbb{H}$ being the inside information and $\mathbb{F}$ being the one generated by the noise events. The optimization portfolio studied is again

$$u(x) := \sup_{\pi \in \mathcal{A}_{\mathbb{H}}} E\left[\log X_\pi^x(T)\right] = E\left[\log X_{\pi^*}^x(T)\right], \tag{16.185}$$

where the wealth $X_\pi(t) = X_\pi^x(t)$, $t \in [0,T]$, of the insider is described by the equation

$$dX_\pi(t) = X_\pi(t^-)\left[\{\rho(t)(1 - \pi(t)) + \pi(t)\mu(t)\}\,dt + \pi(t)\sigma(t)d^-W(t)\right.$$

$$\left. + \int_{\mathbb{R}_0} \pi(t)\theta(t,z)\widetilde{N}(d^-t,dz)\right], \tag{16.186}$$

and $X_\pi(0) = X_\pi^x(0) = 0$. The set of admissible portfolios $\mathcal{A}_{\mathbb{H}}$ is given in Definition 16.42 with the addition of the requirement $\pi(t)\sigma(t)$, $t \in [0,T]$, to be forward integrable with respect to W.

Proceeding in the same line and arguments as for Theorem 16.34 and Theorem 16.43 we have the corresponding following result – see [65].

Theorem 16.48. *Define*

$$M_\pi(t) = \int_0^t \left\{ \mu(s) - \rho(s) - \sigma^2(s)\pi(s) - \int_{\mathbb{R}_0} \frac{\pi(s)\theta^2(s,z)}{1 + \pi(s)\theta(s,z)} \nu_{\mathcal{F}}(dz) \right\} ds$$

$$+ \int_0^t \sigma(s)dW(s) + \int_0^t \int_{\mathbb{R}_0} \frac{\theta(s,z)}{1 + \pi(s)\theta(s,z)} \tilde{N}(d^-s, dz). \quad (16.187)$$

Suppose $\pi = \pi^*$ is optimal for problem (16.185). Then $M_\pi(t)$, $t \in [0, T]$, is a martingale with respect to the filtration $\mathbb{H}$.

Further, we see that the orthogonal decomposition of $M_\pi(t)$, $t \in [0, T]$, into a continuous part $M_\pi^c(t)$ and a discontinuous part $M_\pi^d(t)$, $t \in [0, T]$, is given by

$$M_\pi^c(t) = \int_0^t \sigma(s)dW(s) + \int_0^t \sigma(s)\alpha(s)ds, \quad (16.188)$$

$$M_\pi^d(t) = \int_0^t \int_{\mathbb{R}_0} \frac{\theta(s,z)}{1 + \pi(s)\theta(s,z)} \tilde{N}(d^-s, dz) + \int_0^t \gamma(s)ds, \quad (16.189)$$

where $\alpha(s)$ and $\gamma(s)$ are unique $\mathbb{H}$-adapted processes such that

$$\int_0^t \sigma(s)\alpha(s)ds + \int_0^t \gamma(s)ds = \int_0^t \left\{ \mu - \rho - \sigma^2\pi - \int_{\mathbb{R}_0} \frac{\pi(s)\theta^2(s,z)}{1 + \pi(s)\theta(s,z)} \nu_{\mathcal{F}}(dz) \right\} ds.$$

So the proof of Theorem 16.47, together with the fact that $\int_0^t \frac{1}{\sigma(s)}dM_\pi^c(s) = W(t) + \int_0^t \alpha(s)ds$ also is a $\mathbb{H}$-martingale, gives the following result – [65].

Theorem 16.49. *Suppose there exists an optimal portfolio for problem (16.185). Then the underlying processes*

$$\int_0^t \int_{\mathbb{R}_0} \theta(s,z)\tilde{N}(ds, dz) \quad \text{and} \quad W(t), \qquad t \in [0, T],$$

are $\mathbb{H}$-semimartingales.

Finally, we get as an analog to Theorem 16.45 – see [65].

Theorem 16.50. *Suppose $\pi \in \mathcal{A}_{\mathbb{H}}$ is optimal for problem (16.185). Then π solves the equation*

$$\int_0^t \left\{ \mu(s) - \rho(s) - \sigma^2(s)\pi(s) - \int_{\mathbb{R}_0} \frac{\pi(s)\theta^2(s,z)}{1 + \pi(s)\theta(s,z)} \nu(dz) \right\} ds$$

$$= \int_0^t \sigma(s)\alpha(s)ds + \int_0^t \int_{\mathbb{R}_0} \frac{\theta(s,z)}{1 + \pi(s)\theta(s,z)} (\nu_{\mathbb{F}} - \nu_{\mathbb{H}})(ds, dz), \quad (16.190)$$

where α is the process from (16.188) and $\nu_{\mathbb{H}}$ is the $\mathbb{H}$ compensator of N and, as usual, we have set $\nu_{\mathbb{F}}(dt, dz) = \nu(dz)dt$.

Corollary 16.51. *Suppose* $\mathbb{H} = \mathbb{F}$. *Then a necessary condition for* π *to be optimal is that, for a.a. s,*

$$\mu(s) - \rho(s) - \sigma^2(s)\pi(s) - \int_{\mathbb{R}_0} \frac{\pi(s)\theta^2(s,z)}{1 + \pi(s)\theta(s,z)}\nu(dz) = 0. \tag{16.191}$$

16.6.3 Examples: Enlargement of Filtration

Let us consider a financial market (16.129) and (16.130) where the underlying driving jump process of the risky asset is a pure jump Lévy process $\eta(t)$, $t \in [0,T]$, i.e.,

$$\int_0^t \int_{\mathbb{R}_0} \theta(t,z)\widetilde{N}(dt,dz) = \int_0^t \int_{\mathbb{R}_0} z\widetilde{N}(dt,dz) =: \eta(t).$$

In this series of examples we want to analyze the optimization problem (16.185) in which the insider has at most knowledge about the value of the underlying driving processes $W(T_0)$ and $\eta(T_0)$ at some time $T_0 \geq T$. This means that the insider filtration $\mathbb{H}$ is such that $\mathcal{F}_t \subseteq \mathcal{H}_t \subseteq \mathcal{K}_t$, where $\mathbb{K} := \{\mathcal{K}_t := \mathcal{F}_t \vee \sigma\left(W(T_0),\eta(T_0)\right) : t \in [0,T]\}$. We refer to [65] for the following result.

Proposition 16.52. *Let* $\mathbb{H}$ *be an insider filtration such that* $\mathcal{F}_t \subseteq \mathcal{H}_t \subseteq \mathcal{K}_t$. *Then*

$$W(t) - \int_0^t \frac{E\left[W(T_0)|\mathcal{H}_s\right] - W(s)}{T_0 - s}ds \tag{16.192}$$

and

$$\eta(t) - \int_0^t \frac{E\left[\eta(T_0)|\mathcal{H}_s\right] - \eta(s)}{T_0 - s}ds = \eta(t) - \int_0^t E\left[\int_s^{T_0} \int_{\mathbb{R}} \frac{z}{T_0 - s}\widetilde{N}(dr,dz)\Big|\mathcal{H}_s\right]ds \tag{16.193}$$

are $\mathbb{H}$*-martingales.*

Proof We know by an extension of a result of Itô [122] (see also [204] p. 356) that for a general Lévy process $\Lambda(t)$, $t \in [0,T]$, with respect to some filtration $\widehat{\mathcal{F}}_t$, $t \in [0,T]$, the process

$$\Lambda(t) - \int_0^t \frac{\Lambda(T_0) - \Lambda(s)}{T_0 - s}ds, \quad t \in [0,T],$$

is a $\left\{\widehat{\mathcal{F}}_t \vee \sigma\left(\Lambda(T_0)\right)\right\}$-martingale for $0 \leq t \leq T_0$. Using this result and the fact that $W(t)$, $t \in [0,T]$, and $\eta(t)$, $t \in [0,T]$, are independent we obtain that

$$W(t) - \int_0^t \frac{W(T_0) - W(s)}{T_0 - s}ds \quad \text{and} \quad \eta(t) - \int_0^t \frac{\eta(T_0) - \eta(s)}{T_0 - s}ds, \quad t \in [0,T],$$

are $\mathbb{K}$-martingales. So we have

$$E\left[W(t) - \int_0^t \frac{E\left[W(T_0)|\mathcal{H}_s\right] - W(s)}{T_0 - s}ds\Big|\mathcal{H}_r\right]$$

$$= E\left[W(t) - W(r) - \int_r^t \frac{E\left[W(T_0)|\mathcal{H}_s\right] - W(s)}{T_0 - s}ds\Big|\mathcal{H}_r\right]$$

$$+ W(r) - \int_0^r \frac{E\left[W(T_0)|\mathcal{H}_s\right] - W(s)}{T_0 - s}ds$$

$$= E\left[E\left[W(t) - W(r) - \int_r^t \frac{W(T_0) - W(s)}{T_0 - s}ds\Big|\mathcal{K}_r\right]\Big|\mathcal{H}_r\right] + W(r)$$

$$- \int_0^r \frac{E\left[W(T_0)|\mathcal{H}_s\right] - W(s)}{T_0 - s}ds$$

$$= 0 + W(r) - \int_0^r \frac{E\left[W(T_0)|\mathcal{H}_s\right] - W(s)}{T_0 - s}ds.$$

For $\eta(t) - \int_0^t \frac{E[\eta(T_0)|\mathcal{H}_s]-\eta(s)}{T_0-s}ds$, $t \in [0,T]$, the reasoning is analogous. $\square$

Proposition 16.52 tells us that in the present situation of enlargement of filtration the process α from (16.188) is of the form $-\frac{E[W(T_0)|\mathcal{H}_t]-W(t)}{T_0-t}$, $t \in [0,T]$. Moreover, we can easily deduce the $\mathbb{H}$ compensator $\nu_{\mathbb{H}}$ of N from Proposition 16.52. In fact, we have the following result – see [65].

Proposition 16.53. *The compensating measure $\nu_{\mathbb{H}}$ of the jump measure N with respect to $\mathbb{H}$ is given by*

$$\nu_{\mathbb{H}}(ds, dz) = \nu(dz)ds + E\left[\frac{1}{T_0 - s}\int_s^{T_0} \tilde{N}(dr, dz)\Big|\mathcal{H}_s\right]ds \qquad (16.194)$$

$$= E\left[\frac{1}{T_0 - s}\int_s^{T_0} N(dr, dz)\Big|\mathcal{H}_s\right]ds \qquad (16.195)$$

Proof We know (see [123] p. 80) that it is sufficient to show that if $\hat{\nu}_{\mathbb{H}}$ is the right-hand side of (16.194) then

$$\int_0^t \int_{\mathbb{R}_0} f(z)(N - \hat{\nu}_{\mathbb{H}})(ds, dz)$$

is a $\mathbb{H}$-martingale for all f, which are bounded deterministic functions on $\mathbb{R}$, zero around zero, and that determine a measure on $\mathbb{R}$ with weight zero in zero. The same argument holds for f invertible functions that are integrable with respect to $\hat{\nu}_{\mathbb{H}}$ (note that this implies also integrability with respect to ν). Let $f(z)$, $\omega \in \Omega$, be such a function. Then $\bar{\eta}(t)$, $t \in [0,T]$, given by

$$\bar{\eta}(t) := \int_0^t \int_{\mathbb{R}_0} f(z) \tilde{N}(ds, dz)$$

is a pure jump Lévy process (see, e.g., Sect. 2.3.2 in [8]). Consider the Lévy process

$$W(t) + \bar{\eta}(t), t \in [0, T],$$

whose filtration is denoted by $\bar{\mathcal{F}}_t$, $t \in [0, T]$. Since f is invertible, we have $\bar{\mathcal{F}}_t = \mathcal{F}_t$ and $\bar{\mathcal{K}}_t = \mathcal{K}_t$, where $\bar{\mathcal{K}}_t = \bar{\mathcal{F}}_t \vee \sigma(W(T_0), \bar{\eta}(T_0))$. From Proposition 16.52 we then get that

$$\bar{M}(t) := \int_0^t \int_{\mathbb{R}_0} f(z) \tilde{N}(ds, dz) - \int_0^t E\left[\int_s^{T_0} \int_{\mathbb{R}} \frac{f(z)}{T_0 - s} \tilde{N}(dr, dz)\Big|\mathcal{G}_s\right] ds$$

is a $\mathbb{H}$-martingale. (16.195) is then a straight forward algebraic transformation.

□

Using the measure given by (16.194) (note that here $\theta(t, z) = z$), we see that the necessary condition for an optimal portfolio given by equation (16.190) in our present context becomes

$$\int_0^t \left\{ \mu(s) - \rho(s) - \sigma^2(s)\pi(s) - \int_{\mathbb{R}_0} \frac{\pi(s)z^2}{1 + \pi(s)z} \nu(dz) \right\} ds$$

$$= \int_0^t \left\{ -\sigma(s) \frac{E\left[W(T_0)|\mathcal{G}_s\right] - W(s)}{T_0 - s} \right.$$

$$\left. - E\left[\int_s^{T_0} \int_{\mathbb{R}} \frac{z}{(1 + \pi(s)z)(T_0 - s)} \tilde{N}(dr, dz)\Big|\mathcal{G}_s\right] \right\} ds. \quad (16.196)$$

When $\eta(t)$, $t \in [0, T]$, is of finite variation this can be rewritten as

$$\int_0^t \left\{ \mu(s) - \rho(s) - \sigma^2(s)\pi(s) - \int_{\mathbb{R}_0} z\nu(dz) \right\} ds$$

$$= \int_0^t \left\{ -\sigma(s) \frac{E\left[W(T_0)|\mathcal{H}_s\right] - W(s)}{T_0 - s} \right.$$

$$\left. - E\left[\int_s^{T_0} \int_{\mathbb{R}} \frac{z}{(1 + \pi(s)z)(T_0 - s)} N(dr, dz)\Big|\mathcal{H}_s\right] \right\} ds. \quad (16.197)$$

Given some additional assumptions, this is also a *sufficient* condition for a portfolio $\pi \in \mathcal{A}_{\mathbb{H}}$ to be optimal. In fact, we have the following result – see [65].

Theorem 16.54. *Assume that $\eta(t)$, $t \in [0, T]$, is of finite variation. The portfolio $\pi = \pi(\omega, s)$, $\omega \in \Omega$, $s \in [0, T]$, is optimal for the insider if and only if $\pi \in \mathcal{A}_{\mathbb{H}}$ and for a.a. (ω, s) π solves the equation*

$$\mu(s) - \rho(s) - \sigma^2(s)\pi(s) - \int_{\mathbb{R}_0} z\nu(dz)$$

$$= -\sigma(s)\frac{E\left[W(T_0)|\mathcal{H}_s\right]^- - W(s)}{T_0 - s} - E\left[\int_s^{T_0}\int_{\mathbb{R}} \frac{z}{(1 + \pi(s)z)(T_0 - s)}N(dr, dz)\Big|\mathcal{H}_s\right]^-$$

$$(16.198)$$

where the notation $E[...]^-$ *denotes the left limit in* s.

Proof By Proposition 16.52 and Proposition 16.53, the solution to equation (16.186) becomes

$$E\left[\log\frac{X^\pi(T)}{x}\right] = E\left[\int_0^T\left\{\rho(s) + (\mu(s) - \rho(s))\pi(s) - \frac{1}{2}\sigma(s)^2\pi^2(s)\right.\right.$$

$$\left. - \int_{\mathbb{R}_0}\pi(s)z\nu(dz)\right\}ds + \int_0^T \sigma(s)\pi(s)d(W(s) + \alpha(s)ds)$$

$$- \int_0^T \sigma(s)\pi(s)\alpha(s)ds + \int_0^T\int_{\mathbb{R}_0}\log(1 + \pi(s)z)(N - \nu_{\mathbb{H}})(ds, dz)$$

$$\left. + \int_0^T\int_{\mathbb{R}_0}\log(1 + \pi(s)z)\nu_{\mathbb{H}}(ds, dz)\right] \qquad (16.199)$$

$$= E\left[\int_0^T\left\{\rho(s) + (\mu(s) - \rho(s))\pi(s) - \frac{1}{2}\sigma(s)^2\pi^2(s)\right.\right.$$

$$- \int_{\mathbb{R}_0}\pi(s)z\nu(dz)\right\}ds + \int_0^T\left\{\sigma(s)\pi(s)\frac{E\left[W(T_0)|\mathcal{H}_s\right] - W(s)}{T_0 - s}\right.$$

$$\left.\left. + E\left[\int_s^{T_0}\int_{\mathbb{R}_0}\frac{\log(1 + \pi(s)z)}{(T_0 - s)}N(dr, dz)\Big|\mathcal{H}_s\right]\right\}ds\right].$$

We can maximize this pointwise for each fixed (ω, s). Define

$$H(\pi) = \rho(s) + (\mu(s) - \rho(s))\pi - \frac{1}{2}\sigma(s)^2\pi^2 - \int_{\mathbb{R}_0}\pi z\nu(dz)$$

$$+ \sigma(s)\pi\frac{E\left[W(T_0)|\mathcal{H}_s\right] - W(s)}{T_0 - s} + E\left[\int_s^{T_0}\int_{\mathbb{R}_0}\frac{\log(1 + \pi z)}{(T_0 - s)}N(dr, dz)\Big|\mathcal{H}_s\right].$$

Then a stationary point π of H is given by

$$0 = H'(\pi) = \mu(s) - \rho(s) - \sigma^2(s)\pi(s) - \int_{\mathbb{R}_0} z\nu(dz)$$

$$+ \sigma(s)\frac{E\left[W(T_0)|\mathcal{H}_s\right] - W(s)}{T_0 - s}$$

$$+ E\left[\int_s^{T_0}\int_{\mathbb{R}}\frac{z}{(1 + \pi(s)z)(T_0 - s)}N(dr, dz)\Big|\mathcal{H}_s\right].$$

Since H is concave, a stationary point of H is a maximum point of H. But, since for a given ω the set of discontinuities has Lebesgue measure zero, the equation

$$\mu(s) - \rho(s) - \sigma^2(s)\pi(s) - \int_{\mathbb{R}_0} z\nu(dz)$$

$$= -\sigma(s)\frac{E\left[W(T_0)|\mathcal{H}_s\right]^- - W(s)}{T_0 - s} - E\left[\int_s^{T_0}\int_{\mathbb{R}}\frac{z}{(1+\pi(s)z)(T_0-s)}N(dr,dz)\Big|\mathcal{H}_s\right]^-$$

also describes an optimal portfolio. □

Example 16.55. In order to get explicit expressions for π, we now apply this to the case when $\eta(t)$ is a compensated Poisson process of intensity $\lambda > 0$. In this case the corresponding Lévy measure is

$$\nu(dz)ds = \lambda\delta_1(dz)ds,$$

where $\delta_1(dz)$ is the unit point mass at 1, and we can write

$$\eta(t) = Q(t) - \lambda t, \quad t \in [0,T],$$

Q being a Poisson process of intensity λ. Since in this case

$$l(\pi) := \log(1+\pi)\frac{E\left[Q(T_0)|\mathcal{H}_s\right] - Q(s)}{T_0 - s}$$

is concave in π, we get by Theorem 16.54 that a necessary and sufficient condition for an optimal insider portfolio π for a.a (ω, s) is given by the equation

$$0 = \mu(s) - \rho(s) - \lambda - \sigma^2(s)\pi(s) + \sigma(s)\frac{E\left[W(T_0)|\mathcal{H}_s\right]^- - W(s)}{T_0 - s}$$

$$+ \frac{E\left[Q(T_0) - Q(s)|\mathcal{H}_s\right]^-}{(1+\pi(s))(T_0 - s)}. \tag{16.200}$$

If we deal with the market (16.129) and (16.130), with the Lévy measure given above and $\sigma \equiv 0$, then we have the following result – see [65].

Theorem 16.56. *Assume that $\rho(s)$ and $\mu(s)$ are bounded and $\lambda + \rho(s) - \mu(s) > 0$ and bounded away from 0. Then*

(1) There exists an optimal insider portfolio if and only if

$$E\left[Q(T_0)|\mathcal{H}_s\right] - Q(s) > 0$$

for a.a (ω, s). In this case

$$\pi^*(s) = \frac{E\left[Q(T_0) - Q(s)|\mathcal{H}_s\right]^-}{(T_0 - s)(\lambda + \rho(s) - \mu(s))} - 1 \tag{16.201}$$

is the optimal portfolio for the insider.

(2) Assume there exists an optimal insider portfolio π. Then the value function
$u(x)$, $x \in \mathbb{R}$, for the insider is finite for all $T_0 \geq T$.

Proof Part (1) follows from equation (16.200) setting $\sigma \equiv 0$. It remains to prove (2). We substitute the value (16.201) for π^* into the expression (16.199) and get

$$E\left[\log \frac{X^\pi(T)}{x}\right] = E\left[\int_0^T \left\{2\rho(s) - \mu(s) + \lambda + \left(\frac{E\left[Q(T_0)|\mathcal{H}_s\right] - Q(s)}{(T_0 - s)}\right)\right.\right.$$

$$\left.\left. + \log\left(\frac{E\left[Q(T_0)|\mathcal{H}_s\right] - Q(s)}{(T_0 - s)(\lambda + \rho(s) - \mu(s))}\right)\left(\frac{E\left[Q(T_0)|\mathcal{H}_s\right] - Q(s)}{(T_0 - s)}\right)\right\} ds\right].$$
$$(16.202)$$

By means of the value of the moments of the Poisson distribution and of the Jensen inequality in its conditional form, we obtain

$$E\left[\int_0^T \frac{E\left[Q(T_0)|\mathcal{H}_s\right] - Q(s)}{T_0 - s} ds\right] = \int_0^T \frac{E[Q(T_0) - Q(s)]}{T_0 - s} ds = \lambda T < \infty$$
$$(16.203)$$

and

$$E\left[\int_0^T \log\left(E\left[Q(T_0)|\mathcal{H}_s\right] - Q(s)\right) \frac{E\left[Q(T_0)|\mathcal{H}_s\right] - Q(s)}{T_0 - s} ds\right]$$

$$\leq E\left[\int_0^T \frac{\left(E[Q_{T_0}|\mathcal{H}_s] - Q(s)\right)^2}{T_0 - s} ds\right]$$

$$\leq \int_0^T \frac{E\left[E[(Q(T_0) - Q(s))^2|\mathcal{H}_s]\right]}{T_0 - s} ds$$

$$\leq \int_0^T \frac{E[(Q(T_0) - Q(s))^2]}{T_0 - s} ds$$

$$= \int_0^T \left(\lambda^2(T_0 - s) + \lambda\right) ds < \infty$$
$$(16.204)$$

and also

$$E\left[\int_0^T \log\left(\frac{1}{T_0 - s}\right) \frac{E\left[Q(T_0)|\mathcal{H}_s\right] - Q(s)}{T_0 - s} ds\right] = \lambda \int_0^T \log\left(\frac{1}{T_0 - s}\right) ds < \infty.$$
$$(16.205)$$

Using (16.203)–(16.205), we see that (16.202) is finite. $\square$

Remark 16.57. In the pure Poisson jump case, Theorem 16.56 shows that if the insider filtration is $\mathcal{K}_t = \mathcal{F}_t \vee \sigma(Q(T_0))$, $t \in [0, T]$, then there is no optimal portfolio since $E\left[Q(T_0)|\mathcal{H}_s\right] - Q(s) = 0$ for all ω such that $Q(T_0) = 0$. This is contrary to the pure Brownian motion case with the enlargement of filtration $\mathcal{K}_t = \mathcal{F}_t \vee \sigma(W(T_0))$, $t \in [0, T]$, where we have an optimal portfolio (see [191]). The reason is that the insider has an arbitrage opportunity as soon as he knows where $Q(t)$, $t \in [0, T]$, does not jump. On the other hand, as soon as there exists an optimal portfolio in the Poisson pure jump market for an insider filtration $\mathcal{F}_t \subseteq \mathcal{H}_t \subseteq \mathcal{K}_t$, the value function $u(x)$, $x > 0$, for the insider is finite also for $T_0 = T$, which again is contrary to the pure Brownian motion case.

Example 16.58. If again we deal with market (16.129) and (16.130), with the Lévy measure given by the Poisson process, then we have the following result – [65].

Theorem 16.59. *Set*

$$\alpha(s) = -\frac{E\left[W(T_0) - W(s) | \mathcal{H}_s\right]^-}{T_0 - s} \quad and \quad \gamma(s) = -\frac{E\left[Q(T_0) - Q(s) | \mathcal{H}_s\right]^-}{T_0 - s}.$$

Then

(1) For all insider filtrations $\mathcal{F}_t \subseteq \mathcal{H}_t \subseteq \mathcal{K}_t$ and $T_0 > T$ there exists an optimal insider portfolio given by

$$\pi^*(s) = \frac{1}{2\sigma^2(s)} \left(\mu(s) - \rho(s) - \lambda - \sigma(s)\alpha(s) - \sigma^2(s)\right.$$

$$\left. + \sqrt{(\mu(s) - \rho(s) - \lambda - \sigma(s)\alpha(s) + \sigma^2(s))^2 - 4\sigma^2(s)\gamma(s)}\right). \quad (16.206)$$

(2) The value function $u(x)$, $x > 0$, for the insider is finite for all $T_0 > T$.

Proof Part (1) follows by solving (16.200) for π. Here, the condition $\pi(s) > -1$ is not fulfilled $\nu(dz)dt$ a.e., but $N(dz, dt)$ a.e., which is sufficient in the our situation of the Poisson process. Concerning part (2), it is sufficient to consider the largest insider filtration $\mathcal{K}_t = \mathcal{F}_t \vee \sigma(W(T_0), Q(T_0))$, $t \in [0, T]$. Then

$$\alpha(s) = -\frac{W(T_0) - W(s)}{T_0 - s} \quad and \quad \gamma(s) = -\frac{Q(T_0) - Q(s)}{T_0 - s}.$$

Using the fact that

$$E[\int_0^T \alpha^2(s)ds] = \int_0^T \frac{1}{T_0 - s}ds = \log\left(\frac{T_0}{T_0 - T}\right)$$

in addition to (16.203)–(16.205) and Jensen inequality, one can show the finiteness of $u(x)$, $x > 0$, with the same techniques as in the proof of Theorem 16.56. $\qquad\square$

Remark 16.60. Contrary to the pure Poisson jump case, the mixed case gives rise to an optimal insider portfolio for all insider filtrations $\mathbb{H}$ such that $\mathcal{F}_t \subseteq \mathcal{H}_t \subseteq \mathcal{K}_t$, $t \in [0,T]$. The reason is that while it is possible to introduce arbitrage possibilities for the insider through an enlargement of filtration in the pure jump case (see e.g., [91]), this is no longer the case in the mixed market. But in contrast to the pure jump case, the finiteness of the value function $u(x)$, $x > 0$, is only ensured for T_0 strictly bigger than T. However, choosing the filtration "small enough with respect to the information $W(T_0)$," one can generate a finite value function also for $T_0 = T$. The most obvious example would be $\mathcal{H}_t = \mathcal{F}_t \vee \sigma(Q(T_0))$, in which case $\alpha(s) \equiv 0$. This example is treated in [77].

16.7 Exercises

Problem 16.1. (*) Suppose $U(x) = -\frac{1}{x}$, $x > 0$. Let

$$F_\pi(T) = U'(X_\pi(T))X_\pi(T) = \frac{1}{X_\pi(T)}$$

as in (16.10), where $X_\pi(T)$ is given in (16.4) and (16.5). Assume that the coefficients $\rho, \mu, \sigma, \theta$ and the control π are deterministic.

(a) Find $D_t F_\pi(T)$ and $D_{t,z} F_\pi(T)$;
(b) Find, among the deterministic portfolios, the portfolio π that maximizes

$$E\left[-\frac{1}{X_\pi(T)}\right]$$

in the market (16.1) and (16.2).

Problem 16.2. Consider a market that is a combination of the markets (16.36)–(16.37) and (16.70)–(16.71), i.e.,

(Bond price) $dS_0(t) = \rho(t)S_0(t)dt$, $S_0(0) = 1$

(Stock price) $dS_1(t) = S_1(t)\left[\mu(t)dt + \sigma(t)d^-W(t)\right.$

$$\left. + \int_{\mathbb{R}_0} \theta(t,z)\tilde{N}(d^-t, dz)\right],   S_1(0) > 0$$

(with the described appropriate conditions for the coefficients). Define $\mathcal{A}_{\mathbb{E}}^{W,\tilde{N}}$ to be the set of portfolios π satisfying the conditions in both Definition 16.14 and Definition 16.19. Suppose $\tilde{\pi} \in \mathcal{A}_{\mathbb{E}}^{W,\tilde{N}}$ is optimal, in the sense that

$$\sup_{\pi \in \mathcal{A}_{\mathbb{E}}^{W,\tilde{N}}} E\left[\log X_\pi(T)\right] = E\left[\log X_{\tilde{\pi}}(T)\right],$$

where $X_\pi(T)$ is the final value of the wealth process corresponding to portfolio π. Prove that $\widetilde{\pi}$ satisfies the equation

$$E\Bigg[\mu(t) - \rho(t) + D_{t^+}\sigma(t) - \sigma^2(t)\widetilde{\pi}(t) + \int_{\mathbb{R}_0} \bigg(\frac{\theta(t,z) + D_{t,z}\theta(t,z)}{1 + \widetilde{\pi}(t)\big(\theta(t,z) + D_{t^+,z}\theta(t,z)\big)}$$

$$-\,\theta(t,z)\bigg)\nu(dz)\bigg|\mathcal{E}_t\Bigg] = 0.$$

[*Hint*. Combine the proofs of Theorem 16.15 and Theorem 16.20.]

Regularity of Solutions of SDEs Driven by Lévy Processes

In this chapter, we consider strong solutions of Itô diffusions driven by Lévy noise and we study their smoothness related to Malliavin differentiability. The next chapter will be devoted to the study of the smoothness of the probability distributions associated to the solutions. This application of the Malliavin calculus gave origin to its introduction [158]. Within the study of Itô diffusions driven by Brownian noise, we can refer to the recent monographies, e.g., [53, 160, 169, 212] and the references there in. As for the study within the Lévy type of noise, we can refer to, e.g., [35, 118, 221] and references therein. In this short chapter, we present only some fundamental results.

17.1 The Pure Jump Case

As before, consider a compensated Poisson random measure $\widetilde{N}(dt, dz) = N(dt, dz) - \nu(dz)dt$ defined on the complete probability space equipped with filtration

$$(\Omega, \mathcal{F}, P), \quad \mathbb{F} := \{\mathcal{F}_t : 0 \leq t \leq T\},$$

where $\mathcal{F}_t$, $0 \leq t \leq T$, is the P-augmented filtration generated by the values of the compensated Poisson random measure $\widetilde{N}(dt, dz)$ associated to a pure jump Lévy process and $\mathcal{F} = \mathcal{F}_T$. We require that the Lévy measure ν fulfills the integrability condition

$$\int_{\mathbb{R}_0} z^2 \nu(dz) < \infty.$$

We are interested in studying the regularity of solutions of the pure jump Lévy stochastic differential equation (SDE)

$$X(t) = x + \int_0^t \int_{\mathbb{R}_0} \gamma(s, X(s^-), z)\widetilde{N}(ds, dz), \quad 0 \leq t \leq T, \tag{17.1}$$

G.Di Nunno et al., *Malliavin Calculus for Lévy Processes with Applications to Finance*,
© Springer-Verlag Berlin Heidelberg 2009

for $x \in \mathbb{R}$, where $\gamma : [0, T] \times \mathbb{R} \times \mathbb{R}_0 \longrightarrow \mathbb{R}$ satisfies the linear growth condition

$$\int_{\mathbb{R}_0} |\gamma(t, x, z)|^2 \, \nu(dz) \le C(1 + |x|^2), \quad 0 \le t \le T, x \in \mathbb{R}, \qquad (17.2)$$

for a constant $C < \infty$ as well as the Lipschitz condition

$$\int_{\mathbb{R}_0} |\gamma(t, x, z) - \gamma(t, y, z)|^2 \, \nu(dz) \le K \, |x - y|^2, \quad 0 \le t \le T, \ x, y \in \mathbb{R}, \quad (17.3)$$

for a constant $K < \infty$.

Under the assumptions (17.2) and (17.3), one can employ Picard iteration (just as in the Brownian motion case, see, e.g., [179]) to construct a unique (global) strong solution $X(t), t \in [0, T]$, of (17.1), that is, there exists a unique càdlàg adapted solution $X(t), t \in [0, T]$, such that

$$E\left[\sup_{0 \le t \le T} |X(t)|^2 \right] < \infty. \qquad (17.4)$$

We aim at showing that $X(t), t \in [0, T]$, is regular in the sense that

$$X(t) \in \mathbb{D}_{1,2} \qquad (17.5)$$

for all $0 \le t \le T$.

Before proceeding to the main result, we need the following useful auxiliary lemma.

Lemma 17.1. *Let $F_n, n \ge 1$, be a sequence in $\mathbb{D}_{1,2}$ such that*

$$F_n \longrightarrow F, \quad n \longrightarrow \infty,$$

in $L^2(P)$. Further, we require that

$$\sup_{n \ge 1} E\left[\int_0^T \int_{\mathbb{R}_0} |D_{s,z} F_n|^2 \, \nu(dz) ds \right] < \infty.$$

Then $F \in \mathbb{D}_{1,2}$ and $D_{\cdot,\cdot} F_n, n \ge 1$ converges to $D_{\cdot,\cdot} F$ in the sense of the weak topology of $L^2(P \times \lambda \times \nu)$.

Proof By the boundedness assumption there exists a subsequence $D_{\cdot,\cdot} F_{n_k}$ such that

$$D_{\cdot,\cdot} F_{n_k} \longrightarrow \alpha, \quad k \longrightarrow \infty,$$

weakly in $L^2(P \times \lambda \times \nu)$. Using the duality relation we conclude that

$$E\left[(\alpha, u)_{L^2(\lambda \times \nu)}\right] = E\left[F\delta(u)\right]$$

for all u in the domain $Dom(\delta)$ of the Skorohod integral δ. So we see that $\alpha(t, z) = D_{t,z}F$. $\square$

The next result shows the Malliavin differentiability of solutions to the SDE (17.1). It was first proved in [35]. Our presentation is different from the original in the sense that we use different operators: our Malliavin derivative is in fact a difference operator (see, e.g., Theorem 12.8).

Theorem 17.2. *Suppose that the conditions (17.2) and (17.3) hold. Then there exists a unique strong solution $X(t)$ to the SDE (17.1) such that $X(t)$ belongs to $\mathbb{D}_{1,2}$ for all $0 \le t \le T$.*

Proof The idea of the proof is first to show that the Picard approximations $X_n(t)$, $n \ge 0$, to $X(t)$ given by

$$X_{n+1}(t) = x + \int_0^t \int_{\mathbb{R}_0} \gamma(s, X_n(s^-), z)\widetilde{N}(ds, dz), \quad X_0(t) = x \qquad (17.6)$$

are in $\mathbb{D}_{1,2}$ and then to perform the limit $n \longrightarrow \infty$ (See, e.g., [169] for the Brownian motion case). Let us first prove by induction on n that

$$X_n(t) \in \mathbb{D}_{1,2} \qquad (17.7)$$

and that

$$\phi_{n+1}(t) \le k_1 + k_2 \int_0^t \phi_n(s)ds \qquad (17.8)$$

for all $0 \le t \le T$, $n \ge 0$, where k_1, k_2 are constants and

$$\phi_n(t) := \sup_{0 \le r \le t} E\left[\int_{\mathbb{R}_0} \sup_{r \le s \le t} (D_{r,\cdot}X_n(s))^2 \, \nu(dz)\right] < \infty. \qquad (17.9)$$

One can check that (17.7) and (17.8) are fulfilled for $n = 0$, since

$$D_{t,z} \int_0^T \int_{\mathbb{R}_0} \gamma(s, x, \zeta)\widetilde{N}(ds, d\zeta) = \gamma(t, x, z)$$

by Theorem 12.15. We assume that (17.7) and (17.8) hold n. Then, the closability of the Malliavin derivative $D_{t,z}$ (see Theorem 12.6) and Theorem 12.8 imply that

$$D_{r,z}\gamma(t, X_n(t^-), z) = \gamma(t, X_n(t^-) + D_{r,z}X_n(t^-), z) - \gamma(t, X_n(t^-), z) \quad (17.10)$$

for $r \leq t$ a.e. and $\nu-$a.e. Hence, Theorem 12.15 gives that $X_{n+1}(t) \in \mathbb{D}_{1,2}$ and

$$D_{r,z}X_{n+1}(t) = \int_0^t \int_{\mathbb{R}_0} D_{r,z}\gamma(s, X_n(s^-), \zeta)\tilde{N}(ds, d\zeta) + \gamma(r, X_n(r^-), z)$$

$$= \int_0^t \int_{\mathbb{R}_0} \Big(\gamma(s, X_n(s^-) + D_{r,z}X_n(s^-), \zeta) - \gamma(s, X_n(s^-), \zeta)\Big)\tilde{N}(ds, d\zeta)$$

$$+\gamma(r, X_n(r^-), z)$$

$$= \int_r^t \int_{\mathbb{R}_0} \Big(\gamma(s, X_n(s^-) + D_{r,z}X_n(s^-), \zeta) - \gamma(s, X_n(s^-), \zeta)\Big)\tilde{N}(ds, d\zeta)$$

$$+\gamma(r, X_n(r^-), z),$$

for $r \leq t$ a.e. and $\nu-$a.e. So it follows from (17.2), (17.3), Doob maximal inequality, Fubini theorem, and the Itô isometry that

$$E\Big[\int_{\mathbb{R}_0} \sup_{r \leq s \leq t} (D_{r,\cdot}X_{n+1}(s))^2\, \nu(dz)\Big]$$

$$\leq 8K \int_r^t E\Big[\int_{\mathbb{R}_0} |D_{r,z}X_n(u^-)|^2\, \nu(dz)\Big]du + 2C\left(1 + E\Big[|X_n(r^-)|^2\Big]\right)$$

$$\leq 8K \int_r^t E\Big[\int_{\mathbb{R}_0} |D_{r,z}X_n(u^-)|^2\, \nu(dz)\Big]du + 2C(1 + \lambda), \tag{17.11}$$

for all $0 \leq r \leq t$, where

$$\lambda := \sup_{n \geq 0} E\Big[\sup_{0 \leq s \leq T} |X_n(s)|^2\Big] < \infty.$$

Note that
$$E\Big[\sup_{0 \leq s \leq T} |X_n(s) - X(s)|^2\Big] \longrightarrow 0. \quad n \longrightarrow \infty,$$

by the Picard iteration scheme. Thus (17.11) shows that (17.7) and (17.8) are valid for $n + 1$.

Finally, a discrete version of Gronwall inequality (see, e.g., [20, Lemma 4.1] or Problem 17.1) applied to (17.11) yields

$$\sup_{n \geq 0} E\Big[\int_0^T \int_{\mathbb{R}_0} |D_{s,z}X_n(t)|^2\, \nu(dz)ds\Big] < \infty$$

for all $0 \leq t \leq T$. Then it follows from Lemma 17.1 that $X(t) \in \mathbb{D}_{1,2}$. $\square$

Remark 17.3. In [35] it is shown that the solution in Theorem 17.2 even satisfies

$$X(t) \in \mathbb{D}_{1,\infty} \tag{17.12}$$

for all $0 \leq t \leq T$, where the space $\mathbb{D}_{1,\infty} \subset L^2(P)$ is introduced as

$$\mathbb{D}_{1,\infty} := \bigcap_{p \geq 1} \mathbb{D}_{1,p}.$$

Here, the space $\mathbb{D}_{1,p}$ is defined by completion on a class of smooth random variables (e.g., finite linear combinations of $I_n(f_n), n \geq 0$) with respect to the seminorm $\|\cdot\|_{1,p}$ that is given by

$$\|F\|_{1,p}^2 = E\Big[|F|^p\Big] + E\Big[\|D.F\|_{L^2(\lambda \times \nu)}^p\Big], \quad p \geq 1. \tag{17.13}$$

17.2 The General Case

Similar arguments in the proof of Theorem 17.2 can be employed to extend the latter result to the case of solutions $X(t)$, $t \in [0,T]$, to stochastic differential equations of the type

$$X(t) = x + \int_0^t \alpha(s, X(s))ds + \int_0^t \sigma(s, X(s))dW(s)$$

$$+ \int_0^t \int_{\mathbb{R}_0} \gamma(s, X(s^-), z)\widetilde{N}(ds, dz),$$

$$0 \leq t \leq T, \ x \in \mathbb{R}^n, \tag{17.14}$$

where $\alpha : [0,T] \times \mathbb{R}^n \longrightarrow \mathbb{R}^n, \sigma : [0,T] \times \mathbb{R}^n \longrightarrow \mathbb{R}^{n \times m}$, and $\gamma : [0,T] \times \mathbb{R}^n \times \mathbb{R}_0^n \longrightarrow \mathbb{R}^{n \times l}$ are Borel measurable functions. Here

$$W(t) = (W_1(t), ..., W_m(t)) \tag{17.15}$$

is an m-dimensional Brownian motion and

$$\widetilde{N}(dt, dz)^T = (\widetilde{N}_1(dt, dz_1), ..., \widetilde{N}_l(dt, dz_l))$$

$$= (N_1(dt, dz_1) - \nu_1(dz_1), ..., N_l(dt, dz_l) - \nu_l(dz_l)), \tag{17.16}$$

where $N_j, j = 1, ..., l$ are independent Poisson random measures with Lévy measures $\nu_j, j = 1, ..., l$ coming from l independent one-dimensional Lévy processes. Let $\alpha(t, x) = (\alpha_i(t, x))_{1 \leq i \leq n}$, $\sigma(t, x) = (\sigma_{ij}(t, x))_{1 \leq i \leq n, 1 \leq j \leq m}$, and $\gamma(t, x, z) = (\gamma_{ij}(t, x, z_j))_{1 \leq i \leq n, 1 \leq j \leq l}$ be the coefficients of (17.14) in the component form. Then $X(t) = (X_i(t))_{1 \leq i \leq n}$ in (17.14) can be equivalently written as

$$X_i(t) = x_i + \int_0^t \alpha_i(s, X(s))ds + \sum_{j=1}^n \int_0^t \sigma_{ij}(s, X(s))dW_j(s)$$

$$+ \sum_{j=1}^l \int_0^t \int_{\mathbb{R}_0} \gamma_{ij}(s, X(s^-), z_j)\widetilde{N}_j(ds, dz), \quad 0 \le t \le T, \ x_i \in \mathbb{R}, \quad (17.17)$$

for $x = (x_i)_{1 \le i \le n}$. Let us assume that $W(t)$ and $\widetilde{N}(dt, dz)$, $t \in [0, T]$, $z \in \mathbb{R}_0$, are constructed on the probability space $(\Omega, \mathcal{F}, P)$ given by (13.66) in Sect. 12.5. To guarantee a unique strong solution to (17.14), we assume that the coefficients of (17.14) satisfy linear growth and Lipschitz continuity, i.e.,

$$\|\sigma(t, x)\|^2 + |\alpha(t, x)|^2 + \sum_{j=1}^l \sum_{i=1}^n \int_{\mathbb{R}_0} |\gamma_{ij}(t, x, z_j)|^2 \nu_j(dz_j) \le C(1 + |x|^2) \quad (17.18)$$

for all $x \in \mathbb{R}^n$ and

$$\|\sigma(t, x) - \sigma(t, y)\|^2 + |\alpha(t, x) - \alpha(t, y)|^2$$

$$+ \sum_{j=1}^l \sum_{i=1}^n \int_{\mathbb{R}_0} |\gamma_{ij}(t, x, z_j) - \gamma_{ij}(t, y, z_j)|^2 \nu_j(dz_j) \le K |x - y|^2 \quad (17.19)$$

for all $x, y \in \mathbb{R}^n, 0 \le t \le T$, where $C, K < \infty$ are constants and $\| \cdot \|$ is a matrix norm.

Recall that a random variable F is Malliavin differentiable if and only if $F \in \mathbb{D}_{1,2}$, where the space $\mathbb{D}_{1,2} \subset L^2(P)$ is defined by completion with respect to the seminorms $\|\cdot\|_{1,2}$ that are given by $(l, m = 1, 2, \dots)$

$$\|F\|_{1,2}^2 = E\Big[|F|^2\Big] + \sum_{i=1}^m E\Big[\|D_{\cdot,i}F\|_{L^2(\lambda)}^2\Big] + \sum_{j=1}^l E\Big[\|D_{\cdot,\cdot,j}F\|_{L^2(\lambda \times \nu)}^2\Big]. \quad (17.20)$$

Here $D_{t,i}$ and $D_{t,z,j}$ denote the Malliavin derivatives with respect to the $W_i(t)$ and $\widetilde{N}_j(ds, dz)$, respectively. See Sect. 12.5. The next result is a generalization of Theorem 17.2.

Theorem 17.4. *Under the conditions of (17.18) and (17.19), the values $X(t)$ of the solution to (17.14) are Malliavin differentiable for all $0 \le t \le T$.*

Proof The proof is similar to the one of Theorem 17.2. For details see [35, Chap. 10], where the Peano approximation scheme is used (instead of Picard iteration). □

Remark 17.5. As already indicated in Remark 17.3, the regularity of the solutions to (17.14) can be improved in various ways. For example, it is proved in [115] for $m = n = 1$ and $l = 0$ in (17.14) that if α, σ are time-independent and in $C_b^\infty(\mathbb{R})$ (space of infinitely differentiable functions with bounded derivatives of all orders) then $X(t)$ belongs to the *Meyer–Watanabe test function space* $\mathbb{D}_\infty$, that is,

$$X(t) \in \mathbb{D}_\infty$$

for all $0 \le t \le T$, where

$$\mathbb{D}_\infty := \bigcap_{k \in \mathbb{N}_0, p \ge 1} \mathbb{D}_{k,p}.$$

Here the spaces $\mathbb{D}_{k,p}$ are defined by completion with respect to the seminorms $\|\cdot\|_{k,p}$, that is given by

$$\|F\|_{k,p}^2 = E\Big[|F|^p\Big] + \sum_{j=1}^{k} E\Big[\big\|D_\cdot^j F\big\|_{L^2(\lambda^j)}^p\Big], \quad k \in \mathbb{N}_0, \ p \ge 1, \qquad (17.21)$$

where $D_\cdot^j F$ means

$$D_{t_1,\ldots,t_j,}^j F = D_{t_1} D_{t_2} \ldots D_{t_j} F \qquad (17.22)$$

for a j-times Malliavin differentiable random variable F (see Definition 3.1).

17.3 Exercises

Problem 17.1. Let $m \in \mathbb{N}$. Assume that there exist constants C, D and numbers $\alpha_0, \ldots, \alpha_m$ such that

$$\alpha_0 \le C$$

and

$$\alpha_k \le C + \frac{D}{m} \sum_{j=0}^{k-1} \alpha_j, \quad k = 1, \ldots, m.$$

Then

$$\alpha_k \le C\Big(1 + \frac{D}{m}\Big)^k, \quad \text{for all } k = 0, 1, \ldots, m.$$

Problem 17.2. Consider the SDE (17.18) under conditions (17.19) and (17.20) for $m = l = n = 1$. Suppose that $\alpha(t, \cdot), \sigma(t, \cdot)$, and $\gamma(t, \cdot, z)$ are continuously differentiable for all $t \in [0, T]$, $z \in \mathbb{R}_0$. Use the Itô formula to represent $D_s X_t$ explicitly.

18

Absolute Continuity of Probability Laws

One of the original applications of the Malliavin calculus or the stochastic calculus of variations pertains to the analysis of the existence and smoothness of densities of random vectors on the Wiener space. In this short chapter we want to provide some useful criteria based on Malliavin calculus for Lévy processes to ensure the absolute continuity of probability laws with respect to Lebesgue measure. Finally, we discuss the smoothness of densities of strong solutions to stochastic differential equations driven by Lévy processes. See [35, 38] for further details. See also recent related developments in, e.g., [13, 14, 57, 82, 83, 137].

18.1 Existence of Densities

In this section we consider a one-dimensional Brownian motion $W(t)$, $t \in [0, T]$, and a compensated Poisson random measure $\widetilde{N}(dt, dz) = N(dt, dz) - \nu(dz)dt$, $t \in [0, T]$, $z \in \mathbb{R}_0$, on the probability space $(\Omega, \mathcal{F}, P)$ that is given by (13.66) in Sect. 12.5 for $N = 1$ and $R = 1$. Recall that this space was called the Wiener–Poisson space. We denote D_t, respectively, by $D_{t,z}$ the Malliavin derivative with respect to W, respectively, $\widetilde{N}$.

The main idea of the study of the absolute continuity of laws of random vectors F is based on integration by parts, that is, on the duality relation between D_t and δ. To develop some intuition for this principle let us first analyze the real-valued case.

Theorem 18.1. *Let the random variable $F : \Omega \longrightarrow \mathbb{R}$ be Malliavin differentiable with respect to W. Assume that the process*

$$u(t) := \frac{D_t F}{\|D.F\|_{L^2(\lambda)}^2}, \quad t \in [0, T],$$

is Skorohod integrable. Then the law of F is absolutely continuous with respect to Lebesgue measure. Furthermore, its density f is continuous and bounded *and can be explicitly represented as*

$$f(x) = E\Big[\chi_{\{F>x\}}\delta\Big(\frac{D.F}{\|D.F\|^2_{L^2(\lambda)}}\Big)\Big], \quad x \in \mathbb{R}. \tag{18.1}$$

Proof Let $x_1 < x_2$. Define

$$\alpha(y) = \chi_{[x_1,x_2]}(y)$$

and

$$\beta(y) = \int_{-\infty}^{y} \alpha(u)du.$$

Using the chain rule for D_t, one can show that $\beta(F)$ is Malliavin differentiable with respect to W and that

$$\Big(D.\beta(F), D.F\Big)_{L^2(\lambda)} = \Big(\alpha(F)D.F, D.F\Big)_{L^2(\lambda)} = \alpha(F)\|D.F\|^2_{L^2(\lambda)}.$$

So

$$\alpha(F) = \Big(D.\beta(F), \frac{D.F}{\|D.F\|^2_{L^2(\lambda)}}\Big)_{L^2(\lambda)}.$$

Then the duality relation between D_t and the Skorohod integral δ (on the Wiener–Poisson space) and the Fubini theorem give

$$P(x_1 \le F \le x_2) = E\Big[\alpha(F)\Big] = E\Big[\Big(D.\beta(F), \frac{D.F}{\|D.F\|^2_{L^2(\lambda)}}\Big)_{L^2(\lambda)}\Big]$$

$$= E\Big[\beta(F)\delta\Big(\frac{D.F}{\|D.F\|^2_{L^2(\lambda)}}\Big)\Big]$$

$$= E\Big[\int_{-\infty}^{F} \alpha(u)du \, \delta\Big(\frac{D.F}{\|D.F\|^2_{L^2(\lambda)}}\Big)\Big]$$

$$= \int_{x_1}^{x_2} E\Big[\chi_{\{F>u\}}\delta\Big(\frac{D.F}{\|D.F\|^2_{L^2(\lambda)}}\Big)\Big]du,$$

which completes the proof. □

We wish to derive a multidimensional version of Theorem 18.1. To this end, we need the following auxiliary result from finite dimensional real analysis.

Lemma 18.2. *Let m be a finite measure on the Borel sets of $\mathbb{R}^n$. Suppose that*

$$\Big|\int_{\mathbb{R}^n} \frac{\partial}{\partial x_i}\phi(x)m(dx)\Big| \le C\|\phi\|_\infty \tag{18.2}$$

for all test functions $\phi \in C_b^\infty(\mathbb{R}^n)$ (space of infinitely differentiable functions with bounded partial derivatives of all orders), where $C < \infty$ is a constant not dependent of ϕ and $\|\cdot\|_\infty$ is the supremum-norm. Then m is absolutely continuous with respect to the Lebsegue measure.

Proof A proof of this result can be found, e.g., in [158], where techniques from harmonic analysis are invoked. □

Let us consider a random vector $F = (F_1, ..., F_n)$, whose components are Malliavin differentiable with respect to W. We introduce the *Malliavin matrix of F* as the random symmetric nonnegative definite matrix

$$\sigma_F = \left(\sigma_F^{ij}\right)_{1 \leq i,j \leq n} \tag{18.3}$$

with entries given by

$$\sigma_F^{ij} = \left(D.F_i, D.F_j\right)_{L^2(\lambda)}.$$

The next result is essentially due to [158].

Theorem 18.3. *Let $F = (F_1, ..., F_n)$ be a random vector such that*
(1) F belongs to $\mathbb{D}_{2,4}$, where the space $\mathbb{D}_{2,4}$ is defined as in (17.21) on the Wiener–Poisson space.
(2) The Malliavin matrix σ_F in (18.3) has an inverse a.e.
Then the probability law of F is absolutely continuous with respect to Lebesgue measure.

Proof Let $\phi \in C_b^\infty(\mathbb{R}^n)$. Then by the chain rule (on the Wiener–Poisson space) we see that $\phi(F) \in \mathbb{D}_{1,4}$ and that

$$D_t\phi(F) = \sum_{i=1}^n \frac{\partial}{\partial x_i}\phi(F)D_t F_i.$$

The latter implies

$$\left(D.\phi(F), D.F_j\right)_{L^2(\lambda)} = \sum_{i=1}^n \frac{\partial}{\partial x_i}\phi(F)\sigma_F^{ij}.$$

So using the invertibility of the Malliavin matrix of F we have

$$\frac{\partial}{\partial x_i}\phi(F) = \sum_{j=1}^n \left(D.\phi(F), D.F_j\right)_{L^2(\lambda)}\left(\sigma_F^{-1}\right)^{ij}. \tag{18.4}$$

Let us introduce the function $\Phi : \mathbb{R}^{n \times n} \longrightarrow \mathbb{R}^{n \times n}$ given by

$$\Phi(x) = \begin{cases} x^{-1}\exp\left(-\left(\det(x)\right)^{-2}\right), & \det(x) \neq 0, \\ 0, & \det(x) = 0, \end{cases}$$

where $\det(x)$ denotes the determinant of $x \in \mathbb{R}^{n \times n}$. One can see that $\Phi \in C_b^\infty(\mathbb{R}^{n \times n})$. Then using (18.4) gives

$$E\left[\exp\left(-\left(\det(\sigma_F)\right)^{-2}\right)\frac{\partial}{\partial x_i}\phi(F)\right]$$

$$=\sum_{j=1}^{n}E\left[\left(D.\phi(F),D.F_j\right)_{L^2(\lambda)}\exp\left(-\left(\det(\sigma_F)\right)^{-2}\right)\left(\sigma_F^{-1}\right)^{ij}\right]$$

$$=\sum_{j=1}^{n}E\left[\left(D.\phi(F),D.F_j\right)_{L^2(\lambda)}\left(\Phi(\sigma_F)\right)^{ij}\right]$$

$$=\sum_{j=1}^{n}E\left[\left(D.\phi(F),\left(\Phi(\sigma_F)\right)^{ij}D.F_j\right)_{L^2(\lambda)}\right].\tag{18.5}$$

We want to apply the duality relation between D_t and the Skorohod integral δ to the last expression in (18.5). Thus, we have to verify that $\left(\Phi(\sigma_F)\right)^{ij}D.F_j$ belongs to the domain of δ. To justify this, let us recall from Chap. 3 the integration by parts formula (which can be extended to the Wiener–Poisson space setting by the same proof). For Skorohod integrable processes $u(t)$, $t \in [0,T]$, and random variables $X \in \mathbb{D}_{1,2}$ with $E\left[X^2 \|u\|^2_{L^2(\lambda)}\right] < \infty$ we have

$$\delta(Xu) = X\delta(u) - \left(D.X,u\right)_{L^2(\lambda)},\tag{18.6}$$

provided that the right-hand side belongs to $L^2(P)$ (see Theorem 3.15 and the remark that follows). Because of condition (1) and $\Phi \in C_b^{\infty}(\mathbb{R}^{n \times n})$, we observe that $\left(\Phi(\sigma_F)\right)^{ij} \in \mathbb{D}_{1,2}$ is bounded and that $\left(D.\left(\Phi(\sigma_F)\right)^{ij},D.F_j\right)_{L^2(\lambda)}$ is in $L^2(P)$. So if we choose $u(t) = D_tF_j$ and $X = \left(\Phi(\sigma_F)\right)^{ij}$, we conclude that the right-hand side of (18.6) is square integrable. Thus Xu is in the domain of δ. Hence, by applying the duality relation to (18.5) we obtain

$$E\left[\exp\left(-\left(\det(\sigma_F)\right)^{-2}\right)\frac{\partial}{\partial x_i}\phi(F)\right] = \sum_{j=1}^{n}E\left[\phi(F)\delta\left(\left(\Phi(\sigma_F)\right)^{ij}D.F_j\right)\right]$$

$$\leq \sum_{j=1}^{n}E\left[\left|\delta\left(\left(\Phi(\sigma_F)\right)^{ij}D.F_j\right)\right|\right]\|\phi\|_{\infty}$$

$$= C\|\phi\|_{\infty}.$$

So it follows from Lemma 18.2 that the measure m defined by

$$m(B) = \int_{F^{-1}(B)}\exp\left(-\left(\det(\sigma_F)\right)^{-2}\right)P(d\omega), \quad B \in \mathcal{B}(\mathbb{R}^n)$$

is absolutely continuous with respect to Lebesgue measure λ^n. So if $\lambda^n(B) = 0$ then $m(B) = 0$. Thus we get

$$P(F \in B) = 0,$$

which gives the desired result.   □

Remark 18.4. We mention that the regularity assumption on the random vector F in Theorem 18.3 can be considerably relaxed. See [42], where the authors use a different approach based on a criterion for absolute continuity in finite dimesional spaces.

18.2 Smooth Densities of Solutions to SDE's Driven by Lévy Processes

The density result for random vectors in Theorem 18.3 in concert with Theorem 17.2 (or Remark 17.3) can be actually used to establish criteria for the existence of densities of laws of (strong) solutions to SDEs of the type

$$X(t) = x + \int_0^t \alpha(s, X(s))ds + \int_0^t \sigma(s, X(s))dW(s)$$

$$+ \int_0^t \int_{\mathbb{R}_0} \gamma(s, X(s^-), z)\widetilde{N}(ds, dz),$$

$$0 \le t \le T, x \in \mathbb{R}^n, \tag{18.7}$$

where $\alpha : [0, T] \times \mathbb{R}^n \longrightarrow \mathbb{R}^n, \sigma : [0, T] \times \mathbb{R}^n \longrightarrow \mathbb{R}^{n \times m}$ and $\gamma : [0, T] \times \mathbb{R}^n \times \mathbb{R}_0^n \longrightarrow \mathbb{R}^{n \times l}$ are Borel measurable functions.

An important problem in view of applications refers to the existence of smooth densities of distributions of solutions of SDEs. We want to state a useful result that provides a criterion for the existence of C^∞-densities of solutions to (18.7). The latter criterion ensures that for each fixed $t > 0$ there exists a function $f \in C^\infty(\mathbb{R}^n)$ such that

$$P(X(t) \in B) = \int_B f(x)dx, \quad B \in \mathcal{B}(\mathbb{R}^n).$$

The statement of this result, which goes back to [38], requires the following conditions and notation. In the sequel we confine ourselves to the case $l = 1$ in (18.7). Further, we assume that the Brownian motion $W(t) = (W_1(t), ..., W_m(t))$ and the compensated Poisson random measure $\widetilde{N}(dt, dz) = N(dt, dz) - \nu(dz)dt$ are defined on the Wiener–Poisson space (13.66).

Let us denote by $L_+^{p,\infty}$ the space given by

$$L_+^{p,\infty} := \bigcap_{q \ge p} L_+^q, \tag{18.8}$$

where
$$L_+^q := \{f : \mathbb{R}_0 \longrightarrow [0, \infty) : f \in L^q(\nu)\}.$$

We impose the following conditions on the coefficients of (18.7):

(C.1) α, σ, and γ are time-independent.
(C.2) $\alpha, \sigma_1, ..., \sigma_m \in C_b^\infty(\mathbb{R}^n)$, where $\sigma = (\sigma_1, ..., \sigma_m)$.
(C.3) There exists some function $\rho \in L_+^{2,\infty}$ such that

$$\sup_{z \in \mathbb{R}_0, x \in \mathbb{R}^n} \frac{1}{\rho(z)} \|D_1^m \gamma(x, z)\| < \infty$$

for all $m \in \mathbb{N}$. Here, D_1 is the derivative with respect to $x \in \mathbb{R}^n$.
(C.4)

$$\sup_{z \in \mathbb{R}_0, x \in \mathbb{R}^n} \left\| \left(I + D_1\gamma(x, z)\right)^{-1} \right\| < \infty,$$

where $I \in \mathbb{R}^{n \times n}$ is the unit matrix.

Under these conditions one shows (see, e.g., [168]) that there exist unique predictable processes $Y(t), Z(t) \in \mathbb{R}^{n \times n}$, such that

$$\sup_{0 \leq s \leq t} \|Y(s)\|, \ \sup_{0 \leq s \leq t} \|Z(s)\| \in L^p(P) \tag{18.9}$$

for all $t \geq 0$ and $p < \infty$ as well as

$$Y(t) = I + \int_0^t D\alpha(X(s^-))Y(s^-)ds + \int_0^t D\sigma(X(s^-))Y(s^-)dW(s)$$

$$+ \int_0^t \int_{\mathbb{R}_0} D_1\gamma(X(s^-), z)Y(s^-)\tilde{N}(ds, dz) \tag{18.10}$$

and

$$Z(t) = I - \int_0^t Z(s^-)\left(D\alpha(X(s^-)) - \sum_{i=1}^m D\sigma_i(X(s^-))^2\right.$$

$$- \int_{\mathbb{R}_0} \left(I + D_1\gamma(X(s^-), z)\right)^{-1} D_1\gamma(X(s^-), z)^2 \nu(dz)\right)ds$$

$$- \int_0^t D\sigma(X(s^-))Z(s^-)dW(s)$$

$$- \int_0^t \int_{\mathbb{R}_0} Z(s^-)\left(I + D_1\gamma(X(s^-), z)\right)^{-1} D_1\gamma(X(s^-), z)\tilde{N}(ds, dz). \tag{18.11}$$

Using Itô formula one verifies that

$$Z(t) = Y^{-1}(t) \text{ for all } t \geq 0 \text{ a.e.}$$

If, e.g., (18.7) has no jump part then

$$D_s X(t) = Y(t)Y^{-1}(s)\sigma(X_s)\chi_{[0,t]}(s) \text{ a.e.}$$

Finally, define the reduced Malliavin covariance matrix C_t by

$$C_t = \int_0^t \sum_{i=1}^m Z(s^-)\sigma_i(X(s^-))Z(s^-)\sigma_i(X(s^-))ds.$$

The next result gives a sufficient condition for the existence of a smooth density of $X(t)$ in terms of a moment condition of the inverse of C_t.

Theorem 18.5. *Let $t > 0$ and require that*

$$\|C_t^{-1}\| \in L^p(P)$$

for all $p \geq 2$. Then $X(t)$ has a C^∞-density with respect to Lebesgue measure.

Proof See [35]. $\square$

18.3 Exercises

Problem 18.1. Use Theorem 18.1 to derive a formula for the density of the (risk neutral) log-prices $X(t)$, $t \in [0, T]$, in the Barndorff–Nielsen and Shephard model (12.47), (12.48).

Appendix A: Malliavin Calculus on the Wiener Space

In this book we have, for several reasons, chosen to present the Malliavin calculus via chaos expansions. In the Brownian motion case this approach is basically equivalent to the construction given in the setting of the Hida white noise probability space $(\Omega, \mathcal{F}, \mathcal{P})$, where $\Omega = S'(\mathbb{R})$ is the Schwartz space of tempered distributions. In the Brownian case there is an alternative setting, namely the Wiener space $\Omega = C_0[0, T]$ of continuous functions $\omega : [0, T] \to \mathbb{R}$ with $\omega(0) = 0$. We now present this approach.

Malliavin calculus was originally introduced to study the regularity of the law of functionals of the Brownian motion, in particular, of the solution of stochastic differential equations driven by the Brownian noise [158].

Shortly, the idea is as follows. Let f be a smooth function on $\mathbb{R}^d$. The crucial idea for proving the regularity of the law of an $\mathbb{R}^d$-valued functional X of the Wiener process is to express the partial derivative of f at X as a derivative of the functional $f(X)$ with respect to a *new derivation* on the Wiener space. Based on some *integration by parts formula*, this derivation should exhibit the property of fulfilling the following relation:

$$E\left[\frac{\partial^{|\alpha|} f}{\partial x^\alpha}(X)\right] = E\left[f(X) L_\alpha(X)\right],$$

where $L_\alpha(X)$ is a functional of the Wiener process not depending on f and where $\frac{\partial^{|\alpha|}}{\partial x^\alpha}$ is the partial derivative of order $|\alpha| = \alpha_1 + ... + \alpha_d$, $\alpha = (\alpha_1, ..., \alpha_d)$. Provided $L_\alpha(X)$ is sufficiently integrable, the law of X should be smooth.

Hereafter we outline the classical presentation of the Malliavin derivative on the Wiener space. For further reading, we refer to, for example, [53, 169, 212].

A.1 Preliminary Basic Concepts

Let us first recall some basic concepts from classical analysis, see, for example, [79].

Definition A.1. *Let U be an open subset of $\mathbb{R}^n$ and let $f : U \longrightarrow \mathbb{R}^m$.*

(1) We say that f has a directional derivative *at the point $x \in U$ in the direction $y \in \mathbb{R}^n$ if*

$$D_y f(x) := \lim_{\varepsilon \to 0} \frac{f(x + \varepsilon y) - f(x)}{\varepsilon} = \frac{d}{d\varepsilon}[f(x + \varepsilon y)]_{|\varepsilon=0} \qquad (A.1)$$

exists. If this is the case we call the vector $D_y f(x) \in \mathbb{R}^m$ the directional derivative *at x in the direction y. In particular, if we choose y to be the jth unit vector $e_j = (0, \ldots, 1, \ldots, 0)$, with 1 on jth place, we get*

$$D_{e_j} f(x) = \frac{\partial f}{\partial x_j}(x),$$

the jth partial derivative of f.

(2) We say that f is differentiable *at $x \in U$ if there exists a matrix $A \in \mathbb{R}^{m \times n}$ such that*

$$\lim_{\substack{h \to 0 \\ h \in \mathbb{R}^n}} \frac{|f(x + h) - f(x) - Ah|}{|h|} = 0. \qquad (A.2)$$

If this is the case we call A the derivative *of f at x and we write*

$$A = f'(x).$$

Proposition A.2. *The following relations between the two concepts hold true.*

(1) If f is differentiable at $x \in U$, then f has a directional derivative in all directions $y \in \mathbb{R}^n$ and

$$D_y f(x) = f'(x)y = Ay. \qquad (A.3)$$

(2) Conversely, if f has a directional derivative at all $x \in U$ in all the directions $y = e_j$, $j = 1, \ldots, n$, and all the partial derivatives

$$D_{e_j} f(x) = \frac{\partial f}{\partial x_j}(x)$$

are continuous functions of x, then f is differentiable at all $x \in U$ and

$$f'(x) = \left[\frac{\partial f_i}{\partial x_j}(x)\right]_{\substack{1 \le i \le m \\ 1 \le j \le n}} = A \in \mathbb{R}^{m \times n}, \qquad (A.4)$$

where f_i is component number i of f, that is, $f = (f_1, \ldots, f_m)^T$.

We define similar operations in a more general context. First let us recall some basic concepts from functional analysis.

Definition A.3. *Let X be a Banach space, that is, a complete, normed vector space over $\mathbb{R}$, and let $\|x\|$ denote the norm of the element $x \in X$. A linear functional on X is a linear map*

$$T : X \to \mathbb{R}.$$

Recall that T is called linear *if $T(ax + y) = aT(x) + T(y)$ for all $a \in \mathbb{R}$, $x, y \in X$. A linear functional T is called* bounded *(or* continuous*) if*

$$\||T\|| := \sup_{\|x\| \le 1} |T(x)| < \infty$$

Sometimes we write $\langle T, x \rangle$ or Tx instead of $T(x)$ and call $\langle T, x \rangle$ "the action of T on x". The set of all bounded linear functionals is called the dual *of X and is denoted by X^*. Equipped with the norm $\|| \cdot \||$, the space X^* is a Banach space.*

Example A.4. $X = \mathbb{R}^n$ with the Euclidean norm $|x| = \sqrt{x_1^2 + \cdots + x_n^2}$ is a Banach space. In this case it is easy to see that we can identify X^* with R^n.

Example A.5. Let $X = C_0([0, T])$ be the space of continuous real functions ω on $[0, T]$ such that $\omega(0) = 0$. Then

$$\|\omega\|_\infty := \sup_{t \in [0,T]} |\omega(t)|$$

is a norm on X called the *uniform norm*. With this norm, X is a Banach space and its dual X^* can be identified with the space $M([0, T])$ of all signed measures ν on $[0, T]$, with norm

$$\||\nu\|| = \sup_{|f| \le 1} \int_0^T f(t) d\nu(t) = |\nu|([0, T]).$$

Example A.6. Let $X = L^p([0, T]) = \{f : [0, T] \to \mathbb{R}; \int_0^T |f(t)|^p dt < \infty\}$ be equipped with the norm

$$\|f\|_p = \left[\int_0^T |f(t)|^p dt \right]^{1/p} \qquad (1 \le p < \infty).$$

Then X is a Banach space and its dual can be identified with $L^q([0, T])$, where

$$\frac{1}{p} + \frac{1}{q} = 1.$$

In particular, if $p = 2$, then $q = 2$, so $L^2([0, T])$ is its own dual.

We now extend the definitions of derivative and differentiability we had for $\mathbb{R}^n$ to arbitrary Banach spaces.

Definition A.7. *Let U be an open subset of a Banach space X and let f be a function from U into $\mathbb{R}^m$.*

(1) We say that f has a directional derivative *(or Gateaux derivative) $D_y f(x)$ at $x \in U$ in the direction $y \in X$ if*

$$D_y f(x) := \frac{d}{d\varepsilon}[f(x + \varepsilon y)]_{\varepsilon=0} \in \mathbb{R}^m \tag{A.5}$$

exists.

(2) We say that f is Fréchet-differentiable *at $x \in U$, if there exists a bounded linear map*

$$A : X \to \mathbb{R}^m,$$

that is, $A = (A_1, ..., A_m)^T$, with $A_i \in X^*$ for $i = 1, ..., m$, such that

$$\lim_{\substack{h \to 0 \\ h \in X}} \frac{|f(x + h) - f(x) - A(h)|}{\|h\|} = 0. \tag{A.6}$$

We write

$$f'(x) = \begin{bmatrix} f'(x)_1 \\ \vdots \\ f'(x)_m \end{bmatrix} = A \quad \in (X^*)^m. \tag{A.7}$$

for the Fréchet derivative of f at x.

Similar to the Euclidean case (see Proposition A.2) we have the following result.

Proposition A.8.

(1) *If f is Fréchet-differentiable at $x \in U \subset X$, then f has a directional derivative at x in all directions $y \in X$ and*

$$D_y f(x) = \langle f'(x), y \rangle \in \mathbb{R}^m, \tag{A.8}$$

where

$$\langle f'(x), y \rangle = (\langle f'(x)_1, y \rangle, \ldots, \langle f'(x)_m, y \rangle)^T$$

is the m-vector whose ith component is the action of the ith component $f'(x)_i$ of $f'(x)$ on y.

(2) *Conversely, if f has a directional derivative at all $x \in U$ in all directions $y \in X$ and the linear map*

$$y \to D_y f(x), \qquad y \in X$$

is continuous for all $x \in U$, then there exists an element $\nabla f(x) \in (X^*)^m$ such that

$$D_y f(x) = \langle \nabla f(x), y \rangle .$$

If this map $x \rightarrow \nabla f(x) \in (X^*)^m$ is continuous on U, then f is Fréchet differentiable and

$$f'(x) = \nabla f(x) . \tag{A.9}$$

A.2 Wiener Space, Cameron–Martin Space, and Stochastic Derivative

We now apply these operations to the Banach space $\Omega = C_0([0, T])$ considered in Example A.5 above. This space is called the *Wiener space*, because we can regard each path

$$t \rightarrow W(t, \omega)$$

of the Wiener process starting at 0 as an element ω of $C_0([0, T])$. Thus we may identify $W(t, \omega)$ with the value $\omega(t)$ at time t of an element $\omega \in C_0([0, T])$:

$$W(t, \omega) = \omega(t).$$

The space $\Omega = C_0([0, T])$ is naturally equipped with the Borel σ-algebra generated by the topology of the uniform norm. One can prove that this σ-algebra coincides with the σ-algebra generated by the cylinder sets (see, e.g., [36]). This measurable space is equipped with the probability measure P, which is given by the probability law of the Wiener process:

$$P\{W(t_1) \in F_1, \ldots, W(t_k) \in F_k\}$$

$$= \int_{F_1 \times \cdots \times F_k} \rho(t_1, x, x_1)\rho(t_2 - t_1, x, x_2) \cdots \rho(t_k - t_{k-1}, x_{k-1}, x_k)dx_1, \cdots dx_k,$$

where $F_i \subset \mathbb{R}$, $0 \leq t_1 < t_2 < \cdots < t_k \leq T$, and

$$\rho(t, x, y) = (2\pi t)^{-1/2} \exp(-\frac{1}{2}|x - y|^2), \qquad t \in [0, T], \quad x, y \in \mathbb{R}.$$

The measure P is called the *Wiener measure* on Ω.

Just as for Banach spaces, we now give the following definition.

Definition A.9. *Let $F : \Omega \rightarrow \mathbb{R}$ be a random variable, choose $g \in L^2([0, T])$, and consider*

$$\gamma(t) = \int_0^t g(s)ds \quad \in \Omega. \tag{A.10}$$

Then we define the directional derivative of F at the point $\omega \in \Omega$ in direction $\gamma \in \Omega$ by

$$D_\gamma F(\omega) = \frac{d}{d\varepsilon}[F(\omega + \varepsilon\gamma)]_{|\varepsilon=0}, \tag{A.11}$$

if the derivative exists in some sense (to be made precise later).

Note that we consider the derivative only in special directions, namely in the directions of elements γ of the form (A.10). The set of $\gamma \in \Omega$, which can be written on the form (A.10) for some $g \in L^2([0,T])$, is called the *Cameron–Martin space* and it is hearafter denoted by H. It turns out that it is difficult to obtain a tractable theory involving derivatives in all directions. However, the derivatives in the directions $\gamma \in H$ are sufficient for our purposes.

Definition A.10. *Assume that $F : \Omega \to \mathbb{R}$ has a directional derivative in all directions γ of the form $\gamma \in H$ in the strong sense, that is,*

$$\mathbf{D}_\gamma F(\omega) := \lim_{\varepsilon \to 0} \frac{F(\omega + \varepsilon\gamma) - F(\omega)}{\varepsilon} \tag{A.12}$$

exists in $L^2(P)$. Assume in addition that there exists $\psi(t,\omega) \in L^2(P \times \lambda)$ such that

$$\mathbf{D}_\gamma F(\omega) = \int_0^T \psi(t,\omega)g(t)dt, \quad \text{for all } \gamma \in H. \tag{A.13}$$

Then we say that F is differentiable *and we set*

$$\mathbf{D}_t F(\omega) := \psi(t,\omega). \tag{A.14}$$

We call $\mathbf{D}.F \in L^2(P \times \lambda)$ the stochastic derivative *of F. The set of all differentiable random variables is denoted by $\mathcal{D}_{1,2}$.*

Example A.11. Suppose $F = \int_0^T f(s)dW(s) = \int_0^T f(s)d\omega(s)$, where $f(s) \in L^2([0,T])$. Then if $\gamma \in H$, we have

$$F(\omega + \varepsilon\gamma) = \int_0^T f(s)(d\omega(s) + \varepsilon d\gamma(s))$$

$$= \int_0^T f(s)d\omega(s) + \varepsilon \int_0^T f(s)g(s)ds,$$

and hence

$$\frac{F(\omega + \varepsilon\gamma) - F(\omega)}{\varepsilon} = \int_0^T f(s)g(s)ds$$

for all $\varepsilon > 0$. Comparing with (A.13), we see that $F \in \mathcal{D}_{1,2}$ and

$$\mathbf{D}_t F(\omega) = f(t), \quad t \in [0,T], \ \omega \in \Omega. \tag{A.15}$$

In particular, choosing

$$f(t) = \mathcal{X}_{[0,t_1]}(t)$$

we get

$$F = \int_0^T \mathcal{X}_{[0,t_1]}(s)dW(s) = W(t_1)$$

and hence

$$\mathbf{D}_t(W(t_1)) = \mathcal{X}_{[0,t_1]}(t). \tag{A.16}$$

Let $\mathbb{P}$ denote the family of all random variables $F : \Omega \to \mathbb{R}$ of the form

$$F = \varphi(\theta_1, \ldots, \theta_n),$$

where $\varphi(x_1, \ldots, x_n) = \sum_\alpha a_\alpha x^\alpha$, with $x^\alpha = x_1^{\alpha_1} \ldots x_n^{\alpha_n}$ and $\alpha = (\alpha_1, ..., \alpha_n)$, is a polynomial and $\theta_i = \int_0^T f_i(t)dW(t)$ for some $f_i \in L^2([0,T])$, $i = 1, ..., n$. Such random variables are called *Wiener polynomials*. Note that $\mathbb{P}$ is dense in $L^2(P)$.

Lemma A.12. Chain rule. *Let $F = \varphi(\theta_1, \ldots, \theta_n) \in \mathbb{P}$. Then $F \in \mathcal{D}_{1,2}$ and*

$$\mathbf{D}_t F = \sum_{i=1}^n \frac{\partial \varphi}{\partial \theta_i}(\theta_1, \ldots, \theta_n) \cdot f_i(t). \tag{A.17}$$

Proof Let $\psi(t)$ denote the right-hand side of (A.17). Since

$$\sup_{s \in [0,T]} E[|W(s)|^N] < \infty \qquad \text{for all } N \in \mathbb{N},$$

we see that

$$\frac{1}{\varepsilon}[F(\omega + \varepsilon\gamma) - F(\omega)] = \frac{1}{\varepsilon}\Big[\varphi(\theta_1 + \varepsilon(f_1, g), \ldots, \theta_n + \varepsilon(f_n, g)) - \varphi(\theta_1, \ldots, \theta_n)\Big]$$

$$\longrightarrow \sum_{i=1}^n \frac{\partial \varphi}{\partial \theta_i}(\theta_1, \ldots, \theta_n) \cdot \mathbf{D}_\gamma(\theta_i), \qquad \varepsilon \to 0,$$

in $L^2(P)$. Hence F has a directional derivative in direction γ in the strong sense and by (A.15) we have

$$\mathbf{D}_\gamma F = \int_0^T \psi(t)g(t)dt.$$

By this we end the proof. □

We now introduce the norm $\| \cdot \|_{1,2}$, on $\mathcal{D}_{1,2}$:

$$\|F\|_{1,2}^2 := \|F\|_{L^2(P)}^2 + \|\mathbf{D}_t F\|_{L^2(P \times \lambda)}^2, \qquad F \in \mathcal{D}_{1,2}. \tag{A.18}$$

Unfortunately, it is not clear if $\mathcal{D}_{1,2}$ is closed under this norm. To avoid this difficulty we work with the following family.

Definition A.13. *We define* $\mathbb{D}_{1,2}$ *to be the closure of the family* $\mathbb{P}$ *with respect to the norm* $\| \cdot \|_{1,2}$.

Thus $\mathbb{D}_{1,2}$ consists of all $F \in L^2(P)$ such that there exists $F_n \in \mathbb{P}$ with the property that

$$F_n \longrightarrow F \quad \text{in } L^2(P) \quad \text{as } n \to \infty \qquad (A.19)$$

and

$$\{\mathbf{D}_t F_n\}_{n=1}^{\infty} \quad \text{is convergent in } L^2(P \times \lambda).$$

If this is the case, it is tempting to *define*

$$D_t F := \lim_{n \to \infty} \mathbf{D}_t F_n .$$

However, for this to work we need to know that this defines $D_t F$ *uniquely*. In other words, if there is another sequence $G_n \in \mathbb{P}$ such that

$$G_n \to F \quad \text{in } L^2(P) \text{ as } n \to \infty \qquad (A.20)$$

and

$$\{\mathbf{D}_t G_n\}_{n=1}^{\infty} \quad \text{is convergent in } L^2(P \times \lambda), \qquad (A.21)$$

does it follow that $\lim_{n\to\infty} \mathbf{D}_t F_n = \lim_{n\to\infty} \mathbf{D}_t G_n$?

By considering the difference $H_n = F_n - G_n$, we see that the answer to this question is positive, in view of the following theorem.

Theorem A.14. Closability of the derivative. *The operator* $\mathbf{D}_t$ *is closable, that is, if the sequence* $\{H_n\}_{n=1}^{\infty} \subset \mathbb{P}$ *is such that*

$$H_n \to 0 \quad \text{in } L^2(P) \text{ as } n \to \infty \qquad (A.22)$$

and

$$\{\mathbf{D}_t H_n\}_{n=1}^{\infty} \quad \text{converges in } L^2(P \times \lambda) \text{ as } n \to \infty,$$

then

$$\lim_{n \to \infty} \mathbf{D}_t H_n = 0. \qquad (A.23)$$

The proof is based on the following useful result.

Lemma A.15. Integration by parts formula. *Suppose* $F, \varphi \in \mathcal{D}_{1,2}$ *and* $\gamma \in H$ *with* $g \in L^2([0,T])$. *Then*

$$E[\mathbf{D}_\gamma F \cdot \varphi] = E\left[F \cdot \varphi \cdot \int_0^T g(t) dW(t)\right] - E[F \cdot \mathbf{D}_\gamma \varphi]. \qquad (A.24)$$

Proof By the Cameron–Martin theorem (see, e.g., [159]) we have

$$\int_\Omega F(\omega + \varepsilon\gamma) \cdot \varphi(\omega) P(d\omega) = \int_\Omega F(\omega)\varphi(\omega - \varepsilon\gamma)Q(d\omega),$$

where

$$Q(d\omega) = \exp\left\{\varepsilon \int_0^T g(t) dW(t) - \frac{1}{2}\varepsilon^2 \int_0^T g^2(t) dt\right\} P(d\omega),$$

being $\omega(t) = W(t,\omega)$, $t \geq 0$, $\omega \in \Omega$, a Wiener process on the Wiener space $\Omega = C_0([0,T])$. This gives

$$E[\mathbf{D}_\gamma F \cdot \varphi] = \int_\Omega \lim_{\varepsilon \to 0} \frac{1}{\varepsilon} \Big[F(\omega + \varepsilon\gamma) - F(\omega) \Big] \cdot \varphi(\omega) P(d\omega)$$

$$= \lim_{\varepsilon \to 0} \frac{1}{\varepsilon} \int_\Omega F(\omega + \varepsilon\gamma)\varphi(\omega) - F(\omega)\varphi(\omega) P(d\omega)$$

$$= \lim_{\varepsilon \to 0} \frac{1}{\varepsilon} \int_\Omega F(\omega) \Big[\varphi(\omega - \varepsilon\gamma) \exp \Big\{ \varepsilon \int_0^T g(t) d\omega(t)$$

$$- \frac{1}{2}\varepsilon^2 \int_0^T g^2(t) dt \Big\} - \varphi(\omega) \Big] P(d\omega)$$

$$= \int_\Omega F(\omega) \cdot \frac{d}{d\varepsilon} \Big[\varphi(\omega - \varepsilon\gamma) \exp(\varepsilon \int_0^T g(t) d\omega(t)$$

$$- \frac{1}{2}\varepsilon^2 \int_0^T g^2(t) dt) \Big]_{|\varepsilon=0} P(d\omega)$$

$$= E\Big[F\varphi \cdot \int_0^T g(t) dW(t) \Big] - E[F\mathbf{D}_\gamma \varphi].$$

By this we end the proof. □

Proof of Theorem A.14. By Lemma A.15 we get

$$E\Big[\mathbf{D}_\gamma H_n \cdot \varphi \Big] = E\Big[H_n \varphi \cdot \int_0^T g dW \Big] - E\Big[H_n \cdot \mathbf{D}_\gamma \varphi \Big] \longrightarrow 0, \qquad n \to \infty,$$

for all $\varphi \in \mathbb{P}$. Since $\{\mathbf{D}_\gamma H_n\}_{n=1}^\infty$ converges in $L^2(P)$ and $\mathbb{P}$ is dense in $L^2(P)$, we conclude that $\mathbf{D}_\gamma H_n \to 0$ in $L^2(P)$ as $n \to \infty$. Since this holds for all $\gamma \in H$, we obtain that $\mathbf{D}_t H_n \to 0$ in $L^2(P \times \lambda)$. □

In view of Theorem A.14 and the discussion preceding it, we can now make the following unambiguous definition.

Definition A.16. *Let $F \in \mathbb{D}_{1,2}$, so that there exists $\{F_n\}_{n=1}^\infty \subset \mathbb{P}$ such that*

$$F_n \to F \qquad in\ L^2(P)$$

and $\{\mathbf{D}_t F_n\}_{n=1}^\infty$ is convergent in $L^2(P \times \lambda)$. Then we define

$$D_t F = \lim_{n \to \infty} \mathbf{D}_t F_n \qquad in\ L^2(P \times \lambda) \tag{A.25}$$

and

$$D_\gamma F = \int\limits_0^T D_t F \cdot g(t)\,dt$$

for all $\gamma(t) = \int\limits_0^t g(s)\,ds \in H$, *with* $g \in L^2([0,T])$. *We call* $D_t F$ *the* Malliavin *derivative of* F.

Remark A.17. Strictly speaking we now have two apparently different definitions of the derivative of F:

1. The stochastic derivative $\mathbf{D}_t F$ of $F \in \mathcal{D}_{1,2}$ given by Definition A.10.
2. The Malliavin derivative $D_t F$ of $F \in \mathbb{D}_{1,2}$ given by Definition A.16.

However, the next result shows that if $F \in \mathcal{D}_{1,2} \cap \mathbb{D}_{1,2}$, then the two derivatives coincide.

Lemma A.18. *Let* $F \in \mathcal{D}_{1,2} \cap \mathbb{D}_{1,2}$ *and suppose that* $\{F_n\}_{n=1}^\infty \subset \mathbb{P}$ *has the properties*

$$F_n \to F \quad in\ L^2(P) \qquad and \qquad \{\mathbf{D}_t F_n\}_{n=1}^\infty \quad converges\ in\ L^2(P \times \lambda). \tag{A.26}$$

Then

$$\mathbf{D}_t F = \lim_{n\to\infty} \mathbf{D}_t F_n \quad in\ L^2(P \times \lambda). \tag{A.27}$$

Hence

$$D_t F = \mathbf{D}_t F \qquad for \quad F \in \mathcal{D}_{1,2} \cap \mathbb{D}_{1,2}. \tag{A.28}$$

Proof By (A.26) we get that $\{\mathbf{D}_\gamma F_n\}_{n=1}^\infty$ converges in $L^2(P)$ for each $\gamma(t) = \int\limits_0^t g(s)\,ds$ with $g \in L^2([0,T])$. By Lemma A.15 and (A.26) we get

$$E\Big[(\mathbf{D}_\gamma F_n - \mathbf{D}_\gamma F)\cdot\varphi\Big] = E\Big[(F_n - F)\cdot\varphi\cdot\int\limits_0^t g\,dW\Big] - E\Big[(F_n - F)\cdot \mathbf{D}_\gamma\varphi\Big] \longrightarrow 0$$

for all $\varphi \in \mathbb{P}$. Hence $\mathbf{D}_\gamma F_n \to \mathbf{D}_\gamma F$ in $L^2(P)$ and (A.27) follows. $\square$

In view of Lemma A.18 we now use the same symbol $D_t F$ for the derivative and $D_\gamma F$ for the directional derivative of all the elements $F \in \mathcal{D}_{1,2} \cap \mathbb{D}_{1,2}$.

Remark A.19. Note that from the definition of $\mathbb{D}_{1,2}$ follows that, if $\{F_n\}_{n=1}^\infty \in \mathbb{D}_{1,2}$ with $F_n \to F$ in $L^2(P)$ and $\{D_t F_n\}_{n=1}^\infty$ converges in $L^2(P \times \lambda)$, then

$$F \in \mathbb{D}_{1,2} \quad and \quad D_t F = \lim_{n\to\infty} D_t F_n.$$

A.3 Malliavin Derivative via Chaos Expansions

Since an arbitrary $F \in L^2(P)$ can be represented by its chaos expansion

$$F = \sum_{n=0}^{\infty} I_n(f_n),$$

where $f_n \in \widetilde{L}^2([0,T]^n)$ for all n, it is natural to ask if we can express the derivative of F (if it exists) by means of this. See Chap. 1 for the definition and properties of the Itô iterated integrals $I_n(f_n)$. Hereafer, we consider derivation according to Definition A.16 and Lemma A.18.
Let us first look at a special case.

Lemma A.20. *Suppose* $F = I_n(f_n)$ *for some* $f_n \in \widetilde{L}^2([0,T]^n)$. *Then* $F \in \mathbb{D}_{1,2}$ *and*

$$D_t F = n I_{n-1}(f_n(\cdot, t)), \tag{A.29}$$

where the notation $I_{n-1}(f_n(\cdot, t))$ *means that the* $(n-1)$-*iterated Ito integral is taken with respect to the* $n-1$ *first variables* $t_1, \ldots, t_{n-1}$ *of* $f_n(t_1, \ldots, t_{n-1}, t)$, *that is,* t *is fixed and kept outside the integration.*

Proof First consider the special case when

$$f_n = f^{\otimes n}$$

for some $f \in L^2([0,T])$, that is, when

$$f_n(t_1, \ldots, t_n) = f(t_1) \ldots f(t_n), \qquad (t_1, \ldots, t_n) \in [0,T]^n.$$

Then exploiting the definition and properties of Hermite polynomials h_n (see (1.15)), we have

$$I_n(f_n) = \|f\|^n h_n\left(\frac{\theta}{\|f\|}\right), \tag{A.30}$$

where $\theta = \int_0^T f(t) dW(t)$. Moreover, by the chain rule (A.17) we have

$$D_t I_n(f_n) = \|f\|^n h_n'\left(\frac{\theta}{\|f\|}\right) \cdot \frac{f(t)}{\|f\|}.$$

Recall that a basic property of the Hermite polynomials is that

$$h_n'(x) = n h_{n-1}(x). \tag{A.31}$$

This gives (A.29) in this case:

$$D_t I_n(f_n) = n\|f\|^{n-1} h_{n-1}\left(\frac{\theta}{\|f\|}\right) f(t) = n I_{n-1}(f^{\otimes(n-1)}) f(t) = n I_{n-1}(f_n(\cdot, t)).$$

Next, suppose f_n has the form

$$f_n = \xi_1^{\otimes \alpha_1} \widehat{\otimes} \xi_2^{\otimes \alpha_2} \widehat{\otimes} \cdots \widehat{\otimes} \xi_k^{\otimes \alpha_k}, \qquad \alpha_1 + \cdots + \alpha_k = n, \qquad (A.32)$$

where $\widehat{\otimes}$ denotes symmetrized tensor product and $\{\xi_j\}_{j=1}^{\infty}$ is an orthonormal basis for
$L^2([0,T])$. Then by an extension of (1.15) we have (see [120])

$$I_n(f_n) = h_{\alpha_1}(\theta_1) \cdots h_{\alpha_k}(\theta_k) \qquad (A.33)$$

with

$$\theta_j = \int_0^T \xi_j(t) dW(t)$$

and again (A.29) follows by the chain rule (A.17). Since any $f_n \in \widetilde{L}^2([0,T]^n)$ can be approximated in $L^2([0,T]^n)$ by linear combinations of functions of the form given by (A.32), the general result follows.   □

Lemma A.21. *Let $\mathbb{P}_0 \subseteq \mathbb{P}$ denote the set of Wiener polynomials of the form*

$$p_k\Big(\int_0^T \xi_1(t) dW(t), \ldots, \int_0^T \xi_k(t) dW(t) \Big),$$

where $p_k(x_1, \ldots, x_k)$ is an arbitrary polynomial in k variables and $\{\xi_1, \xi_2, \ldots\}$ is a given orthonormal basis for $L^2([0,T])$. Then $\mathbb{P}_0$ is dense in $\mathbb{P}$ in the norm $\|\cdot\|_{1,2}$.

Proof Let $q := p\Big(\int_0^T f_1(t) dW(t), \ldots, \int_0^T f_k(t) dW(t) \Big) \in \mathbb{P}$. We approximate q by

$$q^{(m)} := p\Big(\int_0^T \sum_{j=0}^m (f_1, \xi_j)_{L^2([0,T])} \xi_j(t) dW(t), \ldots, \int_0^T \sum_{j=0}^m (f_k, \xi_j)_{L^2([0,T])} \xi_j(t) dW(t) \Big).$$

Then $q^{(m)} \to q$ in $L^2(P)$ and

$$D_t q^{(m)} = \sum_{i=1}^k \frac{\partial p}{\partial x_i} \cdot \sum_{j=1}^m (f_i, \xi_j)_{L^2([0,T])} \xi_j(t) \to \sum_{i=1}^k \frac{\partial p}{\partial x_i} \cdot f_i(t)$$

in $L^2(P \times \lambda)$ as $m \to \infty$.   □

Theorem A.22. *Let* $F = \sum\limits_{n=0}^{\infty} I_n(f_n) \in L^2(P)$. *Then* $F \in \mathbb{D}_{1,2}$ *if and only if*

$$\sum_{n=1}^{\infty} n \, n! \|f_n\|_{L^2([0,T]^n)}^2 < \infty \tag{A.34}$$

and if this is the case we have

$$D_t F = \sum_{n=1}^{\infty} n I_{n-1}(f_n(\cdot, t)). \tag{A.35}$$

Proof Define $F_m = \sum\limits_{n=0}^{m} I_n(f_n)$. Then $F_m \in \mathbb{D}_{1,2}$ and $F_m \to F$ in $L^2(P)$. Moreover, if $m > k$ we have

$$\|D_t F_m - D_t F_k\|_{L^2(P \times \lambda)}^2 = \Big\| \sum_{n=k+1}^{m} n I_{n-1}(f_n(\cdot, t)) \Big\|_{L^2(P \times \lambda)}^2$$

$$= \int_0^T E\Big[\Big(\sum_{n=k+1}^{m} n I_{n-1}(f_n(\cdot, t)) \Big)^2\Big] dt$$

$$= \int_0^T \sum_{n=k+1}^{m} n^2 (n-1)! \|f_n(\cdot, t)\|_{L^2([0,T]^{n-1})}^2 dt$$

$$= \sum_{n=k+1}^{m} n \, n! \|f_n\|_{L^2([0,T]^n)}^2, \tag{A.36}$$

by Lemma A.20 and the orthogonality of the iterated Itô integrals (see (1.12)). Hence if (A.34) holds then $\{D_t F_n\}_{n=1}^{\infty}$ is convergent in $L^2(P \times \lambda)$ and hence $F \in \mathbb{D}_{1,2}$ and

$$D_t F = \lim_{m \to \infty} D_t F_m = \sum_{n=0}^{\infty} n I_{n-1}(f_n(\cdot, t)).$$

Conversely, if $F \in \mathbb{D}_{1,2}$ then, thanks to Lemma A.21, there exist polynomials $p_k(x_1, \ldots, x_{n_k}) = \sum\limits_{m_i: \sum m_i \leq k} a_{m_1,\ldots,m_{n_k}} \prod\limits_{i=1}^{n_k} h_{m_i}(x_i)$ of degree k for some $a_{m_1,\ldots,m_{n_k}} \in \mathbb{R}$, such that if we put $F_k = p_k(\theta_1, \ldots, \theta_{n_k})$ then $F_k \in \mathbb{P}$ and $F_k \to F$ in $L^2(P)$ and

$$D_t F_k \to D_t F \quad \text{in } L^2(P \times \lambda) \text{ as } k \to \infty.$$

By applying (A.33) we see that there exist $f_j^{(k)} \in \widetilde{L}^2([0,T]^j)$, $1 \leq j \leq k$, such that

$$F_k = \sum_{j=0}^{k} I_j(f_j^{(k)}).$$

Since $F_k \to F$ in $L^2(P)$ we have

$$\sum_{j=0}^{k} j! \|f_j^{(k)} - f_j\|_{L^2([0,T]^j)}^2 \le \|F_k - F\|_{L^2(P)}^2 \to 0 \qquad \text{as } k \to \infty.$$

Therefore, $\|f_j^{(k)} - f_j\|_{L^2([0,T]^j)} \longrightarrow 0$ as $k \to \infty$, for all j. This implies that

$$\|f_j^{(k)}\|_{L^2([0,T]^j)} \longrightarrow \|f_j\|_{L^2([0,T]^j)} \qquad \text{as } k \to \infty, \text{ for all } j. \qquad (A.37)$$

Similarly, since $D_t F_k \to D_t F$ in $L^2(P \times \lambda)$, we get by the Fatou lemma combined with the calculation, leading to (A.36) that

$$\sum_{j=0}^{\infty} j \cdot j! \|f_j\|_{L^2([0,T]^j)}^2 = \sum_{j=0}^{\infty} \lim_{k \to \infty} \left(j \cdot j! \|f_j^{(k)}\|_{L^2([0,T]^j)}^2 \right)$$

$$\le \varliminf_{k \to \infty} \sum_{j=0}^{\infty} j \cdot j! \|f_j^{(k)}\|_{L^2([0,T]^j)}^2$$

$$= \varliminf_{k \to \infty} \|D_t F_k\|_{L^2(P \times \lambda)}^2$$

$$= \|D_t F\|_{L^2(P \times \lambda)}^2 < \infty,$$

where we have put $f_j^{(k)} = 0$ for $j > k$. Hence (A.34) holds and the proof is complete. $\square$

Solutions

In this chapter we present a solution to the exercises marked with (*) in the book. The level of the exposition varies from fully detailed to just sketched.

Problems of Chap. 1

1.1 Solution

(a) Consider the following equalities:

$$\exp\left\{tx - \frac{t^2}{2}\right\} = \exp\left\{\frac{1}{2}x^2\right\}\exp\left\{-\frac{1}{2}(x-t)^2\right\}$$

$$= \exp\left\{\frac{1}{2}x^2\right\}\sum_{n=0}^{\infty}\frac{t^n}{n!}\frac{d^n}{dt^n}\exp\left\{-\frac{1}{2}(x-t)^2\right\}|_{t=0}$$

$$= \exp\left\{\frac{1}{2}x^2\right\}\sum_{n=0}^{\infty}\left\{\frac{(-1)^n t^n}{n!}\frac{d^n}{du^n}\exp\left\{-\frac{1}{2}u^2\right\}\right\}|_{u=x}$$

$$= \exp\left\{\frac{1}{2}x^2\right\}\sum_{n=0}^{\infty}\frac{(-1)^n t^n}{n!}\frac{d^n}{dx^n}\exp\left\{-\frac{1}{2}x^2\right\}$$

$$= \sum_{n=0}^{\infty}\frac{t^n}{n!}h_n(x),$$

with the substitution $u = x - t$.

(b) Set $u = t\sqrt{\lambda}$. Using the result in (a), we get

$$\exp\left\{tx - \frac{t^2\lambda}{2}\right\} = \exp\left\{u\frac{x}{\sqrt{\lambda}} - \frac{u^2}{2}\right\}$$

$$= \sum_{n=0}^{\infty}\frac{u^n}{n!}h_n\left(\frac{x}{\sqrt{\lambda}}\right)$$

$$= \sum_{n=0}^{\infty}\frac{t^n\lambda^{n/2}}{n!}h_n\left(\frac{x}{\sqrt{\lambda}}\right).$$

(c) If we choose $x = \theta$, $\lambda = \|g\|^2$, and $t = 1$ in (b), we get

$$\exp\left\{\int_0^T gdW - \frac{1}{2}\|g\|^2\right\} = \sum_{n=0}^{\infty} \frac{\|g\|^n}{n!} h_n\left(\frac{\theta}{\|g\|}\right).$$

(d) In particular, if we choose $g(s) = \chi_{[0,t]}(s)$, $s \in [0,T]$, we get

$$\exp\left\{W(t) - \frac{1}{2}t\right\} = \sum_{n=0}^{\infty} \frac{t^{n/2}}{n!} h_n\left(\frac{W(t)}{\sqrt{t}}\right). \quad \square$$

1.3 Solution

(a) $\xi = W(t) = \int_0^T \chi_{[0,t]}(s)dW(s)$, so $f_0 = 0$, $f_1 = \chi_{[0,t]}$, and $f_n = 0$ for $n \geq 2$.

(b) $\xi = \int_0^T g(s)dW(s)$, so $f_0 = 0$, $f_1 = g$, and $f_n = 0$ for $n \geq 2$.

(c) Since

$$\int_0^t \int_0^{t_2} 1\, dW(t_1)dW(t_2) = \int_0^t W(t_2)dW(t_2) = \frac{1}{2}W^2(t) - \frac{1}{2}t,$$

we get that

$$W^2(t) = t + 2\int_0^t \int_0^{t_2} 1\, dW(t_1)dW(t_2)$$

$$= t + 2\int_0^T \int_0^{t_2} \chi_{[0,t]}(t_1)\chi_{[0,t]}(t_2)dW(t_1)dW(t_2) = t + I_2[f_2].$$

Thus $f_0 = t$,

$$f_2(t_1, t_2) = \chi_{[0,t]}(t_1)\chi_{[0,t]}(t_2) =: \chi_{[0,t]}^{\otimes 2},$$

and $f_n = 0$ for $n \neq 2$.

(d) By Problem 1.1 (c) and (1.15), we have

$$\xi = \exp\left\{\int_0^T g(s)dW(s)\right\}$$

$$= \exp\left\{\frac{1}{2}\|g\|^2\right\} \sum_{n=0}^{\infty} \frac{\|g\|^n}{n!} h_n\left(\frac{\theta}{\|g\|}\right)$$

$$= \exp\left\{\frac{1}{2}\|g\|^2\right\} \sum_{n=0}^{\infty} J_n[g^{\otimes n}] = \sum_{n=0}^{\infty} \frac{1}{n!} \exp\left\{\frac{1}{2}\|g\|^2\right\} I_n[g^{\otimes n}].$$

Hence

$$f_n = \frac{1}{n!} \exp(\frac{1}{2}\|g\|^2) g^{\otimes n}, \qquad n = 0, 1, 2, \ldots,$$

where

$$g^{\otimes n}(x_1, \ldots, x_n) := g(x_1)g(x_2) \cdots g(x_n).$$

(e) We have the following equalities:

$$\begin{aligned}
\xi &= \int_0^T g(s)W(s)\,ds \\
&= \int_0^T g(s) \int_0^s 1\,dW(t)\,ds \\
&= \int_0^T \int_t^T g(s)\,ds\,dW(t) \\
&= I_1(f_1),
\end{aligned}$$

where $f_1(t) := \int_t^T g(s)ds$, $t \in [0, T]$. $\square$

1.4 Solution

(a) Since $\int\limits_0^T W(t)dW(t) = \frac{1}{2}W^2(T) - \frac{1}{2}T$, we have

$$F = W^2(T) = T + 2\int\limits_0^T W(t)dW(t).$$

Hence $E[F] = T$ and $\varphi(t) = 2W(t)$, $t \in [0, T]$.

(b) Define $M(t) = \exp\left\{W(t) - \frac{1}{2}t\right\}$, $t \in [0, T]$. Then by the Itô formula

$$dM(t) = M(t)dW(t)$$

and therefore

$$M(T) = 1 + \int\limits_0^T M(t)dW(t).$$

Moreover,

$$F = \exp\{W(T)\} = \exp\left\{\frac{T}{2}\right\} + \exp\left\{\frac{T}{2}\right\}\int\limits_0^T \exp\left\{W(t) - \frac{1}{2}t\right\}dW(t).$$

Hence

$$E[F] = \exp\left\{\frac{T}{2}\right\} \quad \text{and} \quad \varphi(t) = \exp\left\{W(t) + \frac{T-t}{2}\right\}, \quad t \in [0, T].$$

(c) Integration by parts (application of the Itô formula) gives

$$F = \int_0^T W(t)dt = TW(T) - \int_0^T tdW(t) = \int_0^T (T - t)dW(t).$$

Hence, $E[F] = 0$ and $\varphi(t) = T - t$, $t \in [0, T]$.

(d) By the Itô formula

$$dW^3(t) = 3W^2(t)dW(t) + 3W(t)dt.$$

Hence

$$F = W^3(T) = 3\int_0^T W^2(t)dW(t) + 3\int_0^T W(t)dt.$$

Therefore, by (c) we get

$$E[F] = 0 \quad \text{and} \quad \varphi(t) = 3W^2(t) + 3T(1 - t), \quad t \in [0, T].$$

(e) Put $X(t) = e^{\frac{1}{2}t}$, $Y(t) = \cos W(t)$, $N(t) = X(t)Y(t)$, $t \in [0, T]$. Then we have

$$dN(t) = X(t)dY(t) + Y(t)dX(t) + dX(t)dY(t)$$

$$= e^{\frac{1}{2}t}[-\sin W(t)dW(t) - \frac{1}{2}\cos W(t)dt] + \cos W(t) e^{\frac{1}{2}t}\frac{1}{2}dt$$

$$= -e^{\frac{1}{2}t}\sin W(t)dW(t).$$

Hence

$$e^{\frac{1}{2}T}\cos W(T) = 1 - \int_0^T e^{\frac{1}{2}t}\sin W(t)dW(t)$$

and also

$$F = \cos W(T) = e^{-\frac{1}{2}T} - e^{-\frac{1}{2}T}\int_0^T e^{\frac{1}{2}t}\sin W(t)dW(t).$$

Hence $E[F] = e^{-\frac{1}{2}T}$ and $\varphi(t) = -e^{\frac{1}{2}(t-T)}\sin W(t)$, $t \in [0, T]$. $\square$

1.5 Solution

(a)   By Itô formula and Kolmogorov backward equation we have

$$dY(t) = \frac{\partial g}{\partial t}(t, X(t))dt + \frac{\partial g}{\partial x}(t, X(t))dX(t) + \frac{1}{2}\frac{\partial^2 g}{\partial x^2}(t, X(t))(dX(t))^2$$

$$= \frac{\partial}{\partial t}[P_{T-t}f(\xi)]_{|\xi=X(t)}dt + \sigma(X(t))\frac{\partial}{\partial \xi}[P_{T-t}f(\xi)]_{|\xi=X(t)}dW(t)$$

$$+\{b(X(t))\frac{\partial}{\partial \xi}[P_{T-t}f(\xi)]_{|\xi=X(t)} + \frac{1}{2}\sigma^2(X(t))\frac{\partial^2}{\partial \xi^2}[P_{T-t}f(\xi)]_{|\xi=X(t)}\}dt$$

$$= \frac{\partial}{\partial t}[P_{T-t}f(\xi)]_{|\xi=X(t)}dt + \sigma(X(t))\frac{\partial}{\partial \xi}[P_{T-t}f(\xi)]_{|\xi=X(t)}dW(t)$$

$$+\frac{\partial}{\partial u}[P_u f(\xi)]_{\substack{|\xi=X(t)\\|u=T-t}}dt$$

$$= \sigma(X(t))\frac{\partial}{\partial \xi}[P_{T-t}f(\xi)]_{|\xi=X(t)}dW(t).$$

Hence

$$Y(T) = Y(0) + \int_0^T [\sigma(x)\frac{\partial}{\partial \xi}P_{T-t}f(\xi)]_{|\xi=X(t)}dW(t).$$

Since $Y(T) = g(T, X(T)) = [P_0 f(\xi)]_{|\xi=X(T)} = f(X(T))$ and $Y(0) = g(0, X(0)) = P_T f(X)$, (1.29) follows.

(b.1) If $F = W^2(T)$, we apply (a) to the case when $f(\xi) = \xi^2$ and $X(t) = x + W(t)$ (assuming $W(0) = 0$ as before). This gives

$$P_s f(\xi) = E^\xi[f(X(x))] = E^\xi[X^2(s)] = \xi^2 + s$$

and hence

$$E[F] = P_T f(x) = x^2 + T$$

and

$$\varphi(t) = [\frac{\partial}{\partial \xi}(\xi^2 + s)]_{|\xi=x+W(t)} = 2W(t) + 2x.$$

(b.2) If $F = W^3(T)$, we choose $f(\xi) = \xi^3$ and $X(t) = x + W(t)$ and get

$$P_s f(\xi) = E^\xi[X^3(s)] = \xi^3 + 3s\xi.$$

Hence

$$E[F] = P_T f(x) = x^3 + 3Tx$$

and

$$\varphi(t) = [\frac{\partial}{\partial \xi}(\xi^3 + 3(T-t)\xi)]_{|\xi=x+W(t)} = 3(x + W(t))^2 + 3(T-t).$$

(b.3) In this case $f(\xi) = \xi$, so

$$P_s f(\xi) = E^\xi[X(s)] = \xi e^{\rho s}$$

and so

$$E[F] = P_T f(x) = x e^{\rho T}$$

and

$$\varphi(t) = [\alpha\xi\frac{\partial}{\partial\xi}(\xi e^{\rho(T-t)})]_{|\xi=X(t)}$$
$$= \alpha X(t)\exp(\rho(T-t))$$
$$= \alpha x\exp\{\rho T - \frac{1}{2}\alpha^2 t + \alpha W(t)\}.$$

(c)   We proceed as in (a) and put

$$Y(t) = g(t, X(t)), \quad t \in [0, T], \quad \text{with} \quad g(t, x) = P_{T-t}f(x)$$

and

$$dX(t) = b(X(t))dt + \sigma(X(t))dW(t); \qquad X(0) = x \in \mathbb{R}^n,$$

where

$$b: \mathbb{R}^n \to \mathbb{R}^n, \quad \sigma: \mathbb{R}^n \to \mathbb{R}^{n\times m} \quad \text{and} \quad W(t) = (W_1(t), \ldots, W_m(t))$$

is the m-dimensional Wiener process. Then by Itô formula and (1.31), we have

$$dY(t) = \frac{\partial g}{\partial t}(t, X(t))dt + \sum_{i=1}^{n}\frac{\partial g}{\partial x_i}(t, X(t))dX_i(t)$$

$$+\frac{1}{2}\sum_{i,j=1}^{n}\frac{\partial^2 g}{\partial x_i\partial x_j}(t, X(t))dX_i(t)dX_j(t)$$

$$= \frac{\partial}{\partial t}[P_{T-t}f(\xi)]_{|\xi=X(t)}dt + [\sigma^T(\xi)\nabla_\xi(P_{T-t}f(\xi))]_{|\xi=X(t)}dW(t)$$

$$+[L_\xi(P_{T-t}f(\xi))]_{|\xi=X(t)}dt,$$

where

$$L_\xi = \sum_{i=1}^{n}b_i(\xi)\frac{\partial}{\partial\xi_i} + \frac{1}{2}\sum_{i,j=1}^{n}(\sigma\sigma^T)_{ij}(\xi)\frac{\partial^2}{\partial\xi_i\partial\xi_j}$$

is the generator of the Itô diffusion $X(t)$, $t \geq 0$. So by the Kolmogorov backward equation we get

$$dY(t) = [\sigma^T(\xi)\nabla_\xi(P_{T-t}f(\xi))]_{|\xi=X(t)}dW(t)$$

and hence, as in (a),

$$Y(T) = f(X(T)) = P_T f(x) + \int_0^T [\sigma^T(\xi)\nabla_\xi(P_{T-t}f(\xi))]_{|\xi=X(t)}dW(t),$$

which gives, with $F = f(X(T))$,

$$E[F] = P_T f(x) \quad \text{and} \quad \varphi(t) = [\sigma^T(\xi)\nabla_\xi(P_{T-t}f(\xi))]_{|\xi=X(t)}, \quad t \in [0, T]. \quad \square$$

Problems of Chap. 2

2.4 Solution

(a) Since $W(t)$, $t \in [0, T]$, is $\mathbb{F}$-adapted, we have

$$\int_0^T W(t)\delta W(t) = \int_0^T W(t)dW(t) = \frac{1}{2}W^2(T) - \frac{1}{2}T.$$

(b) $\int_0^T \left(\int_0^T g(s)dW(s) \right) \delta W(t) = \int_0^T I_1[f_1(\cdot, t)]\delta W(t) = I_2[\widetilde{f}_1]$, where $f_1(t_1, t) = g(t_1)$, $t \in [0, T]$. This gives

$$\widetilde{f}_1(t_1, t) = \frac{1}{2}[g(t_1) + g(t)], \quad t \in [0, T],$$

and hence

$$
\begin{aligned}
I_2[\widetilde{f}_1] &= 2 \int_0^T \int_0^{t_2} \widetilde{f}_1(t_1, t_2)dW(t_1)dW(t_2) \\
&= \int_0^T \int_0^{t_2} g(t_1)dW(t_1)dW(t_2) + \int_0^T \int_0^{t_2} g(t_2)dW(t_1)dW(t_2) \qquad \text{(S.1)} \\
&= \int_0^T \int_0^{t_2} g(t_1)dW(t_1)dW(t_2) + \int_0^T W(t_2)g(t_2)dW(t_2).
\end{aligned}
$$

Using integration by parts (i.e., the Itô formula), we see that

$$
\begin{aligned}
\left(\int_0^T g(t_1)dW(t_1) \right) W(T) &= \int_0^T \int_0^{t_2} g(t_1)dW(t_1)dW(t_2) \\
&+ \int_0^T g(t_2)W(t_2)dW(t_2) + \int_0^T g(t)dt.
\end{aligned}
\qquad \text{(S.2)}
$$

Combining (S.1) and (S.2), we get

$$\int_0^T \left(\int_0^T g(s)dW(s) \right) \delta W(t) = \left(\int_0^T g(t)dW(t) \right) W(T) - \int_0^T g(t)dt.$$

(c) By Problem 1.3 (c) we have

$$\int_0^T W^2(t_0)\delta W(t) = \int_0^T \left(t_0 + I_2[f_2(\cdot,t)]\right)\delta W(t),$$

where

$$f_2(t_1,t_2,t) = \chi_{[0,t_0]}(t_1)\chi_{[0,t_0]}(t_2), \quad t \in [0,T].$$

Now

$$\widetilde{f}_2(t_1,t_2,t) = \frac{1}{3}\left[f_2(t_1,t_2,t) + f_2(t,t_2,t_1) + f_2(t_1,t,t_2)\right]$$

$$= \frac{1}{3}\left[\chi_{[0,t_0]}(t_1)\chi_{[0,t_0]}(t_2) + \chi_{[0,t_0]}(t)\chi_{[0,t_0]}(t_2) + \chi_{[0,t_0]}(t_1)\chi_{[0,t_0]}(t)\right]$$

$$= \frac{1}{3}\left[\chi_{\{t_1,t_2<t_0\}} + \chi_{\{t,t_2<t_0\}} + \chi_{\{t_1,t<t_0\}}\right]$$

$$= \chi_{\{t,t_1,t_2<t_0\}} + \frac{1}{3}\chi_{\{t_1,t_2<t_0<t\}} + \frac{1}{3}\chi_{\{t,t_2<t_0<t_1\}} + \frac{1}{3}\chi_{\{t,t_1<t_0<t_2\}}$$

and hence, using (1.15),

$$\int_0^T W^2(t_0)\delta W(t) = t_0 W(T) + \int_0^T I_2[f_2(\cdot,t)]\delta W(t)$$

$$= t_0 W(T) + I_3[\widetilde{f}_2] = t_0 W(T) + 6J_3[\widetilde{f}_2]$$

$$= t_0 W(T)$$

$$+ 6\int_0^T\int_0^{t_3}\int_0^{t_2} \chi_{[0,t_0]}(t_1)\chi_{[0,t_0]}(t_2)\chi_{[0,t_0]}(t_3)\,dW(t_1)dW(t_2)dW(t_3)$$

$$+ 6\int_0^T\int_0^{t_3}\int_0^{t_2} \frac{1}{3}\chi_{\{t_1,t_2<t_0<t_3\}}\,dW(t_1)dW(t_2)dW(t_3)$$

$$= t_0 W(T) + t_0^{3/2}h_3\left(\frac{W(t_0)}{\sqrt{t_0}}\right)$$

$$+ 2\int_{t_0}^T\int_0^{t_0}\int_0^{t_2} dW(t_1)dW(t_2)dW(t_3)$$

$$= t_0 W(T) + t_0^{3/2}\left(\frac{W^3(t_0)}{t_0^{3/2}} - 3\frac{W(t_0)}{\sqrt{t_0}}\right)$$

$$+ 2\int_{t_0}^T \left(\frac{1}{2}W^2(t_0) - \frac{1}{2}t_0\right)dW(t_3)$$

$$= t_0 W(T) + W^3(t_0) - 3t_0 W(t_0) + \left(W^2(t_0) - t_0\right)\left(W(T) - W(t_0)\right)$$

$$= W^2(t_0)W(T) - 2t_0 W(t_0).$$

(d) By Problem 1.3 (d) and (1.15) we get

$$\int_0^T \exp(W(T))\delta W(t) = \int_0^T \sum_{n=0}^\infty \frac{1}{n!} e^{T/2} I_n[1]\, \delta W(t)$$

$$= \sum_{n=0}^\infty \frac{1}{n!} e^{T/2} I_{n+1}[1]$$

$$= e^{T/2} \sum_{n=0}^\infty \frac{1}{n!} T^{\frac{n+1}{2}} h_{n+1}\Big(\frac{W(T)}{\sqrt{T}}\Big).$$

(e) Using Problem 1.3 (e) we have that

$$F = \int_0^T F\delta W(t) = I_1(f_1),$$

with $f_1(t) := \int_t^T g(s)ds$, $t \in [0, T]$. Hence

$$\int_0^T F\delta W(t) = I_2(\widetilde{f_1}),$$

where

$$\widetilde{f_1}(t_1, t_2) = \frac{1}{2}\Big(\int_{t_1}^T g(s)ds + \int_{t_2}^T g(s)ds\Big).$$

This gives

$$F = \int_0^T F\delta W(t)$$

$$= J_2\Big(\int_{t_1}^T g(s)ds\Big) + J_2\Big(\int_{t_2}^T g(s)ds\Big)$$

$$= \int_0^T \int_0^{t_2} \Big(\int_{t_1}^T g(s)ds\Big) dW(t_1)dW(t_2)$$

$$+ \int_0^T \Big(\int_0^{t_2} 1 dW(t_1)\Big)\Big(\int_{t_2}^T g(s)ds\Big)dW(t_2)$$

$$= \int_0^T \Big(\int_0^{t_2} g(s)W(s)ds\Big)dW(t_2) + \int_0^T 2W(t_2)\Big(\int_{t_2}^T g(s)ds\Big)dW(t_2). \quad \square$$

Problems of Chap. 3

3.2 Solution

(a) $D_t W(T) = \chi_{[0,T]}(t) = 1$, $t \in [0, T]$, by (3.8).

(b) By (3.8) we get

$$D_t \int_0^T s^2 dW(s) = t^2.$$

(c) By (3.2) we have

$$D_t \int_0^T \int_0^{t_2} \cos(t_1 + t_2) dW(t_1) dW(t_2) = D_t\left(\frac{1}{2} I_2[\cos(t_1 + t_2)]\right)$$

$$= \frac{1}{2} 2 I_1[\cos(\cdot + t)]$$

$$= \int_0^T \cos(t_1 + t) dW(t_1).$$

(d) By the chain rule, we get

$$D_t\big(3W(s_0)W^2(t_0) + \log(1 + W^2(s_0))\big) = \left[3W^2(t_0) + \frac{2W(s_0)}{1 + W^2(s_0)}\right] \chi_{[0,s_0]}(t)$$

$$+ 6W(s_0)W(t_0)\chi_{[0,t_0]}(t).$$

(e) By Problem 2.4 (b) we have

$$D_t \int_0^T W(t_0)\delta W(t) = D_t\big(W(t_0)W(T) - t_0\big)$$

$$= W(t_0) \chi_{[0,T]}(t) + W(T)\chi_{[0,t_0]}(t)$$

$$= W(t_0) + W(T)\chi_{[0,t_0]}(t). \quad \square$$

3.3 Solution

(a) By Problem 1.3 (d) and (3.2), we have

$$D_t \exp\left\{ \int_0^T g(s)dW(s) \right\} = D_t \sum_{n=0}^{\infty} I_n[f_n] = \sum_{n=1}^{\infty} nI_{n-1}[f_n(\cdot, t)]$$

$$= \sum_{n=1}^{\infty} n \frac{1}{n!} \exp\left\{\frac{1}{2}\|g\|^2\right\} I_{n-1}\big[g(t_1) \ldots g(t_{n-1})g(t)\big]$$

$$= g(t) \sum_{n=1}^{\infty} \frac{1}{(n-1)!} \exp\left\{\frac{1}{2}\|g\|^2\right\} I_{n-1}\big[g^{\otimes(n-1)}\big]$$

$$= g(t) \exp\left\{ \int_0^T g(s)dW(s) \right\}.$$

(b) The suggested chain rule and (3.8) give

$$D_t \exp\left\{ \int_0^T g(s)dW(s) \right\} = \exp\left\{ \int_0^T g(s)dW(s) \right\} D_t\left(\int_0^T g(s)dW(t) \right)$$

$$= g(t)\exp\left\{ \int_0^T g(s)dW(s) \right\}.$$

(c) The points above together with Corollary 3.13 give

$$D_t \exp\{W(t_0)\} = \exp\{W(t_0)\}\, \chi_{[0,t_0]}(t). \quad \square$$

Problems of Chap. 4

4.1 Solution

(a) If $s > t$ we have

$$
\begin{aligned}
E_Q[\widetilde{W}(s)|\mathcal{F}_t] &= \frac{E[Z(T)\widetilde{W}(s)|\mathcal{F}_t]}{E[Z(T)|\mathcal{F}_t]} \\
&= \frac{E[Z(T)\widetilde{W}(s)|\mathcal{F}_t]}{Z(t)} \qquad\qquad\qquad (\mathrm{S.3}) \\
&= Z^{-1}(t)E[E[Z(T)\widetilde{W}(s)|\mathcal{F}_s]|\mathcal{F}_t] \\
&= Z^{-1}(t)E[\widetilde{W}(s)E[Z(T)|\mathcal{F}_s]|\mathcal{F}_t] \\
&= Z^{-1}(t)E[\widetilde{W}(s)Z(s)|\mathcal{F}_t]. \qquad\qquad (\mathrm{S.4})
\end{aligned}
$$

Applying Itô formula to $Y(t) := Z(t)\widetilde{W}(t)$, we get

$$
\begin{aligned}
dY(t) &= Z(t)d\widetilde{W}(t) + \widetilde{W}(t)dZ(t) + d\widetilde{W}(t)dZ(t) \\
&= Z(t)[\theta(t)dt + dW(t)] + \widetilde{W}(t)[-\theta(t)Z(t)dW(t)] - \theta(t)Z(t)dt \\
&= Z(t)[1 - \theta(t)\widetilde{W}(t)]dW(t),
\end{aligned}
$$

and hence $Y(t)$ is an $\mathcal{F}_t$-martingale (with respect to P). Therefore, by (S.3),

$$E_Q[\widetilde{W}(s)|\mathcal{F}_t] = Z^{-1}(t)E[Y(s)|\mathcal{F}_t] = Z^{-1}(t)Y(t) = \widetilde{W}(t).$$

(b) We apply the Girsanov theorem to the case with $\theta(t) = a$. Then X is a Wiener process with respect to the measure Q defined by

$$Q(d\omega) = Z(T,\omega)P(d\omega) \qquad \text{on } \mathcal{F}_T,$$

where

$$Z(t) = \exp\left\{ -aW(t) - \frac{1}{2}a^2t \right\}, \qquad 0 \le t \le T.$$

(c) In this case we have

$$\beta(t) = bY(t), \quad \alpha(t) = aY(t), \quad \gamma(t) = cY(t)$$

and hence we put

$$\theta = \frac{\beta(t) - \alpha(t)}{\gamma(t)} = \frac{b - a}{c}$$

and

$$Z(t) = \exp\left\{ -\theta W(t) - \frac{1}{2}\theta^2 t \right\}, \qquad 0 \le t \le T.$$

Then

$$\widetilde{W}(t) := \theta t + W(t), \quad 0 \le t \le T,$$

is a Wiener process with respect to the measure Q defined by $Q(d\omega) = Z(T, \omega)P(d\omega)$ on $\mathcal{F}_T$ and

$$dY(t) = bY(t)dt + cY(t)[d\widetilde{W}(t) - \theta dt] = aY(t)dt + cY(t)d\widetilde{W}(t). \quad \square$$

4.2 Solution

(a) $F = W(T)$ implies $D_t F = \chi_{[0,T]}(t) = 1$, for $t \in [0, T]$, and hence

$$E[F] + \int_0^T E[D_t F | \mathcal{F}_t] dW(t) = \int_0^T 1 \, dW(t) = W(T) = F.$$

(b) $F = \int_0^T W(s) ds$ implies $D_t F = \int_0^T D_t W(s) ds = \int_0^T \chi_{[0,s]}(t) ds = \int_t^T ds = T - t$,
which gives

$$E[F] + \int_0^T E[D_t F | \mathcal{F}_t] dW(t) = \int_0^T (T - t) dW(t)$$

$$= \int_0^T W(s) dW(s) = F,$$

using integration by parts.

(c) $F = W^2(T)$ implies $D_t F = 2W(T) D_t W(T) = 2W(T)$. Hence

$$E[F] + \int_0^T E[D_t F | \mathcal{F}_t] dW(t) = T + \int_0^T E[2W(T) | \mathcal{F}_t] dW(t)$$

$$= T + 2 \int_0^T W(t) dW(t)$$

$$= T + W^2(T) - T = W^2(T) = F.$$

(d) $F = W^3(T)$ implies $D_t F = 3W^2(T)$. Hence, by Itô formula,

$$E[F] + \int_0^T E[D_t F|\mathcal{F}_t]dW(t) = \int_0^T E[3W^2(T)|\mathcal{F}_t]dW(t)$$

$$= 3\int_0^T E[(W(T) - W(t))^2 + 2W(t)W(T) - W^2(t)|\mathcal{F}_t]dW(t)$$

$$= 3\int_0^T (T - t)dW(t) + 6\int_0^T W^2(t)dW(t) - 3\int_0^T W^2(t)dW(t)$$

$$= 3\int_0^T W^2(t)dW(t) - 3\int_0^T W(t)dt = W^3(T).$$

(e) $F = \exp\{W(T)\}$ implies $D_t F = \exp\{W(T)\}$. Hence

$$RHS = E[F] + \int_0^T E[D_t F|\mathcal{F}_t]dW(t)$$

$$= e^{T/2} + \int_0^T E[\exp\{W(T)\}|\mathcal{F}_t]dW(t)$$

$$= e^{T/2} + \int_0^T E[\exp\left\{W(T) - \frac{1}{2}T\right\} e^{T/2}|\mathcal{F}_t]dW(t)$$

$$= e^{T/2} + \exp\{\frac{1}{2}T\}\int_0^T \exp\left\{W(t) - \frac{1}{2}t\right\}dW(t). \qquad (S.5)$$

Here we have used that

$$M(t) := \exp\left\{W(t) - \frac{1}{2}t\right\}$$

is a martingale. In fact, by Itô formula we have $dM(t) = M(t)dW(t)$. Combined with (S.5) this gives

$$RHS = \exp\left\{\frac{1}{2}T\right\} + \exp\left\{\frac{1}{2}T\right\}(M(T) - M(0)) = \exp W(T) = F.$$

(f) $F = (W(T) + T)\exp\left\{ - W(T) - \frac{1}{2}T\right\}$ implies $D_t F = \exp\left\{ - W(T) - \frac{1}{2}T\right\}[1 - W(T) - T]$. Note that

$$Y(t) := (W(t) + t)N(t), \quad \text{with} \quad N(t) = \exp\left\{ - W(t) - \frac{1}{2}t\right\}$$

is a martingale, since

$$
\begin{aligned}
dY(t) &= (W(t)+t)N(t)(-dW(t)) + N(t)(dW(t)+dt) - N(t)dt \\
&= N(t)[1-t-W(t)]dW(t).
\end{aligned}
$$

Therefore,

$$
\begin{aligned}
E[F] + \int_0^T E[D_t F|\mathcal{F}_t]dW(t) &= \int_0^T E[N(T)(1-(W(T)+T))|\mathcal{F}_t]dW(t) \\
&= \int_0^T N(t)(1-(W(t)+t))dW(t) \\
&= \int_0^T 1\,dY(t) = Y(T) - Y(0) \\
&= (W(T)+T)\exp\left\{-W(T)-\frac{1}{2}T\right\} = F. \quad \square
\end{aligned}
$$

4.3 Solution

(a) $\tilde{\varphi}(t) = E_Q[D_t F - F\int_t^T D_t\theta(s)d\widetilde{W}(s)|\mathcal{F}_t]$. If $\theta(s)$, $t \in [0,T]$, is deterministic, then $D_t\theta = 0$ and hence

$$
\begin{aligned}
\tilde{\varphi}(t) &= E_Q[D_t F|\mathcal{F}_t] = E_Q[2W(T)|\mathcal{F}_t] \\
&= E_Q[2\widetilde{W}(T) - 2\int_0^T \theta(s)ds|\mathcal{F}_t] \\
&= 2\widetilde{W}(t) - 2\int_0^T \theta(s)ds \\
&= 2W(t) - 2\int_t^T \theta(s)ds.
\end{aligned}
$$

(b) By application of the generalized Clark–Ocone formula, we have

$$
\begin{aligned}
\tilde{\varphi}(t) &= E_Q[D_t F|\mathcal{F}_t] \\
&= E_Q\left[\exp\left\{\int_0^T \lambda(s)dW(s)\right\}\lambda(t)|\mathcal{F}_t\right] \\
&= \lambda(t)E_Q\left[\exp\left\{\int_0^T \lambda(s)d\widetilde{W}(s) - \int_0^T \lambda(s)\theta(s)ds\right\}|\mathcal{F}_t\right]
\end{aligned}
$$

$$= \lambda(t) \exp \left\{ \int_0^T (\frac{1}{2}\lambda^2(s) - \lambda(s)\theta(s))ds \right\} E_Q \left[\exp \left\{ \int_0^T \lambda(s)d\widetilde{W}(s) \right. \right.$$

$$\left. \left. - \frac{1}{2} \int_0^T \lambda^2(s)ds \right\} | \mathcal{F}_t \right]$$

$$= \lambda(t) \exp \left\{ \int_0^T \lambda(s)(\frac{1}{2}\lambda(s) - \theta(s))ds \right\} \exp \left\{ \int_0^t \lambda(s)d\widetilde{W}(s) - \frac{1}{2} \int_0^t \lambda^2(s)ds \right\}$$

$$= \lambda(t) \exp \left\{ \int_0^t \lambda(s)dW(s) + \int_t^T \lambda(s)(\frac{1}{2}\lambda(s) - \theta(s))ds \right\}.$$

$$\text{(S.6)}$$

(c) By application of the generalized Clark–Ocone formula, we have

$$\widetilde{\varphi}(t) = E_Q[D_t F - F \int_t^T D_t \theta(s)d\widetilde{W}(s)|\mathcal{F}_t]$$

$$\text{(S.7)}$$

$$= E_Q[\lambda(t)F|\mathcal{F}_t] - E_Q[F \int_t^T d\widetilde{W}(s)|\mathcal{F}_t]$$

$$=: A - B.$$

Now $\widetilde{W}(t) = W(t) + \int_0^t \theta(s)ds = W(t) + \int_0^t W(s)ds$ or

$$dW(t) + W(t)dt = d\widetilde{W}(t).$$

We solve this equation for $W(t)$ by multiplying by the "integrating factor" e^t and get

$$d(e^t W(t)) = e^t d\widetilde{W}(t).$$

Hence

$$W(u) = e^{-u} \int_0^u e^s d\widetilde{W}(s)$$

$$\text{(S.8)}$$

or

$$dW(u) = -e^{-u} \int_0^u e^s d\widetilde{W}(s)du + d\widetilde{W}(u).$$

$$\text{(S.9)}$$

Using (S.9) we may rewrite F as follows:

$$F = \exp\left\{ \int_0^T \lambda(s) dW(s) \right\}$$

$$= \exp\left\{ \int_0^T \lambda(s) d\widetilde{W}(s) - \int_0^T \lambda(u) e^{-u} \left(\int_0^u e^s d\widetilde{W}(s) \right) du \right\}$$

$$= \exp\left\{ \int_0^T \lambda(s) d\widetilde{W}(s) - \int_0^T \left(\int_0^T \lambda(u) e^{-u} du \right) e^s d\widetilde{W}(s) \right\}$$

$$= K(T) \exp\left\{ \frac{1}{2} \int_0^T \xi^2(s) ds \right\},$$

where

$$\xi(s) = \lambda(s) - e^s \int_s^T \lambda(u) e^{-u} du \qquad\qquad (S.10)$$

and

$$K(t) = \exp\left\{ \int_0^t \xi(s) d\widetilde{W}(s) - \frac{1}{2} \int_0^t \xi^2(s) ds \right\}, \qquad 0 \le t \le T. \qquad (S.11)$$

Hence

$$A = E_Q[\lambda(t) F | \mathcal{F}_t] \qquad\qquad (S.12)$$

$$= \lambda(t) \exp\left\{ \frac{1}{2} \int_0^T \xi^2(s) ds \right\} E[K(T) | \mathcal{F}_t]$$

$$= \lambda(t) \exp\left\{ \frac{1}{2} \int_0^T \xi^2(s) ds \right\} K(t). \qquad\qquad (S.13)$$

Moreover, if we put

$$H := \exp\left\{ \frac{1}{2} \int_0^T \xi^2(s) ds \right\}, \qquad\qquad (S.14)$$

we get

$$B = E_Q[F(\widetilde{W}(T) - \widetilde{W}(t)) | \mathcal{F}_t]$$
$$= H E_Q[K(T)(\widetilde{W}(T) - \widetilde{W}(t)) | \mathcal{F}_t]$$

$$= H\,E_Q\Big[K(t)\exp\Big\{\int_t^T \xi(s)d\widetilde{W}(s) - \frac{1}{2}\int_t^T \xi^2(s)ds\Big\}(\widetilde{W}(T) - \widetilde{W}(t))|\mathcal{F}_t\Big]$$

$$= H\,K(t)E_Q\Big[\exp\Big\{\int_t^T \xi(s)d\widetilde{W}(s) - \frac{1}{2}\int_t^T \xi^2(s)ds\Big\}(\widetilde{W}(T) - \widetilde{W}(t))\Big]$$

$$= H\,K(t)E\Big[\exp\Big\{\int_t^T \xi(s)dW(s) - \frac{1}{2}\int_t^T \xi^2(s)ds\Big\}(W(T) - W(t))\Big]. \quad \text{(S.15)}$$

This last expectation can be evaluated by using the Itô formula. Put

$$X(t) = \exp\Big\{\int_{t_0}^t \xi(s)dW(s) - \frac{1}{2}\int_{t_0}^t \xi^2(s)ds\Big\}$$

and

$$Y(t) = X(t)\,(W(t) - W(t_0)).$$

Then

$$dY(t) = X(t)dW(t) + (W(t) - W(t_0))dX(t) + dX(t)dW(t)$$
$$= X(t)[1 + (W(t) - W(t_0))\xi(t)]dW(t) + \xi(t)X(t)dt$$

and hence

$$E[Y(T)] = E[Y(t_0)] + E[\int_{t_0}^T \xi(s)X(s)ds]$$

$$= \int_{t_0}^T \xi(s)E[X(s)]ds \qquad \text{(S.16)}$$

$$= \int_{t_0}^T \xi(s)ds. \qquad \text{(S.17)}$$

Combining (S.7) and (S.10)–(S.16), we conclude that

$$\widetilde{\varphi}(t) = \lambda(t)HK(t) - H\,K(t)\int_t^T \xi(s)ds$$

$$= \exp\Big\{\frac{1}{2}\int_0^T \xi^2(s)ds\Big\}\exp\Big\{\int_0^t \xi(s)d\widetilde{W}(s)$$

$$-\frac{1}{2}\int_0^t \xi^2(s)ds\Big\}\Big[\lambda(t) - \int_t^T \xi(s)ds\Big]. \quad \square$$

4.4 Solution

(a) Since $u = \frac{\mu-\rho}{\sigma}$ is constant we get using (4.27)

$$\theta_1(t) = e^{\rho t}\sigma^{-1}S^{-1}(t)E_Q[e^{-\rho T}D_t W(T)|\mathcal{F}_t] = e^{\rho(t-T)}\sigma^{-1}S^{-1}(t).$$

(b) Here $\mu = 0$, $\sigma(s) = cS^{-1}(s)$ and hence

$$u(s) = \frac{\mu-\rho}{\sigma} = -\frac{\rho}{c}S(s) = -\rho(W(s) + S(0)).$$

Hence

$$\int\limits_t^T D_t u(s)d\widetilde{W}(s) = \rho\big[\widetilde{W}(t) - \widetilde{W}(T)\big].$$

Therefore,

$$B := E_Q\Big[F\int\limits_t^T D_t u(s)d\widetilde{W}(s)|\mathcal{F}_t\Big] = \rho E_Q[e^{-\rho T}W(T)(\widetilde{W}(t) - \widetilde{W}(T))|\mathcal{F}_t].$$

$$(S.18)$$

To proceed further, we need to express W in terms of $\widetilde{W}$: since

$$\widetilde{W}(t) = W(t) + \int\limits_0^t u(s)ds = W(t) - \rho S(0)t - \rho\int\limits_0^t W(s)ds,$$

we have

$$d\widetilde{W}(t) = dW(t) - \rho W(t)dt - \rho S(0)dt$$

or

$$e^{-\rho t}dW(t) - e^{-\rho t}\rho W(t)dt = e^{-\rho t}(d\widetilde{W}(t) + \rho S(0)dt)$$

or

$$d(e^{-\rho t}W(t)) = e^{-\rho t}d\widetilde{W}(t) + \rho e^{-\rho t}S(0)dt.$$

Hence

$$W(t) = S(0)[e^{\rho t} - 1] + e^{\rho t}\int\limits_0^t e^{-\rho s}d\widetilde{W}(s). \qquad (S.19)$$

Substituting this in (S.18) we get

$$B = \rho E_Q[\int\limits_0^T e^{-\rho s}d\widetilde{W}(s)\,(\widetilde{W}(t) - \widetilde{W}(T))|\mathcal{F}_t]$$

$$= \rho E_Q[\int\limits_0^t e^{-\rho s}d\widetilde{W}(s)\,(\widetilde{W}(t) - \widetilde{W}(T))|\mathcal{F}_t]$$

$$+\rho E_Q[\int_t^T e^{-\rho s} d\widetilde{W}(s)\,(\widetilde{W}(t) - \widetilde{W}(T))|\mathcal{F}_t]$$

$$= \rho E_Q[\int_t^T e^{-\rho s} d\widetilde{W}(s)\,(\widetilde{W}(t) - \widetilde{W}(T))]$$

$$= \rho \int_t^T e^{-\rho s}(-1)ds = e^{-\rho T} - e^{-\rho t}.$$

Hence

$$\theta_1(t) = e^{\rho t}c^{-1}(E_Q[D_t(e^{-\rho T}W(T))|\mathcal{F}_t] - B)$$
$$= e^{\rho t}c^{-1}(e^{-\rho T} - e^{-\rho T} + e^{-\rho t}) = c^{-1},$$

as expected.

(c) Here $\sigma = c\,S^{-1}(t)$ and hence

$$u(s) = \frac{\mu - \rho}{c}S(s) = \frac{\mu - \rho}{c}\left[e^{\mu s}S(0) + c\int_0^s e^{\mu(s-r)}dW(r)\right].$$

So

$$D_t u(s) = (\mu - \rho)e^{\mu(s-t)}\chi_{[0,s]}(t).$$

Hence

$$\theta_1(t) = e^{\rho t}c^{-1}E_Q[(D_t(e^{-\rho T}W(T)) - e^{-\rho T}W(T)\int_t^T D_t u(s)d\widetilde{W}(s)|\mathcal{F}_t]$$

$$= e^{\rho(t-T)}c^{-1}(1 - (\mu - \rho)E_Q[W(T)\int_t^T e^{\mu(s-t)}d\widetilde{W}(s)|\mathcal{F}_t]). \quad \text{(S.20)}$$

Again we try to express W in terms of $\widetilde{W}$: since

$$d\widetilde{W}(t) = dW(t) + u(t)dt$$

$$= dW(t) + \frac{\mu - \rho}{c}[c^{\mu t}S(0) + c\int_0^t c^{\mu(t-r)}dW(r)]dt,$$

we have

$$e^{-\mu t}d\widetilde{W}(t) = e^{-\mu t}dW(t) + [\frac{\mu - \rho}{c}S(0) + (\mu - \rho)\int_0^t e^{-\mu r}dW(r)]dt. \quad \text{(S.21)}$$

If we put

$$X(t) = \int_0^t e^{-\mu r} dW(r), \qquad \widetilde{X}(t) = \int_0^t e^{-\mu r} d\widetilde{W}(r),$$

(S.21) can be written as

$$d\widetilde{X}(t) = dX(t) + \frac{\mu - \rho}{c} S(0) dt + (\mu - \rho) X(t) dt$$

or

$$d(e^{(\mu-\rho)t} X(t)) = e^{(\mu-\rho)t} d\widetilde{X}(t) - \frac{\mu - \rho}{c} S(0) e^{(\mu-\rho)t} dt$$

or

$$X(t) = e^{(\rho-\mu)t} \int_0^t e^{-\rho s} d\widetilde{W}(s) - \frac{\mu - \rho}{c} S(0) e^{(\rho-\mu)t} \int_0^t e^{(\mu-\rho)s} ds$$

$$= e^{(\rho-\mu)t} \int_0^t e^{-\rho s} d\widetilde{W}(s) - \frac{S(0)}{c} [1 - e^{(\rho-\mu)t}].$$

From this we get

$$e^{-\mu t} dW(t) = e^{(\rho-\mu)t} e^{-\rho t} d\widetilde{W}(t) + (\rho - \mu) e^{(\rho-\mu)t} \left(\int_0^t e^{-\rho s} d\widetilde{W}(s) \right) dt$$

$$+ \frac{S(0)}{c} (\rho - \mu) e^{(\rho-\mu)t} dt$$

or

$$dW(t) = d\widetilde{W}(t) + (\rho - \mu) e^{\rho t} \left(\int_0^t e^{-\rho s} d\widetilde{W}(s) \right) dt + \frac{S(0)}{c} (\rho - \mu) e^{\rho t} dt.$$

In particular,

$$W(T) = \widetilde{W}(T) + (\rho - \mu) \int_0^T e^{\rho s} \left(\int_0^s e^{-\rho r} d\widetilde{W}(r) \right) ds + \frac{S(0)}{\rho c} (\rho - \mu)(e^{\rho T} - 1).$$

$$\text{(S.22)}$$

Substituted in (S.20) this gives

$$\theta_1(t) = e^{\rho(t-T)} c^{-1} \Big\{ 1 - (\mu - \rho) E_Q [\widetilde{W}(T) \int_t^T e^{\mu(s-t)} d\widetilde{W}(s) | \mathcal{F}_t]$$

$$+ (\mu - \rho)^2 E_Q [\int_0^T e^{\rho s} \left(\int_0^s e^{-\rho r} d\widetilde{W}(r) \right) ds \int_t^T e^{\mu(s-t)} d\widetilde{W}(s) | \mathcal{F}_t] \Big\}$$

$$= e^{\rho(t-T)}c^{-1}\{1 - (\mu - \rho)\int_t^T e^{\mu(s-t)}ds$$

$$+ (\mu - \rho)^2 \int_t^T e^{\rho s} E_Q[(\int_t^s e^{-\rho r}d\widetilde{W}(r))(\int_t^T e^{\mu(r-t)}d\widetilde{W}(r))|\mathcal{F}_t]ds\}$$

$$= e^{\rho(t-T)}c^{-1}\{1 - \frac{\mu - \rho}{\mu}(e^{\mu(T-t)} - 1) + (\mu - \rho)^2 \int_t^T e^{\rho r}(\int_t^s e^{-\rho r}e^{\mu(r-t)}dr)ds\}$$

$$= e^{\rho(t-T)}c^{-1}\{1 - \frac{\mu - \rho}{\rho}(e^{\rho(T-t)} - 1)\}. \quad \square$$

Problems of Chap. 5

5.5 Solution

(a) We have the following equations:

$$\int_0^T W(T)\delta W(t) = \int_0^T W(T) \diamond \dot{W}(t)dt$$

$$= W(T) \diamond \int_0^T \dot{W}(t)dt$$

$$= W(T) \diamond W(T) = W^2(T) - T,$$

by (5.65).

(b) We have the following equations:

$$\int_0^T (\int_0^T gdW) \diamond \dot{W}(t)dt = (\int_0^T gdW) \diamond \int_0^T \dot{W}(t)dt$$

$$= (\int_0^T gdW) \diamond W(T)$$

$$= (\int_0^T gdW)W(T) - \int_0^T g(s)ds,$$

by (5.62).

(c) We have the following equations:

$$\int_0^T W^2(t_0)\delta W(t) = \int_0^T (W^{\diamond 2}(t_0) + t_0)\delta W(t)$$

$$= W^{\diamond 2}(t_0) \diamond W(T) + t_0 W(T)$$

$$= W^{\diamond 2}(t_0) \diamond (W(T) - W(t_0))$$

$$+ W^{\diamond 2}(t_0) \diamond W(t_0) + t_0 W(T)$$

$$= W^{\diamond 2}(t_0)(W(T) - W(t_0)) + W^{\diamond 3}(t_0) + t_0 W(T)$$

$$= (W^2(t_0) - t_0)(W(T) - W(t_0)) + W^3(t_0)$$

$$- 3t_0 W(t_0) + t_0 W(T)$$

$$= W^2(t_0)W(T) - 2t_0 W(t_0),$$

where we have used (5.40) and (5.65).

(d) We have the following equations:

$$\int_0^T \exp(W(T))\delta W(t) = \exp(W(T)) \diamond \int_0^T \dot{W}(t)dt$$

$$= \exp(W(T)) \diamond W(T)$$

$$= \exp^\diamond(W(T) + \frac{1}{2}T) \diamond W(T)$$

$$= \exp(\frac{1}{2}T) \sum_{n=0}^\infty \frac{1}{n!} W(T)^{\diamond(n+1)}$$

$$= \exp(\frac{1}{2}T) \sum_{n=0}^\infty \frac{T^{\frac{n+1}{2}}}{n!} h_{n+1}\left(\frac{W(T)}{\sqrt{T}}\right). \quad \Box$$

Problems of Chap. 6

6.4 Solution

Since $\dot{W}(s) = \sum_{i=1}^\infty e_i(s)H_{\epsilon^{(i)}}$, the expansion (6.8) for $D_t\dot{W}(s)$ is

$$D_t\dot{W}(s) = \sum_{i,k=1}^\infty e_i(s)e_k(t)H_{\epsilon^{(i)}-\epsilon^{(k)}}\chi_{\{i=k\}}$$

$$= \sum_{i=1}^\infty e_i(s)e_i(t),$$

which is not convergent. Hence, for all $s \in \mathbb{R}$, we have $\dot{W}(s) \notin Dom(D_t)$. $\quad \Box$

Problems of Chap. 7

7.2 Solution

By Proposition 7.2 we have

$$g(e^Y) = \int_{\mathbb{R}} g(t^y) \frac{1}{\sqrt{2\pi v}} \exp^\diamond \left\{ -\frac{(y-Y)^{\diamond 2}}{2v} \right\} dy.$$

Substituting $e^y = z$ in the integral, we obtain

$$g(Z) = g(e^Y) = \int_0^\infty g(z) \frac{1}{\sqrt{2\pi v}} \exp^\diamond \left\{ -\frac{(\log z - \log Z)^{\diamond 2}}{2v} \right\} \frac{dz}{z}.$$

Hence the Donsker delta function of Z is

$$\delta_Z(z) = \frac{1}{\sqrt{2\pi v}} \frac{1}{z} \exp^\diamond \left\{ -\frac{(\log z - \log Z)^{\diamond 2}}{2v} \right\} \chi_{(0,\infty)}(z),$$

as claimed.   □

Problems of Chap. 8

8.2 Solution

(a) $\int_0^T W(T) d^- W(t) = W(T) \int_0^T dW(t) = W^2(T)$

(b) Consider the following equations:

$$\int_0^T W(t) [W(T) - W(t)] d^- W(t)$$

$$= \int_0^T W(t) W(T) d^- W(t) - \int_0^T W^2(t) d^- W(t)$$

$$= W(T) \int_0^T W(t) dW(t) - \left[\frac{1}{3} W^3(T) - \int_0^T W(t) dt \right]$$

$$= W(T) \left[\frac{1}{2} W^2(T) - \frac{1}{2} T \right] - \left[\frac{1}{3} W^3(T) - \int_0^T W(t) dt \right]$$

$$= \frac{1}{6} W^3(T) - \frac{1}{2} TW(T) + \int_0^T W(t) dt.$$

(c) Consider the following equations:

$$\int_0^T \left(\int_0^T g(s) dW(s) \right) d^- W(t)$$

$$= \int_0^T g(s) dW(s) \int_0^T d^- W(t)$$

$$= W(T) \int_0^T g(s) dW(s).   □$$

Problems of Chap. 9

9.1 Solution

(a) By Theorem 9.4 we have

$$
\begin{aligned}
dY(t) =\,&2Y(t)\big[\alpha(t)dt + \beta(t)dW(t)\big] + \beta^2(t)dt \\
&+ \int_{\mathbb{R}_0} \big[(X(t)+\gamma(t,z))^2 - X^2(t) - 2X(t)\gamma(t,z)\big]\nu|(dz)dt \\
&+ \int_{\mathbb{R}_0} \big[(X(t^-)+\gamma(t,z))^2 - X^2(t^-)\big]\widetilde{N}(dt,dz) \\
=\,&(2\alpha(t)Y(t) + \beta^2(t) + \int_{\mathbb{R}_0} \gamma^2(t,z)\nu(dz))dt + 2\beta(t)Y(t)dW(t) \\
&+ \int_{\mathbb{R}_0} \big[2X(t^-)\gamma(t,z) + \gamma^2(t,z)\big]\widetilde{N}(dt,dz).
\end{aligned}
$$

(b) By Theorem 9.4 we have

$$
\begin{aligned}
dY(t) =\,&Y(t)\big[\alpha(t)dt + \beta(t)dW(t)\big] + \tfrac{1}{2}Y(t)\beta^2(t)dt \\
&+ \int_{\mathbb{R}_0} \big[\exp\{X(t)+\gamma(t,z)\} - \exp\{X(t)\} - \exp\{X(t)\}\gamma(t,z)\big]\nu(dz)dt \\
&+ \int_{\mathbb{R}_0} \big[\exp\{X(t^-)+\gamma(t,z)\} - \exp\{X(t^-)\}\big]\widetilde{N}(dt,dz) \\
=\,&Y(t^-)\Big[\Big(\alpha(t) + \tfrac{1}{2}\beta^2(t) + \int_{\mathbb{R}_0} \big[\exp\{\gamma(t,z)\} - 1 - \gamma(t,z)\big]\nu(dz)\Big)dt\Big] \\
&+ \beta(t)dW(t) + \int_{\mathbb{R}_0} \big[\exp\{\gamma(t,z) - 1\}\big]\widetilde{N}(dt,dz).
\end{aligned}
$$

(c) By Theorem 9.4 we have

$$
\begin{aligned}
dY(t) =\,& -\sin X(t)\big[\alpha(t)dt + \beta(t)dW(t)\big] - \tfrac{1}{2}\cos X(t)\,\beta^2(t)dt \\
&+ \int_{\mathbb{R}_0} \big[\cos\big(X(t)+\gamma(t,z)\big) - \cos X(t) + \sin X(t)\,\gamma(t,z)\big]\nu(dz)dt \\
&+ \int_{\mathbb{R}_0} \big[\cos\big(X(t^-)+\gamma(t,z)\big) - \cos X(t^-)\big]\widetilde{N}(dt,dz) \\
=\,& \big[-\alpha(t)\sin X(t) - \tfrac{1}{2}\beta^2(t)\cos X(t)\big] \\
&+ \cos X(t)\int_{\mathbb{R}_0} \big[\cos\gamma(t,z) - 1\big]\nu(dz) \\
&+ \sin X(t)\int_{\mathbb{R}_0} \big[\gamma(t,z) - \sin\gamma(t,z)\big]\nu(dz)dt - \beta(t)\sin X(t)dW(t) \\
&+ \int_{\mathbb{R}_0} \big[\cos X(t^-)\big(\cos\gamma(t,z)-1\big) - \sin X(t^-)\sin\gamma(t,z)\big]\widetilde{N}(dt,dz). \quad \square
\end{aligned}
$$

9.2 Solution

Applying Problem 9.1 (b) to the case

$$\alpha(t) = -\int_{\mathbb{R}_0} \left[e^{h(t)z} - 1 - h(t)z\right]\nu(dz),$$

$$\beta(t) = 0$$

$$\gamma(t, z) = h(t)z,$$

we obtain

$$dY(t) = Y(t)\left\{\left(-\int_{\mathbb{R}_0} \left[e^{h(t)z} - 1 - h(t)z\right]\nu(dz) + \int_{\mathbb{R}_0} \left[e^{h(t)z} - 1 - h(t)z\right]\nu(dz)\right)dt\right.$$

$$\left. + \int_{\mathbb{R}_0} \left[e^{h(t)z} - 1big\right]\widetilde{N}(dt, dz)\right\}$$

$$= Y(t)\int_{\mathbb{R}_0} \left[e^{h(t)z} - 1big\right]\widetilde{N}(dt, dz). \quad \square$$

9.6 Solution

(a) Since $d(t\eta(t)) = td\eta(t) + \eta(t)dt$, we have

$$F = \int_0^T \eta(t)dt$$

$$= T\eta(T) - \int_0^T td\eta(t)$$

$$= T\int_0^T \int_{\mathbb{R}_0} z\widetilde{N}(dt, dz) - \int_0^T t\int_{\mathbb{R}_0} z\widetilde{N}(dt, dz)$$

$$= \int_0^T \int_{\mathbb{R}_0} (T - t)z\widetilde{N}(dt, dz).$$

Hence F is replicable, with replicating portfolio $\varphi(t) = T - t$, $t \in [0, T]$.
(c) Define

$$Y(t) = \exp\left\{\eta(t) - \int_0^t \int_{\mathbb{R}_0} (e^z - 1 - z)\nu(dz)ds\right\}.$$

Then, by Problem 9.2 we have

$$dY(t) = Y(t^-)\int_{\mathbb{R}_0} (e^z - 1)\widetilde{N}(dt, dz).$$

Hence

$$Y(T) = 1 + \int_0^T \int_{\mathbb{R}_0} Y(t^-)(e^z - 1)\widetilde{N}(dt, dz).$$

Therefore,

$$e^{\eta(T)} = Y(T)M$$

$$= M + \int_0^T \int_{\mathbb{R}_0} MY(t^-)(e^z - 1)\widetilde{N}(dt, dz),$$

where

$$M = \exp\left\{T \int_{\mathbb{R}_0} (e^z - 1 - z)\nu(dz)\right\}.$$

Hence $F = \exp\{\eta(T)\}$ is not replicable unless $\nu(dz) = \lambda\delta_{z_0}(dz)$ is a point mass at some point $z_0 \neq 0$. In this case, the process $\eta(t)$, $t \in [0, T]$, corresponds to the compensated Poisson process with jump size z_0 and intensity $\lambda > 0$. $\square$

Problems of Chap. 10

10.1 Solution

(a) By the Itô formula we have

$$\eta^3(T) = \int_0^T \int_{\mathbb{R}_0} \left[(\eta(t) + z)^3 - \eta^3(t) - 3\eta^2(t)z\right]\nu(dz)dt$$

$$+ \int_0^T \int_{\mathbb{R}_0} \left[(\eta(t) + z)^3 - \eta^3(t)\right]\widetilde{N}(dt, dz)$$

$$= \int_0^T \int_{\mathbb{R}_0} \left[3\eta(t)z^2 + z^3\right]\nu(dz)dt$$

$$+ \int_0^T \int_{\mathbb{R}_0} \left[3\eta^2(t)z + 3\eta(t)z^2 + z^3\right]\widetilde{N}(dt, dz)$$

$$= \int_0^T \int_{\mathbb{R}_0} z^3\nu(dz)dt + \int_0^T \int_{\mathbb{R}_0} 3\eta(t)z^2\nu(dz)dt$$

$$+ \int_0^T \int_{\mathbb{R}_0} z^3\widetilde{N}(dz, dt) + \int_0^T \int_{\mathbb{R}_0} 3\eta(t)z^2\widetilde{N}(dt, dz)$$

$$+ \int_0^T \int_{\mathbb{R}_0} 3\eta^2(t)z\widetilde{N}(dt, dz)$$

$$= \int_0^T \int_{\mathbb{R}_0} z^3\nu(dz)dt + \int_0^T \int_{\mathbb{R}_0} 3\left(\int_0^t \int_{\mathbb{R}_0} z_1\widetilde{N}(ds, dz_1)\right)z^2\nu(dz)dt$$

$$+ \int_0^T \int_{\mathbb{R}_0} z^3\widetilde{N}(dt, dz) + \int_0^T \int_{\mathbb{R}_0} \left(\int_0^t \int_{\mathbb{R}_0} 3z_1\widetilde{N}(dt, dz_1)\right)z^2\widetilde{N}(dt, dz)$$

$$+ \int_0^T \int_{\mathbb{R}_0} 3\eta^2(t)z\widetilde{N}(dt, dz).$$

Now we also have

$$\int_0^T \int_{\mathbb{R}_0} 3\eta^2(t) z \widetilde{N}(dt, dz) = \int_0^T \int_{\mathbb{R}_0} 3\Big(t \int_{\mathbb{R}_0} \varsigma^2 \nu(d\varsigma)\Big) z \widetilde{N}(dt, dz)$$

$$+ \int_0^T \int_{\mathbb{R}_0} 3\Big(\int_0^t \int_{\mathbb{R}_0} z_1^2 \widetilde{N}(dt, dz_1)\Big) z \widetilde{N}(dt, dz)$$

$$+ \int_0^T \int_{\mathbb{R}_0} 6\Big(\int_0^t \int_{\mathbb{R}_0} \int_0^{t_2} \int_{\mathbb{R}_0} z_1 z_2 \widetilde{N}(dt_1, dz_1) \widetilde{N}(dt_2, dz_2)\Big) z \widetilde{N}(dt, dz)$$

$$= I_1\Big(3t_1 z_1 \int_{\mathbb{R}_0} \varsigma^2 \nu(d\varsigma)\Big) + J_2\big(3z_1^2 z_2\big) + J_3\big(6 z_1 z_2 z_3\big).$$

Summing up we get

$$\eta^3(T) = T m_3 + I_1\big(3 T m_2 z_1 + z_1^3\big) + J_2\big(3 z_1 z_2^2 + 3 z_1^2 z_2\big) + J_3\big(6 z_1 z_2 z_3\big)$$

$$= T m_3 + I_1\big(3 T m_2 z_1 + z_1^3\big) + I_2\Big(\frac{3}{2}\big(z_1 z_2^2 + z_1^2 z_2\big)\Big) + I_3\big(z_1 z_2 z_3\big),$$

where $m_1 = \int_{\mathbb{R}_0} \varsigma^i \nu(d\varsigma)$, $i = 2, 3, \ldots$.

(b) By Example 10.4 we have that

$$F_0 := \exp\Big\{\int_0^T \int_{\mathbb{R}_0} z \widetilde{N}(dt, dz) - \int_0^T \int_{\mathbb{R}_0} (e^z - 1 - z)\nu(dz) dt\Big\}$$

has chaos expansion $F_0 = \sum_{n=0}^{\infty} I_n(f_n)$ given by (10.7):

$$f_n = \frac{1}{n!}\big(e^z - 1\big)^{\otimes n}(t_1, z_1, \ldots, t_n, z_n).$$

It follows that F has the expansion

$$F = \sum_{n=0}^{\infty} I_n(K f_n) \quad \text{where} \quad K := \exp\Big\{T \int_{\mathbb{R}_0} (e^z - 1 - z)\nu(dz)\Big\}.$$

(c) We have the following equalities:

$$F = \int_0^T g(s) d\eta(s) = \int_0^T \int_{\mathbb{R}_0} g(s) z \widetilde{N}(ds, dz) = I_1(f_1),$$

where $f_1(s, z) = g(s) z$, $s \in [0, T]$, $z \in \mathbb{R}_0$.

(d) We have the following equalities:

$$F = \int_0^T g(s)\eta(s) = \int_0^T g(s) \int_0^s \int_{\mathbb{R}_0} z \widetilde{N}(ds, dz)$$

$$= \int_0^T \int_{\mathbb{R}_0} \Big(\int_t^T g(s) ds\Big) z \widetilde{N}(dt, dz) = I_1(f_1)$$

with $f_1(t, z) = z \int_t^T g(s) ds$, $t \in [0, T]$, $z \in \mathbb{R}_0$. $\square$

Problems of Chap. 11

11.2 Solution

(a) Since

$$\int_0^T g(s)d\eta(s) = \int_0^T \int_{\mathbb{R}_0} g(s)z\widetilde{N}(dt,dz) = I_1(g(t_1)z_1),$$

we get

$$\int_0^T \left(\int_0^T g(s)d\eta(s) \right) f(t)\delta\eta(t) = \int_0^T \int_{\mathbb{R}_0} I_1(g(t_1)z_1)f(t_z)z_2\widetilde{N}(dt_2,dz_2)$$

$$= I_2\left(\frac{1}{2}(g(t_1)f(t_2) + g(t_2)f(t_1))z_1z_2 \right)$$

$$= \int_0^T \int_{\mathbb{R}_0} \left(int_0^{t_2^-} \int_{\mathbb{R}_0} \frac{1}{2}(g(t_1)f(t_2) + g(t_2)f(t_1))z_1z_2\widetilde{N}(dt_1,dz_1) \right) \widetilde{N}(dt_2,dz_2)$$

$$= \int_0^T \int_{\mathbb{R}_0} \left[f(t_2) \int_0^{t_2^-} g(t_1)d\eta(t_1) + g(t_2) \int_0^{t_2^-} f(t_1)d\eta(t_1) \right] z_2\widetilde{N}(dt_2,dz_2)$$

$$= \int_0^T \left(\int_0^{t_2^-} g(t_1)d\eta(t_1) \right) f(t_2)d\eta(t_2) + \int_0^T \left(\int_0^{t_2^-} f(t_1)d\eta(t_1) \right) g(t_2)d\eta(t_2).$$

$$(S.23)$$

(b) Using the computation in (S.23), but with f and g interchanged, we get

$$\int_0^T \left(\int_0^T f(t)d\eta(t) \right) g(s)\delta\eta(s) = I_2\left(\frac{1}{2}(f(t_1)g(t_2) + f(t_2)g(t_1))z_1z_2 \right),$$

which is the same as we obtained in (a).

(c) This is a direct consequence of (a) and (b).   □

Problems of Chap. 12

12.1 Solution

(a) By Problem 10.1 we have the expansion

$$\eta^3(T) = Tm_3 + I_1\left(3Tm_2z_1 + z_1^3\right) + I_2\left(\frac{3}{2}(z_1z_2^2 + z_1^2z_2) \right) + I_3\left(z_1z_2z_3 \right),$$

where $m_i = \int_{\mathbb{R}_0} \zeta^i \nu(d\zeta)$, $i = 1, 2, \dots$. This gives

$$D_{t,z}\eta^3(T) = 3Tm_2z + z^3 + 3I_1(z_1z^2 + z_1^2z) + 3I_2(z_1z_2z)$$

$$= 3Tm_2z + z^3 + 3z^2\eta(T) + 3zI_1(z_1^2) + 3zI_2(z_1z_2).$$

If we use that

$$\eta^2(T) = Tm_2 + I_1(z^2) + I_2(z_1 z_2)$$

(see Example 12.9), we can see that the above expression can be written

$$D_{t,z}\eta^3(T) = 3\eta^2(T)z + 3\eta(T)z^2 + z^3.$$

(b) By Problem 10.1 we have the expansion

$$e^{\eta(T)} = \sum_{n=0}^{\infty} I_n(g_n),$$

where

$$g_n = K f_n \quad \text{with} \quad f_n = \frac{1}{n!}(e^z - 1)^{\otimes n}, \quad n = 1, 2, \dots$$

$$\text{and} \quad K = \exp\left\{T \int_{\mathbb{R}_0} (e^z - 1 - z)\nu(dz)\right\}.$$

This gives

$$D_{t,z}e^{\eta(T)} = \sum_{n=1}^{\infty} n I_{n-1}(g_n(\cdot, t.z))$$

$$= \sum_{n=1}^{\infty} n I_{n-1}\left(\frac{K}{n!}(e^z - 1)^{\otimes n-1}\right)(e^z - 1)$$

$$= \sum_{n=1}^{\infty} I_{n-1}\left(\frac{K}{(n-1)!}(e^z - 1)^{\otimes n-1}\right)(e^z - 1)$$

$$= e^{\eta(T)}(e^z - 1). \quad \square$$

12.2 Solution
The direct application of Theorem 12.8 yields

(a) $D_{t,z}\eta^3(T) = (\eta(T) + z)^3 - \eta^3(T) = 3\eta^2(T)z + 3\eta(T)z^2 + z^3.$
(b) $D_{t,z}e^{\eta(T)} = e^{\eta(T)+z} - e^{\eta(T)} = e^{\eta(T)}(e^z - 1).$

Compare with the solution of Problem 12.1. $\square$

Problems of Chap. 13

13.1 Solution
Recall that

$$\eta(t) = m_2 \sum_{i=1}^{\infty} \left(\int_0^t e_i(s)ds\right) K_{\varepsilon(i,1)}, \quad t \in \mathbb{R},$$

and

$$\overset{\bullet}{\eta}(t) = m_2 \sum_{i=1}^{\infty} e_i(t) K_{\varepsilon^{(i,1)}}, \quad t \in \mathbb{R},$$

where $\varepsilon^{(i,j)} = \varepsilon^{(\kappa(i,j))}$. Hence

$$\frac{\eta(t+h) - \eta(t)}{h} - \overset{\bullet}{\eta}(t) = m_2 \sum_{i=1}^{\infty} \left(\frac{1}{h} \int_t^{t+h} \left[e_i(s) - e_i(t) \right] ds \right) K_{\varepsilon^{(i,1)}}.$$

By (13.11) we have

$$\kappa(i, 1) = 1 + \frac{i(i-1)}{2}.$$

Therefore, if we put

$$a_i(h) := \frac{1}{h} \int_t^{t+h} \left[e_i(s) - e_i(t) \right] ds,$$

we have

$$\left\| \frac{\eta(t+h) - \eta(t)}{h} - \overset{\bullet}{\eta}(t) \right\|_{-q}^2 = m_2^2 \sum_{i=1}^{\infty} |a_i(h)|^2 \varepsilon^{(i,1)}! (2\mathbb{N})^{-q\varepsilon^{(i,1)}}$$

$$= m_2^2 \sum_{i=1}^{\infty} |a_i(h)|^2 (2\kappa(i,1))^{-q}$$

$$= m_2^2 \sum_{i=1}^{\infty} |a_i(h)|^2 (2 + i(i-1))^{-q}$$

by (13.25). Since

$$\sup_{t \in \mathbb{R}} |e_k(t)| = \mathcal{O}(k^{-1/2})$$

(see [105]), we see that

$$\sup \left\{ |a_i(h)| \; h \in [0,1], i = 1, 2, ... \right\} < \infty.$$

Moreover, since

$$a_i(h) \longrightarrow 0, \quad h \to 0 \quad (i = 1, 2, ...),$$

we can conclude that

$$\left\| \frac{\eta(t+h) - \eta(t)}{h} - \overset{\bullet}{\eta}(t) \right\|_{-q}^2 \longrightarrow 0, \quad h \to 0,$$

for all $q \geq 1$, by bounded convergence. This implies that

$$\frac{d}{dt}\eta(t) = \overset{\bullet}{\eta}(t) \quad \text{in } (\mathcal{S})^*. \quad \square$$

Problems of Chap. 15

15.1 Solution

(a)   By Lemma 15.5 we have

$$
\begin{aligned}
X(t) &= \int_0^t \int_{\mathbb{R}_0} \eta(T) z \widetilde{N}(\delta s, dz) \\
&= \int_0^t \int_{\mathbb{R}_0} \big(\eta(T) z + D_{t+,z} \eta(T) z \big) \widetilde{N}(\delta s, dz) - \int_0^t \int_{\mathbb{R}_0} z^2 \widetilde{N}(ds, dz) \\
&= \int_0^t \int_{\mathbb{R}_0} \eta(T) z \widetilde{N}(d^- s, dz) - \int_0^T \int_{\mathbb{R}_0} z^2 \nu(dz) ds - \int_0^t \int_{\mathbb{R}_0} z^2 \widetilde{N}(ds, dz) \\
&= \eta(t) \eta(T) - \int_0^t \int_{\mathbb{R}_0} z^2 N(ds, dz), \quad t \in [0, T].
\end{aligned}
$$

(b)   Since

$$
D_{t+,z} \eta(t) = D_{t+,z} \left(\int_0^t \int_{\mathbb{R}_0} z^2 N(ds, dz) \right) = 0,
$$

by applications of the chain rule we get that $D_{t+,z} X(t) = \eta(t) z$.

(c)   By (15.8), we have that

$$
\theta(t, z) := S_{t,z} \gamma(t, z) = z S_{t,z} \eta(T) = z(\eta(T) - z).
$$

(d)   First note that by the above we have

$$
\begin{aligned}
A(t, z) :=& D_{t+,z} \Big(2X(t^-) z (\eta(T) - z) + z^2 (\eta(T) - z)^2 \Big) \\
=& 2X(t^-) z^2 + z(\eta(T) - z) 2\eta(t^-) z + 2\eta(t^-) z^3 \\
& + z^2 (\eta^2(T) - (\eta(T) - z)^2) \\
=& 2X(t^-) z^2 + 2\eta(t^-) \eta(T) z^2 + 2\eta(T) z^3 - z^4.
\end{aligned}
$$

Therefore, the Itô formula for Skorohod integrals gives

$$
\begin{aligned}
\delta X^2(t) =& \int_{\mathbb{R}_0} \Big[(X(t^-) + \theta(t, z))^2 - X^2(t^-) + A(t, z) \Big] \widetilde{N}(\delta t, dz) \\
& + \int_{\mathbb{R}_0} \Big[(X(t) + \theta(t, z))^2 - X^2(t) - 2X(t)\theta(t, z) - A(t, z) \\
& \quad - f'(X(t)) D_{t+,z} \theta(t, z) \Big] \nu(dz) dt \\
=& \int_{\mathbb{R}_0} \Big[2X(t^-) z (\eta(T) - z) + z^2 (\eta(T) - z)^2 + A(t, z) \Big] \widetilde{N}(\delta t, dz) \\
& + \int_{\mathbb{R}_0} \Big[z^2 (\eta(T) - z)^2 + A(t, z) - 2X(t) z^2 \Big] \nu(dz) dt. \quad \square
\end{aligned}
$$

Problems of Chap. 16

16.1 Solution

(a)   By the chain rule and (16.5) we have

$$D_t F_\pi(T) = D_t \frac{1}{X_\pi(T)} = -\frac{1}{X_\pi^2(T)} D_t X_\pi(T) = -\frac{1}{X_\pi(T)} \sigma(t)\pi(t)$$

and

$$
\begin{aligned}
D_{t,z} F_\pi(T) &= \frac{1}{X_\pi(T) + D_{t,z} X_\pi(T)} - \frac{1}{X_\pi(T)} \\
&= \frac{1}{X_\pi(T)(1 + \pi(t)\theta(t,z))} - \frac{1}{X_\pi(T)} \\
&= \frac{-\pi(t)\theta(t,z)}{X_\pi(T)(1 + \pi(t)\theta(t,z))}.
\end{aligned}
$$

(b)   The equation for the optimal deterministic portfolio π is obtained by choosing $\mathcal{E}_t = \mathcal{F}_0$ for all $t \in [0, T]$ in (16.12). This gives

$$\mu(s) - \rho(s) - 2\sigma^2(s)\pi(s) - \int_{\mathbb{R}_0} \frac{\pi(s)\theta^2(s,z)(2 + \pi(s)\theta(s,z))}{(1 + \pi(s)\theta(s,z))^2} \nu(dz) = 0. \quad \square$$

References

1. K. Aase, T. Bjuland, and B. Øksendal. Transfinite mean value interpolation. Eprint in Mathematics, University of Oslo, 2007.
2. K. Aase, B. Øksendal, N. Privault, and J. Ubøe. White noise generalizations of the Clark-Haussmann-Ocone theorem with application to mathematical finance. *Finance Stoch.*, 4(4):465–496, 2000.
3. K. Aase, B. Øksendal, and J. Ubøe. Using the Donsker delta function to compute hedging strategies. *Potential Analysis*, 14(4):351–374, 2001.
4. S. Albeverio, Y. Hu, and X.Y. Zhou. A remark on non-smoothness of the self-intersection local time of planar Brownian motion. *Statist. Probab. Lett.*, 32(1):57–65, 1997.
5. E. Alòs and D. Nualart. An extension of Itô's formula for anticipating processes. *J. Theoret. Probab.*, 11(2):493–514, 1998.
6. J. Amendinger, P. Imkeller, and M. Schweizer. Additional logarithmic utility of an insider. *Stochastic Process. Appl.*, 75(2):263–286, 1998.
7. D. Applebaum. Covariant Poisson fields in Fock space. In *Analysis, geometry and probability*, volume 10 of *Texts Read. Math.*, pages 1–15. Hindustan Book Agency, Delhi, 1996.
8. D. Applebaum. *Lévy Processes and Stochastic Calculus*, volume 93 of *Cambridge Studies in Advanced Mathematics*. Cambridge University Press, Cambridge, 2004.
9. J. Asch and J. Potthoff. Itô's lemma without nonanticipatory conditions. *Probab. Theory Related Fields*, 88(1):17–46, 1991.
10. M. Avellaneda and R. Gamba. Conquering the Greeks in Monte Carlo: efficient calculation of the market sensitivities and hedge-ratios of financial assets by direct numerical simulation. In *Mathematical finance—Bachelier Congress, 2000 (Paris)*, Springer Finance, pages 93–109. Springer, Berlin, 2002.
11. K. Back. Insider trading in continuous time. *Review of Financial Studies*, 5:387–409, 1992.
12. V. Bally. On the connection between the Malliavin covariance matrix and Hörmander's condition. *J. Funct. Anal.*, 96(2):219–255, 1991.
13. V. Bally. Integration by parts formula for locally smooth laws and applications to equations with jumps I. Technical report, Mittag-Leffler Institut, 2007.
14. V. Bally. Integration by parts formula for locally smooth laws and applications to equations with jumps II. Technical report, Mittag-Leffler Institut, 2007.

15. V. Bally, M.-P. Bavouzet, and M. Messaoud. Integration by parts formula for locally smooth laws and applications to sensitivity computations. *Ann. Appl. Probab.*, 17(1):33–66, 2007.

16. V. Bally, L. Caramellino, and A. Zanette. Pricing and hedging American options by Monte Carlo methods using a Malliavin calculus approach. *Monte Carlo Methods Appl.*, 11(2):97–133, 2005.

17. O.E. Barndorff-Nielsen. Processes of normal inverse Gaussian type. *Finance Stoch.*, 2(1):41–68, 1998.

18. O.E. Barndorff-Nielsen and N. Shephard. Non-Gaussian Ornstein-Uhlenbeck-based models and some of their uses in financial economics. *J. R. Stat. Soc. Ser. B Stat. Methodol.*, 63(2):167–241, 2001.

19. R.F. Bass and M. Cranston. The Malliavin calculus for pure jump processes and applications to local time. *Ann. Probab.*, 14(2):490–532, 1986.

20. D.R. Bell. *The Malliavin Calculus*, volume 34 of *Pitman Monographs and Surveys in Pure and Applied Mathematics*. Longman Scientific & Technical, Harlow, 1987.

21. E. Benhamou. Optimal Malliavin weighting function for the computation of the Greeks. *Math. Finance*, 13(1):37–53, 2003.

22. A. Bensoussan. *Stochastic Control of Partially Observable Systems*. Cambridge University Press, Cambridge, 1992.

23. F.E. Benth. Integrals in the Hida distribution space $(\mathcal{S})^*$. In *Stochastic analysis and related topics (Oslo, 1992)*, volume 8 of *Stochastics Monogr.*, pages 89–99. Gordon and Breach, Montreux, 1993.

24. F.E. Benth, L.O. Dahl, and K.H. Karlsen. Quasi Monte-Carlo evaluation of sensitivities of options in commodity and energy markets. *Int. J. Theor. Appl. Finance*, 6(8):865–884, 2003.

25. F.E. Benth, G. Di Nunno, A. Løkka, B. Øksendal, and F. Proske. Explicit representation of the minimal variance portfolio in markets driven by Lévy processes. *Math. Finance*, 13(1):55–72, 2003.

26. F.E. Benth and J. Gjerde. A remark on the equivalence between Poisson and Gaussian stochastic partial differential equations. *Potential Anal.*, 8(2):179–193, 1998.

27. F.E. Benth, M. Groth, and P.C. Kettler. A quasi-Monte Carlo algorithm for the normal inverse Gaussian distribution and valuation of financial derivatives. *Int. J. Theor. Appl. Finance*, 9(6):843–867, 2006.

28. F.E. Benth, M. Groth, and O. Wallin. Derivative-free Greeks for the Barndorff-Nielsen and Shephard stochastic volatility model. Eprint in Mathematics, University of Oslo, 2007.

29. F.E. Benth and A. Løkka. Anticipative calculus for Lévy processes and stochastic differential equations. *Stoch. Stoch. Rep.*, 76(3):191–211, 2004.

30. M.A. Berger. A Malliavin-type anticipative stochastic calculus. *Ann. Probab.*, 16(1):231–245, 1988.

31. M.A. Berger and V.J. Mizel. An extension of the stochastic integral. *Ann. Probab.*, 10(2):435–450, 1982.

32. J. Bertoin. *Lévy Processes*, volume 121 of *Cambridge Tracts in Mathematics*. Cambridge University Press, Cambridge, 1996.

33. F. Biagini and B. Øksendal. A general stochastic integral approach to insider trading. *Appl. Math. Optim.*, 52(4):167–181, 2005.

34. P. Biane. Calcul stochastique non-commutatif. In *Lectures on probability theory (Saint-Flour, 1993)*, volume 1608 of *Lecture Notes in Math.*, pages 1–96. Springer, Berlin, 1995.

35. K. Bichteler, J.-B. Gravereaux, and J. Jacod. *Malliavin Calculus for Processes with Jumps*, volume 2 of *Stochastics Monographs*. Gordon and Breach Science Publishers, New York, 1987.

36. P. Billingsley. *Convergence of Probability Measures*. John Wiley & Sons Inc., New York, second edition, 1999.

37. J.-M. Bismut. Martingales, the Malliavin calculus and hypoellipticity under general Hörmander's conditions. *Z. Wahrsch. Verw. Gebiete*, 56(4):469–505, 1981.

38. J.-M. Bismut. Calcul des variations stochastique et processus de sauts. *Z. Wahrsch. Verw. Gebiete*, 63(2):147–235, 1983.

39. F. Black and M. Scholes. The pricing of options and corporate liabilities. *Journal of Political Economy*, 81:637–654, 1973.

40. C. Blanchet-Scalliet, N. El Karoui, M. Jeanblanc, and L. Martinelli. Optimal investment and consumption decisions when time-horizon is uncertain. Technical report, 2002.

41. B. Bouchard and H. Pham. Wealth-path dependent utility maximization in incomplete markets. *Finance Stoch.*, 8(4):579–603, 2004.

42. N. Bouleau and F. Hirsch. Propriétés d'absolue continuité dans les espaces de Dirichlet et application aux équations différentielles stochastiques. In *Séminaire de Probabilités, XX, 1984/85*, volume 1204 of *Lecture Notes in Math.*, pages 131–161. Springer, Berlin, 1986.

43. M. Broadie and P. Glasserman. Estimating security price derivatives using simulation. *Management Science*, 42:169–285, 1996.

44. E.A. Carlen and É. Pardoux. Differential calculus and integration by parts on Poisson space. In *Stochastics, algebra and analysis in classical and quantum dynamics (Marseille, 1988)*, volume 59 of *Math. Appl.*, pages 63–73. Kluwer Acad. Publ., Dordrecht, 1990.

45. N. Chen and P. Glasserman. Malliavin Greeks without Malliavin calculus. Eprint, Columbia University, 2006.

46. J.M.C. Clark. The representation of functionals of Brownian motion by stochastic integrals. *Ann. Math. Statist.*, 41:1282–1295, 1970.

47. J.M.C. Clark. Correction to: "The representation of functionals of Brownian motion by stochastic integrals" (Ann. Math. Statist. **41** (1970), 1282–1295). *Ann. Math. Statist.*, 42:1778, 1971.

48. R. Cont and P. Tankov. *Financial Modelling with Jump Processes*. Chapman & Hall/CRC Financial Mathematics Series. Chapman & Hall/CRC, Boca Raton, FL, 2004.

49. J.M. Corcuera, P. Imkeller, A. Kohatsu-Higa, and D. Nualart. Additional utility of insiders with imperfect dynamical information. *Finance Stoch.*, 8(3):437–450, 2004.

50. R. Coviello and F. Russo. Modeling financial assets without semimartingales. Technical report, BiBoS, Bielefeld, 2006.

51. D. Cuoco and J. Cvitanić. Optimal consumption choices for a "large" investor. *J. Econom. Dynam. Control*, 22(3):401–436, 1998.

52. M. Curran. Strata gems. *RISK*, 7:70–71, 1994.

53. G. Da Prato. *Introduction to Stochastic Analysis and Malliavin Calculus*, volume 6 of *Appunti*. *Scuola Normale Superiore di Pisa*. Edizioni della Normale, Pisa, 2007.

54. M.H.A. Davis and M.P. Johansson. Malliavin Monte Carlo Greeks for jump diffusions. *Stochastic Process. Appl.*, 116(1):101–129, 2006.

55. V. Debelley and N. Privault. Sensitivity analysis of European options in jump diffusion models via the Malliavin calculus on the Wiener space. Eprint, Université de la Rochelle, 2004.

56. F. Delbaen and W. Schachermayer. A general version of the fundamental theorem of asset pricing. *Math. Ann.*, 300(3):463–520, 1994.

57. L. Denis. A criterion of density for solutions of Poisson-driven SDEs. *Probab. Theory Related Fields*, 118(3):406–426, 2000.

58. A. Dermoune, P. Krée, and L. Wu. Calcul stochastique non adapté par rapport à la mesure aléatoire de Poisson. In *Séminaire de Probabilités, XXII*, volume 1321 of *Lecture Notes in Math.*, pages 477–484. Springer, Berlin, 1988.

59. G. Di Nunno. Random fields evolution: non-anticipating integration and differentiation. *Teor. Ĭmovīr. Mat. Stat.*, 66:82–94, 2002.

60. G. Di Nunno. Stochastic integral representations, stochastic derivatives and minimal variance hedging. *Stoch. Stoch. Rep.*, 73(1-2):181–198, 2002.

61. G. Di Nunno. On orthogonal polynomials and the Malliavin derivative for Lévy stochastic measures. *Séminaires et Congrès*, 16:55–69, 2007.

62. G. Di Nunno. Random fields: non-anticipating derivative and differentiation formulas. *Infin. Dimens. Anal. Quantum Probab. Relat. Top.*, 10(3):465–481, 2007.

63. G. Di Nunno, A. Kohatsu-Higa, T. Meyer-Brandis, B. ksendal, F. Proske, and A Sulem. Anticipative stochastic control for Lévy processes with application to insider trading. *Handbook in Mathematical Sciences, Publisher: ELSEVIER, Editors: Alain Bensoussan and Qiang Zhang*, 2008.

64. G. Di Nunno, T. Meyer-Brandis, B. Øksendal, and F. Proske. Malliavin calculus and anticipative Itô formulae for Lévy processes. *Infin. Dimens. Anal. Quantum Probab. Relat. Top.*, 8(2):235–258, 2005.

65. G. Di Nunno, T. Meyer-Brandis, B. Øksendal, and F. Proske. Optimal portfolio for an insider in a market driven by Lévy processes. *Quant. Finance*, 6(1):83–94, 2006.

66. G. Di Nunno and B. Øksendal. Optimal portfolio, partial information and Malliavin calculus. *Eprint in Mathematics*, University of Oslo, 2006.

67. G. Di Nunno and B. Øksendal. The Donsker delta function, a representation formula for functionals of a Lévy process and application hedging in incomplete markets. *Séminaires et Congrès*, 16:71–82, 2007.

68. G. Di Nunno and B. Øksendal. A representation theorem and a sensitivity result for functionals of jump diffusions. In *Mathematical analysis of random phenomena*, pages 177–190. World Sci. Publ., Hackensack, NJ, 2007.

69. G. Di Nunno, B. Øksendal, and F. Proske. White noise analysis for Lévy processes. *J. Funct. Anal.*, 206(1):109–148, 2004.

70. G. Di Nunno and Yu.A. Rozanov. On stochastic integration and differentiation. *Acta Appl. Math.*, 58(1-3):231–235, 1999.

71. G. Di Nunno and Yu.A. Rozanov. Stochastic integrals and adjoint derivatives. In *Stochastic Analysis and its Applications*, volume 2 of *Abel Symposia*, pages 265–307. Springer, Heidelberg, 2007.

72. A.A. Dorogovtsev. Elements of stochastic differential calculus. In *Mathematics today '88 (Russian)*, pages 105–131. "Vishcha Shkola", Kiev, 1988.

73. D. Duffie. *Dynamic Asset Pricing Theory*. Princeton University Press, Princeton, second edition, 1992.

74. E.B. Dynkin. *Markov Processes. Vols. I, II*. Academic Press Inc., Publishers, New York, 1965.

75. E. Eberlein and S. Raible. Term structure models driven by general Lévy processes. *Math. Finance*, 9(1):31–53, 1999.

76. R.J. Elliott, H. Geman, and B.M. Korkie. Portfolio optimization and contingent claim pricing with differential information. *Stochastics Stochastics Rep.*, 60(3-4):185–203, 1997.

77. R.J. Elliott and M. Jeanblanc. Incomplete markets with jumps and informed agents. *Mathematical Methods of Operations Research*, 50:475–492, 1998.

78. K.D. Elworthy and X.-M. Li. Formulae for the derivatives of heat semigroups. *J. Funct. Anal.* 125(1):252–286, 1994.

79. G.B. Folland. *Real Analysis*. Pure and Applied Mathematics (New York). John Wiley & Sons Inc., New York, second edition, 1999.

80. E. Fournié, J.-M. Lasry, J. Lebuchoux, and P.-L. Lions. Applications of Malliavin calculus to Monte-Carlo methods in finance. II. *Finance Stoch.*, 5(2):201–236, 2001.

81. E. Fournié, J.-M. Lasry, J. Lebuchoux, P.-L. Lions, and N. Touzi. Applications of Malliavin calculus to Monte Carlo methods in finance. *Finance Stoch.*, 3(4):391–412, 1999.

82. N. Fournier. Smoothness of the law of some one-dimensional jumping S.D.E.s with non-constant rate of jump. *Electron. J. Probab.*, 13:no. 6, 135–156, 2008.

83. N. Fournier and J.-S. Giet. Existence of densities for jumping stochastic differential equations. *Stochastic Process. Appl.*, 116(4):643–661, 2006.

84. U. Franz, R. Léandre, and R. Schott. Malliavin calculus for quantum stochastic processes. *C. R. Acad. Sci. Paris Sér. I Math.*, 328(11):1061–1066, 1999.

85. U. Franz, N. Privault, and R. Schott. Non-Gaussian Malliavin calculus on real Lie algebras. *J. Funct. Anal.*, 218(2):347–371, 2005.

86. L. Gawarecki and V. Mandrekar. Itô-Ramer, Skorohod and Ogawa integrals with respect to Gaussian processes and their interrelationship. In *Chaos expansions, multiple Wiener-Itô integrals and their applications (Guanajuato, 1992)*, Probab. Stochastics Ser., pages 349–373. CRC, Boca Raton, FL, 1994.

87. I.M. Gel'fand and N.Ya. Vilenkin. *Generalized Functions. Vol. 4*. Academic Press [Harcourt Brace Jovanovich Publishers], New York, 1964 [1977].

88. H. Gjessing, H. Holden, T. Lindstrøm, B. Øksendal, J. Ubøe, and T.-S. Zhang. The Wick product. In *Frontiers in Pure and Applied Probability*, volume 1, pages 29–67. TVP Publishers, Moscow, 1993.

89. P.W. Glynn. Optimization of stochastic systems via simulation. In *Proceedings of the 1989 Winter Simulation Conference, Society for Computer Simulation*, pages 90–105, New York, 1989. ACM.

90. E. Gobet and A. Kohatsu-Higa. Computation of Greeks for barrier and lookback options using Malliavin calculus. *Electron. Comm. Probab.*, 8:51–62 (electronic), 2003.

91. A. Grorud. Asymmetric information in a financial market with jumps. *Int. J. Theor. Appl. Finance*, 3(4):641–659, 2000.

92. A. Grorud and M. Pontier. Probabilités neutres au risque et asymétrie d'information. *C. R. Acad. Sci. Paris Sér. I Math.*, 329(11):1009–1014, 1999.

93. A. Grorud and M. Pontier. Asymmetrical information and incomplete markets. *Int. J. Theor. Appl. Finance*, 4(2):285–302, 2001.

94. A. Grorud and M. Pontier. Financial market model with influential informed investors. *Int. J. Theor. Appl. Finance*, 8(6):693–716, 2005.

95. M. Grothaus, Y.G. Kondratiev, and L. Streit. Complex Gaussian analysis and the Bargmann-Segal space. *Methods Funct. Anal. Topology*, 3(2):46–64, 1997.

96. M. Grothaus, Yu. G. Kondratiev, and G. F. Us. Wick calculus for regular generalized stochastic functionals. *Random Oper. Stochastic Equations*, 7(3):263–290, 1999.

97. P.R. Halmos. *Measure Theory*. D. Van Nostrand Company, Inc., New York, N. Y., 1950.

98. U.G. Haussmann. On the integral representation of functionals of Itô processes. *Stochastics*, 3(1):17–27, 1979.

99. T. Hida. *Brownian Motion*. Springer-Verlag, New York, 1980.

100. T. Hida. Generalized Brownian functionals. In *Theory and application of random fields (Bangalore, 1982)*, volume 49 of *Lecture Notes in Control and Inform. Sci.*, pages 89–95. Springer, Berlin, 1983.

101. T. Hida and N. Ikeda. Analysis on Hilbert space with reproducing kernel arising from multiple Wiener integral. In *Proc. Fifth Berkeley Sympos. Math. Statist. and Probability (Berkeley, Calif., 1965/66). Vol. II: Contributions to Probability Theory, Part 1*, pages 117–143. Univ. California Press, Berkeley, Calif., 1967.

102. T. Hida, H.-H. Kuo, J. Potthoff, and L. Streit. *White Noise*. Kluwer Academic Publishers Group, Dordrecht, 1993.

103. T. Hida and J. Potthoff. White noise analysis—an overview. In *White noise analysis (Bielefeld, 1989)*, pages 140–165. World Sci. Publishing, River Edge, NJ, 1990.

104. C. Hillairet. Existence of an equilibrium with discontinuous prices, asymmetric information, and nontrivial initial σ-fields. *Math. Finance*, 15(1):99–117, 2005.

105. E. Hille and R.S. Phillips. *Functional Analysis and Semi-Groups*. American Mathematical Society Colloquium Publications, vol. 31. American Mathematical Society, Providence, R. I., 1957.

106. H. Holden and B. Øksendal. A white noise approach to stochastic differential equations driven by Wiener and Poisson processes. In *Nonlinear theory of generalized functions (Vienna, 1997)*, volume 401 of *Chapman & Hall/CRC Res. Notes Math.*, pages 293–313. Chapman & Hall/CRC, Boca Raton, FL, 1999.

107. H. Holden, B. Øksendal, J. Ubøe, and T. Zhang. *Stochastic Partial Differential Equations*. Second Ed. Springer. To appear 2008/2009.

108. Y. Hu. Itô-Wiener chaos expansion with exact residual and correlation, variance inequalities. *J. Theoret. Probab.*, 10(4):835–848, 1997.

109. Y. Hu and B. Øksendal. Wick approximation of quasilinear stochastic differential equations. In *Stochastic analysis and related topics, V (Silivri, 1994)*, pages 203–231. Birkhäuser Boston, Boston, MA, 1996.

110. Y. Hu and B. Øksendal. Chaos expansion of local time of fractional Brownian motions. *Stochastic Anal. Appl.*, 20(4):815–837, 2002.

111. Y. Hu and B. Øksendal. Optimal smooth portfolio selection for an insider. *J. Appl. Probab.*, 44(3):742–752, 2007.

112. S. Huddart, J.S. Hughes, and C.B. Levine. Public disclosure and dissimulation of insider trades. *Econometrica*, 69(3):665–681, 2001.

113. F. Huehne. A Clark-Ocone-Haussmann formula for optimal portfolios under Girsanov transformed pure-jump Lévy processes. Technical report, 2005.

114. N. Ikeda and S. Watanabe. An Introduction to Malliavin's Calculus. In *Stochastic analysis (Katata/Kyoto, 1982)*, volume 32 of *North-Holland Math. Library*, pages 1–52. North-Holland, Amsterdam, 1984.

115. N. Ikeda and S. Watanabe. *Stochastic Differential Equations and Diffusion Processes*, volume 24 of *North-Holland Mathematical Library*. North-Holland Publishing Co., Amsterdam, second edition, 1989.

116. P. Imkeller. Malliavin's calculus in insider models: additional utility and free lunches. *Math. Finance*, 13(1):153–169, 2003.

117. P. Imkeller, M. Pontier, and F. Weisz. Free lunch and arbitrage possibilities in a financial market model with an insider. *Stochastic Process. Appl.*, 92(1):103–130, 2001.

118. Y. Ishikawa and H. Kunita. Malliavin calculus on the Wiener-Poisson space and its application to canonical SDE with jumps. *Stochastic Process. Appl.*, 116(12):1743–1769, 2006.

119. K. Itô. On stochastic processes. I. (Infinitely divisible laws of probability). *Jap. J. Math.*, 18:261–301, 1942.

120. K. Itô. Multiple Wiener integral. *J. Math. Soc. Japan*, 3:157–169, 1951.

121. K. Itô. Spectral type of the shift transformation of differential processes with stationary increments. *Trans. Amer. Math. Soc.*, 81:253–263, 1956.

122. K. Itô. Extension of stochastic integrals. In *Proceedings of the International Symposium on Stochastic Differential Equations (Res. Inst. Math. Sci., Kyoto Univ., Kyoto, 1976)*, pages 95–109, New York, 1978. Wiley.

123. J. Jacod and A.N. Shiryaev. *Limit Theorems for Stochastic Processes*, volume 288 of *Grundlehren der Mathematischen Wissenschaften*. Springer-Verlag, Berlin, second edition, 2003.

124. Ju.M. Kabanov. A generalized Itô formula for an extended stochastic integral with respect to Poisson random measure. *Uspehi Mat. Nauk*, 29(4(178)):167–168, 1974.

125. Ju.M. Kabanov. Extended stochastic integrals. *Teor. Verojatnost. i Primenen.*, 20(4):725–737, 1975.

126. A.B. Kaminsky. Extended stochastic calculus for the Poisson random measures. *Nats. Akad. Nauk Ukraïn. Īnst. Mat. Preprint*, 15:i+16, 1996.

127. A.B. Kaminsky. An integration by parts formula for the Poisson random measures and applications. *Nats. Akad. Nauk Ukrain. Īnst. Mat. Preprint*, 9:i+20, 1996.

128. A.B. Kaminsky. A white noise approach to stochastic integration for a Poisson random measure. *Teor. Īmovīr. Mat. Stat.*, 57:41–50, 1997.

129. I. Karatzas and S.E. Shreve. *Brownian Motion and Stochastic Calculus*, volume 113 of *Graduate Texts in Mathematics*. Springer-Verlag, New York, second edition, 1991.

130. I. Karatzas and S.E. Shreve. *Methods of Mathematical Finance.* Springer-Verlag, New York, 1998.

131. A. Kohatsu-Higa. Enlargement of filtrations and models for insider trading. In *Stochastic processes and applications to mathematical finance*, pages 151–165. World Sci. Publ., River Edge, NJ, 2004.

132. A. Kohatsu-Higa. Models for insider trading with finite utility. In *Paris-Princeton Lectures on Mathematical Finance 2004*, volume 1919 of *Lecture Notes in Math.*, pages 103–171. Springer, Berlin, 2007.

133. A. Kohatsu-Higa and A. Sulem. A large trader-insider model. In *Stochastic processes and applications to mathematical finance*, pages 101–124. World Sci. Publ., Hackensack, NJ, 2006.

134. A. Kohatsu-Higa and A. Sulem. Utility maximization in an insider influenced market. *Math. Finance*, 16(1):153–179, 2006.

135. Yu.G. Kontratiev. *Generalized functions in problems of infinitedimensional analysis*. PhD thesis, Kiev University, 1970.

136. D. Kramkov and W. Schachermayer. Necessary and sufficient conditions in the problem of optimal investment in incomplete markets. *Ann. Appl. Probab.*, 13(4):1504–1516, 2003.

137. O.M. Kulik. Malliavin calculus for Lévy processes with arbitrary Lévy measures. *Teor. Ĭmovīr. Mat. Stat.*, 72:67–83, 2005.

138. H. Kunita. *Stochastic Flows and Stochastic Differential Equations*, volume 24 of *Cambridge Studies in Advanced Mathematics*. Cambridge University Press, Cambridge, 1990.

139. H.-H. Kuo. Donsker's delta function as a generalized Brownian functional and its application. In *Theory and application of random fields (Bangalore, 1982)*, volume 49 of *Lecture Notes in Control and Inform. Sci.*, pages 167–178. Springer, Berlin, 1983.

140. H.-H. Kuo. *White Noise Distribution Theory*. Probability and Stochastics Series. CRC Press, Boca Raton, FL, 1996.

141. S. Kusuoka and D. Stroock. Applications of the Malliavin calculus. I. In *Stochastic analysis (Katata/Kyoto, 1982)*, volume 32 of *North-Holland Math. Library*, pages 271–306. North-Holland, Amsterdam, 1984.

142. S. Kusuoka and D. Stroock. Applications of the Malliavin calculus. II. *J. Fac. Sci. Univ. Tokyo Sect. IA Math.*, 32(1):1–76, 1985.

143. S. Kusuoka and D. Stroock. Applications of the Malliavin calculus. III. *J. Fac. Sci. Univ. Tokyo Sect. IA Math.*, 34(2):391–442, 1987.

144. A.S. Kyle. Continuous auctions and insider trading. *Econometrica*, 53:1315–1335, 1985.

145. A. Lanconelli and F. Proske. On explicit strong solution of Itô-SDE's and the Donsker delta function of a diffusion. *Infin. Dimens. Anal. Quantum Probab. Relat. Top.*, 7(3):437–447, 2004.

146. Y.-J. Lee and H.-H. Shih. Donsker's delta function of Lévy process. *Acta Appl. Math.*, 63(1-3):219–231, 2000.

147. J.A. León, R. Navarro, and D. Nualart. An anticipating calculus approach to the utility maximization of an insider. *Math. Finance*, 13(1):171–185, 2003.

148. J.A. León, J.L. Solé, F. Utzet, and J. Vives. On Lévy processes, Malliavin calculus and market models with jumps. *Finance Stoch.*, 6(2):197–225, 2002.

149. D. Lépingle and J. Mémin. Sur l'intégrabilité uniforme des martingales exponentielles. *Z. Wahrsch. Verw. Gebiete*, 42(3):175–203, 1978.

150. T. Lindstrøm, B. Øksendal, and J. Ubøe. Stochastic differential equations involving positive noise. In *Stochastic analysis (Durham, 1990)*, volume 167 of *London Math. Soc. Lecture Note Ser.*, pages 261–303. Cambridge Univ. Press, Cambridge, 1991.

151. T. Lindstrøm, B. Øksendal, and J. Ubøe. Stochastic modelling of fluid flow in porous media. In *Control theory, stochastic analysis and applications (Hangzhou, 1991)*, pages 156–172. World Sci. Publishing, River Edge, NJ, 1991.

152. T. Lindstrøm, B. Øksendal, and J. Ubøe. Wick multiplication and Itô-Skorohod stochastic differential equations. In *Ideas and methods in mathematical analysis, stochastics, and applications (Oslo, 1988)*, pages 183–206. Cambridge Univ. Press, Cambridge, 1992.

153. M. Loève. *Probability Theory. I and II.* Springer-Verlag, New York, fourth edition, 1977, 1978.

154. A. Løkka. Martingale representation of functionals of Lévy processes. *Stochastic Anal. Appl.*, 22(4):867–892, 2004.

155. A. Løkka, B. Øksendal, and F. Proske. Stochastic partial differential equations driven by Lévy space-time white noise. *Ann. Appl. Probab.*, 14(3):1506–1528, 2004.

156. A. Løkka and F.N. Proske. Infinite dimensional analysis of pure jump Lévy processes on the Poisson space. *Math. Scand.*, 98(2):237–261, 2006.

157. S. Luo and Q. Zhang. Dynamic insider trading. In *Applied probability (Hong Kong, 1999)*, volume 26 of *AMS/IP Stud. Adv. Math.*, pages 93–104. Amer. Math. Soc., Providence, RI, 2002.

158. P. Malliavin. Stochastic calculus of variation and hypoelliptic operators. In *Proceedings of the International Symposium on Stochastic Differential Equations (Res. Inst. Math. Sci., Kyoto Univ., Kyoto, 1976)*, pages 195–263, New York, 1978. Wiley.

159. P. Malliavin. *Integration and Probability*, volume 157 of *Graduate Texts in Mathematics*. Springer-Verlag, New York, 1995.

160. P. Malliavin. *Stochastic Analysis*, volume 313 of *Grundlehren der Mathematischen Wissenschaften [Fundamental Principles of Mathematical Sciences]*. Springer-Verlag, Berlin, 1997.

161. P. Malliavin and A. Thalmaier. *Stochastic Calculus of Variations in Mathematical Finance*. Springer Finance. Springer-Verlag, Berlin, 2006.

162. S. Mataramvura, B. Øksendal, and F. Proske. The Donsker delta function of a Lévy process with application to chaos expansion of local time. *Ann. Inst. H. Poincaré Probab. Statist.*, 40(5):553–567, 2004.

163. M. Mensi and N. Privault. Conditional calculus on Poisson space and enlargement of filtration. *Stochastic Anal. Appl.*, 21(1):183–204, 2003.

164. R. Merton. The theory of rational option pricing. *Bell Journal of Economics and Management Science*, 4:141–183, 1973.

165. P.-A. Meyer and J.A. Yan. Distributions sur l'espace de Wiener (suite) d'après I. Kubo et Y. Yokoi. In *Séminaire de Probabilités, XXIII*, volume 1372 of *Lecture Notes in Math.*, pages 382–392. Springer, Berlin, 1989.

166. T. Meyer-Brandis and F. Proske. On the existence and explicit representability of strong solutions of Lévy noise driven SDE's with irregular coefficients. *Commun. Math. Sci.*, 4(1):129–154, 2006.

167. E. Nicolato and E. Venardos. Option pricing in stochastic volatility models of the Ornstein-Uhlenbeck type. *Math. Finance*, 13(4):445–466, 2003.

168. J. Norris. Simplified Malliavin calculus. In *Séminaire de Probabilités, XX, 1984/85*, volume 1204 of *Lecture Notes in Math.*, pages 101–130. Springer, Berlin, 1986.

169. D. Nualart. *The Malliavin Calculus and Related Topics*. Probability and its Applications (New York). Springer-Verlag, Berlin, second edition, 2006.

170. D. Nualart and É. Pardoux. Stochastic calculus with anticipating integrands. *Probab. Theory Related Fields*, 78(4):535–581, 1988.

171. D. Nualart and W. Schoutens. Chaotic and predictable representations for Lévy processes. *Stochastic Process. Appl.*, 90(1):109–122, 2000.

172. D. Nualart and J. Vives. Anticipative calculus for the Poisson process based on the Fock space. In *Séminaire de Probabilités, XXIV, 1988/89*, volume 1426 of *Lecture Notes in Math.*, pages 154–165. Springer, Berlin, 1990.

173. D. Ocone. Malliavin's calculus and stochastic integral representations of functionals of diffusion processes. *Stochastics*, 12(3-4):161–185, 1984.

174. D.L. Ocone and I. Karatzas. A generalized Clark representation formula, with application to optimal portfolios. *Stochastics Stochastics Rep.*, 34(3-4):187–220, 1991.

175. S. Ogawa. Quelques propriétés de l'intégrale stochastique du type noncausal. *Japan J. Appl. Math.*, 1(2):405–416, 1984.

176. S. Ogawa. The stochastic integral of noncausal type as an extension of the symmetric integrals. *Japan J. Appl. Math.*, 2(1):229–240, 1985.

177. B. Øksendal. Stochastic partial differential equations—a mathematical connection between macrocosmos and microcosmos. In *Analysis, algebra, and computers in mathematical research (Luleå, 1992)*, volume 156 of *Lecture Notes in Pure and Appl. Math.*, pages 365–385. Dekker, New York, 1994.

178. B. Øksendal. An Introduction to Malliavin Calculus with Applications to Economics. Technical report, Norwegian School of Economics and Business Administration, Bergen, 1996.

179. B. Øksendal. *Stochastic Differential Equations.* Universitext. Springer-Verlag, Berlin, sixth edition, 2003.

180. B. Øksendal. A universal optimal consumption rate for an insider. *Math. Finance*, 16(1):119–129, 2006.

181. B. Øksendal and F. Proske. White noise of Poisson random measures. *Potential Anal.*, 21(4):375–403, 2004.

182. B. Øksendal and A. Sulem. Partial observation control in an anticipating environment. *Uspekhi Mat. Nauk*, 59(2(356)):161–184, 2004.

183. B. Øksendal and A. Sulem. *Applied Stochastic Control of Jump Diffusions.* Springer-Verlag, Berlin, second edition, 2007.

184. Y.Y. Okur. White noise generalization of the Clark-Ocone formula under change of measure. Eprint in Mathematics, University of Oslo, 2007.

185. Y.Y. Okur. An extension of the Clark-Ocone formula under change of measure for Lévy processes. Eprint in Mathematics, University of Oslo, 2008.

186. Y.Y. Okur, F. Proske, and H.B. Salleh. SDE solutions in the space of smooth random variables. Eprint in Mathematics, University of Oslo, 2008.

187. É. Pardoux and S.G. Peng. Adapted solution of a backward stochastic differential equation. *Systems Control Lett.*, 14(1):55–61, 1990.

188. K.R. Parthasarathy. *An Introduction to Quantum Stochastic Calculus*, volume 85 of *Monographs in Mathematics*. Birkhäuser Verlag, Basel, 1992.

189. J. Picard. Formules de dualité sur l'espace de Poisson. *Ann. Inst. H. Poincaré Probab. Statist.*, 32(4):509–548, 1996.

190. J. Picard. On the existence of smooth densities for jump processes. *Probab. Theory Related Fields*, 105(4):481–511, 1996.

191. I. Pikovsky and I. Karatzas. Anticipative portfolio optimization. *Adv. in Appl. Probab.*, 28(4):1095–1122, 1996.

192. M. Pontier. [Essai de panorama de la] modélisation et [de la] détection du délit d'initié. *Matapli*, 77:58–75, 2005.

193. J. Potthoff and L. Streit. A characterization of Hida distributions. *J. Funct. Anal.*, 101(1):212–229, 1991.

194. J. Potthoff and M. Timpel. On a dual pair of spaces of smooth and generalized random variables. *Potential Anal.*, 4(6):637–654, 1995.

195. N. Privault. An extension of stochastic caluculus to certain non-Markovian processes. Technical report, 1997.

196. N. Privault. Equivalence of gradients on configuration spaces. *Random Oper. Stochastic Equations*, 7(3):241–262, 1999.

197. N. Privault. Independence of a class of multiple stochastic integrals. In *Seminar on Stochastic Analysis, Random Fields and Applications (Ascona, 1996)*, volume 45 of *Progr. Probab.*, pages 249–259. Birkhäuser, Basel, 1999.

198. N. Privault. Connections and curvature in the Riemannian geometry of configuration spaces. *J. Funct. Anal.*, 185(2):367–403, 2001.

199. N. Privault, J.L. Solé, and J. Vives. Chaotic Kabanov formula for the Azéma martingales. *Bernoulli*, 6(4):633–651, 2000.

200. N. Privault and X. Wei. A Malliavin calculus approach to sensitivity analysis in insurance. *Insurance Math. Econom.*, 35(3):679–690, 2004.

201. N. Privault and J.-L. Wu. Poisson stochastic integration in Hilbert spaces. *Ann. Math. Blaise Pascal*, 6(2):41–61, 1999.

202. F. Proske. The stochastic transport equation driven by Lévy white noise. *Commun. Math. Sci.*, 2(4):627–641, 2004.

203. F. Proske. Stochastic differential equations – some new ideas. *Stochastics*, 79(6):563–600, 2007.

204. P.E. Protter. *Stochastic Integration and Differential Equations*. Springer-Verlag, Berlin, 2005. Second edition.

205. M. Reed and B. Simon. *Methods of Modern Mathematical Physics. I.* Academic Press Inc., New York, second edition, 1980.

206. L.C.G. Rogers and D. Williams. *Diffusions, Markov Processes, and Martingales. Vol. 2.* Cambridge University Press, Cambridge, 2000.

207. Yu.A. Rozanov. *Innovation Processes.* V. H. Winston & Sons, Washington, D. C., 1977. Translated from the Russian, Preface by translation editor A. V. Balakrishnan, Scripta Series in Mathematics.

208. W. Rudin. *Functional Analysis*. International Series in Pure and Applied Mathematics. McGraw-Hill Inc., New York, second edition, 1991.

209. F. Russo and P. Vallois. Forward, backward and symmetric stochastic integration. *Probab. Th. Rel. Fields*, 93(4):403–421, 1993.

210. F. Russo and P. Vallois. The generalized covariation process and Itô formula. *Stoch. Proc. Appl.*, 59(4):81–104, 1995.

211. F. Russo and P. Vallois. Stochastic calculus with respect to continuous finite quadratic variation processes. *Stoch. Stoch. Rep.*, 70(4):1–40, 2000.

212. M. Sanz-Solé. *Malliavin Calculus*. Fundamental Sciences. EPFL Press, Lausanne, 2005.

213. K. Sato. *Lévy Processes and Infinitely Divisible Distributions*, volume 68 of *Cambridge Studies in Advanced Mathematics*. Cambridge University Press, Cambridge, 1999.

214. W. Schachermayer. *Portfolio optimization in incomplete financial markets*. Cattedra Galileiana. Scuola Normale Superiore, Classe di Scienze, Pisa, 2004.

215. W. Schoutens. *Stochastic Processes and Orthogonal Polynomials*, volume 146 of *Lecture Notes in Statistics*. Springer-Verlag, New York, 2000.

216. I. Shigekawa, Stochastic Analysis. Translation of Mathematical Monographs 214. American Mathematical Society, 2004.
217. A.V. Skorohod. On a generalization of the stochastic integral. *Teor. Verojatnost. i Primenen.*, 20(2):223–238, 1975.
218. D.W. Stroock. The Malliavin calculus and its application to second order parabolic differential equations. I. *Math. Systems Theory*, 14(1):25–65, 1981.
219. D.W. Stroock. The Malliavin calculus and its application to second order parabolic differential equations. II. *Math. Systems Theory*, 14(2):141–171, 1981.
220. D. Surgailis. On multiple Poisson stochastic integrals and associated Markov semigroups. *Probab. Math. Statist.*, 3(2):217–239, 1984.
221. A. Takeuchi. The Malliavin calculus for SDE with jumps and the partially hypoelliptic problem. *Osaka J. Math.*, 39(3):523–559, 2002.
222. S. Thangavelu. *Lectures on Hermite and Laguerre expansions*, volume 42 of *Mathematical Notes*. Princeton University Press, Princeton, NJ, 1993.
223. A. S. Üstünel. Representation of the distributions on Wiener space and stochastic calculus of variations. *J. Funct. Anal.*, 70(1):126–139, 1987.
224. A.S. Üstünel. *An Introduction to Analysis on Wiener Space*, volume 1610 of *Lecture Notes in Mathematics*. Springer-Verlag, Berlin, 1995.
225. S. Watanabe. *Lectures on Stochastic Differential Equations and Malliavin Calculus*, volume 73 of *Tata Institute of Fundamental Research Lectures on Mathematics and Physics*. Published for the Tata Institute of Fundamental Research, Bombay, 1984.
226. G.C. Wick. The evaluation of the collision matrix. *Physical Rev. (2)*, 80:268–272, 1950.
227. N. Wiener. The homogeneous chaos. *Amer. J. Math.*, 60(4):897–936, 1938.
228. A.L. Yablonski. The Malliavin calculus for processes with conditionally independent increments. In *Stochastic analysis and applications*, volume 2 of *Abel Symp.*, pages 641–678. Springer, Berlin, 2007.
229. J. Yong and X.Y. Zhou. *Stochastic Controls*, volume 43 of *Applications of Mathematics (New York)*. Springer-Verlag, New York, 1999.
230. M. Zakai. The Malliavin calculus. *Acta Appl. Math.*, 3(2):175–207, 1985.
231. T. Zhang. Characterizations of the white noise test functionals and Hida distributions. *Stochastics Stochastics Rep.*, 41(1-2):71–87, 1992.

Notation and Symbols

Numbers

$\mathbb{N}$	The natural numbers
$\mathbb{Z}$	The integer numbers
$\mathbb{Q}$	The rational numbers
$\mathbb{R}$	The real numbers
$\mathbb{R}_0$	p. 162
$\mathbb{C}$	The complex numbers
$\mathbb{C}^{\mathbb{N}}$	The set of all sequences of complex numbers
T	p. 7
$\mathcal{J}$	p. 66
$(2\mathbb{N})^{\alpha}$	p. 68
$\mathbb{K}_q(R)$	p. 74
$\mathbb{C}_c^{\mathbb{N}}$	p. 75
m_2	p. 217

Measures

P	pp. 7, 64, 215, 238
P^W	pp. 197, 238
$P^{\widetilde{N}}$	pp. 197, 238
$\lambda = dt$	Lebesgue measure
ν	p. 162
ρ	p. 217

Operations

$f \otimes g$	p. 11
$f \hat{\otimes} g$	p. 11
$W^{\otimes(n+1)}$	p. 13
$\langle \omega, \phi \rangle$	p. 64
$X \diamond Y$	pp. 70, 221

Spaces and Norms

$(\Omega, \mathcal{F}, P)$	pp. 7, 64, 161, 197, 215, 238
S_n	p. 8
G_n	p. 178
$C_0([0, T])$	pp. 27, 355
$L^2([0, T]^n)$	p. 8
$\tilde{L}^2([0, T]^n)$	p. 8
$\tilde{L}^2((\lambda \times \nu)^n)$	p. 178
$L^2(S_n)$	p. 8
$L^2(G_n)$	p. 178
$L^2((\lambda \times \nu)^n)$	p. 177
$L^2(([0, T] \times \mathbb{R}_0)^n)$	p. 177
$L^2(P)$	p. 9
$L^2(\mathcal{F}_T, P)$	p. 168
$L^2(P \times \lambda)$	p. 22
$L^2(P \times \lambda \times \nu)$	p. 188
$L^2(S)$	p. 210
$Dom(\delta)$	pp. 20, 183
$\mathbb{D}_{1,2}$	pp. 28, 187, 360
$Dom(D_t)$	p. 89
$\mathcal{S} = \mathcal{S}(\mathbb{R}^d),$	p. 63
$\|\cdot\|_{K,\alpha,}$	p. 63
$\mathcal{S}' = \mathcal{S}'(\mathbb{R}^d)$	p. 64
$(\mathcal{S})_k$	p. 69
$\|f\|_k^2$	p. 69
$(\mathcal{S})$	pp. 69, 219
$(\mathcal{S})_{-q}$	p. 69
$\|F\|_{-q}^2$	p. 69
$(\mathcal{S})^*$	pp. 69, 219
$\mathcal{G}_\lambda$	pp. 77, 229
$\mathcal{G}$	pp. 78, 229
$\mathcal{G}^*$	pp. 78, 229
$\mathbb{D}_0$	p. 140
$\mathbb{D}_{1,2}^{\mathcal{E}}$	p. 190
$\widetilde{\mathbb{D}}_{1,2}$	p. 207
$\mathbb{D}_{1,2}^W$	p. 240
$\mathcal{M}$	p. 268
$\mathbb{M}_{1,2}$	p. 268
$\mathbb{D}_{1,p}$	p. 341
$\mathbb{D}_{1,\infty}$	p. 341
$\mathbb{D}_{k,p}$	p. 343
$\mathbb{D}_\infty$	p. 343
$\mathcal{D}_{1,2}$	p. 358
$\mathbb{P}$	p. 359

Filtrations and σ-Algebras

Functions, Random Variables, and Transforms

Processes and Fields

Integrals and Differentials

$J_n(f)$	pp. 8, 178
$I_n(g)$	pp. 10, 178
$\delta(u)$	pp. 20, 183
$\widetilde{N}(\delta t, dz)$	p. 183
$\delta\eta(t)$	pp. 20, 184
$d^-W(s)$	p. 134
$\widetilde{N}(d^-t, dz)$	p. 267

Derivatives

$\partial^\alpha,$	p. 64
$D_t F$	pp. 28, 88, 89, 360
$D_\gamma F$	pp. 87, 357
$\mathbf{D}_\gamma F$	p. 358
$\mathbf{D}_t F$	p. 358
$D_{t,z} F$	pp. 188, 230
$D_{t+}\varphi(t)$	p. 140
$D_{t+,z}\theta(t,z)$	p. 268
$\mathcal{D}_t F$	p. 240
$D^j_{t_1,\dots,t_j,}$	p. 343
$D_y f$	pp. 354, 356

Admissible Controls

$\mathcal{A}$	p. 170
$\mathcal{A}_\mathbb{F}$	pp. 131, 301,
$\mathcal{A}_\mathbb{G}$	pp. 145, 132
$\mathcal{A}_{\mathbb{G},\mathcal{Q}}$	p. 150
$\mathcal{A}_\mathbb{E}$	pp. 200, 278, 291, 296
$\mathcal{A}_\mathbb{H}$	pp. 305, 311, 322

Notations

M^T	Transpose of a matrix M
$P \sim Q$	Measure P is equivalent to measure Q
$E[F]$	(generalized) Expectation w.r.t. measure P
$E_Q[F]$	Expectation w.r.t. measure Q
$E[F|\mathcal{F}_t]$	(generalized) Conditional expectation
càdlàg	Right continuous with left limits
càglàd	Left continuous with right limits
a.a., a.e., a.s.	Almost all, almost everywhere, almost surely
s.t.	Such that
w.r.t.	With respect to
SDE	Stochastic differential equation
BSDE	Backward stochastic differential equation
:=	Equal to by definition

Index